# Processing of RNA

Editor

**David Apirion, Ph.D.**
Professor of Microbiology and Immunology
Department of Microbiology and Immunology
Washington University School of Medicine
St. Louis, Missouri

CRC Press is an imprint of the
Taylor & Francis Group, an **informa** business

First published 1984 by CRC Press
Taylor & Francis Group
6000 Broken Sound Parkway NW, Suite 300
Boca Raton, FL 33487-2742

Reissued 2018 by CRC Press



A Library of Congress record exists under LC control number: 82022834

Publisher's Note
The publisher has gone to great lengths to ensure the quality of this reprint but points out that some imperfections in the original copies may be apparent.

Disclaimer
The publisher has made every effort to trace copyright holders and welcomes correspondence from those they have been unable to contact.

ISBN 13: 978-1-138-50600-8 (hbk)
ISBN 13: 978-1-138-56148-9 (pbk)
ISBN 13: 978-0-203-71078-4 (ebk)

Visit the Taylor & Francis Web site at http://www.taylorandfrancis.com and the CRC Press Web site at http://www.crcpress.com

# PREFACE

"RNA processing comes of age," as Dr. R. Perry so aptly put it in naming a recent article. However, there is still much to be learned about it. As can be attested from the chapters in this book, all organisms from bacteriophage to man can do it. This book attempts to cover the various aspects of RNA processing in the researched biological world. For this reason, the organization is by organism rather than by the kind of RNA processed (tRNA, rRNA, etc.). This means that certain phenomena such as RNA ligation in wheat germ and in yeast, or RNA splicing in yeast and *Tetrahymena* will have to be discussed in different chapters.

In a fast-moving field it is unlikely that articles written more than a year ago would be completely up to date. The purpose of this book is to bring to the nonspecialist an overall view as well as an update on the state of the art as it existed in the beginning of 1982, and to the specialist the opportunity to have a single source of information for how the other organisms do it, and also to enable him to find out the status of the various aspects of RNA processing with which he might not be too familiar. Even if only some of these goals are achieved, all those who labored so diligently to bring about the publication of this book would be more than gratified.

D. A.

# THE EDITOR

**David Apirion, Ph.D.,** is Professor of Microbiology and Immunology in the Department of Microbiology and Immunology of the Washington University School of Medicine, St. Louis.

Dr. Apirion received his M.Sc. degree from Hebrew University, Jerusalem, in 1960, and subsequently was awarded the Ph.D. degree in Genetics from Glasgow University in 1963. From 1963 to 1965 he was a Post-Doctoral Fellow at Harvard University.

At Glasgow University, Dr. Apirion was an Assistant Lecturer from 1962 to 1963. He came to Washington University as Assistant Professor in the Department of Microbiology in 1965, becoming Associate Professor in 1970. He has held his present position in the Department of Microbiology and Immunology since 1978. In 1973 he was a Visiting Scholar at Cambridge University, England.

Dr. Apirion has received the Tuvia Kushnir Prize (1960), the Alexander Milman Prize (1961), and the Sir Maurice Block Award.

Dr. Apirion serves on the Editorial Board of the *Journal of Bacteriology,* and holds memberships in the Genetics Society of America, the American Society for Microbiology, and the American Association for the Advancement of Science.

Dr. Apirion is the author of more than 100 publications, specializing particularly in molecular genetics and cellular organization.

## CONTRIBUTORS

**David Apirion**
Professor of Microbiology and Immunology
Department of Microbiology and Immunology
Washington University School of Medicine
St. Louis, Missouri

**Giuseppe Attardi**
Professor of Biology
Division of Biology
California Institute of Technology
Pasadena, California

**Glenn Björk**
Professor
Department of Microbiology
University of Umeå
Umeå, Sweden

**Thomas R. Broker**
Senior Scientist
Cold Spring Harbor Laboratory
Cold Spring Harbor, New York

**Robert J. Crouch**
Research Chemist
Laboratory of Molecular Genetics
National Institutes of Health
Bethesda, Maryland

**S. J. Flint**
Associate Professor
Department of Biochemical Sciences
Princeton University
Princeton, New Jersey

**Peter Gegenheimer**
Research Associate
Department of Chemistry
University of Colorado
Boulder, Colorado

**Anita K. Hopper**
Associate Professor
Department of Biological Chemistry
Milton S. Hershey Medical Center
Hershey, Pennsylvania

**Bernard Moss**
Head
Macromolecular Biology Section
Laboratory of Biology of Viruses
National Institute of Allergy and Infectious Diseases
National Institutes of Health
Bethesda, Maryland

**Joseph R. Nevins**
Associate Professor
The Rockefeller University
New York, New York

**Norman R. Pace**
Professor
National Jewish Hospital and Research Center and
Department of Biochemistry, Biophysics, and Genetics
University of Colorado Medical Center
Denver, Colorado

**Francis J. Schmidt**
Assistant Professor of Biochemistry
Department of Biochemistry
University of Missouri-Columbia
School of Medicine and College of Agriculture
Columbia, Missouri

# TABLE OF CONTENTS

Chapter 1

# PROTEIN-POLYNUCLEOTIDE RECOGNITION AND THE RNA PROCESSING NUCLEASES IN PROKARYOTES

**Norman R. Pace**

## TABLE Of CONTENTS

# I. INTRODUCTION

Several recent review articles,[1,2] including some in this volume, consider the properties of the few RNA processing enzymes on which we have information. This paper, as well, will survey the known prokaryotic RNA processing enzymes, but not in exhaustive detail. Rather, the author feels it useful to devote much of the available space to a consideration of the features of polynucleotides with which proteins may specifically interact. The fact is that we know little about the molecular details of any specific protein-polynucleotide complex, and the RNA processing enzymes offer excellent models for exploring these. The collection of references used is not intended to be all-inclusive, but rather, generally, to provide access to this literature.

# II. PROTEIN-POLYNUCLEOTIDE CONTACTS

None of the RNA processing enzymes is sufficiently well characterized to encourage even speculation on the detailed character of substrate recognition. We are, however, accumulating a reasonably detailed picture of the sorts of interactions which probably occur. It seems of use, therefore, to draw the discussion of substrate recognition a bit further than simple consideration of nucleotide sequences, even in some folded form. The paradigm offered by the sequence-specific DNA restriction endonucleases may have lulled us into overconfidence regarding our abilities to recognize in RNA the same complex information that proteins do, so the chemical details of possible protein-nucleic acid interactions must be borne in mind as we attempt to ferret out the targets of the processing nucleases.

Protein *surfaces* contact polynucleotide *surfaces*. We therefore must consider a recognition/manipulation process in terms of matrices of complementary contacts between the interacting molecules; the unique geometry possible among multiple contacts provides the overall specificity of the interaction. The important questions to pose, then, are (1) what chemical groups in the nucleic acids are potential binding contacts for proteins; and (2) what is their relationship to the surface, i.e., to an interacting protein? We consider, in passing, DNA as well as RNA because they often are considered to be informationally equivalent molecules. Although the similarities are considerable, these two nucleic acids also offer some strikingly different structural and chemical aspects which are instructive to consider. Moreover, sometimes our knowledge regarding certain aspects of protein-nucleic acid interactions is limited to DNA, so that is presented as exemplary.

Except for tRNA, the most detailed structural information that we have on polynucleotides involves the regular double helices.[3-5] These are of concern, here, because RNA processing sites often are found in regions of high secondary structure.

In general, duplex DNA adopts the familiar "B-form" helix under physiological conditions. Certain deoxynucleotide sequences can assume other helical forms (A, D, Z, etc.), but the bulk of the cellular DNA probably is in the B-form. Somewhat in contrast, largely because of conformational constraints imposed by the ribose 2′-OH groups, duplex RNA assumes the A-form of helix. Space-filling projections of the canonical A- and B-form helices and the positioning of the base pairs about the axes are shown in Figures 1 and 2. For the purposes of this discussion, important points to note regarding these structures are the following: bear in mind that the local structures of the polynucleotides probably are very mushy and readily molded by interacting proteins.

1. The periodicity and orientation of the negatively charged phosphate groups differ in the two forms. DNA-B offers a somewhat narrower profile and about 10 nucleotide pair phosphates per 34-Å turn. Phosphate groups project outward from the helix cylinder in DNA-B, but are more tucked into the RNA-A helix, partly blocking the wide groove (Figure 1).

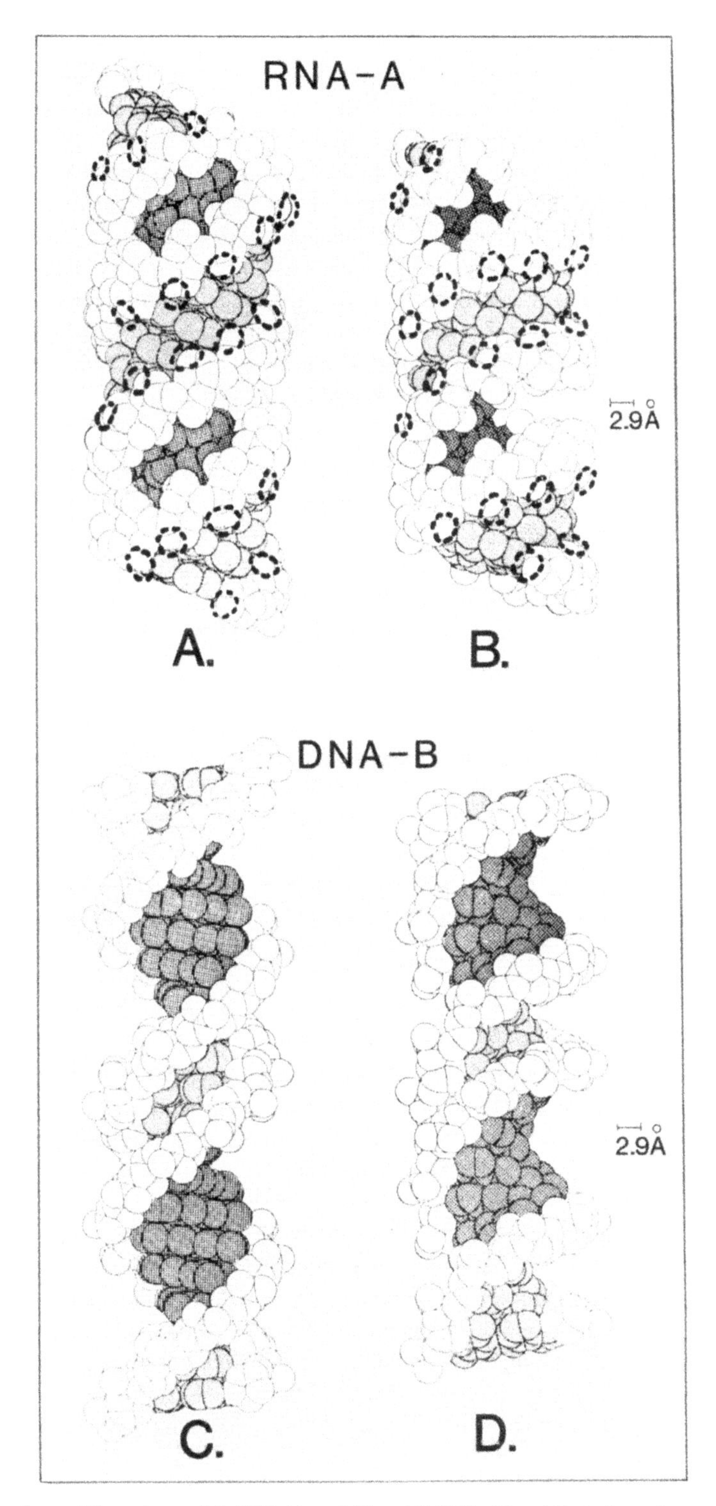


FIGURE 1. Space-filling views of A-RNA (A and B) and B-DNA (C and D) helices. In panels B and D the helices are tilted about 30° to reveal the depths of the helix grooves. Dark shading indicates the wide groove and light shading the narrow groove. In panels A and B, the approximate hydrogen bonding radii of the ribose 2′-OH groups are indicated by dashed circles. The 2.9-Å bar indicates the optimal negative-to-negative center distance for a hydrogen bond contact. (Modified from Alden C. J. and Kim, S.-H., *J. Mol. Biol.*, 132, 411, 1979.)

2. The displacement of the base pairs from the RNA-A helix axis (Figure 2A) defines a very deep, wide groove, which is most evident upon tilting the projection, as seen in Figure 1B. The depth of the wide groove and the barrier offered by the overhanging phosphate groups mean that the base-pair functional groups in the RNA-A wide groove are virtually inaccessible from the surface of the helix. The narrow groove of the RNA-A helix, on the other hand, is superficial (Figures 1 and 2). In DNA-B, because the base pairs are stacked along the helix axis (Figure 2B), both the wide and narrow grooves, in principle, are available to probing groups from an interacting protein.
3. The RNA-A narrow groove is populated by the ribose 2′-OH groups, which are important H-bond donors (Figures 1 and 2). In terms of information content, the 2′-OH groups are a major contrast between RNA and DNA.

Alden and Kim[6] (see also Pullman and Pullman[8]) have provided a more detailed theoretical picture of the information available in the base pairs of the A- and B-form helices by calculating the accessibilities of their various functional groups to "hard-shell" probes of various sizes. Their findings are that, for probe radii greater than 3 Å (amino acids), the helix wide groove contains the most accessible base contacts available in B-form DNA, and only the narrow groove of RNA-A has significant base exposure. The phosphate and 2′-OH groups are freely available.

Considerably less attention has been given to crystallographic analysis of "single-strand" polynucleotides than to the duplexes. It is clear from these and other studies, however, that unpaired sequences are far from disordered. Even without the constraints of a complementary pairing, single-strand sequences adopt ordered, helical arrays, stabilized mostly by intramolecular base stacking. Poly A, for example, crystallizes as a right-hand helical structure with about 9 residues per 25-Å repeat;[9] poly C collapses into a 6 base per 18.5-Å repeat helix.[10] Only poly U seems to be substantially unstacked, but is still highly ordered by the conformational constraints of the phosphodiester backbone.[11] In contrast to the regular duplex helices, all of the potential interaction sites for proteins would seem to be freely available in these "single-strand" structures.

Paired sequences containing RNA processing sites often are imperfect complements, containing unpaired or non-Watson-Crick base pairs (G.U, G.A, etc.), or out-of-register complements. At least these latter presumably would yield structures containing bulge loops or extrahelical bases, but little is known regarding their details and how they reflect into adjacent regions. Since protein contacts on the bases seem to be sterically very limited in the regular RNA-A helix, any irregularities within the helices may be important to focus upon in comparative analyses of processing enzyme substrate sites.

Four somewhat overlapping classes of contacts between proteins and polynucleotides can be envisaged.[12-14] These are (1) electrostatic interactions, (2) hydrogen bonding, (3) hydrophobic interactions, and (4) stacking interactions. The most important contributions to protein-polynucleotide binding energies probably are the electrostatic and hydrogen bonds. Stacking and hydrophobic interactions probably are not substantial as proteins confront duplex polynucleotides, but they offer interesting possibilities in considering irregular (non-duplex) nucleic acid conformations or in cases where the nucleic acid conformation is significantly perturbed. Let us consider, now, how these potential contacts are arranged in space in the nucleic acids, and the protein groups which may interact with them.

### A. Electrostatic Contacts

Electrostatic interactions are diverse in energy and type; they include the strong ionic contacts afforded by basic amino acids countering the negatively charged phosphate groups in the nucleic acids as well as a hierarchy of weaker and less-defined dipole interactions.

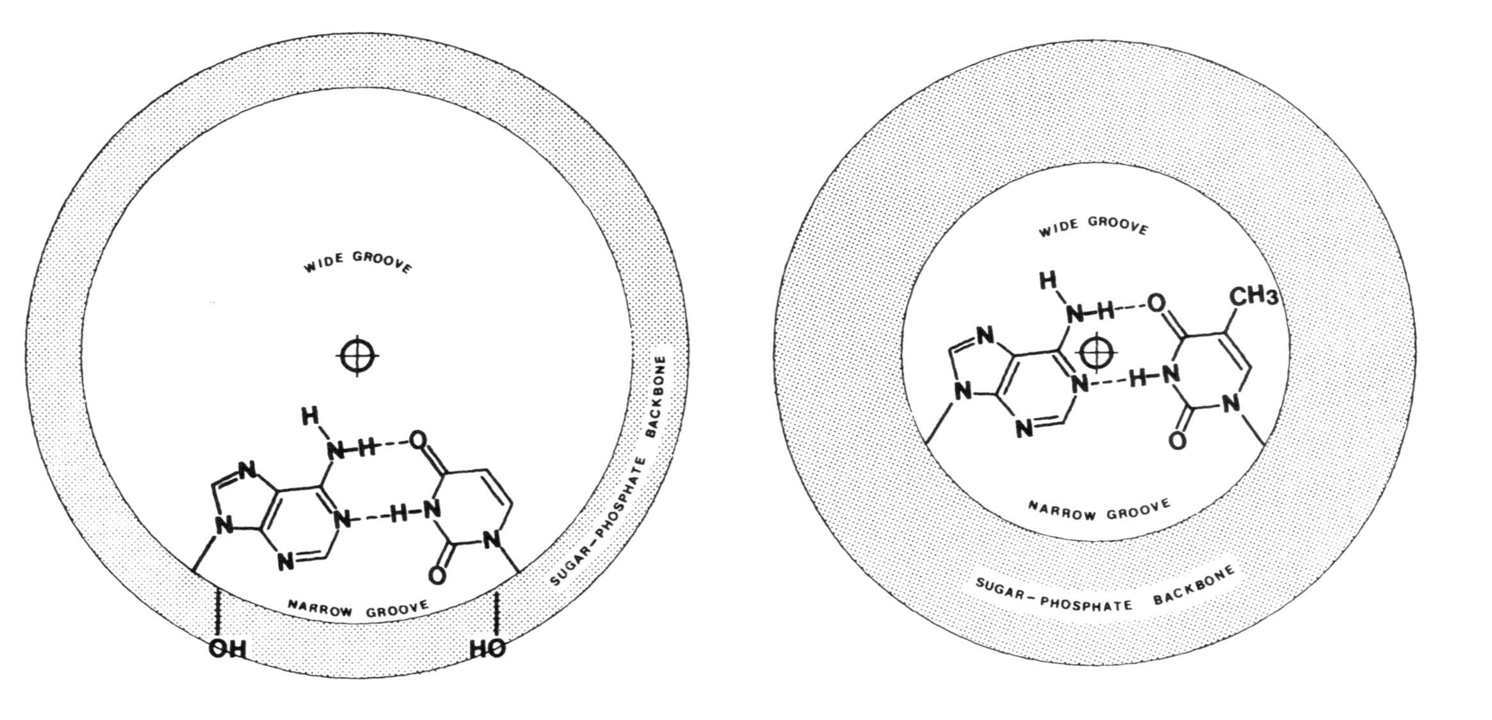


FIGURE 2. The relationships of the base pairs to the RNA-A and DNA-B helix axes. (Modified from Bloomfield, V. A., Crothers, D. M., and Tinoco, I., Jr., *Physical Chemistry of the Nucleic Acids*, Harper & Row, New York, 1974, 125.)

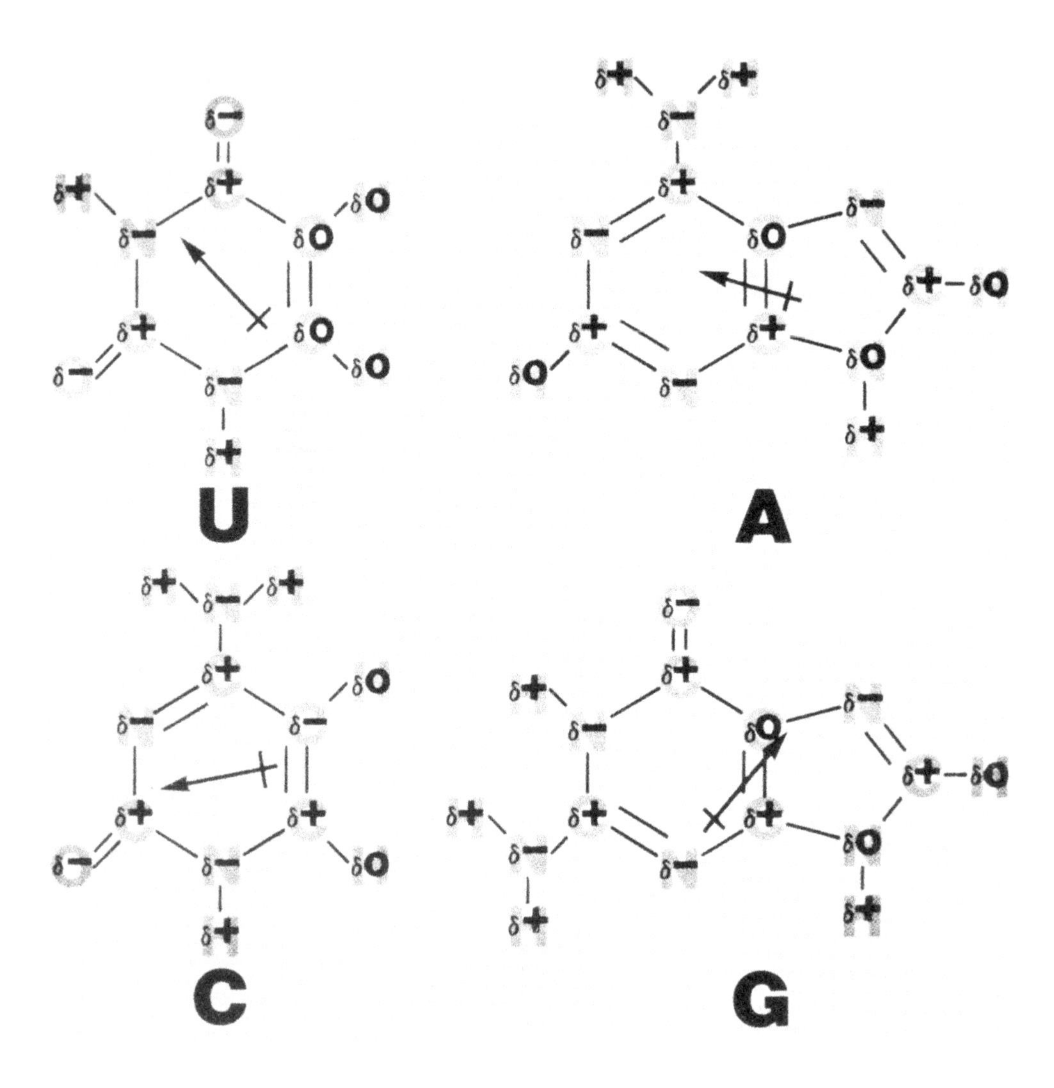


FIGURE 3. The partial charge distributions and permanent dipole moments of the purine and pyrimidine rings. (Modified from Bloomfield, V. A., Crothers, D. M., and Tinoco, I., Jr., *Physical Chemistry of the Nucleic Acids*, Harper & Row, New York, 1974.)

Nucleotide sequences should be viewed not simply as a string of letters, but as some array of partial charges. Figure 3 illustrates the partial charge distribution in the base rings. The intensity of any particular local charge of course is very dependent upon the environment of the base, i.e., its stacking neighbors as well as its association with the solvent, ions, etc.

An important aspect of all the electrostatic contacts is that their interaction energies decay rapidly with increasing distance from the optimum approach. In principle, the interactive energy of two point charges decreases inversely with the square of the distance between them (r); dipole-point charge (e.g., basic amino acid-phosphodiester backbone phosphate) energy decreases with $1/r^3$, dipole-dipole interactions with $1/r^4$, and induced dipole-induced dipole interactions decay with $1/r^6$. In fact, the decay is even more rapid because the interaction energy also is an inverse function of the dielectric constant of the medium. As water molecules are displaced from the contact sites (water occupies a sphere of about 1.4 Å radius), the dielectric constant of the intervening space decreases 20- to 40-fold and the contact affinity increases correspondingly.

It is obvious, then, that interacting arrays of electrostatic contacts must be precisely in register to be most effective. The strong ionic contacts probably provide the first order of interaction as proteins kinetically nestle onto their nucleic acid targets. Following the primary ionic fit, cooperation from many other and weaker interactions would contribute greatly to the binding energy, if the appropriate matrix of contacts between the molecules is present.

It is easy to see how ionic complementarity between a protein and a polynucleotide can provide considerable specificity. For example, the array of phosphate charges on the RNA-A helix is different from that of the DNA-B helix (Figure 1), therefore selective, strong contacts for proteins are available. Other conformations of course could provide more diverse ionic patterns; the arrays of phosphate contacts which alone could result from RNA tertiary structure seem almost without limit. In our searches for "consensus sequences" associated with RNA processing sites, therefore, it is important to remember that quite different nucleotide sequences can result in a similar matrix of electrostatic (e.g., phosphates) or other groups. The tRNAs offer a good example of this. Also, it is noteworthy that enzymes which operate on structurally different substrates (e.g., RNase P with tRNAs or RNase III with a variety of RNA species, see below) need not always use exactly the same matrix or types of contacts. More intensely negative or positive centers can serve as hydrogen bond donor/acceptor sites (below) as well as dipole contacts; when a dipole interaction involves an in-line proton, it is a hydrogen bond.

In the regular duplex helices, because the bases are tightly stacked, their potential dipole or hydrogen binding contact sites for an interacting protein are available only in the helix grooves, and the sites most diagnostic of the local sequence are in the wide groove.[12] Access to both grooves is no particular problem in the case of B-form DNA or helical, single-strand RNA, but the wide groove in duplex RNA essentially is not available to a protein surface unless it is unwound. The charge center-to-center distance in an effective dipole contact is short, say 2 to 3 Å, and the wide-groove groups in the bases are removed from the RNA-A helix surface considerably farther than that (see Figures 1 and 2). Presumably, then, any base-specific contacts in the intact RNA-A helix must originate from the narrow groove, which essentially is flush with the surface.

The importance of ionic contacts in protein-polynucleotide complexes is evident in their general inactivation at elevated salt concentrations, but oftentimes dependence upon low salt concentrations. This is less well documented for RNA processing or binding proteins than for proteins which bind to DNA, but likely will prove to be a common theme. As an illustration of how substantial these effects may be, the monovalent cation dependence of one rather nonspecific protein-RNA complex is shown in Figure 4.[15] The progressive enhancement of the binding constant as the $NA^+$ concentration increases to 0.1 *M* is interpreted as a requirement for counterions to shield charged groups (the protein is somewhat acidic) which otherwise would be repulsive, preventing surface contacts by the macromolecules. Then, as the ionic strength of the solution is further increased, the counterion clouds bury the interacting electrostatic pairs and the binding constant of the protein for the RNA is reduced. Record and colleagues,[16] and others, have made elegant use of this phenomenon to distinguish electrostatic and nonelectrostatic contributions to the thermodynamics of protein-polynucleotide interactions.

Appropriate (physiological) solution ionic strengths are demonstrably important to the specificity of processing nucleases. For example, *Escherichia coli* RNase III, which at physiological or higher ionic strength is sufficiently specific to release the rRNAs from their tandem transcript,[17,18] dramatically relaxes its selectivity at low ionic strength ($<0.05$ *M* $Na^+$). It then is capable of binding to and hydrolyzing a variety of natural and synthetic

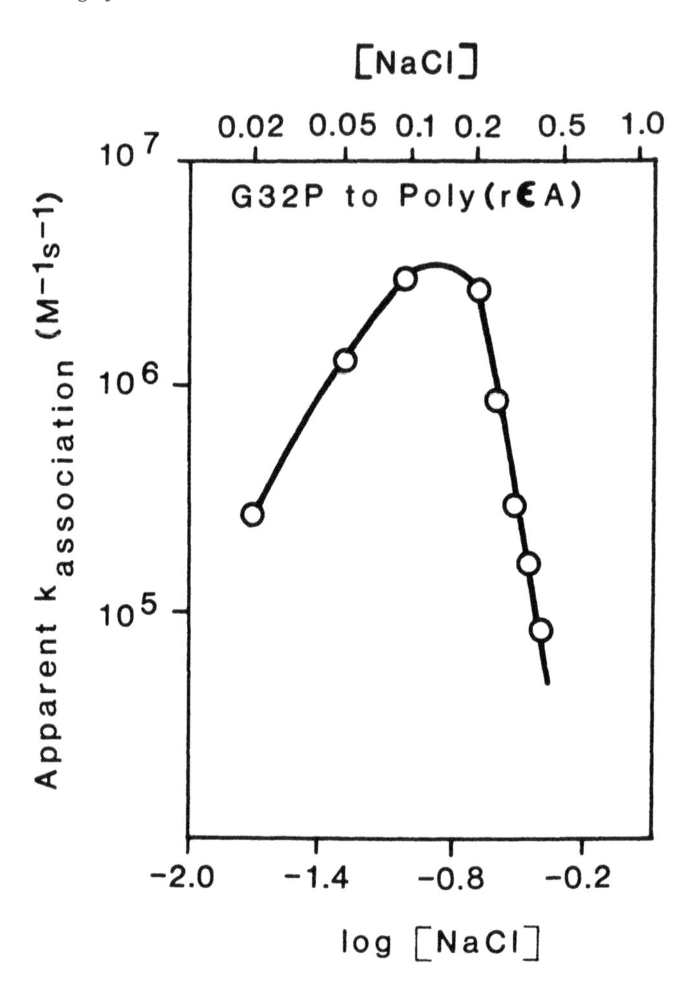


FIGURE 4. The dependence upon salt concentration of the apparent association constant of phage T4 gene 32 protein with poly (1,N[6]-ethenoadenylic acid). These measurements with a variety of nucleic acids are described in Reference 15. (Modified from Lohman, T. M. and Kowalczykowski, S. C., *J. Mol. Biol.*, 152, 67, 1981. With permission.)

RNAs.[19] This loss of specificity conceivably is explained by an insufficient cation concentration to shield ionic contacts on the RNA which are not quite in register with basic groups on the protein surface. At higher ionic strengths these spurious contact sites would not participate in the protein-RNA binding, but at low salt concentrations they could mimic, to some extent, a natural cleavage site.

**B. Hydrogen Bond Contacts**

Hydrogen bonds are the interactions between a proton-containing dipole (H-bond donor) and an electronegative center (acceptor). An important key to thinking of hydrogen bonding interactions, as proteins confront nucleic acids, is their dimension and vectorial nature. The acceptor must be in line (within approximately 20°) with the donor-proton dipole, and the optimal negative center-to-negative center distance for a hydrogen bond is about 2.9 Å. This dimension, relative to the regular helices, is indicated in Figure 1. Much closer approach becomes repulsive, and a bit further is energetically ineffective. Additionally, potential competition from water molecules means that hydrogen bond donor and acceptor pairs must be positioned very precisely in order to contribute to an interaction.

The polynucleotides offer a variety of potential contributions to hydrogen bonding.[12] Some base nitrogens and the exocyclic base amino groups are potential donors, as are the exocyclic oxy groups, if the base is polarized by its environment (nucleotide sequence or presence of a protein or other ligand) such that the lactim (enol) mode of the base prevails over the normal lactam (keto) form. Additionally RNA, but not DNA, contains the 2′-OH group, a highly conspicuous and important H-bond donor (below). Potential hydrogen bond acceptors associated with the bases are the exocyclic oxygens and the ring nitrogens, and conceivably the exocyclic amino groups in their imino tautomeric forms. The charged backbone phosphate groups are also potential acceptors.

As with the electrostatic contacts, an array of H-bond participants in a polynucleotide can provide a highly specific surface for recognition by a protein surface with a complementary donor/acceptor matrix. The contacts on the nucleic acid may be dependent upon the particular bases involved,[12] but, to reiterate, different nucleotide sequences may provide similar matrices. This is not to say that we should throw up our hands at the prospects of finding "consensus sequences" defining protein action sites in nucleic acids. The local nucleotide sequence certainly determines the detailed, local geometry (helix pitch, base tilt, and alignment relative to the helix axis, etc.) of a potential target. However, even sequence-dependent contacts need not present the same nucleotide sequences along their entire length. Since it seems unlikely (and unnecessary) that interacting proteins wrap extensively around the nucleic acid helices, probably only five or six base pairs per helix turn, even using both grooves, would be in register with a "globular" protein face and available for interaction.

Proteins offer a wealth of contributors to hydrogen bonding. Point contacts are available from the hydroxyl groups of Ser, Tyr, and Thr, the sulfhydryl group of Cys, the amino and amide groups of Lys, Asn, and Gln, and the His imidazole. More complex, somewhat base-specific associations involving the carboxylate anions of Asp and Glu or the Arg guanidinium cation also seem possible (Figure 5).[12,14] Of course, the peptide bond itself offers both an H-bond donor (the amide) and an acceptor (the carbonyl).

As with the in-plane dipoles associated with the base pairs, the DNA and RNA duplexes offer strikingly different aspects in their hydrogen bonding groups which are accessible to an interacting protein. Hydrogen bonding possibilities on the base pairs in duplex DNA are available from both the wide and the narrow grooves, while strictly duplex RNA (because of the deeply recessed A-form wide groove) would seem capable of H-bond contact between the base pairs and proteins only in the narrow groove.

The narrow groove of RNA, however, has information not present in DNA in that it is densely populated by the ribose 2′-OH groups. These undoubtedly play a significant role in many protein-RNA contacts. Carter and Kraut[20] have proposed an attractive model by which proteins might coordinate onto the RNA-A narrow groove, utilizing the 2′-OH groups. Their notion derives from the fact that antiparallel polypeptide β-sheets, which are common components of protein foldings, fit remarkably well to the RNA-A helix, such that alternate carbonyl groups in each of the two peptide chains are in appropriate register for hydrogen bond contact with each 2′-OH group of the nucleotide units. It was envisaged by Carter and Kraut[20] that additional contact might derive by the H-bond bridging of a water molecule between alternate amide protons in the polypeptides and the furanose ring oxygen atoms, although the latter is a questionable H-bond acceptor. Kim and colleagues[21] have pointed out that the antiparallel polypeptide β-sheet also can be fitted to the narrow groove of DNA-B, if alternate amide protons are invoked to form H-bonds with each of the 3′-oxygen atoms in the phosphodiester chain.

In the polypeptide β-sheet, alternate amino acid side chains are on alternate sides of the sheet, so in principle those pointing into the narrow groove could interact with the poly-

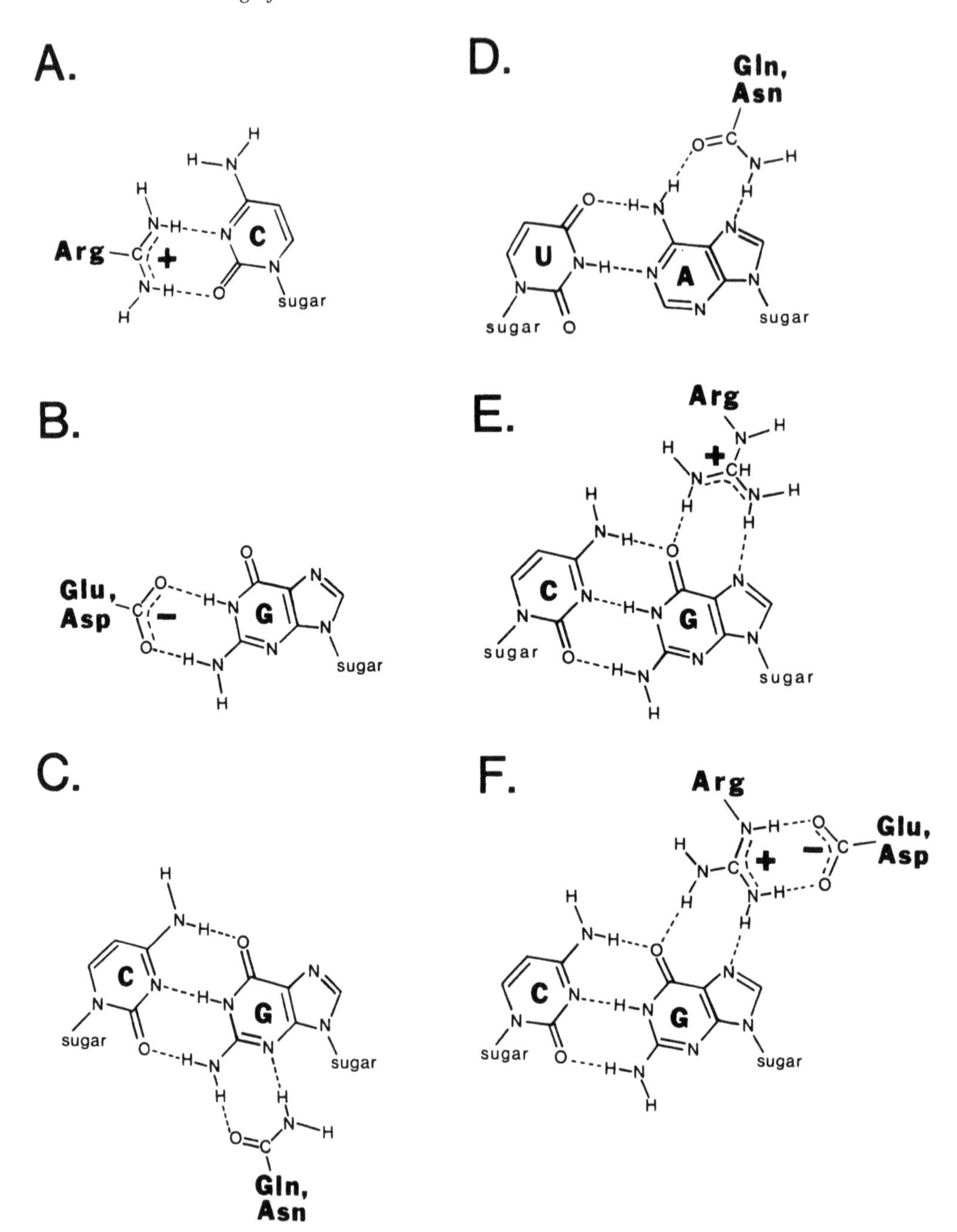


FIGURE 5. Various multiple-point hydrogen bonding associations between amino acids and bases. (Modified from Seeman, N. C. et al.[12] and Helene and Maurizot.[14])

nucleotide base pairs. The narrow groove of DNA, which is somewhat recessed (Figure 1), could accommodate the bulk of most of the amino acid side chains. In RNA-A, however, the narrow groove is essentially flush with the helix surface so would seem to offer steric barrier to the amino acid side chains. Of course, it is possible that the polynucleotide could rearrange somewhat to offer appropriate fit. Imperfections in the Watson-Crick complementarity at or very near the cleaved phosphodiester bonds in double helical processing sites might also offer accommodation to an interacting amino acid side chain.

The recently determined[22] structure of the phage λ cro repressor has features suggesting that these theoretically attractive schemes may reflect reality. The cro dimer (apparently the functional form) has remarkable complementarity to B-DNA. A region of β-sheet is flanked on either side by α-helical segments, which are separated by 34 Å, the DNA-B helix repeat distance. Positioning the cro dimer β-sheet on the DNA narrow groove orients the α-helical regions in good position and geometry for probing successive wide grooves along the DNA surface. This fit is consistent with experiments in which cro protein is shown to protect selected DNA groups from chemical modifying agents, or in which cro binding is reduced by the modifications. The cro repressor-DNA interaction is diagrammed in Figure 6; note the polypeptide β-sheet alignment on the narrow groove of the helix. RNA processing enzyme-substrate complexes probably are not so different in their general aspect. Even if wide groove contacts cannot be established in a helical RNA substrate, such β-sheet alignment could poise the protein well for endonucleolytic attack on the phosphodiester chains.

Each of the nucleotide residues in RNA-A contributes a ribose 2′-OH group to the narrow groove of the helix. It might seem difficult for a protein to abstract site-specific information from such a general array of hydrogen bond donors, but it is possible to imagine that the presentation of this array in space is very dependent upon the local nucleotide sequence for at least three reasons. First, the local sequence defines the exact wrapping periodicity of the helix and therefore the spatial distribution of 2′-OH groups which must be in register with H-bond acceptors on a protein surface. Second, the local sequence could influence substantially the puckering of the furanose ring, which in turn alters the vectorial aspect of the 2′-OH. Since the establishment of a hydrogen bond requires that the proton be pretty much in-line with the electronegative centers (above), this possibly is an important consideration. Third, it probably is the case that not all the 2′-OH groups are equally available for H-bond formation, and their availability can be imagined to be a function of nucleotide sequence. For instance, as diagrammed in Figure 7, the H-bonding capacity of the 2′-OH group associated with a uridylate unit probably is substantially quenched by its proximity to, and hence coordination on, the 2-oxy group of the pyrimidine ring. (Interaction between the dipoles is not subject to the linear constraints of hydrogen bonding.) The cytosine 2-oxy group, because of its pairing with guanine, would not have such significant influence in a duplex. The purine N-3 atoms also possibly affect the 2′-OH availability, but the important point is that the intensities of the 2′-OH groups as hydrogen bond donors are highly dependent upon the associated base. Taken together, then, local geometric and electrostatic influences can be imagined to mold the 2′-OH matrix such that it can contribute very specifically to a contact site for an interacting protein.

### C. Hydrophobic and Stacking Interactions

Polynucleotides are among the most polar of the macromolecules, but at least DNA has an apolar group, the thymine methyl, which is available to the surface of the molecule.[6] This is a prominent feature in the wide groove of DNA and possibly orients on hydrophobic puddles on an interacting protein surface. RNA, of course, generally lacks the methyl group so has virtually no apolar contact sites available. Even if the methyl groups were present, the recessed wide groove in double helical RNA would exclude all but the most disruptive of proteins from contact with them. RNA sometimes contains residues with apolar modifications, including thymine, but these seem to occur mostly in unpaired sequences, for example in the various tRNA loops, so in principle are available for hydrophobic interactions.

Probably the most direct data that hydrophobic contacts with thymine methyl groups in DNA contribute to a specific protein interaction are supplied by the experiments of Goeddel et al.[23] with lac repressor and synthetic lac operators. Replacement of certain thymine residues with less hydrophobic 5-bromothymine or with uracil was seen to substantially reduce the DNA-repressor affinity. That protein hydrophobic sites may lie in close proximity to the

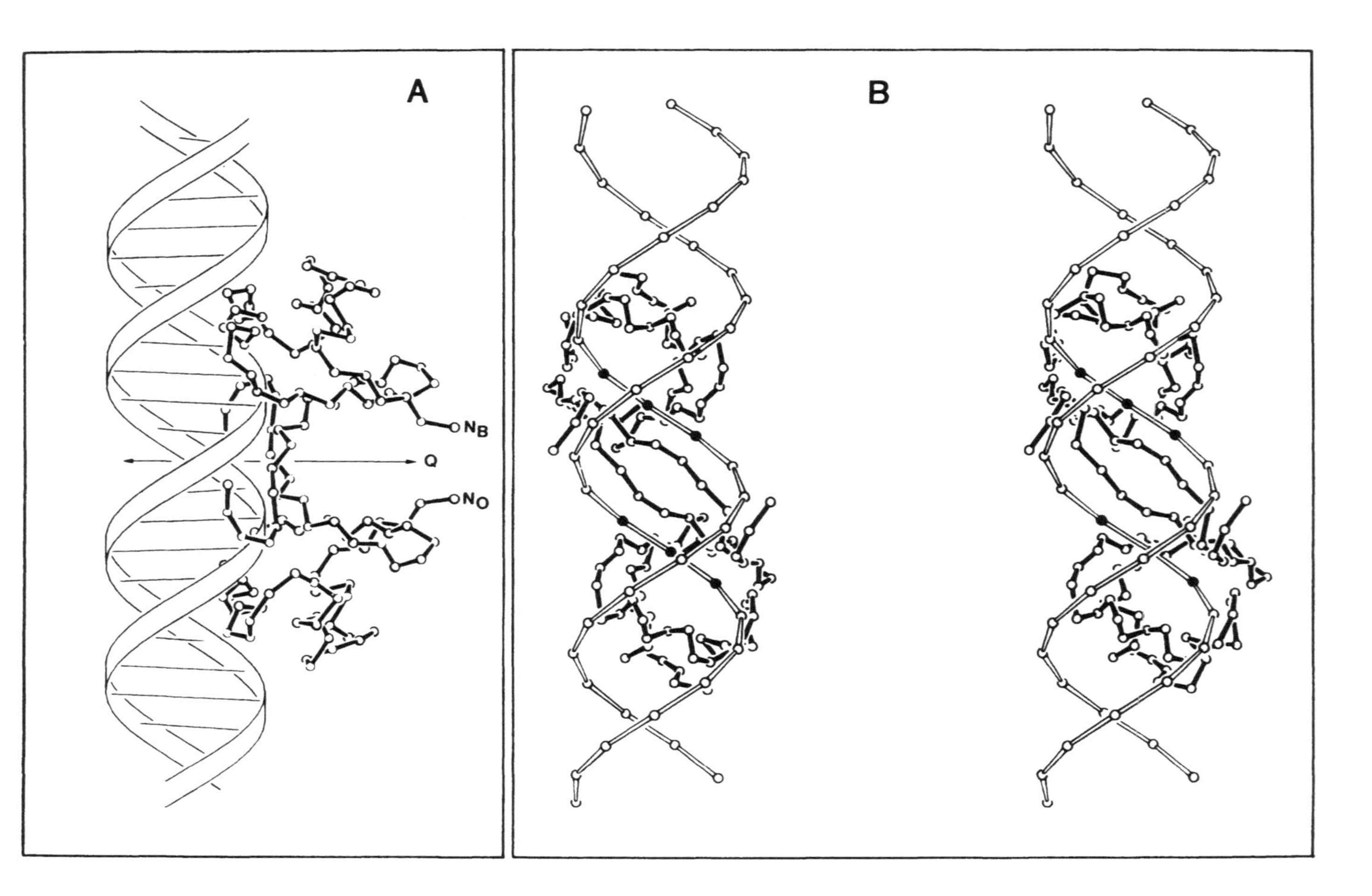


FIGURE 6. The cro repressor-DNA interaction, in profile (A) and stereo pair (B). (From Anderson, W. F., Ohlendorf, D. H., Takeda, Y., and Matthews, B. W., *Nature (London)*, 290, 754, 1981. With permission. Copyright © 1981 Macmillan Journals Limited.)

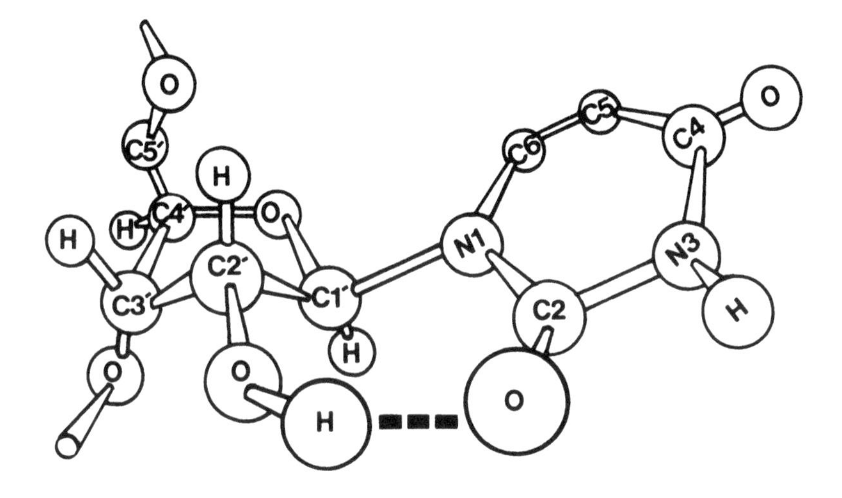


FIGURE 7. "Quenching" of the ribose 2'-OH group as a hydrogen bond donor by the uracil 2-O. The C2'-endo sugar pucker (RNA-A normally is considered to have C3'-endo pucker) brings the ribose 2'-OH very close to the uracil 2-oxy group, reducing by dipole interaction the availability of the 2'-OH group as a hydrogen bond donor.

DNA is suggested by observations of, for example, Gilbert and colleagues,[24] that the susceptibilities of certain DNA residues to hydrophobic alkylating agents (dimethyl sulfate, ethylnitrosourea) are enhanced by bound lac repressor or RNA polymerase. This is interpreted to result from hydrophobic sites on the protein surface serving to increase local concentrations of the agents.

Stacking ("vertical") contacts between the aromatic amino acids (Phe, Trp, Tyr) and the nucleic acid bases in regular duplexes probably are of minimal importance, because intercalation of the apolar amino acid side chains into the helix is required. Energetic barrier to such interaction may derive from the fact that a regular helix must undergo slight unwinding in order to accept an intercalating residue. Additionally, the aromatic amino acid side chains normally are buried in the protein superstructure, so their ability to penetrate the strongly polar shell of the nucleic acid helices would seem limited.

This expectation that stacking interactions are not of major importance in protein-nucleic acid helix complexes is borne out by studies with model peptides, for example the binding of Lys-Trp-Lys to a variety of polynucleotides (reviewed in Reference 14). Such peptides in fact do show some intercalation (as evidenced by quenching of Trp fluorescence), but the strong preference is for single-strand polynucleotides. Even with these, intramolecular base stacking tends to exclude the aromatic amino acid; the efficiency of Trp insertion decreases in the order poly U > poly A > poly C, which follows the increased stacking tendency of the bases.[7]

Although stacking interactions between protein residues and the double- or single-strand nucleic acid helices may not be favorable, *extrahelical* bases would seem to offer possible stacking targets. If nucleotide sequences involved in RNA folding are interrupted by one or a few bases with no complementary partner, collapse of the complementary segments into helical array can pinch out mispaired bases from the base-paired stack. The resulting "bulged" bases potentially provide obvious landmarks for an interacting protein. A variety of chemical and physical data are consistent with the existence of bulged bases in helical stretches, and the regular helix backbone readily accommodates one or two bulged nucleotides (reviewed in Reference 25).

The occurrence of unpaired (and presumably bulged) bases in RNA stems which interact with proteins seems to be a common theme, although certainly not an ubiquitous one. They are seen, for instance, in ribosomal protein binding sites on the rRNAs[26] and in the initial coat protein binding sites in phage R17[27] or tobacco mosaic virus[28] RNAs. In the case of ribosomal protein L18 binding to 5S rRNA, a reasonable case has been made that an unpaired A residue in an otherwise complementary stem in fact is extrahelical and acts as a binding landmark.[29] The adenine ring in question is very susceptible to modification by diethyl-pyrocarbonate, which only attacks unstacked residues, and the modification reduces the affinity of protein L18 for the 5S rRNA. More directly, Carey and Uhlenbeck[112] have synthesized phage R17 coat protein binding sites with altered nucleotide sequences and found that deletion of an ostensibly bulged base from a short hairpin results in an orders-of-magnitude decrease in the binding constant. This is all to say that extrahelical bases in polynucleotides may be important contact/recognition elements for proteins. They conceivably interact by stacking with appropriate amino acid side chains, but of course other associations (e.g., hydrogen bonding) are equally probable.

## III. THE KNOWN PROKARYOTIC PROCESSING NUCLEASES

### A. A Summary of Known and Suspected Processing Events and Enzymes

For perspective, it is useful to tick through briefly the cleavage events and enzymes that we know of, and those which are more nebulous. Table 1 compiles some of the available information; certain of the enzymes are elaborated upon below. All of the better-characterized prokaryotic processing enzymes act on the primary transcript from the ribosomal DNA, so their actions can be summarized as in Figure 8.

During transcription, soon after the completion of the 16S rRNA and formation of the duplex stalk which defines its substrate site, RNase III chips the promoter-proximal sequence and the immediate precursor of 16S rRNA ("p"16 rRNA; "m"16 rRNA denotes the mature form) from the growing transcript.[17,18] It is well established that RNase III carries out this reaction; RNase III$^-$ mutants do not produce the p16 rRNA. Moreover, the purified enzyme is capable of cleaving the entire tandem transcript of the rDNA, p30 rRNA, to generate p16 rRNA among other products. As depicted in Figure 8 these other products include p23 and p5 rRNAs and tRNA-containing spacers.[30,31] RNase III acts on many other cellular substrates, as well (below).

During their synthesis, as the binding sites become available, the growing rRNA molecules are bedecked with ribosomal proteins.[32] By the time that RNase III cleavage occurs, or shortly after, the rRNA precursors contain a complete, or nearly so, complement of ribosomal proteins, but they probably are not capable of engaging in protein synthesis.[33] This association of the rRNA precursors with the ribosomal proteins is a prerequisite for the terminal maturation steps, which remove precursor-specific segments from the 5′ and 3′ ends of all the rRNAs. That is, ribonucleoprotein particles, not the naked rRNAs, are the substrates required by the terminal cleavage enzymes. This is evidenced by the fact that the rRNA precursors accumulate in the absence of protein synthesis (e.g., the presence of an antibiotic such as chloramphenicol).[34] Formally, it is possible that the maturation nucleases themselves are unstable and so require continuous replenishment, but there is no parallel for this in bacteria. The termini of p16 and p23 rRNAs are removed by the putative RNases "M16" and "M23". (The quotation marks indicate the tentative nature of the information on these enzymes.) It is possible, even likely, that both 5′ and 3′ cuts in each of the precursors are made by the same enzyme since the cleavage sites are adjacent in the folded sequences. Not much is known about these enzymes (see below).

In *Escherichia coli*, the maturation of 5S rRNA from the RNase III cleavage product is a bit complex and is not completely worked out. First, RNase E trims off 5′ and 3′ precursor

## Table 1
## PROKARYOTIC PROCESSING NUCLEASES

| Enzyme | Function |
|---|---|
| | **Enzymes with Known Functions** |
| RNase III | Endonuclease. Cleaves "primary" transcripts of rRNA and tRNA to "secondary" precursors. Cleaves some phage (and probably host) mRNAs. Possibly involved in mRNA inactivation during decay. |
| RNase P | Endonuclease. Produces, with one cut, mature 5′ end of tRNA. |
| RNase D | 3′-Exonuclease. Probably produces mature 3′ end of tRNAs containing 3′-CCA. |
| RNase E | Endonuclease. Trims, with two cuts, an RNase III product to a 5S rRNA precursor with 3 extra nucleotides at both 5′ and 3′ ends. Probably involved in the maturation of other RNAs. |
| RNase M5 *(B. subtilis)* | Endonuclease. Makes two cuts in a 5S rRNA precursor, to yield mature 5S rRNA. |
| RNase II | 3′ Exonuclease. Probably responsible for most RNA decay to 5′ = mononucleotides. |
| RNase BN | Analogous to (and possibly related to) RNase D. Probably trims 3′ ends of phage tRNAs which lack 3′-CCA. |
| | **Functions Needing Enzymes** |
| RNase "M16" | Produces mature 5′ and 3′ ends of 16S rRNA. May be two enzymes. |
| RNase "M23" | Produces mature 5′ and 3′ ends of 23S rRNA. May be two enzymes. |
| ? | Trims 3 nucleotides from both 5′ and 3′ ends of 5S rRNA. May be two enzymes. |
| ? | Endonuclease. Cleaves several nucleotides from 3′ end of tRNA, to yield RNase D substrates. |
| ? (Sometimes perhaps RNase III) | Endonuclease. Cleaves mRNAs as prelude to 3′ exonucleolytic decay. |
| | **Enzymes Needing Functions** |
| RNase BN | Trims phage tRNA 3′ ends. Function in normal cell unknown. Conceivably is modified RNase D. |
| RNase F | Endonuclease. |
| RNase N | Endonuclease. |
| Polynucleotide phosphorylase (PNPase) | 3′ Exonuclease. Phosphorylytic release of 5′-nucleoside diphosphates. May serve in RNA decay, in absence of RNase II. |

elements, but the cleavages occur three nucleotides from both ends of the mature domain.[35] Some other so far uncharacterized enzyme(s) must complete the task. The *E. coli* p5 rRNA which accumulates in the absence of protein synthesis is only a few residues longer than the m5 rRNA, and the extra material (one, two, or three nucleotides) is all at the 5′ end.[36] Therefore, RNase E and the nuclease which matures the 3′ end are capable of operating on naked RNA; the 5′ nuclease, which may be exonucleolytic,[37] requires the RNP substrate. The maturation of 5S rRNA in *Bacillus subtilis* is more straightforward. RNase M5 precisely removes the precursor segments to generate the mature termini.[38] The required substrate is p5 rRNA with one 5S-binding ribosomal protein (see below).

The maturation of tRNA requires the action of at least two nucleases and often probably three. RNase P is known to generate the mature 5′ end of the tRNAs.[39] If the excess at the 3′ terminus is not too lengthy, an exonuclease such as RNase D[40] could be responsible for trimming back to the mature 3′ end (see below). If the stretch of nucleotides between the tRNA gene and the next downstream RNase III site is considerable, however, another endonuclease seems to be required to shorten it to a form acceptable to RNase D action.[41-43] RNases P2,[44] O,[45] and F[46] are candidates for this nuclease. RNases P2 and O

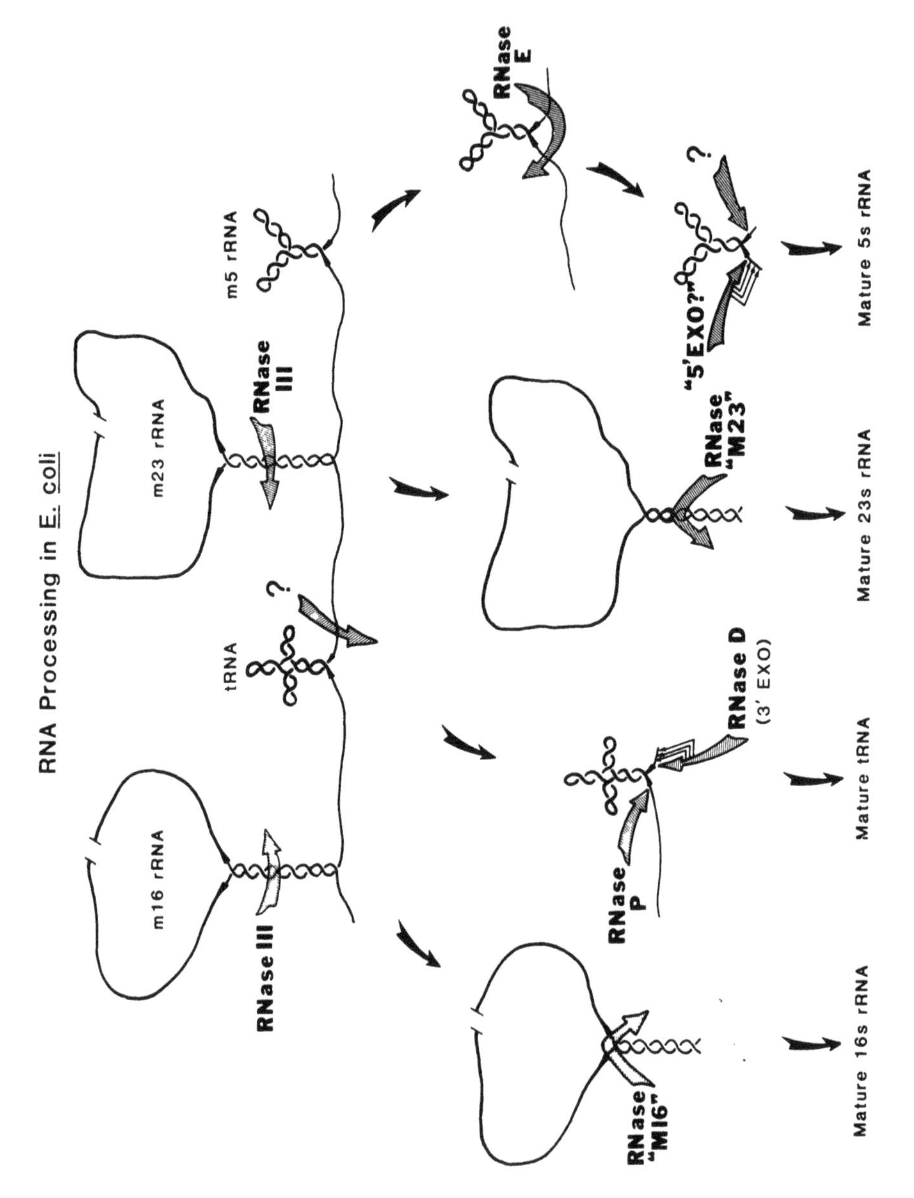


FIGURE 8. Processing of the rDNA primary transcript in *Escherichia coli.*

are not sufficiently characterized to distinguish them from RNase III. RNase F certainly is distinct from RNase III in that it yields 3′ phosphorylated cleavage products and does not have a divalent cation requirement,[47] but it remains to be established that RNase F is a bona fide RNA processing enzyme. All other known prokaryotic processing endonucleases generate 5′-phosphoryl and 3′-hydroxyl termini, although the yeast nuclease which excises introns from tRNA precursors does produce 3′-phosphorylated cleavage intermediates.[48]

Thus, the list of processing nucleases required by the prokaryotic cells is not a long one. Probably not many more than a dozen enzymes are involved, even including degradative activities (Table 1 and below), and the actual number may be less if some of the cleavages listed in Table 1 as distinct are carried out by the same nuclease. The literature contains references to additional processing activities from *E. coli*. Some can be attributed to the more characterized processing enzymes, but others cannot. Particularly, studies with crude extracts are problematic; observed activities may result from the combined actions of two or more of the characterized enzymes, and the results may be obscured by nonspecific nuclease actions. Some probably represent novel activities which require further study.

### B. The Substrate Problem

Probably the most important reason that we know so little about the RNA processing enzymes is the general dearth of substrates for them. The development of an appropriate test substrate is the key to the investigation of any enzyme reaction. Except for RNase III, which has a remarkably relaxed specificity, particularly under certain ionic conditions (see below), the processing enzymes so far defined are highly fastidious, requiring substrates that pretty much resemble their natural precursors. Isotopically labeled, natural precursors have been isolated from pulse-labeled cells (e.g., tRNA precursors), processing mutants (e.g., tRNA, rRNA precursors), or cultures treated with inhibitors of protein synthesis (e.g., rRNA precursors). The latter procedure works because, as pointed out above, the terminal processing enzymes for the rRNAs seek not a naked RNA substrate but rather a ribonucleoprotein (RNP) containing, as a good example, a ribosomal protein. Following inhibition of protein synthesis (e.g., by chloramphenicol) and depletion of the available pool of a required ribosomal protein, the RNA substrate accumulates in the cell and so can be preferentially labeled.

The cleavage of RNA precursors in vitro generally is scored by gel electrophoresis. Although this is a cumbersome assay, it is acceptable for monitoring column eluates during purification schedules and certainly should be used for initial enzyme characterizations. Definition of the correct product on a gel offers reasonable assurance that an observed cleavage activity in fact is the specific one. The ability of an extract to generate an RNA band about the size of a mature product does not prove that correct scission was made, however. This requires sequence or fingerprint analysis of the product to determine whether the termini are those of the cellular product. This rigor is particularly important with the smaller RNAs (e.g., tRNA, 5S rRNA), which are compact in secondary and tertiary structure and so tend to be relatively resistant to nonspecific nuclease action. If precursor-specific segments are less structured, hence less nuclease resistant than the mature domains, subjecting test substrates to limited, nonspecific cleavages can result in fragments of about the same size as the native product.

Natural precursors can be used for identifying and monitoring processing activities, but they are not very desirable for detailed studies, for example, kinetic analyses. Isotopically labeled RNA precursors recovered from cells always are precious, of uncertain specific radioactivity, and often are contaminated by other, nonradioactive cellular RNAs. Some effort, therefore, has gone into developing artificial substrates, although so far this has been done only for the tRNA and 5S rRNA processing enzymes, RNases P, D, and M5 (see below). In each of these cases, the precursor-specific elements contribute little or nothing to the interaction with the enzymes; the key recognition elements are in the mature domains. Synthetic oligonucleotides thus can be added to the 5′ or 3′ termini of the mature RNAs and their release monitored in ways more convenient than gel electrophoresis, for example by release of acid soluble radioactivity or by thin layer chromatography. It remains to be seen how generally useful such fabrications will be.

### C. RNase III

RNase III acts on many cellular substrates. In addition to the tandem rDNA transcript,[17,18] it acts on some multimeric tRNA precursors,[49,50] it cleaves some phage[17] (and probably cellular) polycistronic mRNAs to shorter forms, which sometimes are more active in protein synthesis,[51-53] and it may even participate in the initial stages of mRNA decay.[54] It is remarkable that such a wide-ranging activity is dispensable to the cell, but it is; *E. coli* mutants deficient in the enzyme are viable, albeit limping.[55] In the absence of RNase III, the terminal maturation nucleases (e.g., RNases "M16", "M23", and M5) are capable of producing the required products.[2] Thus, RNase III, while important in much posttranscriptional processing, would seem to offer a kinetic rather than an absolute advantage to cells by providing substrates which are handled with facility by other, and perhaps more discriminating, enzymes.

Although *E. coli* RNase III renders very precise cleavages in the cell, it was first isolated on the basis of its ability to cleave a wide variety of duplex RNAs, including synthetic polymers such as poly(AU) or poly G-poly C, or naturally occurring duplex heteropolymers such as reovirus RNA or RNase A-resistant RNA phage "replicative form".[55,56] The limit reaction products from these sorts of substrates are acid-soluble fragments about 10 to 20 nucleotides in length.[57] As with the other well-documented prokaryotic processing endonucleases, RNase III yields 5′-phosphate and 3′-hydroxyl termini, and the cleavage at both specific and nonspecific sites has an absolute requirement for divalent cations.[57,58] The enzyme does not cleave DNA-RNA hybrids; early observations that it would do so turned out to be due to contaminating RNase H.[57,58] Single-strand substrates (e.g., phage genome RNAs) are not rendered acid soluble by RNase III, but if they contain sufficiently lengthy, local base-paired regions, they are fragmented. This has been shown with diverse RNAs, for instance vesicular stomatitis genome RNA[59] or poliovirus RNA,[60] which in principle have little detailed relationship to RNase III targets in *E. coli*.

Taking advantage of the fact that RNase III has a high affinity for duplex RNAs, even in the absence of divalent cations, where enzymatic activity is not manifested, Dunn used affinity chromatography on (poly I-poly C) agarose to purify the enzyme essentially to homogeneity.[19] The enzyme is about 25,000 in molecular weight by denaturing gel electrophoresis. Sephadex chromatography suggested that it might consist of two of the 25,000-dalton subunits, but a more critical inspection of this association is required.

The specificity of RNase III cleavage is highly dependent upon the monovalent cation concentration in reaction buffers. In the cell, and in vitro at higher ionic strengths, RNase III cleavages of natural substrates are restricted to very specific sites. For instance Dunn observed that the phage T7 early polycistronic RNA, which accumulates in a RNase III$^{-}$ mutant, is cleaved mostly at its five normal sites above approximately 100 m*M* monovalent salt.[19] At decreasing salt concentrations, increasing numbers of "secondary", probably physiologically irrelevant, cleavage sites become available. These are not really nonspecific cleavages, however, because gel electrophoretic analysis shows the secondary cleavage products to be discrete fragments; preferred structures are being attacked by the enzyme. It is not established whether the salt concentration effect is upon the RNA or the enzyme-RNA interaction, but probably both are involved. At low ionic strength the RNA would be structurally more mobile and therefore capable of spuriously fitting to, for example, ionic contacts on the enzyme surface. The ill-fitting substrate could be tolerated and cleaved by the enzyme, because of inadequate ion shielding, as discussed above. Regardless of the mechanism, elevated salt concentrations do not limit RNase III exclusively to physiologically significant cleavage sites; the optimum for poly (AU) duplex solubilization is about 0.3 *M* $NH_4Cl$. The kinetic parameters (e.g., $K_M$, $V_{max}$) of RNase III for specific, in contrast to nonspecific, sites would be of considerable interest in thinking of RNase III recognition/action mechanisms.

Several natural cleavage sites for RNase III have been sequenced.[60-64] All of these are within extensive duplex stalks; illustrative natural substrates sites, two within the early T7 RNA and two associated with rRNA, are shown in Figure 9. The possible recognition elements in such structures have been discussed by several authors,[61,62,65-67] but the upshot is that no outstanding features common to all the substrate sites emerge except their extensive duplex character. Helical imperfections are associated with most, although not all, of the substrate sites, but the character of the imperfections and their relationships to the cleaved bonds are not constant. Some sequences are common to many, although not all, of the cleaved sites in T7 RNA, but the similarities vanish in the rRNA substrates. Possibly this structural divergence is reflected in the fact that the T7 RNA sites undergo single cleavages, whereas the rRNA sites are cleaved twice. It seems equally likely, however, that we cannot recognize the required information in simple terms of nucleotide sequence. Presumably the local helical parameters and the consequent matrix of hydrogen bond and electrostatic contacts define the substrate site and, as belabored above, there may be many nucleotide sequences which can provide the required contact matrix. It also is possible that different ensembles of contact arrays are used by the enzyme in handling the different substrates. Some detailed kinetic studies with defined substrates are definitely merited for this enzyme.

Homologues of the *E. coli* RNase III have not been isolated from other organisms, but they must exist. The rRNA genes in *B. subtilis*, for example, are transcriptionally linked, although a tandem transcript from the rRNA genes does not appear even in cells treated with inhibitors of protein synthesis to inhibit the terminal maturation enzymes. In *B. cereus*, however, the tandem transcript does accumulate in cells poisoned with nucleoside analogues,[68] so the tandem rRNA precursor must be cut by an RNase III homologue. Additionally, certain *B. subtilis* phage mRNAs are cleaved soon after transcription.[69] By analogy with *E. coli* phage T7, these scissions may be effected by the same enzyme.

The initial study of the *B. subtilis* RNase III probably will not be as easy as with *E. coli* because the *B. subtilis* enzyme does not seem to relax its specificity under low-salt reaction conditions in vitro. Our laboratory made a brief, by no means exhaustive, search for a *B. subtilis* RNase III-like activity, using as assay substrates a few duplex homo- and heteropolymers of RNA in various ionic conditions, but observed no acid solubilization of the test substrates. Although further convenient substrates and other reaction environments (e.g., with solvents) should be sought, it may be necessary to turn to natural substrates such as phage mRNAs from pulse-labeled cells. Although such an undertaking would be difficult with *E. coli* because of the high nonspecific nuclease content of cellular extracts, our general experience is that *B. subtilis* extracts are remarkably low in nonspecific nuclease activities, so it should be possible to see specific cleavage products on gels.

## D. RNases "M16" and "M23"

RNase III cleavage of the tandem rDNA transcript produces, among the other products, the immediate precursors of the 16S and 23S rRNAs, p16 and p23 rRNA.[30,31] To reiterate, these precursors accumulate in the absence of cellular protein synthesis; their subsequent maturation requires ribonucleoprotein substrates. Not very much is known about the putative RNases "M16" and "M23"; it is not even known how many enzymes are involved. The fact that the folded sequences spanning the mature ends of the 16S and 23S rRNAs (Figure 9) bring the two cleavage sites in each rRNA near one another might suggest that one enzyme

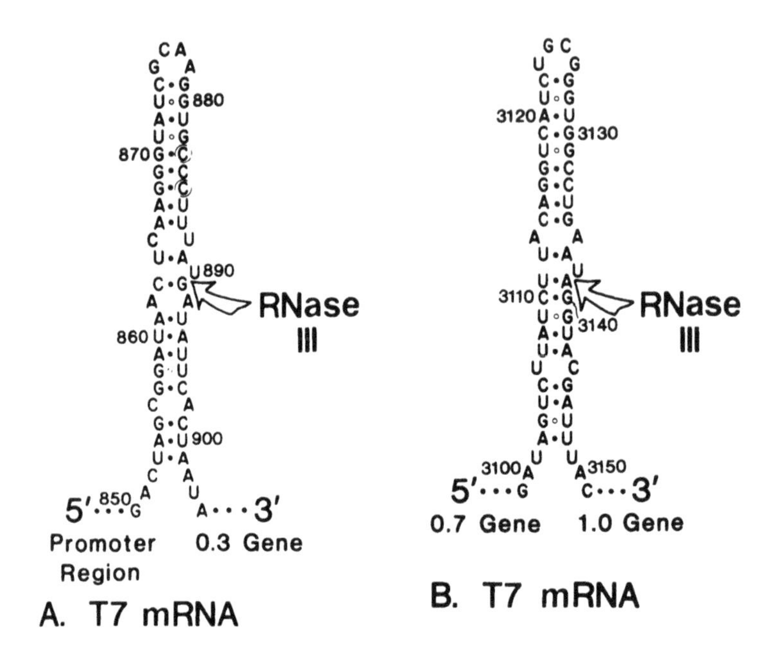


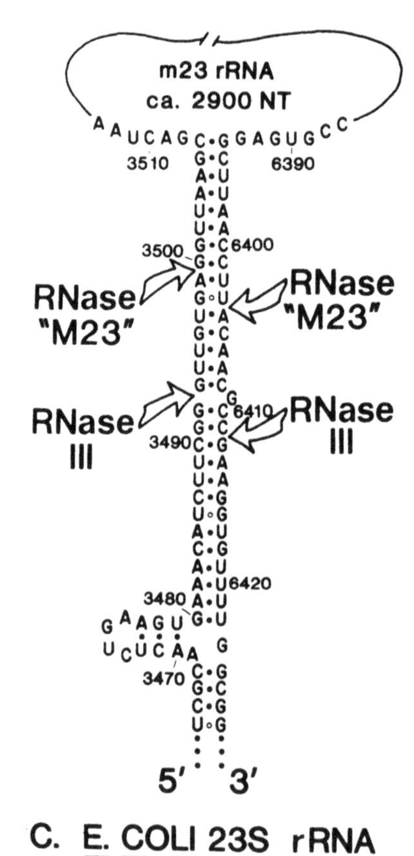


FIGURE 9. Some *Escherichia coli* RNase III and rRNA maturation processing sites. The nucleotide sequences of a few phage T7 and rRNA processing sites are shown, folded into their probable secondary structures. Mutational alteration of the C residues encircled in panel A drastically reduces susceptibility to RNase III.[119] The T7 RNAs are numbered from the genetic left end of T7 DNA. The rRNAs are numbered according to Brosius et al. for the rrnB operon.[61-64]

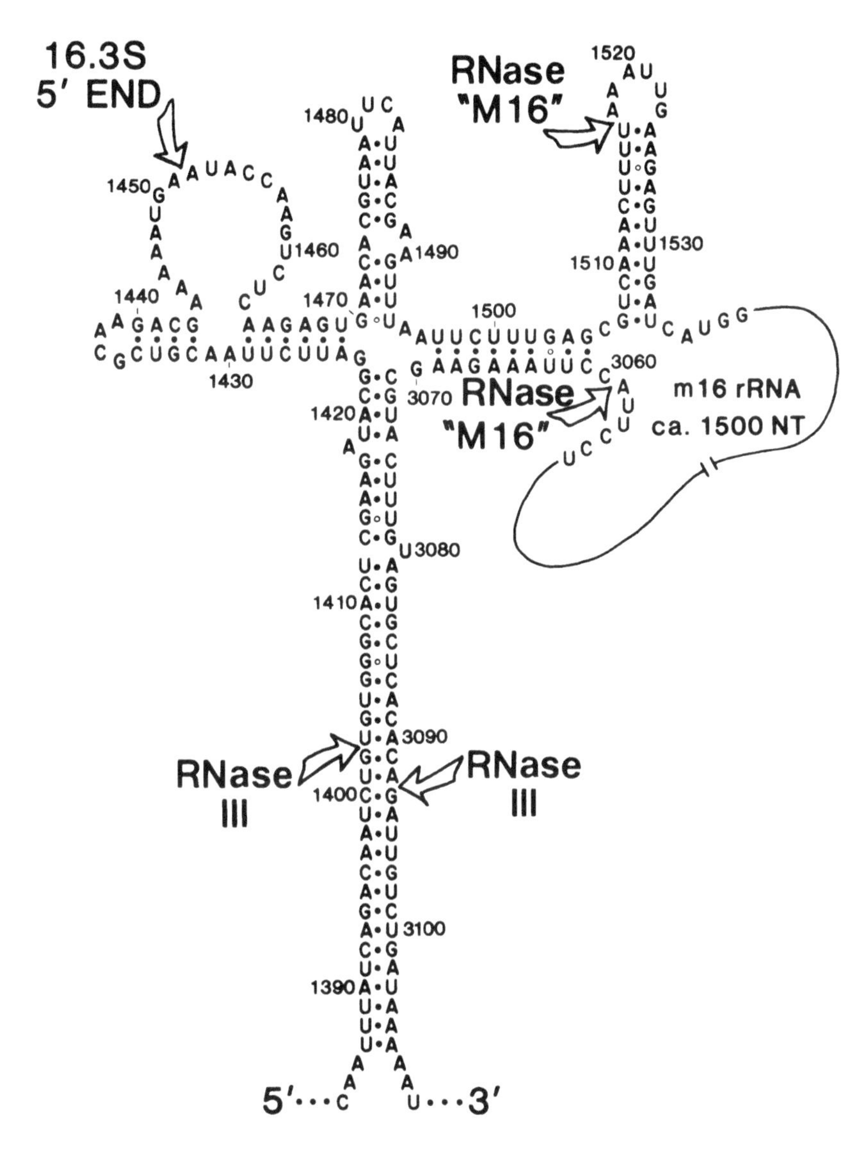


FIGURE 9D

generates both 3′ and 5′ mature termini, especially in the case of 23S rRNA. This would be analogous to the action of the *B. subtilis* RNase M5 (below), but is not necessarily compelling logic; tRNA maturation offers a good counterexample (below).

The concomitant maturation of the 5′ and 3′ ends of p16 rRNA in pulse-labeled ribosome subunit precursors has been demonstrated in vitro, using partially purified extracts from *E. coli.*[70,71] As usual with the specific endonucleases, RNase "M16" requires $Mg^{++}$, and Hayes and Vasseur[71] have reported that it is about 45,000 in molecular weight, as analyzed by sucrose gradient centrifugation. Hayes and Vasseur also thought that the 5′ and 3′ cleavage activities could be separated by DEAE chromatography, but the data do not warrant that conclusion; the question of whether RNase "M16" is one or two independent cleavage activities remains open.

A more convenient assay substrate for RNase "M16" than pulse-labeled precursor particles was discovered by Dahlberg et al.[72] in a temperature-sensitive elongation factor G mutant of *E. coli*. In this mutant, and in a temperature-insensitive revertant, the 16S rRNA is not always processed correctly. Under some growth conditions, about half the 16S rRNA in active ribosomes was found to retain some precursor sequences at the 5′ end. Sequence analysis showed that only 49 of 115 5′ precursor nucleotides had been removed in those "16.3S" rRNA molecules; the aberrant cleavage site is indicated in Figure 9. Slow maturation of the incorrectly trimmed 16S rRNAs during growth was demonstrable, so the mutant certainly was not deficient in the RNase "M16". It remains somewhat uncertain whether the incorrect processing is due to a damaged RNase "M16" or to some altered aspect of the substrate ribonucleoprotein. Ribosomes containing the 16.3S rRNA are a suitable and convenient assay substrate for the enzyme, however, and were used to partially purify an activity that would generate the mature 5′ end of the 16S rRNA.[72] The RNase "M16" was shown to act endonucleolytically, releasing the partial precursor fragment and absolutely required the RNP substrate, presented either as 30S subunits or 70S couples. Naked 16.3S rRNA was not cleaved, nor were ribosome "cores" obtained from CsCl gradients; presumably some essential components of the RNP substrate had been stripped from these latter.

Difficulties in dealing with the complex substrate, the cumbersome assay (polyacrylamide gels), and nonspecific nucleases seem to have discouraged pursuit of the RNase "M16". It would be interesting, at least, to test the partially purified preparations of the activity with p23 rRNA-containing particles. Conceivably the RNases "M16" and "M23" are the same. No information at all is available regarding the RNase "M23" except that we can infer from the mature 23S rRNA[34] that 5′-phosphoryl and 3′-hydroxyl groups result from the cleavages, as also is the case with 16S rRNA.

**E. RNases M5 and E**

5S rRNA is a convenient system for study because of its small size (approximately 120 nucleotides), general lack of modified bases, and its unique structure.[34] Members of the genus *Bacillus* differ from *E. coli* in their terminal processing of 5S rRNA — that of *B. subtilis* is the most straightforward.

Three immediate precursors of 5S rRNA, of lengths about 150, 180, and 240 nucleotides and deriving from different genes, appear in chloramphenicol-treated *B. subtilis*;[73] 5′- and 3′-terminal sequences are cleaved from each precursor by the endonuclease RNase M5, which has been purified substantially.[74] The close relatives of *B. subtilis, B. licheniformis,*[75] *B. stearothermophilus*, and *B. megaterium*[113] also possess RNase M5-like activities, and all in fact will operate on the *B. subtilis* precursors. Figure 10 shows the reaction catalyzed by the *B. subtilis* RNase M5 with one of the precursors. The 3′ segment of this precursor has the structure of a strong transcription termination site, so the 5S gene presumably is the last one in that rRNA transcriptional unit. The 240-nucleotide precursor also contains the transcription stop, but the 150-nucleotide class (3 to 4 species) does not. Probably the transcriptional units for these latter have other genes, perhaps for tRNA, distal to the 5S gene. The enzyme(s) that cleave the 5S precursors from the tandem transcript of the rDNA are not yet known. Probably a RNase III homologue is involved.

The *B. subtilis* RNase M5 consists of two readily separable components, $\alpha$ and $\beta$.[74] The $\beta$ component is capable of binding the precursor to a membrane filter, but no scission of the RNA occurs until $\alpha$, the presumed catalytic component, is added.[114] Component $\beta$ has been purified to homogeneity: it is of molecular weight ca. 15,000, highly basic, and recovered mostly from ribosomes. $\beta$ is electrophoretically indistinguishable from *B. subtilis* ribosomal protein BL16 and undoubtedly is the *B. subtilis* homologue of *E. coli* ribosomal protein EL18, since extremely pure EL18 will substitute for $\beta$ in the reaction catalyzed by

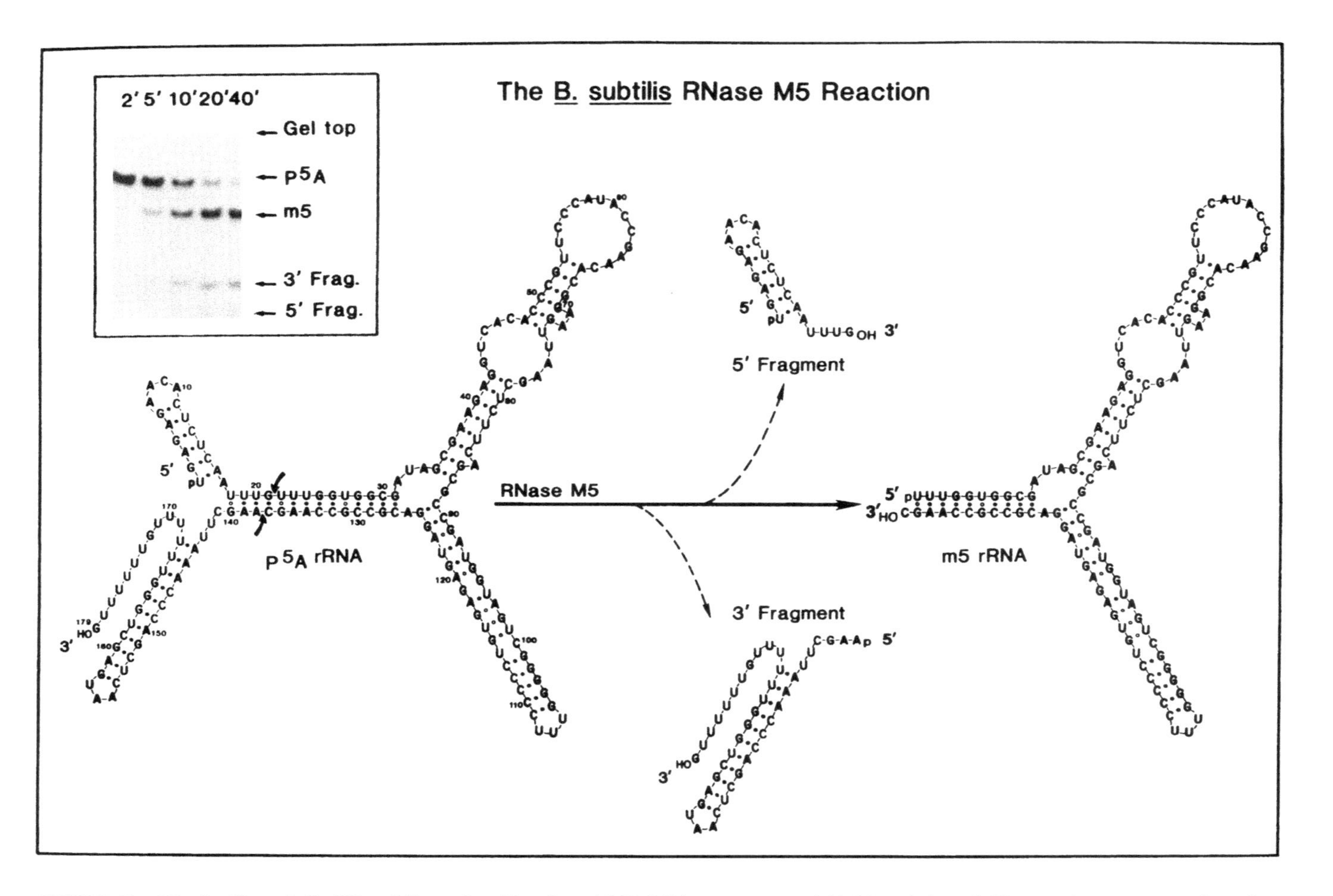


FIGURE 10. The *Bacillus subtilis* RNase M5 reaction. The $p5_A$ and M5 rRNA sequences are folded into their probable secondary structures; the action sites of RNase M5 are shown. The insert shows a polyacrylamide gel electrophoresis assay of the time course of the reaction.[74]

α.[115] RNase M5, then, should be viewed as more akin to the RNases "M16" and "M23" than to RNase III or RNase P (below), in that the recognizable substrate is a ribonucleoprotein rather than naked RNA. It will be interesting to see if the protein contributes to the contact surface (i.e., provides information) required by the highly specific RNase M5, or whether it serves as a scaffold to lock the precursor into a preferred conformation.

The cleavage sites for RNase M5, although 116 nucleotides apart, are immediately adjacent in the folded structure (Figure 10); this is a common feature of processing sites, e.g., two of the RNase III sites and the RNase "M23" site shown in Figure 9. It is not yet known whether RNase M5 removes both precursor segments simultaneously or sequentially, but both of the cleavages probably occur during a single binding event, i.e., without the release of an intermediate, mature at only one end. This is because no such intermediate is observed to accumulate during the RNase M5 reaction. If an intermediate were released, then it would have to be a much better substrate than the native one, or it would be in free competition with uncleaved substrate and must accumulate. Intermediates of the type expected, containing only the 3′ and the 5′ precursor segments, have been produced by annealing "half" molecules of precursor and mature 5S rRNA, and it is clear that neither is cleaved by RNase M5 with more facility than the native precursor. However, such artificial "intermediates" differ substantially in their relative susceptibilities to the enzyme; that lacking the 5′ end is cleaved at only a few percent of the rate of the intermediate lacking only the 3′ end. So, if sequential cleavage occurs, presumably the 3′ precursor sequence is released before the cut at the 5′ end is made.[38]

The components of the 5S precursors which are required for recognition and cleavage by RNase M5 have been explored to some extent by inspecting the ability of the enzyme to act on various precursor fragments.[38] As with RNase P, the precursor-specific segments do not convey substantial information to the enzyme. Rather, a considerable portion of the mature domain is involved; only residues ca. 90 to 120 in the 179-nucleotide precursor in Figure 10 are exempt from requirement by the enzyme, so far as is known.

When it became clear that the most significant RNase M5 recognition elements are in the mature domain of the precursor RNA it was possible to develop a processing assay which was more convenient than the gel electrophoresis of reaction products, as illustrated in Figure 10. Phage T4 RNA ligase was used to append an oligoribonucleotide (normally $U_3G$) to the 5′ end of the mature 5S rRNA. The construct then was labeled at the 5′ end using [$\gamma$-$^{32}$P] ATP and polynucleotide kinase, and the release of [5′-$^{32}$P] $U_3G_{OH}$ from the labeled construct by RNase M5 readily followed by acid-soluble radioactivity or by thin-layer chromatography.[76] The latter procedure is best for quantitative studies, since complete bookkeeping on the product and the residual substrate is possible.

It was pointed out above that, in double-helical RNA (in contrast to duplex DNA), base-specific contacts may be less important than conformation-dependent ones. The cleavage site for RNase M5 offers an interesting example of this.

The main recognition elements for RNase M5 are in the mature domain of the precursor. One precursor residue, however, a G ($G_{21}$ in Figure 10) adjacent to the 5′ cleavage site, significantly enhances the rate of its own cleavage as well as that of the 3′ fragment[76] so it must be an important component of the features recognized by the enzyme. This G residue is opposed in the helical substrate region to a C residue ($C_{137}$, the 3′ end of the mature domain), presenting the question of whether RNase M5 specifically contacts the cleavage site on the basis of nucleotide sequence (the G residue per se) or on the basis of more general aspects of helical conformation. Substrates containing all permutations of complementary and noncomplementary nucleotides at the cleavage site were fabricated to test this, and their susceptibilities to RNase M5 showed that the double-helical aspect, not the detailed sequence, is the important feature adjacent to the cleaved bonds.[77] Unpaired residues in the mature domains adjacent to the cleaved bonds also reduces susceptibility to the enzyme.[75]

The terminal maturation of 5S rRNA in *E. coli* seems more complex than in *B. subtilis*. RNase III cleavage of the primary transcript of the rDNA in *E. coli* releases a transient intermediate of 5S rRNA processing which is about 9S (250 nucleotides) in size (depending somewhat on the rRNA transcriptional unit), and probably analogous to the *B. subtilis* precursors that accumulate in the absence of protein synthesis.[78] The *E. coli* 9S does not appear under those conditions, however, because it is rapidly trimmed by RNase E, an enzyme characterized to some extent by Misra and Apirion.[79,80]

RNase E does not effect the terminal maturation of 5S rRNA; rather, it cleaves the 9S precursor 3 nucleotides from both 3′ and 5′ ends, at least in vitro.[35] 5′-Phosphate and 3′ hydroxyl groups result. In the cell, still further processing occurs in the absence of concomitant protein synthesis; the 3′-terminal precursor residues on the 5S rRNA are removed by some unknown nuclease, and possibly a bit of trimming at the 5′ end occurs. The 5S precursor population which accumulates under those conditions contains a mixture of 5′ ends with one, two, or three additional residues.[36] This heterogeneity evidently is not an aberrancy due to the absence of protein synthesis since the same population can be seen in pulse-labeled cells. In any case, the 5′ maturation nuclease in *E. coli* which acts only on the RNP substrate (analogous to the *B. subtilis* RNase M5), has not been identified. Conceivably it is a 5′ exonuclease,[37] which would be unique among processing enzymes so far examined, but this is not certain. Neither the 5′ nor the 3′ maturation may occur without prior RNase E action; the 9S RNA accumulates in an RNase E temperature-sensitive mutant at the restrictive temperature.[78] It is obscure why *E. coli* would use multiple steps to mature the 5S rRNA while *B. subtilis* does not. It may be that RNase E is involved in more than 5S rRNA processing, which is suggested by experiments with the RNase E temperature-sensitive mutant,[53,81] and so must contain a versatile substrate binding site. The 9S RNA precursor fit to the enzyme surface may not precisely align the mature 5′ and 3′ ends to the catalytic site, so subsequent trimming is required.

Inspection of the features of the 5S-containing precursors which are needed by RNase E will be of interest. There already are suggestions that they will differ from those used by the *B. subtilis* RNase M5. Whereas RNase M5 recognizes mainly the mature domain of its substrates, RNase E may rely at least in part on precursor-specific elements. The 5S rRNA precursors which accumulate in a RNase E temperature-sensitive mutant contain about 80 precursor-specific residues at their 5′ ends, as well as additional residues at their 3′ ends. Roy et al.[35] have shown that precursors lacking only the 5′-terminal 5-7 nucleotide (removed by spurious nuclease action) are not acceptable to the RNase E. The missing residues are partly complementary to another precursor sequence near the 5′ side of the mature domain, so the precursor stem, if it forms, may be part of the information used by RNase E in its selection of substrates.

### F. RNase P

Probably all tRNAs in *E. coli* are derived from RNase P action.[39,82,83] Temperature-sensitive mutants in the enzyme are lethal at restrictive temperatures, and many tRNA precursors accumulate under those conditions. RNase P evidently is involved in the maturation of other RNAs, as well. A precursor of 4.5S RNA (of unknown function) accumulates in RNase P mutants at nonpermissive temperatures and is a substrate for the enzyme.[84] At least one other substrate for the enzyme has been found, a low-molecular-weight RNA from phage $\phi$80-infected cells,[84] but it is not clear that it normally is acted upon in the cell.

The *E. coli* RNase P cleaves precursors of tRNA endonucleolytically to generate their mature 5′ phosphorylated termini.[39] Details regarding the enzyme have been reviewed recently.[2,39] The reason for such interest in RNase P is that it contains, in addition to a protein moiety, an obligatory RNA element and therefore possibly offers an analogy for the ribonucleoproteins apparently involved in RNA processing in eukaryotic cells.

Altman and colleagues[86] first documented that RNase P is a ribonucleoprotein. The key observations were that cleavage activity on a precursor of $tRNA^{Tyr}$ was abolished by treatment with insoluble RNase A or micrococcal nuclease, and the RNase P exhibited a buoyant density of about 1.7 g/mℓ in CsCl gradients. The latter suggests that about 75% of the mass of the enzyme consists of RNA. The fact that the *E. coli* RNase P survives 5 *M* CsCl speaks to the tightness of the protein-RNA aggregate, but it is dissociable by 7 *M* urea or acetic acid extraction. Simple mixing of the inactive separated RNA (here termed "P-RNA") and protein elements restores activity.[87,88] Kole and Altman[89] have purified the RNase P considerably, and suggest that the protein component molecular weight is about 18,000 to 19,000.[89] The P-RNA is about 360 nucleotides in length, although fragments of the RNA, down to about half the size of the intact form, are active upon reconstitution. The relative efficiencies of the intact and fragmented P-RNAs are not known, however.

It is pretty clear that the most important recognition elements for RNase P on the tRNA precursor are within the mature domain.[39,82,83] The 5′-precursor segments which are trimmed off often are only a few nucleotides long and they have highly variable structures. Moreover, oligonucleotides which span RNase P cleavage sites of phage T4 $tRNA^{Pro}$ and $tRNA^{Ser}$ are not substrates,[90] so the simple local nucleotide sequence does not offer sufficient information to define the cleavage site. On the other hand, many tRNA mutants with reduced susceptibility to RNase P have been collected, and they all are associated with the mature domain.[83] In particular, the mutations are at positions such as to disrupt or destabilize the secondary and tertiary folding of the tRNA. RNase P must manipulate tRNA precursors which vary widely in sequence, so it is not unreasonable that the required contact array is oriented by the common tertiary structures of the tRNA mature domains.

tRNA processing enzymes homologous to the *E. coli* RNase P have now been identified from a few eukaryotes, and also contain RNA elements.[91] The only other prokaryotic RNase P to be investigated, however, is that of *B. subtilis.*[92] The *B. subtilis* RNase P seems to be structurally analogous to that of *E. coli,* except that the RNA-protein association does not withstand exposure to CsCl buoyant density centrifugations. It was initially reported that the *B. subtilis* enzymes had at least two protein components, one RNA-associated and a second dissociable, but this now seems to be incorrect since activity can be reassociated from single RNA and protein components.[116]

As with RNase M5, the fact that the main recognition elements for RNase P are within the mature domain of the substrate has made possible the development of semisynthetic substrates, which are more convenient for enzymatic analyses than the native precursors. Some tRNAs are good RNA ligase donors, so may be condensed with a synthetic oligomer, which in turn is end-labeled and RNase P action scored by release of the oligomer.[117,118] The substrate so far most used is AAAC-$tRNA^{fMet}$; RNase P cleaves off the $A_3C$. The $tRNA^{fMet}$ is used because the 5′-terminal A residue is not base-paired in the acceptor stem, so will serve as a RNA ligase substrate. As a variation on the end-labeled substrate, Kline et al.[91] fabricated ApApA$\overset{*}{p}$C-$tRNA^{fMet}$; the $^{32}P$ resides between the A and C in the "precursor" element. The internal phosphate permits assay of RNase P in the presence of confounding amounts of phosphomonoesterase activity.

The interesting question, of course, is why does RNase P need an RNA component? Three possible roles for the P-RNA come quickly to mind.

First, it is possible that the P-RNA is required to establish specific association with the substrate, for example by hydrogen-bonding contacts. This would seem unnecessary, however, as we know perfectly well that proteins do not need auxiliary RNA elements to specifically interact with a substrate RNA molecule, even if the substrate structures vary extensively. The broad specificity of RNase III is witness to that, so a key role for the P-RNA in the primary recognition process seems unlikely.

Second, the P-RNA possibly is required not for binding the substrate, but for adjusting — loosening — its structure slightly, such that intramolecular contact sites in the rather rigid mature domain do not prevent the substrate phosphodiester bond from precise positioning at the active site on the enzyme surface. Although all the tRNAs have the same general tertiary structure, they probably differ slightly, for example, with regard to the helical orientation of the substrate bond in the acceptor stem. The P-RNA perhaps drives the substrate bond into the catalytic site by, as an example, competing for constraining tertiary or secondary structural features in the mature domain. This, in the author's opinion, seems the most likely role for the P-RNA at this time.

Third, it is conceivable that the P-RNA is not involved in the manipulation of the tRNA substrate per se, but rather is the heart of a multienzyme processing complex, so is important to the structural integrity of the RNase P, much as the rRNA molecules are integral to the activities of the ribosome. No other proteins have yet been found associated with the P-RNA, but experience with the system is so far quite limited. It is intriguing that fragments of the P-RNA can reconstitute good activity with the P-protein. Is the rest of the RNA engaged in other functions?

Each of these possibilities has distinctive expectations which undoubtedly will be tested within the coming years.

## G. RNase D

The tRNA genes often are associated with very long transcriptional units, which also contain the rRNAs, messages, or other tRNAs.[39,82,83] However, the immediate tRNA precursors so far characterized all have only a bit of excess beyond their mature 3′ termini, usually less than a dozen residues. The endonuclease which must generate these precursor 3′ ends is not characterized (above), but the exposure of the mature 3′ terminus evidently is effected by an exonuclease which has been purified by Deutscher and colleagues and termed RNase D.[93]

Actually, before the rather recent purification and characterization of RNase D, several laboratories had reported 3′-exonucleolytic activities capable of baring the mature 3′-CCA residues in the tRNA precursors. These exonucleases, observed in rather crude column chromatography cuts, have been termed RNase PIII by Bikoff and Gefter[94] or RNase Q by Shimura and co-workers.[45] However, the credulity of those reports was always a bit clouded by the possibility that the observed exonucleolytic activity was due to the action of RNase II, another 3′ exonuclease. The role of RNase II in the cell is not yet known for certain (it probably scavenges spent RNA, as discussed below), but it almost certainly is not involved in tRNA maturation. It will chip excess residues from the 3′ ends of tRNA precursors; however, it continues on to destroy the mature domain.[40]

RNase D was first isolated by Ghosh and Deutscher[95] on the basis of its ability to digest tRNA which had been "denatured" at the 3′ end by limited venom exonuclease digestion; hence the terminology RNase "D".[95] It soon became evident, however, that RNase D was a good candidate for the 3′ maturation of tRNA.

The essential feature of RNase D is that, in contrast to RNase II, it removes extra residues in a nonprocessive manner, the enzyme dissociating after each cleavage until the mature-CCA end is reached, whereupon the rate of cleavage falls 20- to 30-fold.[96] The simple presence of the -CCA sequence is not the signal to halt digestion, however, since an extra CCA appended onto the native sequence is rapidly removed, the native -CCA remaining relatively refractory. Other aspects of the structural environment of the native -CCA terminus must limit the reaction.

The functional RNase D is a 40K molecular weight protein, probably acting as a monomer; it does not require an RNA element. Cudny et al.[96] employed a variety of partially artificial substrates to characterize the activity, the most useful of these being tRNA to which short oligonucleotide stretches were added by an anomalous reaction of liver nucleotidyl transferase. For example, a few extra radioactive C residues can be added so that exonucleolytic activity is manifested by release of acid-soluble label.

It might seem curious that RNase D will digest, albeit slowly, the mature domains of tRNA precursors during extensive reaction. This need not detract from considering it to be a true maturation nuclease, however. In the cellular millieu, because RNase D is nonprocessive, as soon as the mature -CCA is available, the tRNA would be aminoacylated, associated with elongation factor Tu, and hence removed from the available pool of RNase D substrates. Any tRNAs incapable of undergoing aminoacylation of course would remain susceptible to degradation by RNase D or other scavenging nucleases.

The presence of the 3′-terminal -CCA sequence in tRNA apparently is important to RNase D, but some tRNA precursors, notably those defined by phage T4,[82,83] have sequences replacing this feature. During maturation, these and any additional residues must be removed and the mature -CCA added by nucleotidyl transferase. RNase D probably is not responsible for the final maturation of the phage T4 tRNA 3′ termini, rather, evidence supporting the involvement of still another exonuclease, RNase BN, has been offered by McClain and colleagues.[97,98] They isolated an *E. coli* mutant deficient in processing the 3′ ends of the phage tRNAs lacking the -CCA, but RNase D activity apparently was intact; the mutant grows normally. It still is not clear that RNase D and RNase BN are distinct activities, however. The RNase BN activity conceivably is RNase D, converted by some factor association such that it is capable of handling tRNAs with no resident -CCA sequence. All of the *E. coli* tRNA precursors (and genes) so far known contain the mature -CCA, so the RNase BN mutation could in fact be associated with RNase D, but at a position in the protein which is involved in the factor influence rather than the native enzymatic activity. It has been reported that RNase D undergoes a reversible alteration in size (from 40K to 65K, by gel filtration) during the course of phage T4 infection,[99] but any relationship this has to RNase BN activity is not known. Aspects of the BN mutation are discussed in more detail in Chapter 3 of this volume.

The terminal maturation of tRNA molecules is quite involved, then. In addition to all the cleavages and appendages, many nucleotide residues in the tRNAs are subject to modifications such as methylation, thiolation, etc. The modifications accumulate as the precursors are reduced in size, probably in preferred pathways, as the optimal substrates become available. The modifying enzymes are little studied (see Chapter 11) but many must exist. As with RNase P or RNase D, enzymes responsible for a particular modification probably are capable of handling many different tRNAs. It will be interesting to see whether any of the modifying enzymes contain RNA elements.

### H. The Selective Decay of RNA

Precursor-specific RNA elements are rapidly destroyed following their release, and mRNAs in prokaryotes turn over with half-lives of only a few minutes. We do not know very much about the enzymes responsible for degrading RNA and scavenging the remnants for reuse. Known processing enzymes certainly generate the unstable, precursor-specific fragments, but the early steps in mRNA decay are not at all clear. There are kinetic suggestions that endonuclease action is a prelude to the reduction of at least some mRNAs to nucleotides.[100] RNase III sometimes seems to act here.[54] It is attractive to think that an endonuclease involved in the initial decay of mRNA would attack the 5′ end of a cistron element so as to inactivate ribosome loading, thus avoiding production of abbreviated run-off peptides. Shen et al.,[54] in fact, have shown that RNase III can inactivate lac mRNA in vitro in that manner. RNase III$^-$ strains, however, do not have appreciably slowed mRNA turnover rates.[101] It thus would seem necessary to invoke some other activity, if endonucleolytic fragmentation contributes much kinetic advantage to the degradative pathway. An *E. coli* activity termed RNase N, which seems to be distinct in its properties from other endonucleases, has been suggested for this role,[102] but further characterization of that enzyme definitely is needed.

The reduction of RNA fragments to reusable units seems mostly to occur by 3′-exonuclease action.[103] The major activity involved in *E. coli* evidently is RNase II,[104,105] a processive 3′-exonuclease, which generates 5′-nucleoside monophosphates. Mutants with enhanced RNase II activity have decreased mRNA functional lifetimes and increased rates of unstable RNA turnover;[106] mutants with temperature-sensitive RNase II have more stable messages.[107,108] In the absence of RNase II, the terminal degradation of RNA fragments seems in large part due to the action of polynucleotide phosphorylase (PNPase).[108] This is a phosphorylytic 3′-exonuclease; inorganic phosphate is required in the reaction and 5′-mononucleoside diphosphates result. It still is not certain whether RNase II and PNPase are the end of the story, however, because strains that ostensibly lack both enzymes still display decay of unstable RNA, albeit three or four times more slowly than the wild type.[108] Possibly the mutants retain some residual RNase II and/or PNPase activity in vivo, but conceivably other enzymes so far unidentified can fill their roles.

Consistent with the invoking of RNase II as the major degradative activity in *E. coli* are the experiments of Chaney and Boyer,[109] who interpreted the kinetics of $H_2{}^{18}O$ incorporation into RNA as indicative of hydrolytic, rather than phosphorolytic, degradation.[109] In contrast, Duffy et al.,[110] using the same approach, suggested that in *B. subtilis* RNA turnover is effected by phosphorolysis.[110] This conclusion almost certainly is incorrect, however, because unless special pains are taken, bulk RNA isolated from that organism is heavily contaminated by nonpolynucleotide, phosphate-containing materials. This would have confounded the experimental analysis offered. More directly, inspection of the selective degradation of the precursor-specific fragments released during 5S rRNA maturation in *B. subtilis* extracts suggested the involvement of an RNase II-like activity.[111] Analysis of degradation intermediates and products showed clearly that the process is 3′-exonucleolytic and that 5′-mononucleotides result exclusively. Inclusion of inorganic phosphate inhibited rather than stimulated the fragment degradation.

Further work with RNA scavenging systems definitely is needed; we do not understand the process. Any endonucleases involved offer interesting questions regarding their selection of substrates. Perhaps regulatory aspects are present. In terms of protein-polynucleotide recognitions, however, the scavenging exonuclease(s) may not be very interesting. The author feels that the scavenging exonuclease(s) may not distinguish "stable" (nonsubstrate) from "unstable" (substrate) RNA by a specific recognition process; it needs only to attack all available RNA 3′ termini. The selectivity of the degradative process could result because the stable RNAs are never free from a protein complex for a significant length of time. The rRNAs are well protected in the ribosome (5S rRNA often does get nibbled a bit at its 3′ end) and tRNAs almost always are associated with their cognate synthetase, elongation factor Tu, or other element of the translation apparatus. The tRNA 3′ termini occasionally are attacked, but are reparable by nucleotidyl transferase. The 3′ (growing) termini of the prokaryotic mRNAs are protected until their transcription is complete and they do not survive long after that.

### I. A Comment on RNA Processing Pathways

The notion of metabolic "pathways" with rigid ordering in steps is well ingrained in the biochemist; enzymes commonly have exclusive specificity for the products of the immediately preceding step in their pathways. The order of multiple RNA processing events generally seems to be more relaxed, however.

RNA processing events fall loosely into two classes. "Primary" processing is upon the initial RNA transcript; "secondary" processing generates, sometimes in several steps (Figure 8), mature RNA termini. In normally growing cells, primary processing usually precedes secondary cleavages, but this is not obligatory. For instance, in *E. coli* mutants deficient in RNase III, the key primary processing enzyme, the secondary processing enzymes (e.g., the rRNA maturation endonucleases) are capable of independent function. The collection of precursor forms in such a mutant differs from that in normal cells, of course. Primary cleavages mostly occur rapidly, on nascent RNA chains, soon after the passage of the RNA polymerase. Some secondary scissions also may occur on nascent RNA, but other cleavage enzymes require ribonucleoprotein substrates (e.g., ribosome precursors), whose assembly probably is rate-limiting in the kinetic flow of most processing. Additionally, the products of primary cleavages generally seem to be the best substrates for the secondary processing enzymes, likely because of conformational or steric hindrances in the bulkier, more complex molecules. Nonetheless, the course of processing steps should be viewed as following the kinetic availability of optimal substrate sites in the RNAs, rather than as obligatory pathways of the traditional sort.

## REFERENCES


1. **Abelson, J.,** RNA processing and the intervening sequence problem, *Annu. Rev. Biochem.*, 48, 1035, 1979.
2. **Gegenheimer, P. and Apirion, D.,** Processing of prokaryotic ribonucleic acid, *Microbiol. Rev.*, 45, 502, 1981.
3. **Jack, A.,** Secondary and tertiary structure of nucleic acids, *Int. Rev. Biochem.*, 24, 211, 1979.
4. **Record, M. T., Jr., Mazur, S. J., Melancon, P., Roe, J.-H., Shaner, S. L., and Unger, L.,** Double helical DNA: conformations, physical properties, and interactions with ligands, *Annu. Rev. Biochem.*, 50, 997, 1981.
5. **Leslie, A. G. W., Arnott, S., Chandrasekaran, R., and Ratliff, R. L.,** Polymorphism of DNA double helices, *J. Mol. Biol.*, 143, 49, 1980.
6. **Alden, C. J. and Kim, S.-H.,** Solvent-accessible surfaces of nucleic acids, *J. Mol. Biol.*, 132, 411, 1979.
7. **Bloomfield, V. A., Crothers, D. M., and Tinoco, I., Jr.,** *Physical Chemistry of the Nucleic Acids*, Harper & Row, New York, 1974, 125.
8. **Pullman, A. and Pullman, B.,** Molecular electrostatic potentials of the nucleic acids, *Quart. Rev. Biophys.*, 14, 289, 1981.
9. **Saenger, W., Riecke, J., and Suck, D.,** A structural model for the polyadenylic acid single helix, *J. Mol. Biol.*, 93, 529, 1975.
10. **Arnott, S., Chandrasekaran, R., and Leslie, A. G. W.,** Structure of the single-strand polyribonucleotide polycytidylic acid, *J. Mol. Biol.*, 106, 735, 1976.
11. **Inners, L. D. and Felsenfeld, G.,** Conformation of polyuridylic acid in solution, *J. Mol. Biol.*, 50, 373, 1970.
12. **Seeman, N. C., Rosenberg, J. M., and Rich, A.,** Sequence-specific recognition of double helical nucleic acids by proteins, *Proc. Natl. Acad. Sci. U.S.A.*, 73, 804, 1976.
13. **Von Hippel, P. H.,** On the molecular basis of the specificity of interaction of transcriptional proteins with genome DNA, in *Biological Regulation and Development*, Vol. 1, Goldberger, R.F., Ed., Plenum Press, New York, 1979, chap. 8.
14. **Helene, C. and Maurizot, J.-C.,** Interactions of oligopeptides with nucleic acids, *CRC Crit. Rev. Biochem.*, 10, 213, 1981.
15. **Lohman, T. M. and Kowalczykowski, S. C.,** Kinetics and mechanism of the association of the bacteriophage T4 gene 32 (helix destabilizing) protein with single stranded nucleic acids: evidence for protein translocation, *J. Mol. Biol.*, 152, 67, 1981.

16. **Record, M. T., Jr., Anderson, C. F., and Lohman, T. M.,** Thermodynamic analysis of ion effects on the binding and conformational equilibria of proteins and nucleic acids: the roles of ion association or release, screening and ion effects on water activity, *Quart. Rev. Biophys.*, 11, 103, 1978.
17. **Dunn, J. J. and Studier, F. W.,** T7 early RNAs and *Escherichia coli* ribosomal RNAs are cut from large precursor RNAs *in vivo* by ribonuclease III, *Proc. Natl. Acad. Sci. U.S.A.*, 70, 3296, 1973.
18. **Nikolaev, N., Silengo, L., and Schlessinger, D.,** A role for ribonuclease III in processing of ribosomal ribonucleic acid and messenger ribonucleic acid precursors in *Escherichia coli*, *J. Biol. Chem.*, 248, 7967, 1973.
19. **Dunn, J. J.,** RNase III cleavage of single-stranded RNA. Effect of ionic strength on the fidelity of cleavage, *J. Biol. Chem.*, 251, 3807, 1976.
20. **Carter, C. W., Jr. and Kraut, J.,** A proposed model for interaction of polypeptides with RNA, *Proc. Natl. Acad. Sci. U.S.A.*, 71, 283, 1974.
21. **Church, G. M., Sussman, J. L., and Kim, S.-H.,** Secondary structural complementarity between DNA and proteins, *Proc. Natl. Acad. Sci. U.S.A.*, 74, 1458, 1977.
22. **Anderson, W. F., Ohlendorf, D. H., Takeda, Y., and Matthews, B. W.,** Structure of the cro repressor from bacteriophage λ and its interaction with DNA, *Nature (London)*, 290, 754, 1981.
23. **Goeddel, D. V., Yansura, D. G., and Caruthers, M. H.,** How lac repressor recognizes lac operator, *Proc. Natl. Acad. Sci. U.S.A.*, 75, 3378, 1978.
24. **Siebenlist, U., Simpson, R. B., and Gilbert, W.,** *E. coli* RNA polymerase interacts homologously with two different promoters, *Cell*, 20, 269, 1980.
25. **Lomant, A. J. and Fresco, J. R.,** Structural and energetic consequences of noncomplementary base oppositions in nucleic acid helices, *Prog. Nucleic Acids Res. Mol. Biol.*, 15, 185, 1975.
26. **Woese, C., Magrum, L. J., Gupta, R., Siegel, R. B., Stahl, D. A., Kop, J., Crawford, N., Brosius, J., Gutell, R., Hogan, J. J., and Noller, H. F.,** Secondary structure model for bacterial 16S ribosomal RNA: phylogenetic, enzymatic, and chemical evidence, *Nucleic Acids Res.*, 8, 2275, 1980.
27. **Gralla, J., Steitz, J. A., and Crothers, D. M.,** Direct physical evidence for secondary structure in an isolated fragment of R17 bacteriophage mRNA, *Nature (London)*, 248, 204, 1974.
28. **Zimmern, D.,** The nucleotide sequence at the origin for assembly on tobacco mosaic virus RNA, *Cell*, 11, 463, 1977.
29. **Peattie, D. A., Douthwaite, S., Garrett, R. A., and Noller, H. F.,** A "bulged" double helix in a RNA-protein contact site, *Proc. Natl. Acad. Sci. U.S.A.*, 78, 7331, 1981.
30. **Lund, E., Dahlberg, J. E., and Guthrie, C.,** Processing of spacer tRNAs of *Escherichia coli*, in *Transfer RNA: Biological Aspects*, Söll, D., Abelson, J., and Schimmel, P.R., Eds., Cold Spring Harbor Laboratory, Cold Spring Harbor, N.Y., 1980, 123.
31. **Apirion, D., Ghora, B. K., Plautz, G., Misra, T. K., and Gegenheimer, P.,** Processing of rRNA and tRNA in *Escherichia coli:* Cooperation between processing enzymes, in *Transfer RNA: Biological Aspects*, Söll, D., Abelson, J., and Schimmel, P.R., Eds., Cold Spring Harbor Laboratory, Cold Spring Harbor, N.Y., 1980, 139.
32. **Cowgill de Narvaez, C. and Schaup, H. W.,** *In vitro* transcriptionally coupled assembly of *Escherichia coli* ribosomal subunits, *J. Mol. Biol.*, 134, 1, 1979.
33. **Lindahl, L.,** Intermediates and time kinetics of the *in vivo* assembly of *Escherichia coli* ribosomes, *J. Mol. Biol.*, 92, 15, 1975.
34. **Pace, N. R.,** Structure and synthesis of the ribosomal RNA of prokaryotes, *Bacteriol. Rev.*, 37, 562, 1973.
35. **Roy, M. K., Singh, B., Ray, B. K., and Apirion, D.,** Maturation of 5S rRNA: ribonuclease E cleavages and their dependence on precursor sequences, *Eur. J. Biochem.*, 131, 119, 1983.
36. **Feunteun, J., Jordan, B. K., and Monier, R.,** Study of the maturation of 5S RNA precursors, *J. Mol. Biol.*, 70, 465, 1972.
37. **Galibert, F., Tiollais, P., Sanfourche, F., and Boiron, M.,** Coordination de la transcription des RNA 5S et 23S et coordination de la maturation du RNA 5S et de la sous-unité ribosomique 50S chez *Escherichia coli*, *Eur. J. Biochem.*, 20, 381, 1971.
38. **Pace, N. R., Meyhack, B., Pace, B., and Sogin, M. L.,** The interaction of RNase M5 with a 5S rRNA precursor, in *Transfer RNA: Biological Aspects*, Söll, D., Abelson, J., and Schimmel, P.R., Eds., Cold Spring Harbor Laboratory, Cold Spring Harbor, N.Y., 1980, 155.
39. **Altman, S.,** Transfer RNA biosynthesis, *Int. Rev. Biochem.*, 17, 19, 1978.
40. **Cudny, H. and Deutscher, M. P.,** Apparent involvement of ribonuclease D in the 3′ processing of tRNA precursors, *Proc. Natl. Acad. Sci. U.S.A.*, 77, 837, 1980.
41. **Sekiya, T., Contreras, R., Takeya, T., and Khorana, H. G.,** Total synthesis of a tyrosine suppressor transfer RNA gene. XVII. Transcription, *in vitro*, of the synthetic gene and processing of the primary transcript to transfer RNA, *J. Biol. Chem.*, 254, 5802, 1979.

42. **Fukada, K. and Abelson, J.,** DNA sequence of a T4 transfer RNA gene cluster, *J. Mol. Biol.*, 139, 377, 1980.
43. **Hudson, L., Rossi, J., and Landy, A.,** Dual function transcripts specifying tRNA and mRNA, *Nature (London)*, 294, 422, 1981.
44. **Schedl, P., Roberts, J., and Primakoff, P.,** *In vivo* processing of *E. coli* tRNA precursors, *Cell*, 8, 581, 1976.
45. **Shimura, Y., Sakano, H., and Nagawa, F.,** Specific ribonucleases involved in processing of tRNA precursors of *Escherichia coli*. Partial purification and some properties, *Eur. J. Biochem.*, 86, 267, 1978.
46. **Watson, N. and Apirion, D.,** Ribonuclease F, a putative processing endonuclease from *Escherichia coli*, *Biochem. Biophys. Res. Commun.*, 103, 543, 1981.
47. **Gurevitz, M., Watson, N., and Apirion, D.,** A cleavage site of ribonuclease F: a putative processing endoribonuclease from *Escherichia coli*, *Eur. J. Biochem.*, 124, 533, 1982.
48. **Knapp, G., Ogden, R. C., Peebles, C. L., and Abelson, J.,** Splicing of yeast tRNA precursors: Structure of the reaction intermediates, *Cell*, 18, 37, 1979.
49. **McClain, W. H.,** A role for ribonuclease III in synthesis of bacteriophage T4 transfer RNAs, *Biochem. Biophys. Res. Commun.*, 86, 718, 1979.
50. **Pragai, B. and Apirion, D.,** Processing of bacteriophage T4 tRNAs: the role of RNase III, *J. Mol. Biol.*, 153, 619, 1981.
51. **Dunn, J. J. and Studier, F. W.,** Effect of RNase III cleavage on translation of bacteriophage T7 messenger RNAs, *J. Mol. Biol.*, 99, 487, 1975.
52. **Hercules, K., Schweiger, M., and Saurbier, W.,** Cleavage by RNase III converts T3 and T7 early precursor RNA into translatable message, *Proc. Natl. Acad. Sci. U.S.A.*, 71, 840, 1974.
53. **Gitelman, D. R. and Apirion, D.,** The synthesis of some proteins is affected in RNA processing mutants of *Escherichia coli*, *Biochem. Biophys. Res. Commun.*, 96, 1063, 1980.
54. **Shen, V., Cynamon, M., Daugherty, B., Kung, H.-F., and Schlessinger, D.,** Functional inactivation of lac α-peptide mRNA by a factor that purifies with *Escherichia coli* RNase III, *J. Biol. Chem.*, 256, 1896, 1981.
55. **Kindler, P., Keil, T. V., and Hofschneider, P. H.,** Isolation and characterization of a ribonuclease III deficient mutant of *Escherichia coli*, *Mol. Gen. Genet.*, 126, 53, 1973.
56. **Robertson, H. D., Webster, R. E., and Zinder, N. D.,** Purification and properties of ribonuclease III from *Escherichia coli*, *J. Biol. Chem.*, 243, 82, 1968.
57. **Robertson, H. D. and Dunn, J. J.,** Ribonucleic acid processing activity of *Escherichia coli* ribonuclease III, *J. Biol. Chem.*, 250, 3050, 1975.
58. **Crouch, R. J.,** Ribonuclease III does not degrade deoxyribonucleic acid-ribonucleic acid hybrids, *J. Biol. Chem.*, 249, 1314, 1974.
59. **Wertz, G. W. and Davis, N. L.,** RNase III cleaves vesicular stomatitis virus genome-length RNAs but fails to cleave viral mRNAs, *J. Virol.*, 30, 108, 1979.
60. **Harris, T. J., Dunn, J. J., and Wimmer, E.,** Identification of specific fragments containing the 5′ end of poliovirus after ribonuclease III digestion, *Nucleic Acids Res.*, 5, 4039, 1978.
61. **Young, R. A. and Steitz, J. A.,** Complementary sequences 1700 nucleotides apart form a ribonuclease III cleavage site in *Escherichia coli* ribosomal precursor RNA, *Proc. Natl. Acad. Sci. U.S.A.*, 75, 3593, 1978.
62. **Bram, R. J., Young, R. A., and Steitz, J. A.,** The ribonuclease III site flanking 23S sequences in the 30S ribosomal precursor RNA of *E. coli*, *Cell*, 19, 393, 1980.
63. **Dunn, J. J. and Studier, F. W.,** Nucleotide sequence from the genetic left end of bacteriophage T7 DNA to the beginning of gene 4, *J. Mol. Biol.*, 148, 303, 1981.
64. **Brosius, J., Dull, T. J., Sleeter, D. D., and Noller, H. F.,** Gene organization and primary structure of a ribosomal RNA operon from *Escherichia coli*, *J. Mol. Biol.*, 148, 107, 1981.
65. **Robertson, H. D.,** Structure and function of RNA processing signals, in *Nucleic Acid-Protein Recognition*, Vogel, H.J., Ed., Academic Press, New York, 1977, 549.
66. **Robertson, H. D., Dickson, E., and Dunn, J. J.,** A nucleotide sequence from a ribonuclease III processing site in bacteriophage T7 RNA, *Proc. Natl. Acad. Sci. U.S.A.*, 74, 822, 1977.
67. **Rosenberg, M. and Kramer, R. A.,** Nucleotide sequence surrounding a ribonuclease III processing site in bacteriophage T7 RNA, *Proc. Natl. Acad. Sci. U.S.A.*, 74, 984, 1977.
68. **Grünberger, D., Maslova, R. N., and Sorm, F. L.,** Effect of 8-azaguanine on the synthesis of pulse-labeled ribonucleic acid in *Bacillus cereus*, *Collect. Czech. Chem. Commun.*, 29, 152, 1964.
69. **Downard, J. S. and Whitely, H. R.,** Early RNAs in SP82- and SP01-infected *Bacillus subtilis* may be processed, *J. Virol.*, 37, 1075, 1981.
70. **Meyhack, B., Meyhack, I., and Apirion, D.,** Processing of precursor particles containing 17S rRNA in a cell free system, *FEBS Lett.*, 49, 215, 1974.

71. **Hayes, F. and Vasseur, M.,** Processing of 17S *Escherichia coli* precursor RNA in the 27S pre-ribosomal particle, *Eur. J. Biochem.*, 61, 433, 1976.
72. **Dahlberg, A. E., Dahlberg, J. E., Lund, E., Tokimatsu, H., Rabson, A. B., Calvert, P. C., Reynolds, F., and Zahalak, M.,** Processing of the 5′ end of *Escherichia coli* 16S ribosomal RNA, *Proc. Natl. Acad. Sci. U.S.A.*, 75, 3598, 1978.
73. **Pace, N. R., Pato, M. L., McKibbin, J., and Radcliffe, C. W.,** Precursors of 5S ribosomal RNA in *Bacillus subtilis*, *J. Mol. Biol.*, 75, 619, 1973.
74. **Sogin, M. L., Pace, B., and Pace, N. R.,** Partial purification and properties of a ribosomal RNA maturation endonuclease from *Bacillus subtilis*, *J. Biol. Chem.*, 252, 1350, 1977.
75. **Stiekema, W. J., Raue, H. A., Duin, M. M. C., and Planta, R.,** Structural features of *Bacillus* precursor 5S RNA involved in the interaction with RNase M5, *Nucleic Acids Res.*, 8, 5411, 1980.
76. **Meyhack, B., Pace, B., Uhlenbeck, O., and Pace, N. R.,** Use of T4 RNA ligase to construct model substrates for a ribosomal RNA maturation endonuclease, *Proc. Natl. Acad. Sci. U.S.A.*, 75, 3045, 1978.
77. **Stahl, D. A., Meyhack, B., and Pace, N. R.,** Recognition of local nucleotide conformation in contrast to sequence by a rRNA processing endonuclease, *Proc. Natl. Acad. Sci. U.S.A.*, 77, 5644, 1980.
78. **Ghora, B. K. and Apirion, D.,** Structural analysis and *in vitro* processing to p5 rRNA of a 9S RNA molecule isolated from an *rne* mutant of *E. coli*, *Cell*, 15, 1055, 1978.
79. **Misra, T. K. and Apirion, D.,** RNase E, an RNA processing enzyme from *Escherichia coli*, *J. Biol. Chem.*, 254, 11154, 1979.
80. **Misra, T. K. and Apirion, D.,** Gene *rne* affects the structure of the ribonucleic acid processing enzyme ribonuclease E of *Escherichia coli*, *J. Bacteriol.*, 142, 359, 1980.
81. **Ray, B. K. and Apirion, D.,** RNase P is dependent on RNase E action in processing monomeric RNA precursors which accumulate in an RNase E-mutant of *Escherichia coli*, *J. Mol. Biol.*, 149, 599, 1981.
82. **Mazzara, G. P. and McClain, W. H.,** tRNA synthesis, in *Transfer RNA: Biological Aspects*, Söll, D., Abelson, J., and Schimmel, P. R., Eds., Cold Spring Harbor Laboratory, Cold Spring Harbor, N.Y., 1980, 3.
83. **Daniel, V.,** Biosynthesis of transfer RNA, *CRC Crit. Rev. Biochem.*, 9, 253, 1981.
84. **Bothwell, A. L. M., Garber, R. L., and Altman, S.,** Nucleotide sequence and *in vitro* processing of a precursor molecule to *Escherichia coli* 4.5S RNA, *J. Biol. Chem.*, 251, 7709, 1976.
85. **Bothwell, A. L. M., Stark, B. C., and Altman, S.,** Ribonuclease P substrate specificity: cleavage of a bacteriophage ϕ80-induced RNA, *Proc. Natl. Acad. Sci. U.S.A.*, 73, 1912, 1976.
86. **Stark, B. P., Kole, R., Bowman, E. J., and Altman, S.,** Ribonuclease P: an enzyme with an essential RNA component, *Proc. Natl. Acad. Sci. U.S.A.*, 75, 3717, 1978.
87. **Altman, S., Brown, E. J., Garber, R. L., Kole, R., Koski, R. A., and Stark, B. C.,** Aspects of RNase P structure and function, in *Transfer RNA: Biological Aspects*, Söll, D., Abelson, J., and Schimmel, P. R., Eds., Cold Spring Harbor Laboratory, Cold Spring Harbor, N.Y., 1980, 71.
88. **Guthrie, C. and Atchison, R.,** Biochemical characterization of RNase P: a tRNA processing activity with protein and RNA components, in *Transfer RNA: Biological Aspects*, Söll, D., Abelson, J., and Schimmel, P. R., Eds., Cold Spring Harbor Laboratory, Cold Spring Harbor, N.Y., 1980, 83.
89. **Kole, R. and Altman, S.,** Properties of purified RNase P from *Escherichia coli*, *Biochemistry*, 20, 1902, 1981.
90. **Guthrie, C., Seidman, J. G., Comer, M. M., Bock, R. M., Schmidt, F. J., Barrell, B. G., and McClain, W. H.,** The biology of bacteriophage T4 transfer RNAs, *Brookhaven Symp. Biol.*, 26, 106, 1975.
91. **Kline, L., Nishikawa, S., and Söll, D.,** Partial purification of RNase P from *Schizosaccharomyces pombe*, *J. Biol. Chem.*, 256, 5058, 1981.
92. **Gardiner, K. and Pace, N. R.,** RNase P of *Bacillus subtilis* has a RNA component, *J. Biol. Chem.*, 255, 7507, 1980.
93. **Cudny, H., Zaniewski, R., and Deutscher, M. P.,** *Escherichia coli*, RNase D. Purification and structural characterization of a putative processing nuclease, *J. Biol. Chem.*, 256, 5627, 1981.
94. **Bikoff, E. K. and Gefter, M. L.,** *In vitro* synthesis of transfer RNA. I. Purification of required components, *J. Biol. Chem.*, 250, 6240, 1975.
95. **Ghosh, R. K. and Deutscher, M. P.,** Identification of an *Escherichia coli* nuclease acting on structurally altered transfer RNA molecules, *J. Biol. Chem.*, 253, 997, 1978.
96. **Cudny, H., Zaniewski, R., and Deutscher, M. P.,** *Escherichia coli* RNase D catalytic properties and substrate specificity, *J. Biol. Chem.*, 256, 5633, 1981.
97. **Seidman, J. G., Schmidt, F. J., Foss, K., and McClain, W. H.,** A mutant of *Escherichia coli* defective in removing 3′ terminal nucleotides from some transfer RNA molecules, *Cell*, 5, 389, 1975.

98. **Schmidt, F. J. and McClain, W. H.,** An *Escherichia coli* ribonuclease which removes an extra nucleotide from a biosynthetic intermediate of bacteriophage T4 proline transfer RNA, *Nucleic Acids Res.*, 5, 4129, 1978.
99. **Cudny, H., Roy, P., and Deutscher, M. R.,** Alteration of *E. coli* RNase D by infection with bacteriophage T4, *Biochem. Biophys. Res. Commun.*, 98, 337, 1981.
100. **Lim, L. W. and Kennel, D.,** Evidence for random endonucleolytic cleavages between messages in decay of *Escherichia coli* trp mRNA, *J. Mol. Biol.*, 141, 227, 1980.
101. **Apirion, D. and Gitelman, D. R.,** Decay of RNA in RNA processing mutants of *Escherichia coli, Mol. Gen. Genet.*, 177, 339, 1980.
102. **Misra, T. K. and Apirion, D.,** Characterization of an endoribonuclease, RNase N, from *Escherichia coli, J. Biol. Chem.*, 253, 5594, 1978.
103. **Apirion, D.,** The fate of mRNA and rRNA in *Escherichia coli, Brookhaven Symp. Biol.*, 26, 286, 1975.
104. **Sekiguchi, M. and Cohen, S. S.,** The selective degradation of phage-induced ribonucleic acid by polynucleotide phosphorylase, *J. Biol. Chem.*, 238, 349, 1963.
105. **Spahr, P. R. and Schlessinger, D.,** Breakdown of messenger ribonucleic acid by a potassium-activated phosphodiester from *Escherichia coli, J. Biol. Chem.*, 238, PC2251, 1963.
106. **Simon, M. and Apirion, D.,** Increased inactivation of messenger in an *Escherichia coli* ts mutant, *Nature New Biol.*, 239, 79, 1972.
107. **Kivity-Vogel, T. and Elson, D.,** A correlation between ribonuclease II and the *in vivo* inactivation of messenger RNA in *E. coli, Biochem. Biophys. Res. Commun.*, 33, 412, 1968.
108. **Kinscherf, T. G. and Apirion, D.,** Polynucleotide phosphorylase can participate in decay of mRNA in *Escherichia coli* in the absence of RNase II, *Mol. Gen. Genet.*, 139, 357, 1975.
109. **Chaney, S. G. and Boyer, P. D.,** Incorporation of water oxygens into intracellular nucleotides and RNA. II. Predominantly hydrolytic RNA turnover in *Escherichia coli, J. Mol. Biol.*, 64, 581, 1973.
110. **Duffy, J. J., Chaney, S. G., and Boyer, P. D.,** Incorporation of water oxygens into intracellular nucleotides and RNA. I. Predominantly non-hydrolytic RNA turnover in *Bacillus subtilis, J. Mol. Biol.*, 64, 565, 1972.
111. **Schroeder, E., McKibbin, J., Sogin, M. L., and Pace, N. R.,** Mode of degradation of precursor-specific ribonucleic acid fragments by *Bacillus subtilis, J. Bacteriol.*, 130, 1000, 1977.
112. **Carey, J. and Uhlenbeck, O.,** personal communication.
113. **Betz, J.,** unpublished data.
114. **Pace, B.,** unpublished data.
115. **Stahl, D. and Pace, B.,** unpublished data.
116. **Marsh, T.,** unpublished data.
117. **Guthrie, C. and Uhlenbeck, O.,** unpublished data.
118. **Gardiner, K. and Pace, N.,** unpublished data.
119. **Studier, W. F. and Dunn, J. J.,** personal communication.

Chapter 2

# MOLECULAR BIOLOGY OF RNA PROCESSING IN PROKARYOTIC CELLS

**David Apirion and Peter Gegenheimer**

## TABLE OF CONTENTS

# I. INTRODUCTION

Processing of RNA is a feature of RNA metabolism that contributes to the determination of the final population of active RNA molecules in the cell. Processing is the sum of events which converts a primary RNA transcript into a functional molecule.

The immediate transcription products of prokaryotic genes are frequently not identical with the RNA molecules which are functional in the cell. The stable RNA species such as ribosomal and transfer RNAs as well as a few of the messenger RNAs of *Escherichia coli* differ from the primary transcription products of their genes in one or more aspects. For example, ribosomal RNA transcription units of *E. coli* contain sequences that specify the three rRNAs 16S, 23S, and 5S as well as one or more tRNAs. Similarly, transfer RNA genes are frequently clustered and cotranscribed in multimers of up to seven identical or different tRNAs. Moreover, transfer RNA genes are located in the spacer region of every rRNA transcription unit of *E. coli* and in the trailer regions of some of these gene clusters.

In order to produce functional RNA molecules from the immediate products of transcription, prokaryotic cells carry out three basic types of RNA maturation reactions: first, separation of polycistronic transcripts into monocistronic precursor RNAs; second, accurate recognition of the mature 5′ and 3′ termini, removing extraneous nucleotides without altering the terminal sequences themselves; and third, modification of the base or the ribose moiety of the four primary nucleosides in an RNA chain.

Not every RNA molecule is subject to all three of these processes. Certain polycistronic tRNA transcripts do require all three; many — probably most — messenger RNAs require no processing for their functional expression. This article discusses the first two of these RNA maturation reactions, *namely those endo- and exoribonucleolytic events which reduce the size* of primary RNA transcripts.

In the last decade, a number of reviews covering different aspects of RNA maturation have appeared, some of them discussing processing of stable RNA[1] or all RNAs[2-4a] in both prokaryotes and eukaryotes, and others describing rRNA[5] or tRNA processing in *E. coli*.[6-9]

Recently, a combination of genetic and enzymological approaches has provided a great deal of new information on in vivo RNA processing in bacterial cells. Here, we shall review the genetics of RNA processing and then describe the events occurring during endo- and exoribonucleolytic processing of ribosomal, transfer, and messenger RNAs, primarily in *Escherichia coli*. While we shall try to be up-to-date and to include all the relevant studies, we are likely to err in judgment and to neglect important contributions. We hope that these unintentional oversights will not be serious enough to diminish the value of this review to the general reader.

# II. GENETICS OF RNA PROCESSING

## A. Introduction

With the realization that all the three ribosomal RNAs are cotranscribed from a single promoter[5] it became obvious that there must be a way by which a transcript synthesized from a single promoter ends up as three independent molecules. *A priori*, two obvious solutions could be offered: the first involving a single promoter with multiple termination sites, while the second could involve cutting and trimming events, i.e., processing. It is clear that nature preferred the second solution, and to the best of our knowledge, thus far, the first solution has not been encountered in any case.

The realization that processing of rRNA must take place did not help to solve the problem of how to study this process. Primarily, this stemmed from the fact that in the wild-type cell of *E. coli*, or any other bacteria, almost no obvious rRNA precursors are observed even when cells are labeled with radioisotopes of RNA precursors for exceedingly short periods. (The exception in *E. coli* is a 17S rRNA which is a precursor to the 16S mature rRNA found in the 30S ribosomal subunit. This will be dealt with in some detail later.) The reason for this problem became evident only later, when it was realized that most of the rRNA processing during transcription takes place prior to the termination of the transcript.[4] However, some of the RNA processing occurs only after later cleavages have been introduced, and in these cases precursors can be accumulated, albeit transiently, in the wild-type cell.

In order to circumvent the lack of RNA precursors in the wild-type cell, it became obvious that it would be necessary to inhibit the reaction(s) involved in RNA processing. This, of course, can be carried out by finding specific inhibitors or by isolating mutants defective in these reactions. No reports appeared on inhibitors of primary RNA processing, but the second approach, that of mutant isolation, proved to be very rewarding.

### B. Mutants Defective in RNA Processing

Genetic analysis of RNA processing reactions has provided numerous benefits — in part because the role assigned to particular enzymes in vitro can be verified in vivo, and also because isolation of mutant strains defective in a particular processing step has frequently allowed detection of RNA precursor species which are not detectable in wild-type cells, and which can then be used as substrates in the purification of novel processing enzymes and in the investigation of processing reactions.

The first major steps towards delineating pathways and mechanisms of RNA processing were taken in 1973 with the isolation of two mutant strains, one of which lacked ribonuclease III activity[10] and the other of which was defective in RNase P.[11] Subsequently, screening of colonies derived from mutagenized cells led to the detection of several other mutants with defects in rRNA or tRNA biosynthesis. The isolation and properties of each of these mutants will be described below.

### C. *rnp* Mutants

#### *1. Identification*

To identify enzymes required for tRNA biosynthesis in vivo, Schedl and Primakoff[11] screened temperature-sensitive mutants which at the restrictive temperature were unable to express the amber-suppressing su3$^+$ tRNA$_1^{Tyr}$ after infection with bacteriophage φ80 carrying this gene. Several such mutants were isolated, one of which, A49, proved upon further analysis to be defective in RNase P. Cell-free extracts from the mutant strain showed a greatly reduced specific activity of RNase P[12] and could not process the su3$^+$ tRNA$_1^{Tyr}$ precursor to its mature form (at 30 or 42°C), whereas the parental extract was fully active in such reactions.

Ozeki and collaborators[13] chose a different approach, devising an ingenious selection procedure in which expression of the amber-suppressor su$_3^+$ tRNA gene is lethal. This procedure yielded six mutants in which RNase P activity was thermolabile.[13-15] Another mutant defective in RNase P activity was isolated from a strain initially characterized as defective in rRNA processing[16] (see below). The conditional lethality conferred by these RNase P mutations demonstrated that RNase P cleavage is an essential requirement for the functioning of at least some tRNA species.

Genetic analysis indicates that some of these mutations fall into three loci: *rnpA* at minute 82.5 of the *E. coli* chromosome,[17] *rnpB* between minutes 68.5 and 69,[16,18] and a third locus which has not yet been mapped. These loci are shown in Figure 1. Since complementation tests were not carried out one cannot conclude that all the known *rnp* mutants lie in three genes. The four *rnp* mutations that were studied most extensively are *rnpA49*,[11,17] *rnpB3187*,[16] and *ts241* and *ts709*.[12,13-15,18] While the *ts709* mutation maps near *rnpB3187* and could be in the same gene, the *ts241* mutation, which is not allelic with *ts709*,[18] is not

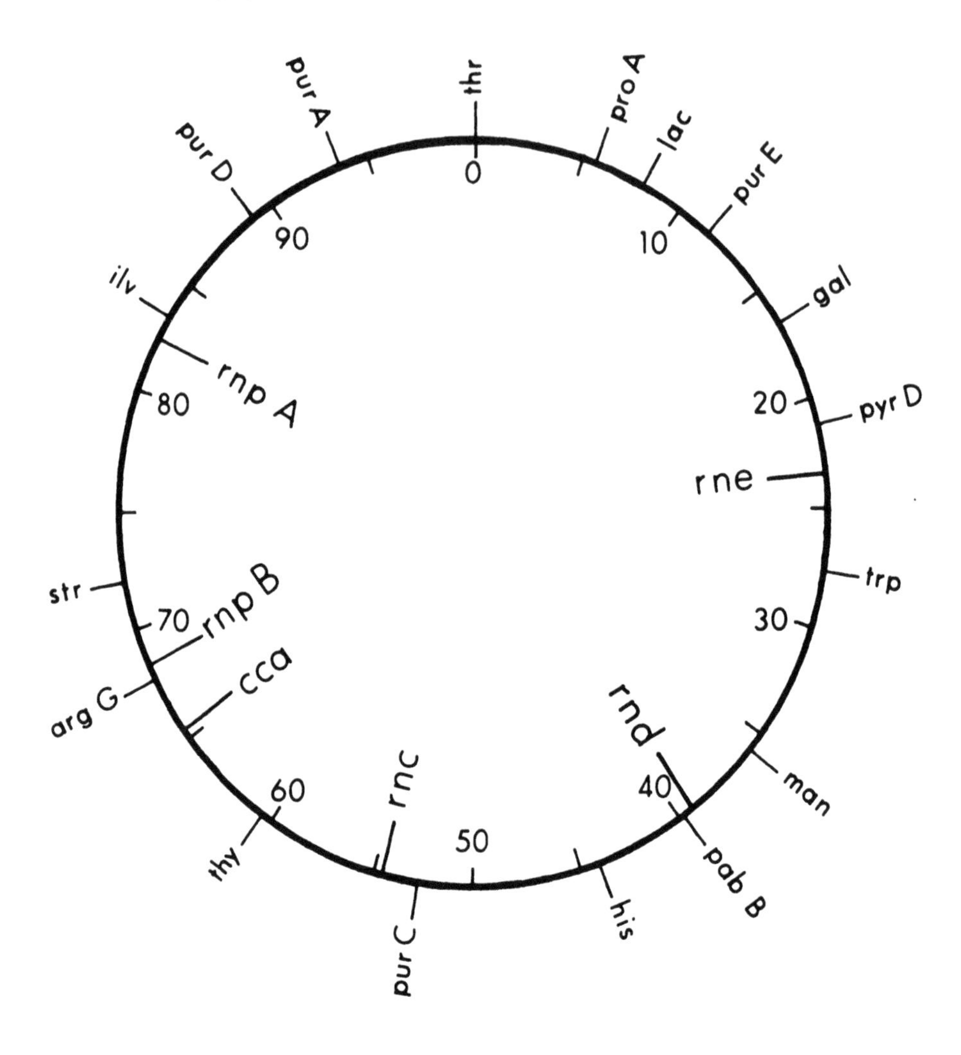


FIGURE 1. Genetic map of RNA processing mutations in *Escherichia coli* K12. The RNA processing genes are inside the circle. The loci are discussed in the text: *rnc*, RNase III;[35,36] *rne*, RNase E;[42] *rnp*, RNase P;[16-18] *cca*, nucleotidyltransferase.[102,103] A mutation which affects RNase D *(rnd)* was mapped at 39.8 min.[104]

complemented by an F′ that complements the *rnpA49* mutation;[18a] therefore the *rnp* mutation *ts241* in all likelihood defines a third gene affecting the RNase P activity *(rnpC)*.

In vitro reconstitution experiments demonstrated that the RNase P enzyme contains an obligatory RNA component.[12,19] The available data are insufficient to determine which part of the enzyme (RNA or protein) is structurally affected in all the available *rnp* mutants. There is evidence to suggest that the *rnpA49* mutation affects the structure of the protein.[12] When RNase P is purified, two relatively small RNAs, M1 and M2, are found to be associated with it.[12,19] The fingerprint of these molecules[12] shows that M2 but not M1 is identical with the 10S RNA identified in *E. coli* cells.[20,21] In vitro RNase P activity requires the presence of M1 but not of M2.[22] A plasmid that complements the *rnpA49* mutation contains *E. coli* DNA sequences that code for a small RNA which is similar to M1 RNA.[18a,22a] Moreover, studies from two laboratories[18a,22a] show that all the *rnpB* mutants tested do not accumulate any detectable levels of the RNA that is being coded by the plasmid. The DNA corresponding to M1 RNA has been sequenced.[23] Obviously, while it is clear that an RNA moiety is required for RNase P activity in vitro, many more studies are required to characterize this RNA, to clarify the relationship between 10S RNA and M1 RNA and to determine their roles in RNase P activity, and to find out whether or not an RNA moiety is required for RNase P activity in vivo.

*2. tRNA Precursors Accumulated in* rnp *Strains*

The RNA species accumulated in various *rnp* mutants have been examined by several groups.[14,15,24-26] The results can be summarized as follows: a large group of RNAs are about 4S in size. These "small monomeric" tRNA precursors bear short stretches of extra nucleotides at their 5′ termini. These species apparently have undergone some processing in vivo as they do not contain 5′ triphosphoryl termini. In addition, only a few residues, or none at all, remain at their 3′ termini. Another group of RNAs (about 4.5S and 5S) are "large monomeric" precursors, some of which do contain 5′ triphosphates and hence represent primary transcripts unprocessed at least at their 5′ ends. Many of these species have also undergone some 3′ cleavages since tRNAs known to be transcribed as much longer precursors, or whose genes are known to include additional 3′ sequences, are found in transcripts bearing only a few extra trailing bases. The prototype of this class is the $tRNA_1^{Tyr}$ precursor, which has 41 extra 5′ bases starting with pppG, and only 1 to 3 extra 3′ nucleotides,[27] although its transcription unit continues for a considerable distance past its 3′ end.[28]

Perhaps the most interesting species produced by *rnp* mutants, however, are the multimeric tRNA transcripts, which may contain up to four or five individual tRNA species.[7,8,14,15,25,26,29] Many of these species contain 5′ triphosphate termini. Some of the monomers and dimers observed in the *rnp* strain are apparently subsets cleaved from these larger species.[15] This phenomenon of partial cleavage of tRNA precursors in temperature-sensitive RNase P cells suggested the involvement of additional endonucleolytic tRNA processing activities apart from RNase P.[24,30] The effects of combining *rnp* mutations with other RNA processing mutations will be discussed later.

Recently transcripts from the *tyrT* locus ($tRNA_1^{Tyr}$) were identified in *rnp* and *rnp rnc* mutant strains of *E. coli* by blotting denatured separated total RNA to a proper matrix and hybridizing it with a specific DNA probe.[30a] Such techniques are potentially capable of revealing all the precursors that are involved in an RNA processing pathway.

**D. An *rnc* Mutant**

As a means of facilitating in vitro studies on the replicative intermediates of RNA bacteriophages, Hofschneider and co-workers[10] screened extracts of heavily mutagenized *E. coli* cells for the inability to solubilize $^3H$-labeled double-stranded replicative forms of phage M13. One such mutant, AB301-105, proved to be unconditionally defective in the activity of ribonuclease III. The ability of extracts from the mutant strain to solubilize labeled synthetic ribohomopolymer duplexes was less than 1% that of the parental strain.[10] The apparent "residual" activity detected by such assays cannot be attributed to RNase III, however, since it has a different pH optimum and different (more stable) thermal inactivation kinetics.[31,32] Furthermore, in RNase $III^-$ strains no RNase III processing activity can be detected in vivo[33] (see below).

Genetic analysis of strain AB301-105 revealed the presence of at least seven mutations other than the RNase III lesion.[34] To study the effect of the RNase III mutation *rnc-105* by itself, the mutation was first mapped at minute 54.5[35,36] (see Figure 1) and isogenic $rnc^+$/*rnc-105* pairs were then constructed by transducing the *rnc-105* mutation into a different genetic background[35] which allowed further study of the physiological role of RNase III (see below).

Electrophoretic analysis of labeled RNAs from an *rnc-105* strain demonstrates a dramatic effect on ribosomal RNA production.[37] The mutant synthesizes a variety of RNAs[37-39] (30S, 25S, and 18S) whose presence is solely dependent upon the *rnc-105* allele.[32] (The derivation of all these new RNAs will be discussed below.)

### E. An *rne* Mutant

In an attempt to obtain mutants defective in rRNA processing enzymes other than RNase III, Apirion and Lassar[40] mutagenized RNase III$^-$ cells and, after enrichment for temperature-sensitive colonies, they screened the mutants for defects in the pattern of radioactively labeled rRNA species by electrophoresis in polyacrylamide gels. One such mutant was detected which at the nonpermissive temperature failed to produce 23S and 5S rRNAs, and accumulated instead a 25S RNA species[40] containing both 23S and 5S.[41] Transduction of the *ts* mutation into an RNase III$^+$ genetic background yielded a strain in which 23S production was normal, but in which 5S rRNA still did not appear at the restrictive temperature whereas several new low-molecular-weight RNAs were detected.[42,43] This mutation, *rne-3071*, is a recessive single point mutation which maps at minute 24 of the *E. coli* chromosome[42] (see Figure 1). A combination of the *rnc-105* and *rne-3071* mutations yields a phenotype similar to that of the original isolate.[42] When an *rne* mutant is shifted to the nonpermissive temperature, growth as well as macromolecule synthesis continues linearily but cell division is promptly blocked.[44]

A major new RNA detected in an *rne* strain is a 9S species (about 250 bases long), containing the entire sequence of p5 rRNA as well as extra sequences at both ends[45] (Figure 2). The 3′ end of 9S contains a stem-and-loop structure[45a,45b] similar to structures near known transcription termination sites.[46,47]

In *rne* cells, at 43°C, no 5S RNA sequences could be detected in the 5S region of polyacrylamide gels.[43] Moreover, RNase E partially purified from an *rne* strain was found to be thermolabile in vitro.[48] The *rne-3071* mutation, therefore, seems to determine a structurally altered RNase E. The *rne*$^+$ gene has been cloned using a λ charon vector.[49]

While 9S RNA contains no stable cellular RNA other than p5 rRNA, an *rne rnp* double mutant strain accumulates novel RNAs consisting of p5 rRNA joined to $tRNA_1^{Asp}$, or p5 plus $tRNA_1^{Asp}$ and $tRNA^{Trp}$.[50] This confirms previous observations that genes coding for $tRNA_1^{Asp}$ and $tRNA^{Trp}$ are found near, and are cotranscribed with, the p5 cistrons of some rRNA gene clusters.[51-53] Another of the novel 8S — 10S RNAs appearing in an *rne* single mutant has been shown to be a dimeric tRNA precursor containing sequences of $tRNA_1^{Leu}$, $tRNA_1^{His}$, and a 100-base-long 3′ trailer segment.[54] Monomeric precursors for $tRNA^{Ser}$ and $tRNA^{Asn}$ have also been detected in an *rne* strain.[55]

Recent studies on T4-directed RNAs that accumulate in an *rne* mutant suggest that in such a mutant RNAs can be accumulated which do not contain cleavage sites for RNase E.[55a] (This problem is discussed in greater detail in Chapter 3 of this volume.) Furthermore, the apparent extreme conservation of the sequences in all the seven rRNA genes where RNase E cleaves (see Figure 2), and the lack of such sequences in the tRNA precursors that accumulate in the *rne* mutant, further strengthen the view that the *rne* mutant can accumulate precursors which do not contain RNase E cleavage sites.

### F. Mutants That Affect Maturation of 16S rRNA

A mutant strain has been described in which maturation of the 5′ terminus of p16 rRNA is altered.[56] This slower-growing but viable strain has a slow rate of p16 maturation, and accumulates two intermediates both of which contain a 3′ terminus identical to that found in m16 rRNA, but whose 5′ ends are longer. One of these, a 16.3S molecule containing 50 extra 5′ nucleotides, is found in 30S ribosomal subunits on polysomes and might be competent in protein synthesis.[56] These 30S particles, or 70S ribosomes containing them, can be matured in vitro. A few extra nucleotides at the 5′ end of m16 rRNA do not affect

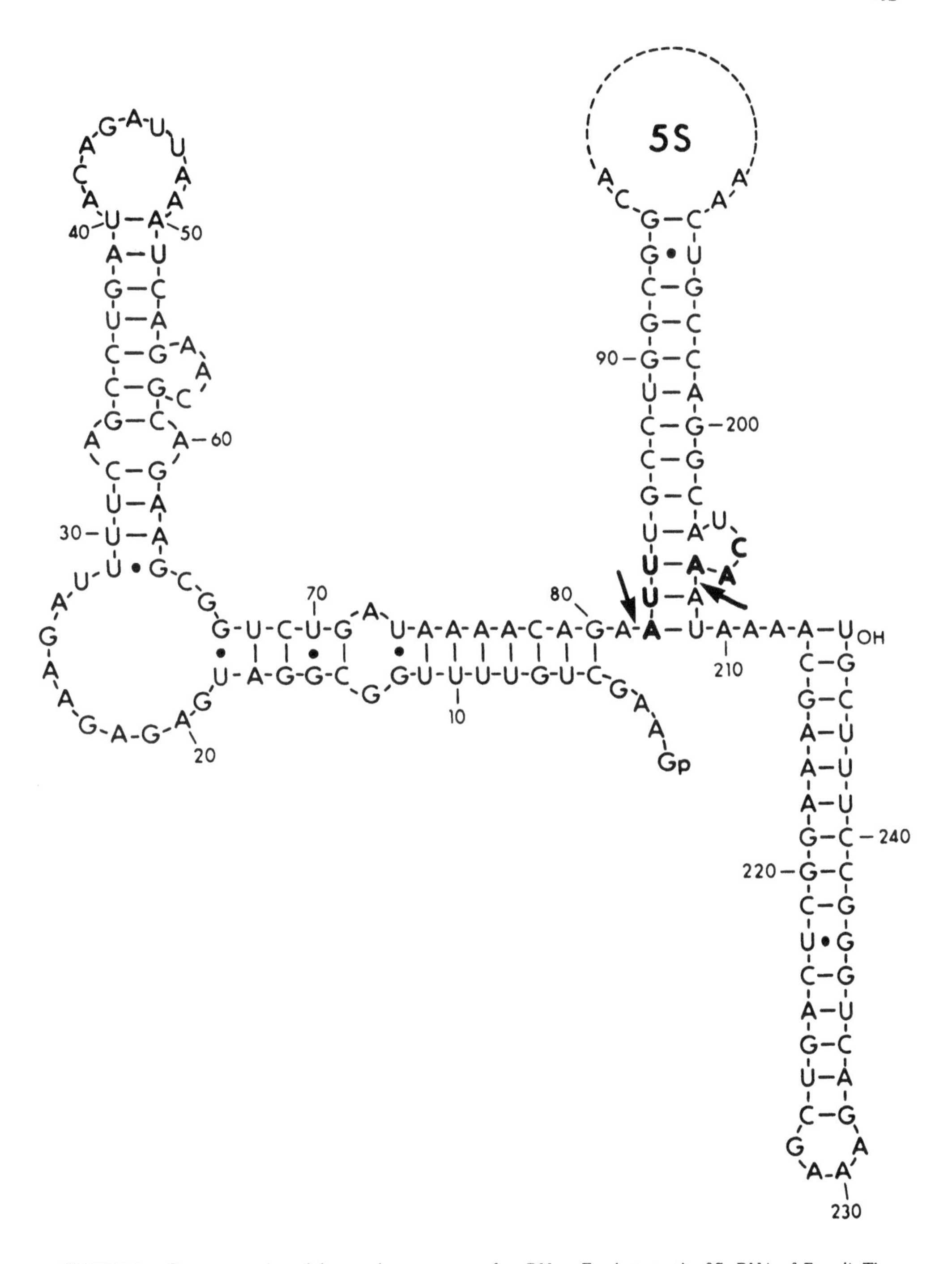


FIGURE 2. Sequence and partial secondary structure of an RNase E substrate, the 9S rRNA of *E. coli*. The sequence shown is from transcripts of the *rrnB* gene cluster[105] (a few base changes are found in the *rrnA*, *rrnE*, and the *rrnG* gene clusters). The last two stems starting with nucleotides 82 and 213, respectively, were verified experimentally.[45a] These stems as well as the structure starting with nucleotide 29 were also indicated by computer analysis of the sequence (R. J. Feldmann, National Institutes of Health). The secondary structure involving sequences from the 5′ end of the molecule was suggested by the finding that molecules missing the first eight nucleotides could not be cleaved by RNase E.[45b] The arrows show the positions of RNase E cleavages.[45,45b] The six bases, near the arrows, indicated with bold letters, are not included in the mature 5S rRNA. The structure shown is transcribed from the four rRNA gene clusters which do not contain trailer tRNAs. An almost identical RNA, up to nucleotide 209, is also transcribed from the other rRNA gene clusters which do contain trailer tRNAs.

ribosome function,[57] but whether the presence of 50 nucleotides is deleterious, or whether this extra fragment is removed before the 30S subunit becomes functional, is not known. Also unknown is the mechanism whereby m16 rRNA is ultimately formed in this mutant strain, which seems to lack an activity capable of p16 maturation in vitro.[56] No genetic analysis of this mutation(s) was carried out.

More recently, a temperature-sensitive mutant was isolated which seems to be defective in RNase M16.[56a] In this mutant, conversion in vivo of 17S to 16S (p16 to 16S) is reduced at permissive temperatures, and is virtually absent at nonpermissive temperatures. Temperature-independent revertants regained the capacity to mature 17S to 16S at all temperatures. In this mutant the maturation of p5 to m5 is normal at all temperatures. (The maturation of p23 to m23 was not tested.) Thus it is likely that different enzymes are involved in the maturation of p16 to m16 and of p5 to m5. The existence of this mutant also suggests that a single enzyme may be responsible for the maturation of p16 to m16. In this mutant, the p16 which accumulates is very similar and possibly identical to p16 accumulating when protein synthesis is blocked. No intermediates between p16 and m16 are observed.

Mutants like the one described above could be defective also in a ribosomal protein, because it is clear that for the maturation of p16 to 16S the p16 has to be in a ribosomal precursor particle. Therefore, it is logical to assume that mutants defective in maturation of p16 to m16 could result from alterations in a soluble factor(s), required for the reaction, as well as in 30S ribosomal proteins.

### G. The BN Mutant

*E. coli* strain BN (a B strain) is a viable mutant[58] which after infection with bacteriophage T4 accumulates several phage-coded tRNAs with extra sequences at their 3′ ends.[59] The mutant strain is apparently defective in an activity which removes these 3′ nucleotides.[59] This activity, denoted RNase "BN", might be related to one of the activities of RNase D.[60] The role of this activity in tRNA maturation will be described later.

### H. *rnd* Mutants

A mutant defective in RNase D was isolated by heavily mutagenizing a culture of *E. coli* and assaying individual colonies specifically for RNase D activity. However, none of these mutants, which are still viable, demonstrated any deficiency in RNA processing. This *rnd* mutation was mapped at 39.8 min on the *E. coli* map.[60a]

## III. PROCESSING OF RIBOSOMAL RNA

The p16 and p23 precursor ribosomal RNAs are the direct products of cleavage by RNase III,[61,62] which has a specificity for double-stranded RNA.[63] RNase III cleavages were proposed to occur in duplex structures formed by base-pairing between regions of the ribosomal RNA transcript flanking the 5′ and 3′ ends of 16S or of 23S sequences.[37,64,65] DNA sequences surrounding 16S and 23S genes in several different rRNA operons have been determined by several groups (cited in Reference 33), and in each case the expected sequence complementarity was found. Ribosomal RNA precursors from RNase III$^-$ cells have been isolated and their duplex regions studied.[33] The sequence and secondary structure of the p16 and p23 RNase III processing sites found in these duplex regions were described in some detail.[33] (For more details about the RNase III cleavage sites in rRNA of *E. coli* see Figure 9 in Chapter 1 of this volume.) Ribosomal RNA transcripts in the cell thus take on the form of giant ribonucleoprotein loops held together by duplex stems; p16 and p23 rRNA precursors arise by cleavage within these stems (Figure 3).

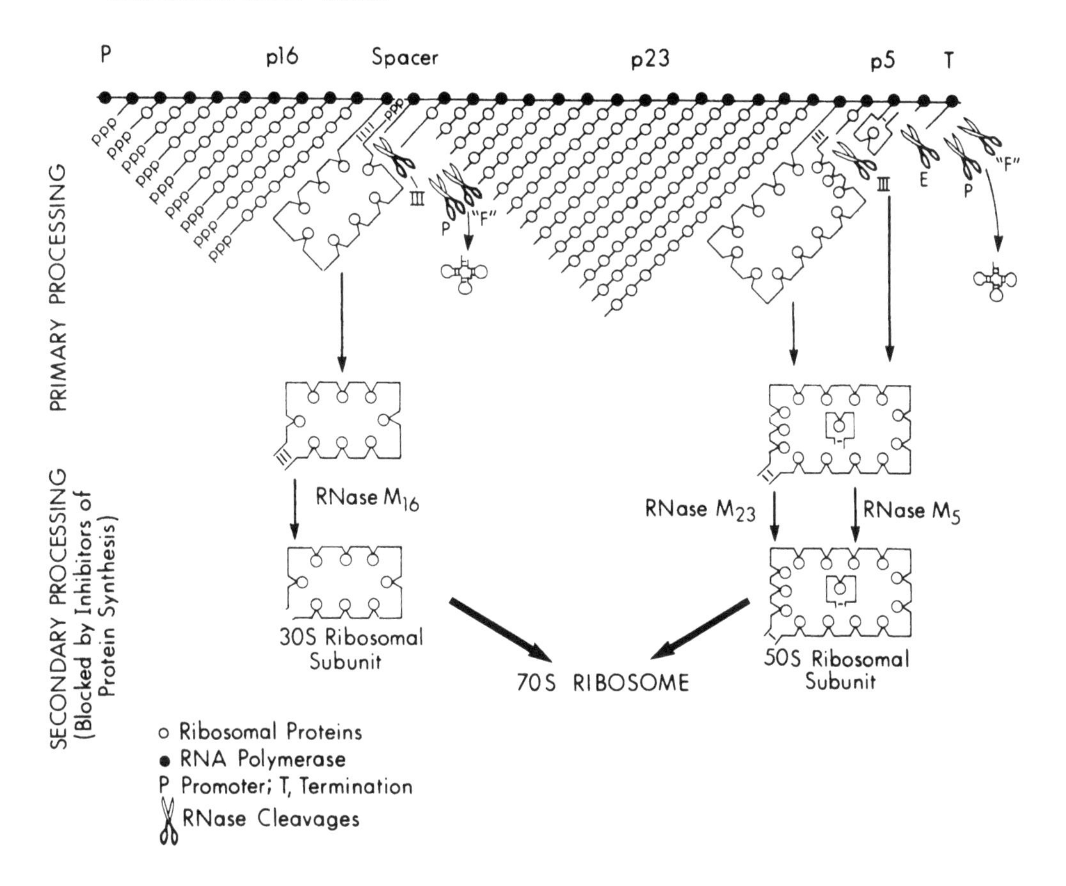


FIGURE 3. Maturation and assembly of ribosomes. Each nascent chain of ribosomal ribonucleoprotein represents a successive timepoint in the transcription of an rRNA operon by an RNA polymerase molecule. Many ribosomal proteins attach sequentially to the nascent rRNA,[71] and endonuclease cleavages by RNases III and E can occur only after the 3′ terminal region of their respective substrates has been synthesized and hydrogen-bonded to the 5′ terminal region. The spacer and trailer tRNAs are not drawn in the nascent transcript, but they too must assume their mature conformation before being cleaved by RNases P and "F". Secondary maturation of p16, p23, and p5 rRNAs occurs only in ribonucleoprotein particles and involves removal of some or all of the duplex formed between 5′ and 3′ termini in the precursor rRNA.

## A. Processing in Wild-Type Cells

The detailed sequence of processing events discussed here is illustrated by the models presented in Figures 3 and 4. An RNA polymerase molecule which initiates transcription of a ribosomal RNA gene cluster (Figures 3 and 4) continues to synthesize the various components of the polycistronic rRNA primary transcript, that is, leader sequences; p16 rRNA; spacer sequences containing tRNAs; p23 rRNA; p5 rRNA; and trailer sequences which in some operons contain tRNAs. Before transcription is terminated, however, processing endonucleases are already acting on the nascent transcript. As polymerase molecules complete synthesis of p16 rRNA, the complementary sequences flanking the m16 transcript anneal to form a double-stranded stem from which m16 sequences loop out. As Figures 3 and 4 show, the stem so formed is susceptible to endonucleolytic processing cleavage by RNase III, which cuts within it (Figure 4, cuts 1A and 1B) to release p16 precursor rRNA plus a 5′ leader fragment from the growing RNA chain. As the spacer region is synthesized,

# PROCESSING MAP OF RIBOSOMAL RNA

## SECONDARY STRUCTURE AND CLEAVAGE SITES

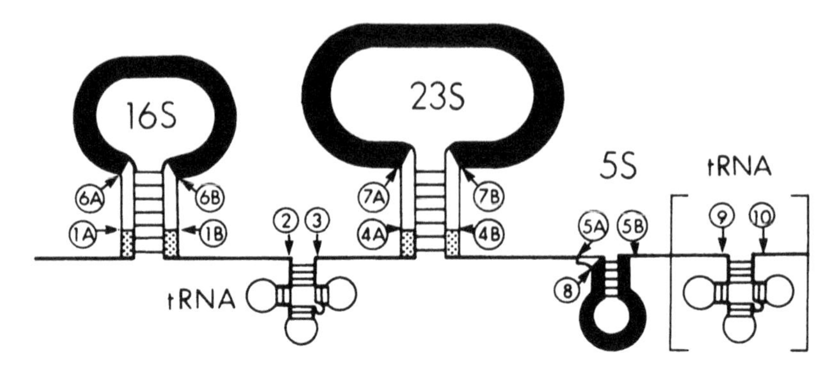


Cuts 1, 4. RNase III 2,9. RNase P 3,10. RNase F 5. RNase E
6. RNase M16 7. RNase "M23" 8. RNase "M5"

A

## PROCESSING IN WILD-TYPE STRAINS

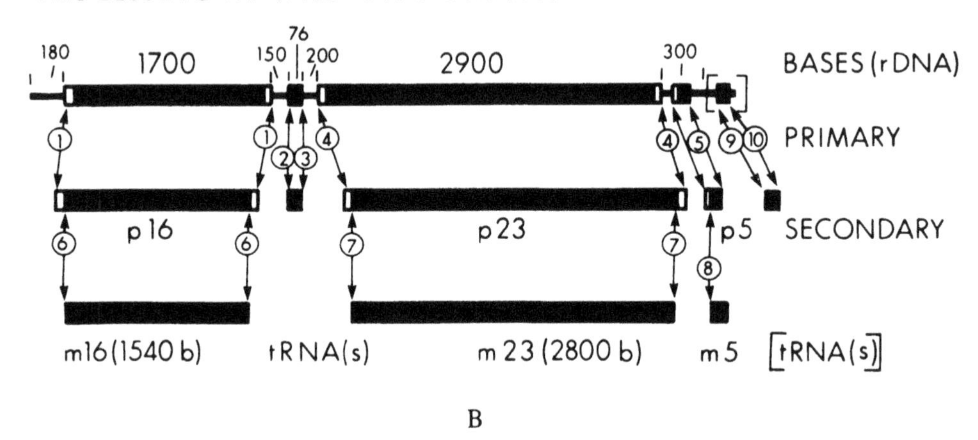


B

FIGURE 4. Structure and processing map of ribosomal RNA transcripts. (A) Structure and cleavage sites of the rRNA primary transcript (not to scale). Derived from data from a number of laboratories, summarized in References 37 and 72. Distal ("trailer") tRNAs are bracketed since not all ribosomal RNA clusters contain them. Transcripts may contain one or two spacer tRNAs, and zero, one, or two trailer RNAs. Arrows indicate endonucleolytic cleavage sites. Each cutting event is given a separate number, referring to the enzyme involved; A and B, where used, indicate that two (or more) separate cuts might be required. (Cut no. 8 is now known to be composed of two cuts or trimming events, on either side of the 5S rRNA.) Thick solid segments represent mature rRNA sequences, thick open segments represent precursor-specific sequences removed during secondary processing steps, stippled segments are sequences found only in p16b and p23b of RNase III$^-$ cells, and thin lines (except for tRNAs) represent nonconserved sequences discarded during primary processing. Enzymes are discussed in the text. (B) Processing in wild-type strains. The first line shows the transcriptional map of a representative rDNA unit, approximately to scale. Distances in bases are between vertical bars above the map. The primary and secondary cuts, numbered as in panel A, are shown above the products they generate. Open and solid segments are as in panel A.

tRNAs are removed by endonucleolytic cleavage (RNase "F") at or near the 3′ side, and at the 5′ end by RNase P (cuts 2 and 3 in Figure 4). Trimming of the 3′ end, perhaps by RNase D, might also be required to produce mature spacer tRNAs. As the RNA polymerase completes transcription of 23S genes, RNase III excises p23 sequences, again by cleaving in the double-stranded stem formed by complementary sequences surrounding m23 (Figure 4, cuts 4A and 4B). Transcription of the distal portion of the gene cluster now proceeds into the 5S gene (Figures 3 and 4). As soon as p5 rRNA sequences are formed and have folded into the appropriate conformation, they are excised by RNase E (Figure 4, cuts 5A and 5B; see also Figure 2), and distal or trailer tRNAs are removed by RNase P (cut 9) and another activity (cut 10) possibly identical with RNase "F".

That p23 and p5 rRNA are excised from rRNA transcripts prior to transcription termination is evidenced by the failure to detect 25S rRNA (p23 plus p5) in wild-type strains, and by the demonstration that no material in the 9S region of a polyacrylamide gel prepared from RNA isolated from wild-type cells contains p5 sequences.[43] It is clear, then, that the primary processing cleavages by RNase III and RNase E are exceedingly rapid events, such that virtually all sites are cleaved within seconds after synthesis of the sequences which comprise them. (The rate of chain elongation of RNA polymerase on rRNA operons is about 85 nucleotides per second at 37°C;[66] the distance between p23 and p5 is 81 bases; and between p5 and the termination signal are 39 bases in four out of the seven rRNA operons which do not contain trailer tRNAs [see Figure 2], and it is somewhat longer in the other three which do contain trailer tRNAs.[53])

It is interesting to mention here a major difference between processing of rRNA in prokaryotic and eukaryotic cells. While, as emphasized here, in prokaryotic cells most of the rRNA processing occurs during transcription, in all the eukaryotic systems studied thus far, most of the rRNA species are transcribed in a single molecule which is processed posttranscriptionally.[3]

### B. Processing in Mutant Strains

The origin of rRNA species seen in mutants defective in RNA processing enzymes (Figure 5) can readily be described by reference to the model shown in Figure 4. In strains lacking the processing endonuclease RNase III, scission of nascent rRNA transcripts is initiated by RNase P and another enzyme(s) which cut in the spacer region to remove the tRNA sequences (cuts 2 and 3). The single-stranded region between cut 2 and the p16 stem is removed, possibly by enzyme(s) such as RNase II or polynucleotide phosphorylase, giving rise to 18S RNA. Subsequently, the 5′ leader sequence is removed from 18S RNA in a slower process by single-strand-specific nucleases which leave intact the duplex stem (see Figure 4). The final rRNA product, a p16b molecule which contains the entire duplex stem and is thus slightly larger than normal p16a,[33] is converted to normal m16 rRNA (see Reference 67) by the maturation enzyme(s) RNase(s) M16 (cuts 6A and B). RNase E cleavage in the distal portion of the nascent rRNA transcript generates p5 rRNA and a p23-like molecule — the latter containing extra single-stranded spacer sequences extending from cleavages 3 to 5A, which could be rapidly and nonspecifically removed, giving rise to the p23b rRNA of the *rnc* strain.[33] This p23b is further processed to mature 23S rRNA via the maturation enzyme(s) RNase(s) "M23" (cuts 7A and B).

When RNase E is inactivated in the *rne* mutant, p5 rRNA is not removed from the distal portion of the transcript, which instead accumulates as 9S RNA. If RNase P is also inactivated, trailer tRNAs are found linked to 5S rRNA[50] (cuts 4B to 10). Failure to perform RNase E and RNase III cleavages, in the *rnc rne* strain at nonpermissive temperatures yields p5 sequences linked to p23 rRNA. A 25S RNA thus accumulates at the expense of p23 and p5 (see Figures 4 and 5). In these *rnc rne* double mutant cells, unlike in *rnc* single mutant strains, no p23 rRNA is detected.

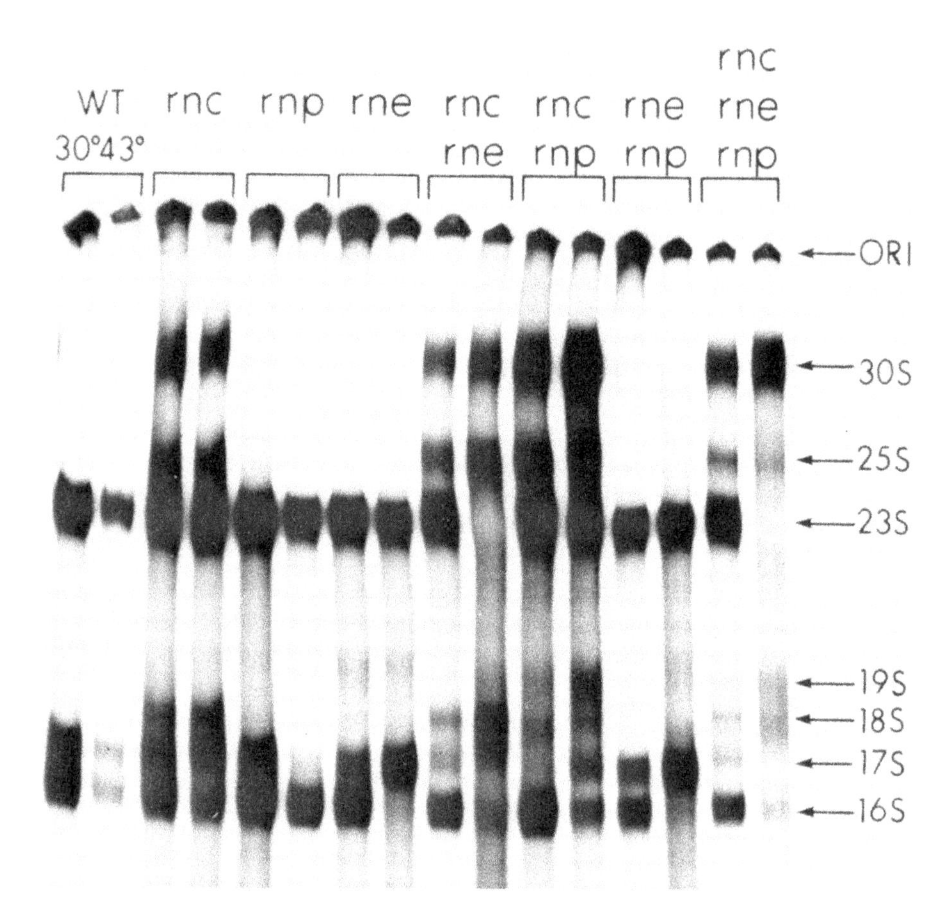


FIGURE 5. Ribosomal RNAs produced in different RNA processing mutants. For each of the genotypes indicated, cultures of the corresponding strain were labeled with $^{32}P_i$ at 30 or at 43°C and nucleic acids were fractionated on a 3% polyacrylamide gel (details as in References 37 and 81). For each strain the left lane represents RNA from cells grown at 30°C, while the right lane represents RNA from cells labeled at 43°C. While the *rne* mutant behaves like a classical *ts* mutant, the *rnc* mutant does not. Strains carrying the *rnc-105* allele are completely missing RNase III[33] and therefore the patterns observed at 30 and 43°C are similar. The *rnpA49* mutation used here is apparently defective in the synthesis but not in the function of RNase P[12] and certain features of this mutation are already expressed at temperatures where the mutant can grow.[17] The *rne* and *rne rnp* strains used here contain an additional *ts* mutation(s) which blocks protein synthesis,[42] and therefore 16S rRNA, the product of secondary rRNA processing, is not observed in these strains. It can be seen that in the *rnc rne* double mutant and in the *rnc rne rnp* triple mutant 16S rRNA appears. While the difference between p16 and m16 is substantial (~150 nucleotides) and can be easily detected in gels, the difference between p23 and m23 is only 15 nucleotides[33] and cannot be observed in the gel (see Figures 3A, 3B, and 5B). The size differences between p16a and p16b, and p23a and p23b (see text) are also not readily observed in gels.

Spacer tRNAs are linked to 18S RNA, giving 19S RNA, if RNase P is inactivated in an *rnc rnp* strain (Figure 4, cut 3). Each 19S molecule (Figure 5) is initiated with a nucleoside 5′ triphosphate and, at its distal end, contains a spacer tRNA which terminates with the mature 3′ $CCA_{OH}$ sequence.[68] Production of 19S RNAs, therefore, requires endonucleolytic cleavage of nascent rRNA transcripts at a site near the 3′ end of each spacer tRNA sequence (cut 3 in Figure 4). In an *rnc rne rnp* mutant some 19S- and 25S-sized species are still produced (Figure 5). It appears that a processing activity(ies), distinct from RNases III, E, or P, does exist which can cut in the rRNA spacer region. At present, we refer to such

**Table 1**
**RIBOSOMAL RNA MOLECULES THAT APPEAR IN VARIOUS RNA PROCESSING MUTANTS**

| Genotype | | | Containing 5S | | | Containing 16S | | | | | Containing 23S | | | | Intact transcript |
|---|---|---|---|---|---|---|---|---|---|---|---|---|---|---|---|
| *rnc* | *rne* | *rnp* | 5S | 9S | 11S | 16S | 17S[a] (p16a) | 17S[a] (p16b) | 18S | 19S | 23S | p23a[b] | p23b[b] | 25S | 30S |
| + | + | + | V | | | V | V | | | | V | V | | | |
| − | + | + | V | | | V | | V | V | | V | | V | V | V |
| + | + | − | V | | | V | V | | | | V | V | | | |
| + | − | + | | V | | V | V | | | | V | V | | | |
| − | − | + | | | | V | | V | V | | | | | V | V |
| − | + | − | V | | | V | | V | (V)[c] | V | V | | V | V | V |
| + | − | − | | | V | V | V | | | | V | V | | | |
| − | − | − | | | | V | | (V)[c] | V | V | | | | V | V |

*Note:* This table shows the rRNA molecules that can be observed (V) in the various strains at elevated temperatures. For further details see text and Figures 4 to 6. Some of the molecules cannot be clearly observed in Figures 5 and 6 and the reader is advised to consult specific references which are mentioned in the text: 11S contains 5S and trailer tRNAs; 18S contains 16S and leader sequences; 19S contains 16S, leader sequences, and spacer tRNAs; 25S contains 23S and 5S; 30S is the whole RNA from the rRNA transcription units, containing leader sequences 16S, spacer tRNA, 23S, 5S, and trailer tRNAs.

[a] p16a is slightly smaller than p16b, the difference being in the double-stranded stem of the molecule[33] (Figure 4).
[b] p23a is slightly smaller than p23b (Figure 4).
[c] Low levels were observed.

activity(ies) as RNase "F". The various rRNA molecules observed in the different strains are listed in Table 1.

### C. Efficiency and Order of Processing Steps

Since RNase III$^-$ cells are viable and form normal, functional, mature m16, m23, and m5 ribosomal RNAs, the physiological usefulness of RNase III might be questioned. One answer may lie in the observation that the doubling time of an RNase III$^-$ cell is 40% longer (at 37°C) than that of an isogenic RNase III$^+$ cell.[31] This observation can best be explained by the lower efficiency of rRNA processing in RNase III$^-$ cells. While in *rnc*$^+$ cells no uncleaved primary transcripts are detected, in *rnc-105* cells a large proportion of the newly synthesized rRNA transcripts are uncleaved 30S and partially cleaved 25S molecules, few if any of which contribute significantly to the pool of mature rRNA species.[37,69]

RNase P cleavage of spacer tRNAs is less efficient when p16 sequences are not previously removed by RNase III, as evidenced by the fact that uncleaved 30S transcripts are detected in RNase III$^-$ RNase P$^+$ strains (see Figure 5), even though RNase P can cleave tRNA sequences in purified 30S RNA.[70] Similarly, tRNA 3′ endonuclease action (RNase "F") is impaired by the presence of p16 sequences linked to spacer tRNA since, as seen in Figure 5, relatively more 30S RNA is found in the RNase III$^-$ RNase P$^-$ strain than in the strain lacking only RNase III. Again, because the 25S species is seen in RNase III$^-$ single mutant strains even though RNase E is active, it can be concluded that the addition of p23 sequences to nascent 9S transcripts severely impairs the efficiency of p5 excision by RNase E.

This pattern of impaired processing might result from a steric hindrance. Indeed, since ribosomal proteins attach to rRNA during transcription[71] it is likely that the uncleaved nascent rRNA exists as a ribonucleoprotein particle which restricts access to the 5S and tRNA processing recognition sites (see Figure 3).

**D. Sequential Nature of Processing Reactions**

Since the presence of p16 sequences impairs the efficiency of RNase P and 3′ endonuclease cleavages in the spacer region of nascent rRNA transcripts, it can be inferred that these tRNA processing cleavages normally occur on RNA chains lacking p16 structures, and that removal of p16 rRNA therefore precedes processing of spacer tRNAs. For the same reason, RNase III-mediated release of p23 must normally precede RNase E excision of p5 rRNA. As previously discussed, RNase III cleavage of a growing transcript is completed during the time required for the polymerization by RNA polymerase of only a few hundred more bases; excision of p16 rRNA will usually have occurred by the time the RNA polymerase has moved into the 23S gene.[71,77a]

Since RNase III and RNase P in vitro can each cleave 19S or 30S RNA independently of the other,[68,70,72] it may be assumed that the order of rRNA processing cleavages in vivo is partly determined by the linearity of rRNA transcription and the relative intrinsic efficiency of competing enzymatic reactions.

**E. Other Processing Enzymes and Secondary Maturation**

All the available data indicate that the four major primary rRNA processing endonucleases of *E. coli* are RNases III, E, "F", and P. (Once a transcript has been cut, "nonconserved" sequences are degraded. The nature of the degradative enzymes is unknown; the nonspecific endonuclease RNase N[73,74] is one candidate for such an activity.)

The primary processing events described here give rise to precursor forms of 16S, 23S, and 5S rRNAs, whose maturation occurs in ribonucleoprotein particles (Figure 3). The processing of p16 to m16 by RNase(s) M16[56,75,76] is probably endonucleolytic (Figure 4, cuts 6A and 6B). Similarly, RNase(s) "M23" (cuts 7A and 7B in Figure 4) is proposed to mature p23 to m23 rRNA, and *E. coli* p5 rRNA might be trimmed to m5 by RNase "M5" (cut 8 in Figure 4).

Since these secondary maturation enzymes are unaffected by the processing mutations described here, the viability of an RNase III$^-$ cell can be ascribed to the ability of the maturation enzymes (RNases M16 and "M23") to recognize and process the abnormal p16b and p23b substrates produced in the absence of RNase III. The short extension of the duplex stem observed in these species (see Figure 4), then, does not alter the accuracy of cleavages which give rise to m16 and m23.

## IV. PROCESSING OF TRANSFER RNA

The immediate transcriptional products of *E. coli* tRNA genes are molecules containing extra nucleotides at both their 5′ and 3′ termini. Reflecting the nature of tRNA gene organization, many transcripts are polycistronic, containing as many as four or five different or identical tRNA species.[14,15,25,26,29] Because these intact transcripts are not detected in wild-type cells, endonucleolytic processing cleavages must begin to occur before transcription is terminated.

**A. Enzymes Involved: a Genetic Approach**

All the information gathered so far on tRNA processing in vivo, including processing of ribosomal spacer tRNAs, leads to the conclusion that transfer RNA species of *E. coli* are the products of posttranscriptional processing by four, and probably only four, endonucleases. These are RNase P, RNase E, RNase III, and at least one distinct 3′ endonuclease (which could be RNase F). (RNase E might be identical with RNase "P2"[24] and RNase III with RNase "O".[30] Since RNase E and III have been defined genetically as well as enzymologically, we shall discuss processing in terms of these activities.) To illustrate the participation of these enzymes in tRNA maturation, the effect of *rnc, rne,* and *rnp* mutations on tRNA

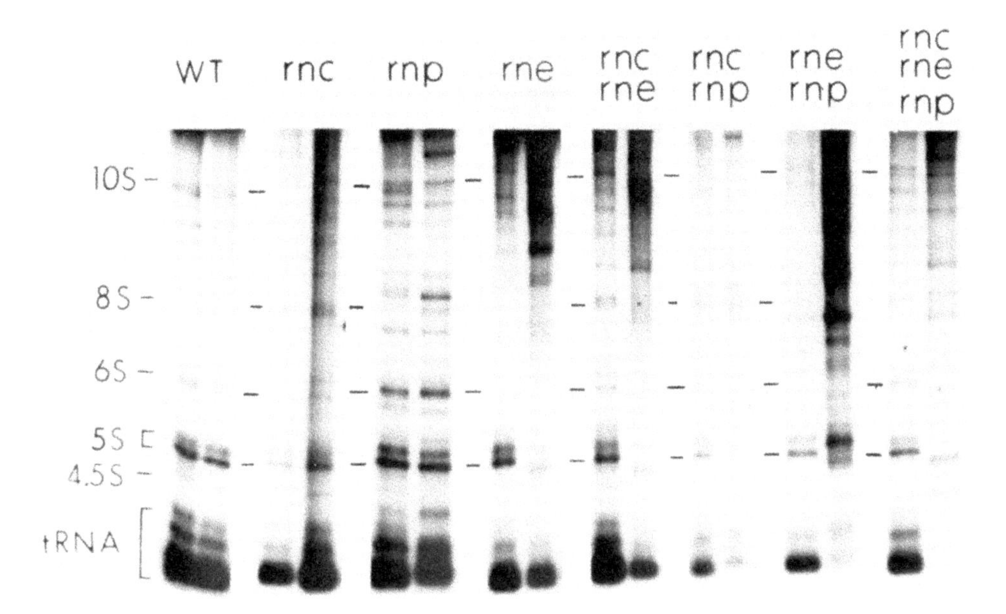


FIGURE 6. Display of small RNAs in various RNA processing mutants of *E. coli*. Cells were labeled with $^{32}P_i$ at 30 or at 43°C. The cells were processed and their contents analyzed on a 5%/10% tandem thin-slab polyacrylamide gel. The 5% portion of the gel was removed from the photograph. The strains used here and in Figure 5 were the same or similar to those described in Reference 82. For each strain the left lane represents RNA from cells grown at 30°C and the right lane represents RNA from cells grown at 43°C.

production in vivo has been examined.[72] Figures 6 and 7 summarize this information, which demonstrates that tRNA maturation is more severely restricted in strains lacking RNase E or RNase E and RNase III, as well as RNase P, compared with the single RNase P mutant.

In Figure 6, small RNAs accumulated in the absence of either RNase III, E, or P, or in mutants lacking combinations of these enzymes, are displayed. It can be seen that the level of tRNA that matures is very much diminished in the absence of these enzymes, and larger RNA precursors are accumulated. In order to analyze this problem more closely, the regions of the gels containing tRNA and 5S rRNA were displayed by analysis in two-dimensional polyacrylamide gels (Figure 7). Figure 7a is a two-dimensional gel pattern of tRNAs from a wild-type strain labeled at 43°C. The tRNAs of the *rnc* strain, which are displayed in Figure 7b, are all apparently normal. The *rne* strain, panel c, shows a fairly normal pattern but lacks mature 5S rRNA and a few tRNAs. The molecules accumulating in the 5S region contain monomeric tRNA precursors.[55] It is noteworthy that the *rnp* strain (Figure 7d) produces an appreciable amount of mature or almost mature-sized tRNA molecules as compared to wild-type cells. Many of these molecules represent the previously described "small monomeric" tRNA precursors[14,15] which are produced by endonucleases other than RNase P. The 4S — 5S RNAs of the *rnc rne* strain are very similar to those of the single *rne* strain panel e). Combination of the *rnc* allele with *rnp* does not significantly change the pattern of tRNA accumulation from that of the single *rnp* strain, as seen in Figure 7f. Most dramatically, however, an *rnp rne* double mutant shows a drastic reduction in mature tRNA species and a number of new 5S-sized species, when compared with the single *rnp* mutant (Figure 7g). A further decrease in tRNA maturation is found upon combining all three processing mutations. The *rnp rnc rne* triple mutant accumulates few if any mature-size tRNA molecules (Figure 7h). Again, although little difference was detected between *rnp* and *rnp rnc* strains, the enhanced deficiency in tRNA maturation observed upon introduction of the *rnc* mutation into an *rnp rne* strain indicates that RNase III is involved in tRNA maturation in *E. coli*.

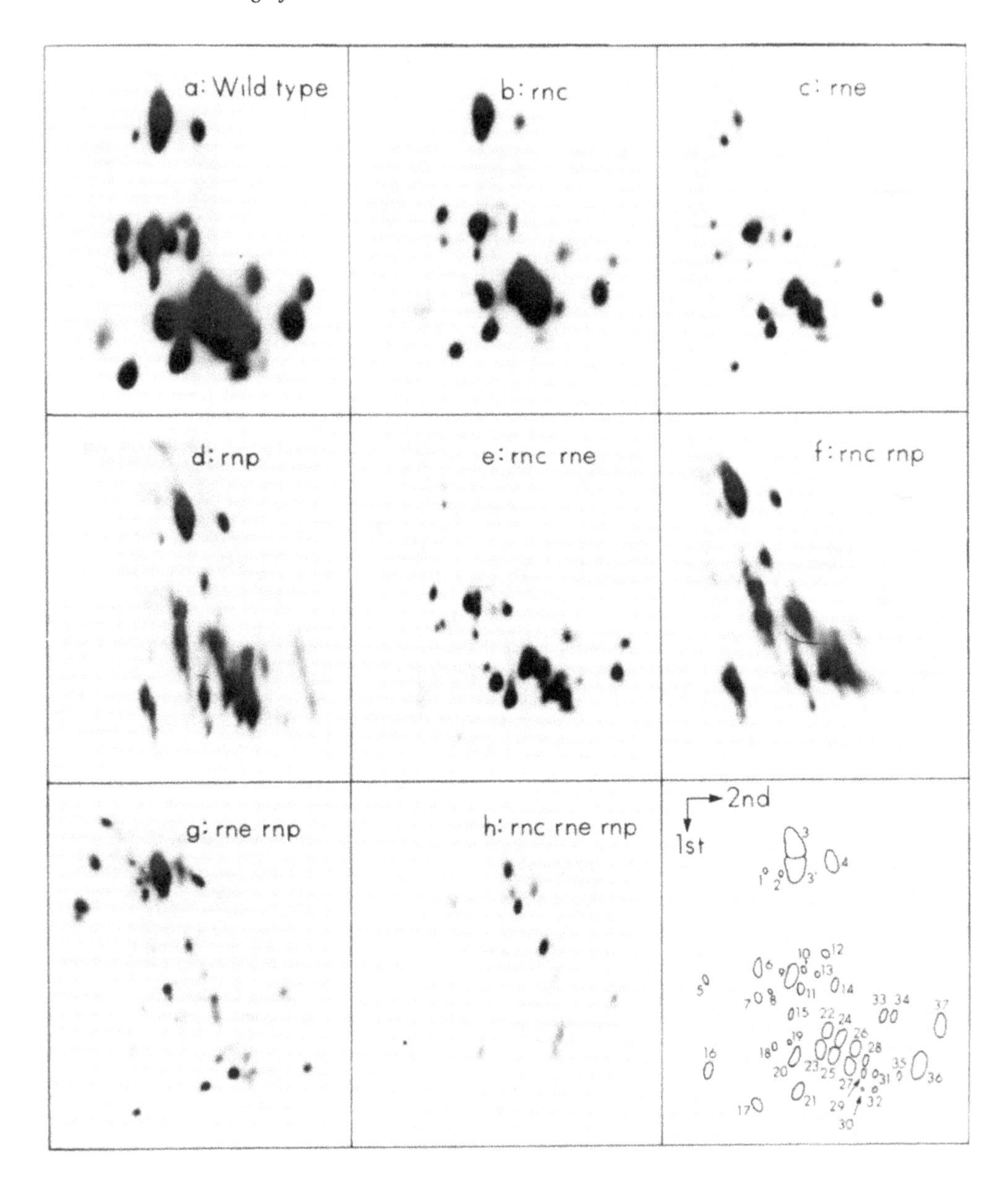


FIGURE 7. Transfer RNAs and 5S rRNA produced in different RNA processing mutants. Cultures of strains with the indicated genotypes were grown at 30°C, shifted to 43°C, and after 4 min they were labeled with $^{32}P_i$ for 60 min. RNA was fractionated on 10% → 20% two-dimensional polyacrylamide gels.[72,81] Spots 3,3′, and 4 are 5S rRNA isomers; the other spots are tRNA, e.g., spot 17 is $tRNA_2^{Glu}$ (see last panel).

Most of the RNA species (16 out of 22 tested) in the 4S to 6S region which accumulate in an *rnc rne rnp* strain at the nonpermissive temperature are primary transcription products initiated with ATP or GTP.[77] The *rnc rne rnp* mutant also accumulates large (>6S) precursors to most cellular tRNAs (see Figure 6) which can be processed with a wild-type cell extract to give a pattern of small RNAs indistinguishable from that shown in Figure 7a.[77]

Apart from RNases III, E, P, and a 3′ endonuclease, one additional activity, presumably an exonuclease, is required for trimming extra sequences from the 3′ termini. Since in all *E. coli* tRNAs examined the CCA terminus of the mature molecule is encoded in the DNA and is present in the RNA precursor, only the activity of an enzyme like RNase D[60] would be required for 3′ trimming. It has been suggested, however, that trimming in vivo or in vitro may require prior RNase P cleavage at the 5′ side of the precursor $tRNA_1^{Tyr}$.[30,60] In an RNase P mutant strain, this and many other pre-tRNAs are found to bear extra sequences past their mature 3′ termini.[7,8] Other tRNA precursors examined, however, possess fully matured 3′ termini, even though extra sequences are present at the 5′ end.[14,78] Thus, either RNase D trimming stops at different sites on different precursors in the absence of RNase P cleavage, or another 3′ exonuclease is also involved in tRNA maturation.

Even though the results observed in Figures 6 and 7 support rather convincingly the notion that RNase E is required for the maturation of tRNA, we consider it more likely that RNase E is not involved directly in tRNA metabolism and that the phenomenon seen in Figures 6 and 7 is a result of interference of the mutant RNase E in strains containing the mutation *rne-3071* with another enzyme that normally participates in tRNA processing, most likely RNase F. Such an inhibition can be explained if RNA processing enzymes exist in a complex[79] or if the mutant RNase E can directly inhibit the activity of the other enzyme.[55a] These two mechanisms are not mutually exclusive. The fact that the RNase E cleavage sites in the seven rRNA genes are very conserved (see Figure 2) and the realization that similar sequences are not found in tRNA precursors that accumulate in the *rne* mutant[54,55] suggest that indeed the observations in Figures 6 and 7 indicate a very close relationship among the RNA processing enzymes, rather than the involvement of the wild-type RNase E in the maturation of tRNA.

**B. Sequence of Processing Events**

After an RNA polymerase moves downstream of a monomeric tRNA gene, the tRNA sequences are released from the nascent transcript by one or more endonucleases. As soon as the mature portion of the molecule has assumed its final secondary and tertiary configuration, it can be recognized by RNase P, whose cleavage generates the mature tRNA 5′ terminus. Trimming by RNase D or a similar enzyme, when necessary, then exposes the mature 3′ terminus $CCA_{OH}$.

Processing of transcripts from polycistronic tRNA gene clusters is somewhat more complex. The processing endonucleases RNase E and RNase III may also cleave between adjacent tRNA sequences to produce smaller multimers and monomeric substrates from which RNase P can rapidly remove 5′ precursor-specific sequences. Again, 3′ trimming by RNase D is probably necessary to expose the terminal sequence of the functional tRNA species.

It is clear that in the absence of RNases E and III (Figure 7e), most tRNAs continue to be processed correctly, but that in the absence of RNase P, very few if any mature 5′ termini are produced. RNase P, therefore, performs the final 5′ maturation of tRNA molecules, and this cleavage cannot be performed by another enzyme — while RNase F or an equivalent enzyme performs most of the 3′ endonucleolytic cleavages.

In cells lacking one or more processing nucleases, various tRNA precursors can be detected, and it is by studying the effects in vivo of inactivating one or more enzymes as well as by comparing their relative efficiencies in vitro that one can begin to establish a hierarchy of processing events. Cleavage by the other endonucleases can clearly precede RNase P processing since monomeric tRNA precursors accumulated in vivo in the absence of RNase P have already been cleaved at or near their 3′ termini. Conversely, RNase P can cleave some multimeric pre-tRNA transcripts in vitro, although inefficiently, without other endonucleolytic cuts.[15,25,26] Three precursors have been identified, however, for which RNase P action is absolutely dependent upon prior cleavage by a 3′ endonuclease. Cleavage between ribosomal spacer tRNAs $tRNA_1^{Ile}$ and $tRNA_{1B}^{Ala}$, by an activity other than RNase III or P, is required before RNase P can cleave at the 5′ side of $tRNA_1^{Ile}$.[68] RNase P processing of monomeric precursors to $tRNA_1^{Ser}$ and $tRNA^{Asn}$ requires prior 3′ processing.[55] Apparently, in all these precursors, the RNase P cleavage site is masked by secondary and/or tertiary structure. The action of RNases III and E is not essential for proper maturation of all tRNAs, since *rnc* strains are viable, and *rnc rne* mutants also accumulate many mature-sized tRNAs (Figure 7e). The role of endonucleases such as RNase III and RNase E might be largely to facilitate RNase P action.[15,25,26]

In the absence of RNases E and III, therefore, RNase P and "F" cleavage can still produce the normal pattern of most mature-sized tRNAs. In the absence of RNases P and III, RNases E and 3′ endonuclease(s) produces "small" tRNA monomers and tRNA precursors. In the absence of RNases P and E, RNase III and 3′ endonuclease action can produce a more limited number of "large" tRNA monomers. Strains lacking RNases III, E, and P accumulate very few 4S to 6S-sized RNA species, almost all of which contain purine nucleoside 5′ triphosphoryl termini and thus represent 5′ unprocessed primary transcripts.[77]

## C. Cooperativity of Processing

Our present understanding of the function, efficiency, and order of RNase III, E, and P cleavages in rRNA processing may be used to assemble a coherent picture of the involvement of these enzymes in tRNA processing. Since most processing of *E. coli* tRNA probably occurs during transcription, the availability of processing sites is governed by the linearity of transcription as well as by the three-dimensional structure assumed by the newly synthesized sequences. The endonucleolytic cleavages of tRNA transcripts are performed primarily by the 5′ endonuclease RNase P which generates the mature 5′ terminus and by a 3′ endonuclease(s) which (alone or followed by the 3′ trimming activity of RNase D or a similar enzyme) exposes the mature 3′ terminus. This basic mode of tRNA processing is graphically visualized in Figure 8A, which illustrates the nucleolytic processing of nascent transcripts from the *E. coli* $tRNA_1^{Tyr}$ doublet gene cluster.[8,28,80] In this simple situation, only a 5′ endonuclease (RNase P), a 3′ endonuclease, and a 3′ exonuclease (RNase D) are required to produce mature-sized tRNAs. The endonuclease cuts are not strictly ordered, but both precede RNase D trimming.

An example of strict ordering of endonucleolytic cleavage events is shown in Figure 8B, which demonstrates processing of $tRNA_1^{Ile}$ and $tRNA_1^{Ala}$ from the spacer region of ribosomal RNA transcripts.[68,70,81] In this transcript, endonuclease cleavage at the 3′ side of the $tRNA_1^{Ile}$ is a prerequisite for cleavage at the 5′ side by RNase P; 5′ and 3′ cleavages of $tRNA_{1B}^{Ala}$ are not ordered.[68,70]

Other enzymes also play roles in tRNA processing. RNase III can participate, but is not absolutely required for production of mature tRNAs in vivo since RNase III$^-$ strains are viable.[35,37] The cellular function of RNases E and III in tRNA biosynthesis is probably to convert large nascent multimeric tRNA transcripts into smaller substrates for which RNase P cleavage is more efficient; this type of tRNA processing is depicted in Figure 8C. (However, the wild-type RNase E might not be involved, as an effect is only observed in an *rne* temperature-sensitive mutant [see above].) In Figure 8C RNase III is shown as cleaving between two tRNAs, but it is conceivable that cleavages by RNase III (or other enzymes) might also be performed in the leader preceding the first tRNA sequence.

From analysis of host tRNA metabolism in vivo and in vitro we can conclude that cooperation exists among all the tRNA processing events and that some but not all of them are ordered. From the available data on tRNA processing, four suggestions can be made: *first,* that most processing of tRNA occurs during transcription or shortly thereafter; *second,* that RNase P performs the final 5′ maturation of tRNA molecules, and this cleavage cannot be performed by any other enzyme; *third,* that some transcripts are cleaved near tRNA 3′ termini by RNase "F" or by yet another endonuclease(s); *fourth,* that (based on genetic evidence) RNase E performs a role in vivo compatible with that attributed to "P2"[24,25]

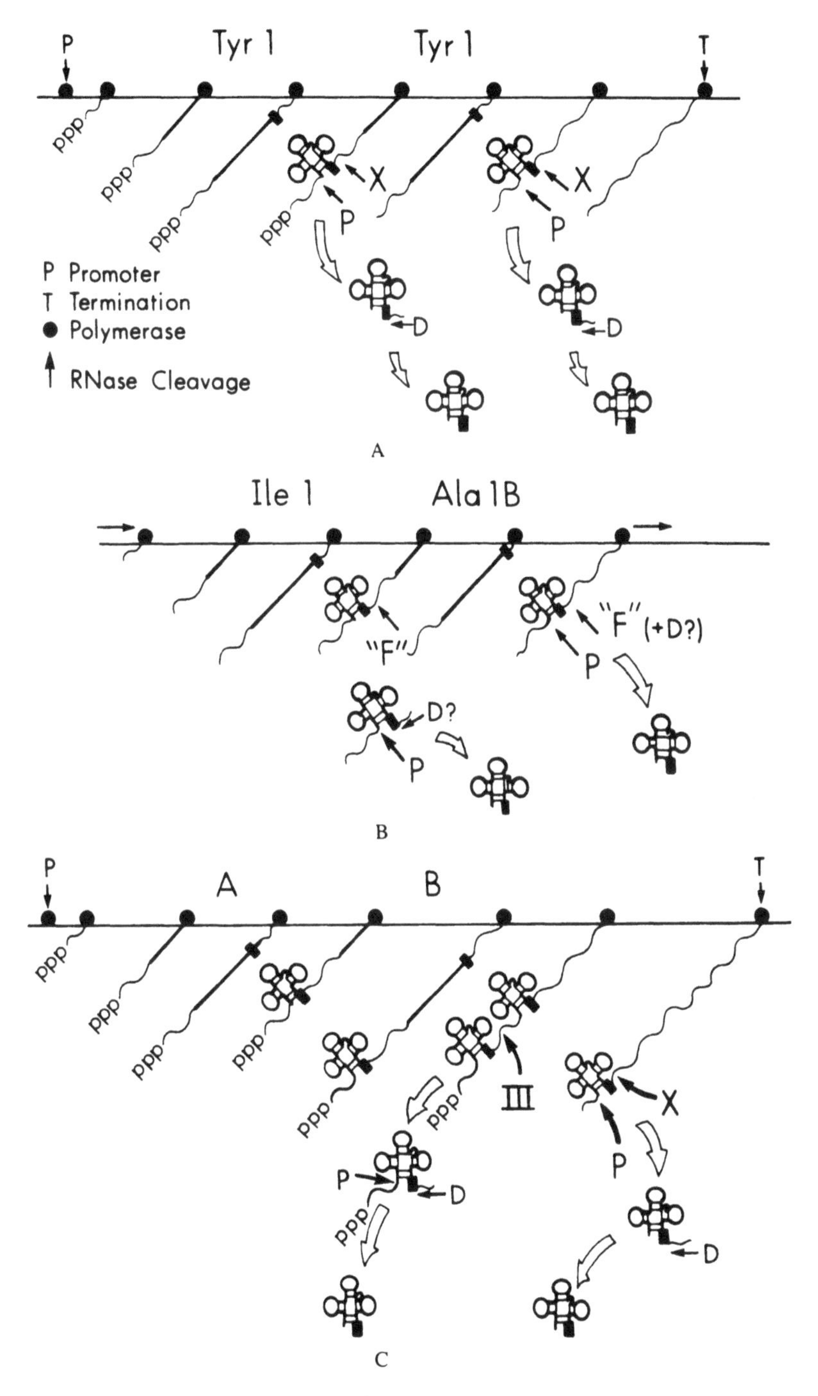


FIGURE 8. Processing of transfer RNA. Several different schema for tRNA processing are illustrated. Successively elongated nascent tRNA transcripts are shown. Thick straight lines represent tRNA sequences not in their mature conformation, and thin wavy lines are leader, spacer, and trailer sequences. The first processing events occur during transcription, but only after the tRNA has assumed its mature conformation, a process which is depicted graphically. Endonuclease cleavages by RNases P, and 3′ endonuclease activities (''F'' and X) are shown, as well as 3′ exonuclease trimming by RNase D. Details are discussed in the text. (A), processing of the *E. coli* $tRNA_1^{Tyr}$ doublet transcript. (B), processing of $tRNA_1^{Ile}$ and $tRNA_{1B}^{Ala}$ from the spacer region of the *E. coli* rRNA genes *rrnA*, *rrnD*, and *rrnF*. RNase ''F'' cleavage must precede RNase P cleavage of $tRNA_1^{Ile}$. The involvement of RNase D is conjectural. (C), processing of a hypothetical dimeric tRNA transcript in *E. coli*. In this instance, RNase P action would require a prior 3′ cleavage by RNase III. RNase X in A and C could be RNase ''F''.

activity (or that the mutant RNase E inhibits RNase "P2" activity; see above), and that "O" activity is performed by RNase III (and is hence a dispensable function). All these endonucleases act synergistically to process mature transfer RNA species from nascent tRNA gene transcripts.

## V. PROCESSING OF MESSENGER RNA

The first clear-cut evidence for processing of messenger RNA in prokaryotes came from a phage rather than a host system. Subsequently, considerable attention was focused on the possibility of processing of host mRNAs. At least one host mRNA appears to be cleaved in vivo, and expression of many genes is altered in cells defective in one or more processing enzymes. So far, however, no host of phage mRNA has been described whose function is absolutely dependent upon a processing event. Additionally, mutations in the enzymes RNase III, E, or P do not seem to affect the stability of host mRNA.[82]

### A. *E. coli* mRNA

Direct evidence has been presented for processing of one host mRNA, the transcript of the *E. coli rplJL-rpoBC* operon which encodes ribosomal proteins L10 and L7/12 and RNA polymerase subunits β and β′. Total cellular RNA was examined for its ability to hybridize with and protect from nuclease S1 cleavage a restriction endonuclease fragment of DNA spanning the *rplL-rpoB* intercistronic region.[83] RNA from wild-type cells allows the probe to be cut into two pieces, whereas RNA from *rnc-105* cells completely protects the probe.[83] In *rnc-105* cells, therefore, mRNA from this intercistronic region is intact, but in wild-type cells it is cleaved in a region about 200 bases beyond the 3′ end of the *rplL* message.[83] The sequences in this region can be depicted in a stem-and-loop structure[83,84] similar to the p23 rRNA stem, and containing a short conserved "core" sequence which might be involved in RNase III recognition. Actual RNase III cleavage of one or both strands in this region, however, remains to be demonstrated in vitro. Again, since the *rnc-105* strain is viable, processing in this region is not required for the qualitative expression of either proximal or distal mRNAs. Processing might, however, play a role in the complex quantitative regulation of the products of this operon.[83,85]

Studies with isogenic *rnc*$^+$/*rnc* strains have failed to detect a requirement for RNase III in *lac* gene expression at 30°C.[86] The RNase III$^-$ strain AB301-105 appears to be defective in initiation of translation of *lac* mRNA,[87] but the involvement of RNase III cannot be easily evaluated because this mutant contains a number of mutations in addition to the RNase III lesion.[34,35]

Most recently, however, both RNase III and RNase E have been found to be required at 43°C for full expression not only of the *lac* operon, but also of a significant number of other proteins in *E. coli*.[88] The differential induction of the *lacZ* gene, as well as the amount of β-galactosidase protein, is reduced in *rnc-105* or *rne-3071* strains at 43°C compared to 30°C. In the double *rnc rne* mutant, this reduction is even more pronounced. Moreover, analysis of total cell protein separated in two-dimensional gels reveals that the synthesis of 21 out of 80 individual proteins examined is markedly decreased in the *rnc rne rnp* strain at 43°C, and that this difference is caused mainly by the *rnc* and the *rne* mutations.[88]

Another interesting and at present unexplained effect of the *rnc-105* mutation is that *rnc-105* strains are nonmotile because they are defective in the production of flagella.[89] Whether this results from a failure to process a specific mRNA, or simply from the known sensitivity of flagellar production to a variety of physiological perturbations, has not been determined. These results, while not demonstrating mRNA processing, do open up such a possibility. While these intriguing observations could be due to processing of mRNA, other possibilities such as effects on synthesis, stability, and utilization of mRNA were not ruled out.

### B. Processing and Attenuation

In two messenger RNA transcripts which are subject to regulated rho-mediated termination of transcription between two cistrons, the site of attenuation is located near a region for RNase III cleavage.[83,90,91] A proposed RNase III cleavage site lies just downstream of an attenuator in the *rplL-rpoB* intercistronic region of the *E. coli rplJL-rpoBC* operon mRNA,[83] and the site of RNase III cleavage following gene N message in the major leftward transcript of bacteriophage λ is in the region where rho protein causes transcription termination in the absence of functional N protein.[91,92] RNA sequences in the vicinity of termination or attenuation sites can form multiple stem-and-loop structures.[46,47,93]

It is noteworthy that these termination signals near RNase III processing regions both occur in intercistronic "spacer" regions of an mRNA transcript, rather than in a 5′ leader region. The mechanism of attenuation in leader RNAs[93] could be somewhat different from that in spacer regions. RNase III cleavage might be a general phenomenon which occurs near attenuation signals in intercistronic spacer regions. Termination or attenuation followed by endonucleolytic cleavage in an intercistronic spacer region might be important in the maintenance of complex transcription and posttranscriptional regulatory patterns. It could, for instance, insure that the size of the messages produced from this region of the genome would remain the same whether or not termination at the attenuation site took place.

Genetic analysis of *rnc rho* double mutants[94] indicated that the presence or absence of RNase III does not influence suppression by *rho* mutants of a polar mutation in the *lacZ* gene. Moreover, no obvious alterations in the RNA molecules were observed in the double mutant.[94] These observations suggest that *rho* and RNase III do not recognize the same sequences.

## VI. GENERAL CONSIDERATIONS

### A. Unity of RNA Processing Mechanisms

The prokaryotic cell, as we have seen, employs a very limited number of nucleolytic activities to accomplish the primary processing of a variety of RNA transcripts and a comparable number of activities for the secondary maturation steps. Indeed, each of the three endoribonucleases of *E. coli* — RNase III, RNase E, and RNase P — plays a role in processing both ribosomal and transfer RNA transcripts. Intracellular parasites — such as the DNA bacteriophages — have evolved so that their RNA transcripts can be processed by host systems. Processing of tRNAs and other stable RNAs in T-even and related phages, as well as of messenger RNAs in T3, T7, and λ, is performed by host enzymes (see Chapter 3).

### B. Function of Polycistronic Transcription and Processing

As Pace originally suggested,[5] one explanation of precursor-specific sequences in monomeric rRNA precursors is that they are remnants of the recognition sites for the enzymes which generate those precursors. This consideration could be true for tRNA precursors as well. The larger question, however, is the physiological necessity or utility of polycistronic transcription of stable RNAs.

Four primary advantages for polycistronic transcription of stable RNAs can be considered. *First*, production of equimolar amounts of functionally interdependent molecules, such as ribosomal RNAs, is achieved. *Second*, cotranscription of rRNAs allows coordinated and orderly sequential assembly of the complex ribonucleoprotein structure of the ribosomal subunits.[71] *Third*, the termini of RNA species are protected during transcription and until the molecule is functional. As an example, endonucleolytic cleavage of rRNA transcripts releases p16 and p23 rRNA precursors[56,62] whose 5′ and 3′ termini are paired in a duplex stem which is resistant to degradative cellular nucleases.[33,95] Since the 3′ terminal portion of m16 RNA, for example, must carry out a crucial function in translation, protection of p16 from exonuclease attack is beneficial until the m16 portion, plus ribosomal proteins, can assume the mature, nuclease-resistant[96] conformation of the ribosome, at which time specific maturation endonuclease(s) can safely expose the functional ends. Likewise, precursor tRNAs may be protected from nonspecific degradation — in particular, irreparable excessive removal of 3′ terminal sequences — by being present in a long (nascent) transcript until the mature portion of the tRNA has assumed its correct (RNase-resistant) configuration, which is recognized by the processing enzymes as well as by (for example) aminoacylating enzymes. *Fourth,* processing of RNA molecules provides the possibility for regulation of stable RNA gene expression at the posttranscriptional level by a physiologically controlled balance between productive processing and decay. Processing vs. degradation of tRNA precursors has been described and discussed elsewhere.[7,8,97,98]

One potentially exciting role for precursor-specific sequences could be to actively regulate RNA conformational folding.[99] Study of *B. subtilis* p5 rRNA by Pace and co-workers[99] has shown that the first six bases of the precursor are hydrogen-bonded to an immediately adjacent precursor-specific sequence. These latter sequences would otherwise base-pair with an internal region of the m5 portion of the precursor, resulting in an abnormal conformation for the RNA. This fact in itself, however, does not explain why any precursor-specific sequences should interact with mature sequences since isolated m5 rRNA can assume an apparently normal conformation without the aid of precursor-specific sequences. Further work, perhaps with mutants affecting RNA structure, will be required to elucidate the roles of processing and precursor-specific sequences. Similar studies of mutations in mature tRNA sequences have yielded important information on tRNA structure and processing (reviewed in References 1,6-8,97,98,100).

### C. Specificity and Efficiency of RNA Processing Reactions

What specific structural elements are recognized by RNA processing enzymes? Since the role of processing events is ultimately to produce mature functional RNA molecules, we would expect that recognition of the mature RNA portion of the precursor transcript would be required. The mature portion contains information indicative not only of the generic species of RNA but also of its "readiness" to be productively processed, as determined by whether or not it has assumed its final and functional configuration. RNase P and *B. subtilis* RNase M5 have been shown to recognize predominantly the mature domain of their respective substrates.[8,101] A similar recognition mechanism can be predicted for *E. coli* RNase E. Likewise, maturation of p16 rRNA precursor by RNase(s) M16 appears to require a properly assembled rRNA-protein complex.[56,57,75,76] Since RNase III cleavage of rRNA transcripts does not generate functional termini, it need not recognize the mature domain; but since it generates base-paired precursor termini, processing by RNase III requires an intact duplex structure.

At present, however, little can be said about such structure-function relationships in processing of phage or host mRNAs. The rationale of processing may well be regulatory, but much more remains to be learned. It is attractive, nonetheless, to imagine that processing enzymes might recognize and cleave a particular configuration, or one of several alternative configurations adopted in a region of an mRNA molecule involved in regulation of its own transcription or translation.

All of the processing endonucleases known thus far — RNases III, E, and P, and *B. subtilis* RNase M5 — recognize their substrates only after the 5′ and 3′ termini are base-paired. In addition, RNase P, which removes tRNA 5′ precursor-specific sequences, requires also the presence of the 3′ terminal CCA sequence of the tRNA. The double-stranded nature of these processing substrates may serve the purpose of ensuring that endonuclease cleavages which could otherwise occur during transcription do not take place until transcription of the entire substrate is completed, and proper RNA folding has occurred. This also applies to some extent to the yeast enzyme which removes intervening sequences from precursor tRNA molecules. In this case the substrate for this enzyme contains the aminoacyl stem which is comprised of nucleotides upstream and downstream from the nucleotides which are excised.[2]

From these considerations we anticipate that cleavages which produce mature termini, or which are required for their stability during processing, cannot be bypassed, whereas other cleavages can be bypassed or performed at slightly different locations by other enzymes. All available experimental evidence supports this view: for example, inactivation of RNase P is a lethal event, since mature tRNA 5′ termini are not formed. Inactivation of RNase III is not lethal, however, since nascent rRNA transcripts are cleaved by the remaining enzymes (RNases E, F, and P), and p16b and p23b rRNA precursors are formed which can be processed successfully to mature m16 and m23 rRNA. Since inactivation of RNase III does, however, drastically reduce the efficiency of the remaining RNA processing steps, a second point becomes clear: cleavages which do not directly yield mature termini, and those which convert polycistronic transcripts to monocistronic precursors, act in concert with other cleavages to enhance the overall efficiency of an RNA processing pathway. The order of such cleavages is largely determined by the order in which their substrates become available, during transcription, in a conformation recognized by the appropriate enzyme.

Another point that should be mentioned is that the transcription machinery of the bacterial cell does not seem to require "feedback" signals from processed RNAs since transcription of precursors for ribosomal and for transfer RNAs continues unabated in the absence of processing by RNase III, P, and E (see Figures 5 and 6 and References 37 and 72).

Bearing these considerations in mind, then, we can offer a general principle governing enzymatic recognition and processing of RNA transcripts. A relatively limited number of processing enzymes, principally endonucleases, acting on a few fundamentally distinctive processing signals, carry out a broad range of distinct, highly specific, and potentially well-ordered processing steps. By their concerted action these enzymes accomplish efficiently the processing of RNA transcripts.

## VII. FINAL POINTS

1. All "stable" RNAs and some messages in *E. coli* are processed.
2. Most "stable" RNA transcripts are processed by more than a single endoribonuclease.
3. Most of the processing of rRNA and tRNA occurs during transcription.
4. Transcription of RNA is independent of its processing.
5. The enzymes responsible for most processing of stable RNA transcripts in *E. coli* are four endoribonucleases, RNases III, E, F, and P, and at least one exonuclease, such as RNase D.
6. A given ribonuclease, by and large, fulfills either a degradative or a processing function.
7. The efficiency but not the specificity of processing cuts is affected by the size of the substrate; multiple cleavages facilitate efficient and accurate processing, but not all the steps are essential.
8. Some processing steps are completely dependent on the occurrence of an earlier step.
9. Processing endonucleases are highly specific and each performs a unique function. Their recognition sites may be composed of unique combinations of secondary and tertiary structure.
10. In some processing events the substrate is not naked RNA, but RNA in ribonucleoprotein particles.
11. The order of processing steps is governed primarily by the order in which substrate recognition signals become available. A certain amount of flexibility exists in the order of initial processing events, but the final steps, which generate the mature RNA termini,

must be preserved. These final cleavages are performed by enzymes which recognize the mature domains of the substrate in its final conformation.

## ACKNOWLEDGMENTS

We wish to thank those colleagues who communicated their results in advance of publication. Research performed in this laboratory and described here was supported by grant GM-19821 from the National Institute of Health.

## REFERENCES

1. **Mazzara, G. P., Plunkett, G., and McClain, W. H.,** Maturation events leading to transfer RNA and ribosomal RNA, in *Cell Biology: A Comprehensive Treatise,* Vol. 3, Goldstein, L. and Prescot, D. H., Eds., Academic Press, New York, 1980, 439.
2. **Abelson, J.,** RNA processing and the intervening sequence problem, *Annu. Rev. Biochem.,* 48, 1035, 1979.
3. **Perry, R. P.,** Processing of RNA, *Annu. Rev. Biochem.,* 45, 608, 1976.
4. **Apirion, D. and Gegenheimer, P.,** Processing of bacterial RNA, *FEBS Lett.,* 12, 1, 1981.
4a. **Gegenheimer, P. and Apirion, D.,** Processing of prokaryotic ribonucleic acid, *Microbiol. Rev.,* 45, 502, 1981.
5. **Pace, N. R.,** Structure and synthesis of the ribosomal RNA of prokaryotes, *Bacteriol. Rev.,* 37, 562, 1973.
6. **Altman, S.,** Biosynthesis of transfer RNA in *Escherichia coli, Cell,* 4, 21, 1975.
7. **Smith, J. D.,** Transcription and processing of transfer RNA precursors, *Prog. Nucleic Acids Res. Mol. Biol.,* 16, 25, 1976.
8. **Altman, S.,** Transfer RNA biosynthesis, in *Biochemistry of Nucleic Acids,* Vol. 2, Clark, B. F. C., Ed., University Park Press, Baltimore, 1978, 19.
9. **Daniel, V.,** Biosynthesis of transfer RNA, *CRC Crit. Rev. Biochem.,* 9, 253, 1981.
**Kindler, P., Keil, T. U., and Hofschneider, P. H.,** Isolation and characterization of a ribonuclease III deficient mutant of *Escherichia coli, Mol. Gen. Genet.,* 126, 53, 1973.
11. **Schedl, P. and Primakoff, P.,** Mutants of *Escherichia coli* thermosensitive for the synthesis of transfer RNA, *Proc. Natl. Acad. Sci. U.S.A.,* 70, 2091, 1973.
12. **Kole, R., Baer, M. F., Stark, B. C., and Altman, S.,** *E. coli* RNase P has a required RNA component *in vivo, Cell,* 19, 881, 1980.
13. **Sakano, H., Yamada, S., Ikemura, T., Shimura, Y., and Ozeki, H.,** Temperature sensitive mutants of *Escherichia coli* for tRNA synthesis, *Nucleic Acids Res.,* 1, 355, 1974.
14. **Ikemura, T., Shimura, Y., Sakano, H., and Ozeki, H.,** Precursor molecules of *Escherichia coli* transfer RNAs accumulated in a temperature-sensitive mutant, *J. Mol. Biol.,* 96, 69, 1975.
15. **Sakano, H. and Shimura, Y.,** Characterization and *in vitro* processing of transfer RNA precursors accumulated in a temperature-sensitive mutant of *Escherichia coli, J. Mol. Biol.,* 123, 287, 1978.
16. **Apirion, D. and Watson, N.,** A second gene which affects the RNA processing enzyme ribonuclease P of *Escherichia coli, FEBS Lett.,* 110, 161, 1980.
17. **Apirion, D.,** Genetic mapping and some characterization of the *rnpA49* mutation of *Escherichia coli* that affects the RNA processing enzyme ribonuclease P, *Genetics,* 94, 291, 1980.
18. **Shimura, Y., Sakano, H., Kubokawa, S., Nagara, F., and Ozeki, H.,** tRNA precursors in RNase P mutants, in *Transfer RNA: Biological Aspects,* Söll, D., Abelson, J., and Schimmel, P., Eds., Cold Spring Harbor Laboratory, Cold Spring Harbor, N.Y., 1980, 43.
18a. **Motamedi, H., Lee, K., Michols, L., and Schmidt, F. J.,** An RNA species involved in *Escherichia coli* ribonuclease P activity: Gene cloning and effect on transfer RNA synthesis *in vivo, J. Mol. Biol.,* 162, 535, 1982.
19. **Stark, B. P., Kole, R., Bowman, E. J., and Altman, S.,** Ribonuclease P: an enzyme with an essential RNA component, *Proc. Natl. Acad. Sci. U.S.A.,* 75, 3717, 1978.
20. **Lee, S. Y., Bailey, S. C., and Apirion, D.,** Small stable RNAs from *Escherichia coli:* evidence for the existence of new molecules and for a new ribonucleoprotein particle containing 6S RNA, *J. Bacteriol,* 133, 1015, 1978.

21. **Ray, B. K. and Apirion, D.**, Characterization of 10S RNA: a new stable RNA molecule from *Escherichia coli, Mol. Gen. Genet.*, 174, 25, 1979.
22. **Kole, R. and Altman, S.**, Properties of purified RNase P from *Escherichia coli, Biochemistry*, 20, 1902, 1981.
22a. **Jain, S. W., Gurewitz, M., and Apirion, D.**, A small RNA that complements mutants in the RNA processing enzyme ribonuclease P, *J. Mol. Biol.*, 162, 515, 1982.
23. **Reed, R. E., Baer, M. F., Guerrier-Takada, C., Donis-Keller, H., and Altman, S.**, Nucleotide sequence of the gene encoding the RNA subunit (M1 RNA) of ribonuclease P from *Escherichia coli, Cell*, 30, 627, 1982.
24. **Schedl, P., Primakoff, P., and Roberts, J.**, Processing of *E. coli* tRNA precursors, *Brookhaven Symp. Biol.*, 26, 53, 1975.
25. **Schedl, P., Roberts, J., and Primakoff, P.**, In vitro processing of *E. coli* tRNA precursors, *Cell*, 8, 581, 1976.
26. **Ilgen, C., Kirk, L. L., and Carbon, J.**, Isolation and characterization of large transfer ribonucleic acid precursors from *Escherichia coli, J. Biol. Chem.*, 251, 922, 1976.
27. **Altman, S. and Smith, J. D.**, Tyrosine tRNA precursor molecule polynucleotide sequence, *Nature New Biol. (London)*, 233, 35, 1971.
28. **Kupper, H., Sekiya, T., Rosenberg, M., Egan, J., and Landy, A.**, A rho-dependent termination site in the gene coding for tyrosine tRNA $su_3$ of *Escherichia coli, Nature (London)*, 272, 423, 1978.
29. **Nakajima, N., Ozeki, H., and Shimura, Y.**, Organization and structure of an *E. coli* tRNA operon containing seven tRNA genes, *Cell*, 23, 239, 1981.
30. **Sakano, H. and Schimura, Y.**, Sequential processing of precursor tRNA molecules in *Escherichia coli, Proc. Natl. Acad. Sci. U.S.A.*, 72, 3369, 1975.
30a. **Rossi, J., Egan, J., Hudson, L., and Landy, A.**, The *tyrT* locus: termination and processing of a complex transcript, *Cell*, 26, 305, 1981.
31. **Apirion, D., Neil, J., and Watson, N.**, Consequences of losing ribonuclease III on the *Escherichia coli* cell, *Mol. Gen. Genet.*, 144, 185, 1976.
32. **Apirion, D., Neil, J., and Watson, N.**, Revertants from RNase III negative strains of *Escherichia coli, Mol. Gen. Genet.*, 149, 201, 1976.
33. **Gegenheimer, P. and Apirion, D.**, Precursors to 16S and 23S ribosomal RNA from a ribonuclease $III^-$ strain of *Escherichia coli* contain intact RNase III processing sites, *Nucleic Acids Res.*, 8, 1873, 1980.
34. **Apirion, D. and Watson, N.**, Analysis of an *Escherichia coli* strain carrying physiologically compensating mutations one of which causes an altered ribonuclease III, *Mol. Gen. Genet.*, 132, 89, 1974.
35. **Apirion, D. and Watson, N.**, Mapping and characterization of a mutation in *Escherichia coli* that reduces the level of ribonuclease III specific for double-stranded ribonucleic acid, *J. Bacteriol.*, 124, 317, 1975.
36. **Studier, F. W.**, Genetic mapping of a mutation that causes ribonuclease III deficiency in *Escherichia coli, J. Bacteriol.*, 124, 307, 1975.
37. **Gegenheimer, P., Watson, N., and Apirion, D.**, Multiple pathways for primary processing of ribosomal RNA in *Escherichia coli, J. Biol. Chem.*, 252, 3064, 1977.
38. **Dunn, J. J. and Studier, F. W.**, T7 early RNAs and *Escherichia coli* ribosomal RNAs are cut from large precursor RNAs *in vivo* by ribonuclease III, *Proc. Natl. Acad. Sci. U.S.A.*, 70, 3296, 1973.
39. **Nikolaev, N., Silengo, L., and Schlessinger, D.**, Synthesis of a large precursor to ribosomal rRNA in a mutant of *Escherichia coli, Proc. Natl. Acad. Sci. U.S.A.*, 70, 3361, 1973.
40. **Apirion, D. and Lassar, A. B.**, A conditional lethal mutant of *Escherichia coli* which affects the processing of ribosomal RNA, *J. Biol. Chem.*, 253, 1738, 1978.
41. **Ghora, B. K. and Apirion, D.**, 5S ribosomal RNA is contained within a 25S ribosomal RNA that accumulates in mutants of *Escherichia coli* defective in processing of ribosomal RNA, *J. Mol. Biol.*, 127, 507, 1979.
42. **Apirion, D.**, Isolation, genetic mapping, and some characterization of a mutation in *Escherichia coli* that affects the processing of ribonucleic acid, *Genetics*, 90, 659, 1978.
43. **Ghora, B. K. and Apirion, D.**, Identification of a novel RNA molecule in a new RNA processing mutant of *Escherichia coli* which contains 5S rRNA sequences, *J. Biol. Chem.*, 254, 1951, 1979.
44. **Goldblum, K. and Apirion, D.**, Inactivation of the RNA processing enzyme ribonuclease E blocks cell division, *J. Bacteriol.*, 146, 128, 1981.
45. **Ghora, B. K. and Apirion, D.**, Structural analysis and *in vitro* processing to p5 rRNA of a 9S RNA molecule isolated from an *rne* mutant of *E. coli, Cell*, 15, 1055, 1978.
45a. **Singh, B. and Apirion, D.**, Primary and secondary structure in a precursor of 5S rRNA, *Biochim. Biophys. Acta*, 698, 252, 1982.
45b. **Roy, M. K., Singh, B., Ray, B. K., and Apirion, D.**, Maturation of 5-S rRNA: ribonuclease E cleavages and their dependence on precursor sequences, *Eur. J. Biochem.*, 131, 119, 1983.
46. **Adhya, S. and Gottesman, M.**, Control of transcription, *Annu. Rev. Biochem.*, 47, 967, 1978.

47. **Rosenberg, M. and Court, D.,** Regulatory sequences involved in the promotion and termination of RNA transcription, *Annu. Rev. Genet.*, 13, 319, 1979.
48. **Misra, T. K. and Apirion, D.,** Gene *rne* affects the structure of the ribonucleic acid processing enzyme ribonuclease E of *Escherichia coli, J. Bacteriol.*, 142, 359, 1980.
49. **Ray, A. and Apirion, D.,** Cloning the gene for ribonuclease E, an RNA processing enzyme, *Gene*, 12, 87, 1980.
50. **Ray, B. K. and Apirion, D.,** Identification of RNA molecules which contain 5S ribosomal RNA and transfer RNA in an RNAase $E^-$ RNAase $P^-$ double mutant strain of *Escherichia coli, J. Mol. Biol.*, 139, 329, 1980.
51. **Morgan, E. A., Ikemura, T., Lindahl, L., Fallon, A. M., and Nomura, M.,** Some rRNA operons in *E. coli* have tRNA genes at their distal ends, *Cell*, 13, 335, 1978.
52. **Morgan, E. A., Ikemura, T., Post, L. E., and Nomura, M.,** tRNA genes in ribosomal RNA operons of *Escherichia coli*, in *Transfer RNA: Biological Aspects*, Söll, D., Abelson, J., and Schimmel, P., Eds., Cold Spring Harbor Laboratory, Cold Spring Harbor, N.Y., 1980, 259.
53. **Young, R. A.,** Transcription termination in the *Escherichia coli* ribosomal RNA operon *rrnC, J. Biol. Chem.*, 254, 12725, 1979.
54. **Ray, B. K. and Apirion, D.,** Transfer RNA precursors are accumulated in *Escherichia coli* in the absence of RNase E, *Eur. J. Biochem.*, 114, 517, 1981.
55. **Ray, B. K. and Apirion, D.,** RNAase P is dependent on RNAase E action in processing monomeric RNA precursors that accumulate in an RNAase $E^-$ mutant of *Escherichia coli, J. Mol. Biol.*, 149, 599, 1981.
55a. **Pragai, B. and Apirion, D.,** Processing of bacteriophage T4 transfer RNAs: structural analysis and *in vitro* processing of precursors that accumulate in RNase $E^-$ strains, *J. Mol. Biol.*, 155, 465, 1982.
56. **Dahlberg, A. E., Dahlberg, J. E., Lund, E., Tokimatsu, H., Rabson, A. B., Calvert, P. C., Reynolds, F., and Zahalak, M.,** Processing of the 5′ end of *Escherichia coli* 16S ribosomal RNA, *Proc. Natl. Acad. Sci. U.S.A.*, 75, 3598, 1978.
56a. **Bhat, B., and Apirion, D.,** unpublished observations.
57. **Feunteun, J., Rosset, R., Ehresmann, C., Stiegler, P., and Fellner, P.,** Abnormal maturation of precursor 16S RNA in a ribosomal assembly defective mutant of *E. coli, Nucleic Acids Res.*, 1, 141, 1974.
58. **Maisurian, A. N. and Buyanovskaya, E. A.,** Isolation of an *Escherichia coli* strain restricting bacteriophage suppressor, *Mol. Gen. Genet.*, 120, 227, 1973.
59. **Seidman, J. G., Schmidt, F. J., Foss, K., and McClain, W. H.,** A mutant of *Escherichia coli* defective in removing 3′ terminal nucleotides from some transfer RNA precursor molecules, *Cell*, 5, 389, 1975.
60. **Ghosh, R. K. and Deutscher, M. P.,** Purification of potential 3′ processing nucleases using synthetic tRNA precursors, *Nucleic Acids Res.*, 5, 3831, 1978.
60a. **Zaniewski, R. and Deutscher, M. P.,** Genetic mapping of a mutation in *Escherichia coli* leading to a temperature-sensitive RNase D, *Mol. Gen. Genet.*, 185, 142, 1982.
61. **Bram, R. J., Young, R. A., and Steitz, J. A.,** The ribonuclease III site flanking 23S sequences in the 30S ribosomal precursor RNA of *E. coli, Cell*, 19, 393, 1980.
62. **Ginsburg, D. and Steitz, J. A.,** The 30S ribosomal precursor RNA from *Escherichia coli*. A primary transcript containing 23S, 16S, and 5S sequences, *J. Biol. Chem.*, 250, 5647, 1975.
63. **Robertson, H. D., Webster, R. D., and Zinder, N. D.,** Purification and properties of ribonuclease III from *Escherichia coli, J. Biol. Chem.*, 243, 82, 1968.
64. **Wu, M. and Davidson, N.,** Use of gene 32 protein straining of single-strand polynucleotides for gene mapping by electron microscopy: application to the $\phi80\ d_3$ *ilv* $su^+7$ system, *Proc. Natl. Acad. Sci. U.S.A.*, 72, 4506, 1975.
65. **Gegenheimer, P., Watson, N., and Apirion, D.,** Processing of ribosomal RNA in *E. coli*, in *ICN/UCLA Symp. Molecular Mechanisms in Control of Gene Expression*, Nierlich, D. P., Rutter, W. J., and Fox, C. J., Eds., Academic Press, New York, 1976, 405.
66. **Dougan, A. H. and Glaser, D. A.,** Rate of chain elongation of ribosomal RNA molecules in *Escherichia coli, J. Mol. Biol.*, 87, 775, 1974.
67. **Sprague, K. U. and Steitz, J. A.,** The 3′ terminal oligonucleotide of *E. coli* 16S ribosomal RNA: the sequence in both wild-type and RNase $III^-$ cells is complementary to the polypurine tracts common to mRNA initiator regions, *Nucleic Acids Res.*, 2, 787, 1975.
68. **Gegenheimer, P. and Apirion, D.,** Structural characterization and *in vitro* processing of *Escherichia coli* ribosomal RNA transcripts containing 5′ triphosphates, leader sequences, 16S RNA and spacer tRNAs, *J. Mol. Biol.*, 143, 227, 1980.
69. **Gegenheimer, P. and Apirion, D.,** *Escherichia coli* ribosomal ribonucleic acids are not cut from an intact precursor molecule, *J. Biol. Chem.*, 250, 2407, 1975.
70. **Lund, E., Dahlberg, J. E., and Guthrie, C.,** Processing of spacer tRNAs from ribosomal RNA transcripts of *E. coli*, in *Transfer RNA: Biological Aspects*, Söll, D., Abelson, J., and Schimmel, P., Eds., Cold Spring Harbor Laboratory, Cold Spring Harbor, N.Y., 1980, 123.

71. **Cowgill de Narvaez, C. and Schaup, H. W.,** *In vivo* transcriptionally coupled assembly of *Escherichia coli* ribosomal subunits, *J. Mol. Biol.*, 134, 1, 1979.
72. **Apirion, D., Ghora, B. K., Plautz, G., Misra, T. K., and Gegenheimer, P.,** Processing of rRNA and tRNA in *Escherichia coli:* Cooperation between processing enzymes, in *Transfer RNA: Biological Aspects*, Söll, D., Abelson, J., and Schimmel, P., Eds., Cold Spring Harbor Laboratory, Cold Spring Harbor, N.Y., 1980, 139.
73. **Misra, T. K. and Apirion, D.,** Characterization of an endoribonuclease, RNase N, from *Escherichia coli*, *J. Biol. Chem.*, 253, 5594, 1978.
74. **Misra, T. K., Rhee, S., and Apirion, D.,** A new endoribonuclease from *Escherichia coli*, ribonuclease N, *J. Biol. Chem.*, 251, 7669, 1976.
75. **Meyhack, B., Meyhack, I., and Apirion, D.,** Processing of precursor particles containing 17S rRNA in a cell free system, *FEBS Lett.*, 49, 215, 1974.
76. **Hayes, F. and Vasseur, M.,** Processing of the 17S *Escherichia coli* precursor RNA in the 27S preribosomal particle, *Eur. J. Biochem.*, 61, 433, 1976.
77. **Plautz, G. and Apirion, D.,** Processing of RNA in *Escherichia coli* is limited in the absence of ribonuclease III, ribonuclease E and ribonuclease P, *J. Mol. Biol.*, 149, 813, 1981.
77a. **Pettijohn, D. E.,** personal communication.
78. **Vogeli, G., Grosjean, H., and Söll, D.,** A method for the isolation of specific tRNA precursors, *Proc. Natl. Acad. Sci. U.S.A.*, 72, 4790, 1975.
79. **Jain, S., Pragai, B., and Apirion, D.,** A possible complex containing RNA processing enzymes, *Biochem. Biophys. Res. Commun.*, 106, 768, 1982.
80. **Eagan, J. and Landy, A.,** Structural analysis of the $tRNA_1^{Tyr}$ gene of *Escherichia coli*. A 178 base pair sequence that is repeated 3.14 times, *J. Biol. Chem.*, 253, 3607, 1978.
81. **Gegenheimer, P. and Apirion, D.,** Processing of rRNA by RNAase P: spacer tRNAs are linked to 16S rRNA in an RNAase P RNAase III mutant strain of *Escherichia coli*, *Cell*, 15, 527, 1978.
82. **Apirion, D. and Gitelman, D.,** Decay of RNA in RNA processing mutants of *E. coli*, *Mol. Gen. Genet.*, 177, 339, 1980.
83. **Barry, G., Squires, C., and Squires, C. L.,** Attenuation and processing of RNA from the *rplJL-rpoBC* transcription unit of *Escherichia coli*, *Proc. Natl. Acad. Sci. U.S.A.*, 77, 3331, 1980.
84. **Post, L. E., Strycharz, G. D., Nomura, M., Lewis, H., and Dennis, P. P.,** Nucleotide sequence of the ribosomal protein gene cluster adjacent to the gene for RNA polymerase subunit β in *Escherichia coli*, *Proc. Natl. Acad. Sci. U.S.A.*, 76, 1697, 1979.
85. **Dennis, P. P. and Fill, N. P.,** Transcriptional and post-transcriptional control of RNA polymerase and ribosomal protein genes cloned on composite colE1 plasmids in the bacterium *Escherichia coli*, *J. Biol. Chem.*, 254, 7540, 1979.
86. **Apirion, D. and Watson, N.,** Unaltered stability of newly-synthesized RNA in strains of *Escherichia coli* missing a ribonuclease specific for double-stranded RNA, *Mol. Gen. Genet.*, 136, 317, 1975.
87. **Talkad, V., Achord, D., and Kennell, D.,** Altered RNA metabolism in ribonuclease III-deficient strains of *Escherichia coli*, *J. Bacteriol.* 135, 528, 1978.
88. **Gitelman, D. R. and Apirion, D.,** The synthesis of some proteins is affected in RNA processing mutants of *Escherichia coli*, *Biochem. Biophys. Res. Commun.*, 96, 1063, 1980.
89. **Apirion, D. and Watson, N.,** Ribonuclease III is involved in motility of *Escherichia coli*, *J. Bacteriol.*, 133, 1543, 1978.
90. **Kourilsky, P., Bourguignon, M. F., and Gros, F.,** Kinetics of viral transcription after induction of prophage, in *The Bacteriophage Lambda*, Hershey, A. D., Ed., Cold Spring Harbor Laboratory, Cold Spring Harbor, N.Y., 1971, 647.
91. **Lozeron, H. A., Anevski, P. J., and Apirion, D.,** Antitermination and absence of processing of the leftward transcript of coliphage lambda in the RNAase III-deficient host, *J. Mol. Biol.*, 109, 359, 1977.
92. **Kourilsky, P., Bourguignon, M. F., Bouquet, M., and Gros, F.,** Early transcription controls after induction of prophage λ, *Cold Spring Harbor Symp. Quant. Biol.*, 35, 305, 1970.
93. **Oxender, D. L., Zurawski, G., and Yanofsky, C.,** Attenuation in the *Escherichia coli* tryptophan operon: role of RNA secondary structure involving the tryptophan codon region, *Proc. Natl. Acad. Sci. U.S.A.*, 76, 5524, 1979.
94. **Apirion, D., Neil, J., Ko, T. S., and Watson, N.,** Metabolism of ribosomal RNA in mutants of *Escherichia coli* doubly defective in ribonuclease III and the transcription termination factor rho, *Genetics*, 90, 19, 1978.
95. **Robertson, H. D., Pelle, E. G., and McClain, W. H.,** RNA processing in an *Escherichia coli* strain deficient in both RNas P and RNase III, in *Transfer RNA: Biological Aspects*, Söll, D., Abelson, J., and Schimmel, P., Eds., Cold Spring Harbor Laboratory, Cold Spring Harbor, N.Y., 1980, 107.
96. **Wireman, J. W. and Sypherd, P. S.,** Properties of 30S ribosomal particles reconstituted from precursor 16S ribonucleic acid, *Biochemistry*, 13, 1215, 1974.

97. **Altman, S., Bothwell, A. L. M., and Stark, B. C.,** Processing of *E. coli* $tRNA^{Tyr}$ precursor *in vitro, Brookhaven Symp. Biol.*, 26, 12, 1975.
98. **Guthrie, C., Seidman, J. G., Comer, M. M., Bock, R. M., Schmidt, F. J., Barrell, B. G., and McClain, W. H.,** The biology of bacteriophage T4 transfer RNAs, *Brookhaven Symp. Biol.*, 26, 106, 1975.
99. **Stahl, D. A., Walker, T. A., Meyhack, B., and Pace, N. R.,** Precursor-specific nucleotides can govern RNA folding, *Cell,* 18, 1133, 1979.
100. **Smith, J. D.,** Mutants which allow accumulation of $tRNA^{Tyr}$ precursor molecules, *Brookhaven Symp. Biol.*, 26, 1, 1975.
101. **Meyhack, B. and Pace, N. R.,** Involvement of the mature domain in the *in vitro* maturation of *Bacillus subtilis* precursor 5S ribosomal RNA, *Biochemistry,* 17, 5804, 1978.
102. **Deutscher, M. P. and Hilderman, R. H.,** Isolation and partial characterization of *Escherichia coli* mutants with low levels of transfer ribonucleic acid nucleotidyltransferase, *J. Bacteriol.*, 118, 621, 1974.
103. **Foulds, J., Hilderman, R. H., and Deutscher, M. P.,** Mapping of the locus for *Escherichia coli* transfer ribonucleic acid nucleotidyltransferase, *J. Bacteriol.*, 118, 628, 1974.
104. **Zaniewski, R. and Deutscher, M. P.,** Genetic mapping of a mutation in *Escherichia coli* leading to a temperature-sensitive RNase D, *Mol. Gen. Genet.*, 185, 142, 1982.
105. **Brosius, J., Dull, T. J., Sleeter, D. D., and Noller, H. F.,** Gene organization and primary structure of a ribosomal RNA operon from *Escherichia coli, J. Mol. Biol.*, 148, 107, 1981.

Chapter 3

# PROCESSING OF BACTERIOPHAGE-CODED RNA SPECIES

**Francis J. Schmidt**

## TABLE OF CONTENTS

# I. INTRODUCTION

Bacteriophage represent nature at its most ruthless efficiency. Within a short time, about half an hour under physiological conditions, a single genome is replicated, not once, as in a bacterial cell, but hundreds of times. Furthermore, the full complement of genes which are present in a phage DNA is maintained at some cost. Deletion mutants of λ often possess increased thermostability over wild-type phage and the Qβ replicase system preferentially replicates smaller RNAs in vitro.[1] The author would argue from these observations that phage genomes, then, come down on the selectionist side of evolutionary genetics and that the presence of a gene product is good evidence that it is useful under at least some circumstances for the maintenance of the virus in the population.

Coupled with the strong selective pressure for minimum size and/or maximum efficiency is the extraordinary plasticity of meeting these requirements. No single strategy has been chosen to ensure phage propagation. No one who has taken a course in molecular biology can fail to be impressed with the variety of approaches to survival, from the extreme compression of the single-stranded coliphage genome to the ability of a T phage to completely appropriate its host's machinery.

Coupled with the variety of phage to study is the relative ease of doing so. Several phage now have saturated or sequenced genetic maps. Furthermore, just about any function can be made conditional for survival in the laboratory. If one wishes to study subtle biological processes (including RNA processing) the ability to amplify the effects of this process at will is more than useful. Likewise, the virus and its host are separately manipulable genetically so that subtle effects can be amplified at two levels. Recent results which illustrate this advantage in the study of RNA processing are described below. For a biochemist, phage offer the advantage that the host genome can be shut off either by virulence or irradiation so that the products one wishes to study are synthesized free of host background.

One pays the price for this efficiency: phage do not carry out many of the functions of cells or even of more complex viruses. Nevertheless, the genetic strategies which phage use often show up in the other systems; we may confidently expect that new insights and tools gained from bacteriophage work will be generally applicable to other RNA processing systems.

This review has two purposes. The first is to enumerate known tRNA and mRNA processing pathways of phage genomes and the second is to point out how these may possibly lead to understanding the utilization of these pathways in regulation of phage functions. Particular aspects of RNA processing in phage systems have been reviewed, either specifically or more generally.[2-6] In addition, the 1974 *Brookhaven Symposium in Biology* remains a valuable reference.[7]

# II. PHAGE-CODED TRANSFER RNA SYNTHESIS

In 1968, subsequent to erroneous reports that Herpes simplex virus coded for tRNA species, Scherberg and co-workers[8] and Daniel et al.[9] showed by hybridization experiments that phage T4 DNA codes for several tRNA isoacceptors. Clearly, tRNA alterations can affect coding of message classes, and since T4 tRNAs complement the coding capacity of *Escherichia coli* tRNA[10] it was a common hope that this phenomenon might prove to be as useful as the T4 assembly pathway in analyzing the molecular biology of development. Unfortunately, that particular hope has not been realized; tRNA alteration as a developmental mechanism is still speculative and the slow pace of data acquisition makes the speculation less and less entrancing. Nevertheless, the study of tRNA metabolism in phage has been the source of substantial insight into the mechanisms of tRNA processing.

These insights were due to an integrated approach involving genetics, sequence analysis of phage-coded RNA products, and biochemical reconstruction of the events of the synthesis pathways.

The genetics of phage-coded tRNA systems depends on the use of chain-termination suppressor forms of the tRNAs. In T4, four independent suppressors of various coding capacities and suppressor-negative point or deletion mutants of T4 suppressors have been isolated by Wilson, McClain, and their co-workers.[11-16] As might be expected, point mutations in tRNA structure can interfere with correct processing of the appropriate precursors, and these altered forms can then be analyzed to give some insight into the structural requirements of tRNA biosynthetic reactions. In addition, unlinked antisuppressor mutations (of $tRNA^{Ser}_{UAG}$ or $tRNA^{Ser}_{UAA}$) which affect the synthesis of T4 $tRNA^{Pro}$, $tRNA^{Ser}$, and $tRNA^{Ile}$ have been isolated. The gene represented by these mutations termed *mb*[15] or *M1*[17] is the only known T4-coded function required for tRNA biosynthesis; all other $su^-$ mutations appear to map within the tRNA operon.

A second advantage of phage systems is their ability to transcribe the viral genome exclusively facilitating analysis of a limited set of RNA species. When bacteriophage T4 infects its host, only eight phage-coded tRNAs are synthesized, along with two stable RNAs of unknown function (see Figure 1).[17]

**A. Sequence Organization of the T4 tRNA Gene Cluster and Its Implications for tRNA Biosynthesis**

Although historically many details of tRNA biosynthesis were known from analysis of individual RNAs before DNA sequence analysis began, the recently determined DNA sequence permits review of the tRNA biosynthetic pathways in a more logical fashion. The sequence analysis has been done by Fukada and Abelson[18] and by Mazzara, Plunkett, and McClain;[19] the data obtained by the two groups using two methodologies agree almost precisely. Prior to the DNA sequencing effort, most of the stable T4 4S RNAs had been sequenced by fingerprint or in vitro transcription[20] techniques.

The T4 tRNAs are clustered in a single operon whose transcription is counterclockwise on the standard T4 map. In the map in Figure 2, the tRNA genes are flanked by gene 57 on the left and gene *e* on the right. Transcription of this region proceeds from a promoter about 900 bp upstream from the 5′ proximal tRNA, $tRNA^{Gln}$.[21] Cloning of the entire operon has not been achieved, arguing that expression of DNA in or from this region of the map is lethal to *E. coli*. Further, when clones were isolated in λgt as minute plaques, spontaneous, faster-growing deletions and insertions arose.[18] These events removed the RNA polymerase binding site of the operon and about 500 bp of DNA downstream from this promoter. It has been suggested that lethality is due to overproduction of T4 internal protein I, a DNA-binding protein found in the phage head, although some of the tRNAs themselves may be lethal. The gene for internal protein I apparently lies downstream from the promoter; if this is so, the T4 tRNA gene cluster will furnish a useful system for study of the processing of composite transcripts leading to both proteins and stable RNA products.[13] Although the upstream portion of the T4 tRNA cluster is difficult to clone, the downstream cluster of $tRNA^{Arg}$, stable RNA Species 1(C) and 2(D), has been isolated several times.[19,21]

**B. Synthesis of tRNAs In Vitro**

Transcription of the T4 gene cluster can be done with purified RNA polymerase holoenzyme, nucleoside triphosphates, and phage DNA; no accessory factors or phage-coded products are necessary for this reaction.[23] Several transcripts have been isolated by Daniel and co-workers;[23] these could then be processed by an S100 supernatant from uninfected *E. coli*.

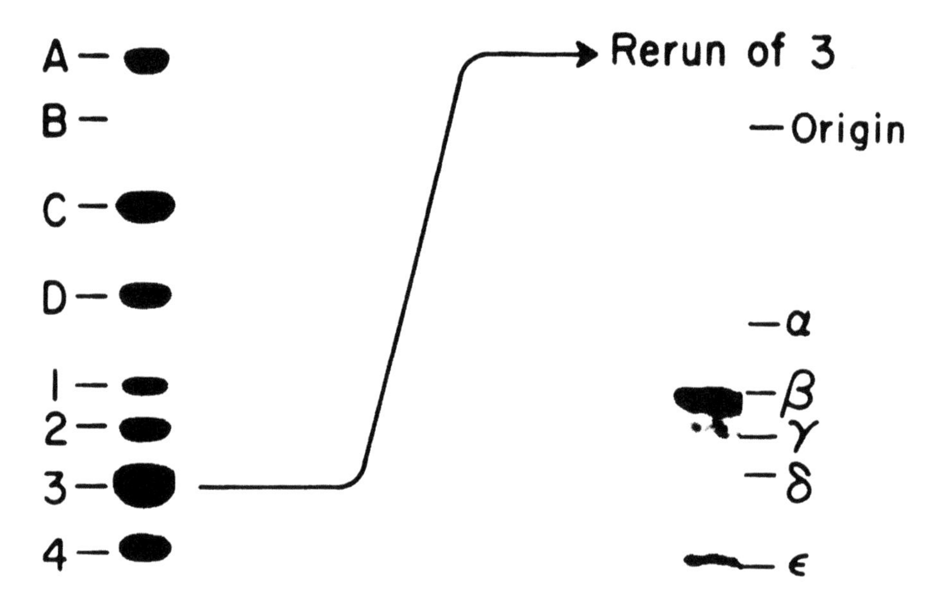


FIGURE 1. Bacteriophage T4 low-molecular-weight RNA species. RNA species A and B are the $tRNA^{Pro}$-$tRNA^{Ser}$ and $tRNA^{Thr}$-$tRNA^{Ile}$ precursors, respectively. RNAs C and D, also called Species 1 and 2, are stable RNAs of unknown function. The identities of individual tRNA species are 1: $tRNA^{Ser}$, 2: $tRNA^{Leu}$, 3α: $tRNA^{Thr}$, 3β: $tRNA^{Pro}$, 3γ: $tRNA^{Glu}$, 3δ: $tRNA^{Ile}$, 3ε: $tRNA^{Arg}$, 4: $tRNA^{Gly}$. (From Guthrie, C., Seidman, J. G., Comer, M. M., Bock, R. M., Schmidt, F. J., Barrell, B. G., and McClain, W. H., *Brookhaven Symp. Biol.*, 26, 106, 1975. With permission.)

In vitro processing of the largest product yielded nine of the ten stable RNA species (all but $tRNA^{Ile}$; see below), showing that the transcription continued through both sets of tRNA genes. In contrast to this work, other workers reported that a coupled in vitro transcription-processing extract was unable to synthesize $tRNA^{Pro}$ or $tRNA^{Ser}$ in addition to $tRNA^{Ile}$.[21] In vivo, synthesis of these three tRNAs requires the T4 *mb (M1)* gene product and it was proposed that failure to synthesize these tRNAs reflected an in vitro requirement as well. An alternative explanation, that these tRNAs are unstable in the processing extract, seems to be favored by the results obtained with isolated precursors.[24] Two other, shorter, transcripts of this region were also isolated; these yielded neither $tRNA^{Ile}$ nor RNAs from subcluster II. These latter RNA species therefore, probably represent the products of termination after (or within) subcluster I.[26]

The nucleotide sequence[18,19] of the DNA corresponding to the stable RNAs (Figure 2) shows that:

1. The individual RNA species are separated by only a few (1 to 10) bp in the DNA; this is true even between different precursor molecules.
2. The sequences of individual RNA species reflect their modes of biosynthesis (see below for a description of the individual pathways).
3. There is a strong potential terminator 70 bp downstream from the Species 1(C) gene; this terminator is preceded by a sequence capable of forming a stable ($\Delta G^{0\prime} \cong -20.7$ kcal $mol^{-1}$) stem and loop. This terminator could explain why in vitro transcription of T4 DNA does not proceed into gene *e* (lysozyme). Since there is evidence that gene *e* is synthesized in phage-infected cells as a 6 kb transcript[24] (i.e., encompassing the entire tRNA operon) it is possible that this termination is not 100% efficient in vivo.[25] Figure 3a shows this terminator.
4. A second potential terminator occurs within the distal half of the $tRNA^{Ile}$ gene, where a somewhat weaker ($\Delta G^{0\prime} \cong -3.3$ kcal $mol^{-1}$) stem and loop precedes a run of 4 to 5 T residues. It is too early to conclude that this terminator is responsible for the rho-dependent termination of transcription that Goldfarb and Daniel[23] observed between tRNA subclusters 1 and 2, but its presence may account for some unusual features of $tRNA^{Ile}$ synthesis (see below). Figure 3b shows this stem and loop and the $tRNA^{Ile}$ structure.

### C. Enzymes Required for tRNA Processing

Once (or while) the primary transcript is synthesized, nucleolytic and biosynthetic functions are required to mature the transcription product into tRNA and stable RNA species. Genetic experiments show several features of this processing:

1. T4 $tRNA^{Pro}$, $tRNA^{Ser}$, $tRNA^{Ile}$, and $tRNA^{Gln}$ genes do not contain the C-C-$A_{OH}$ terminus common to all tRNAs; this must be synthesized posttranscriptionally. Since *E. coli* mutants deficient in tRNA nucleotdyltransferase (the *cca* locus) do not permit expression of $tRNA^{Ser}$ or $tRNA^{Gln}$ suppressors,[26] this enzyme is involved in C-C-A synthesis in vivo.
2. Expression of $tRNA^{Ser}$ (but not $tRNA^{Gln}$, $tRNA,^{Leu}$ or $tRNA^{Arg}$) suppressors is deficient in strain BN, which has been identified as deficient in a cellular exonuclease.[27,28] This is consistent with the DNA and RNA sequence data which show extra nucleotides at the 3′ end of this tRNA.[29]
3. Expression of the suppressor form of $tRNA^{Gln}$ is abolished in hosts deficient in *E.coli* RNase III;[30] therefore this enzyme must act near the proximal part of the tRNA gene cluster. The requirement for other enzymatic activities is indicated by in vivo accumulation of the appropriate tRNA precursors in the absence of an enzyme activity. Temperature-sensitive RNase P[31,32] mutants and RNase E[33] mutants interfere with tRNA synthesis at high temperature and so these enzymes must also be directly or indirectly required for synthesis.

Finally, an enzyme has been purified by Goldfarb and Daniel[34] and termed RNase PC. This enzyme, when allowed to digest primary transcript produced in vitro, produced molecules which are probably the $tRNA^{Pro}$-$tRNA^{Ser}$ and $tRNA^{Gln}$-$tRNA^{Leu}$ dimeric precursors. In addition, $tRNA^{Arg}$ and Species 1(C) and 2(D) precursors were formed, although the cleavages were not 100% faithful.

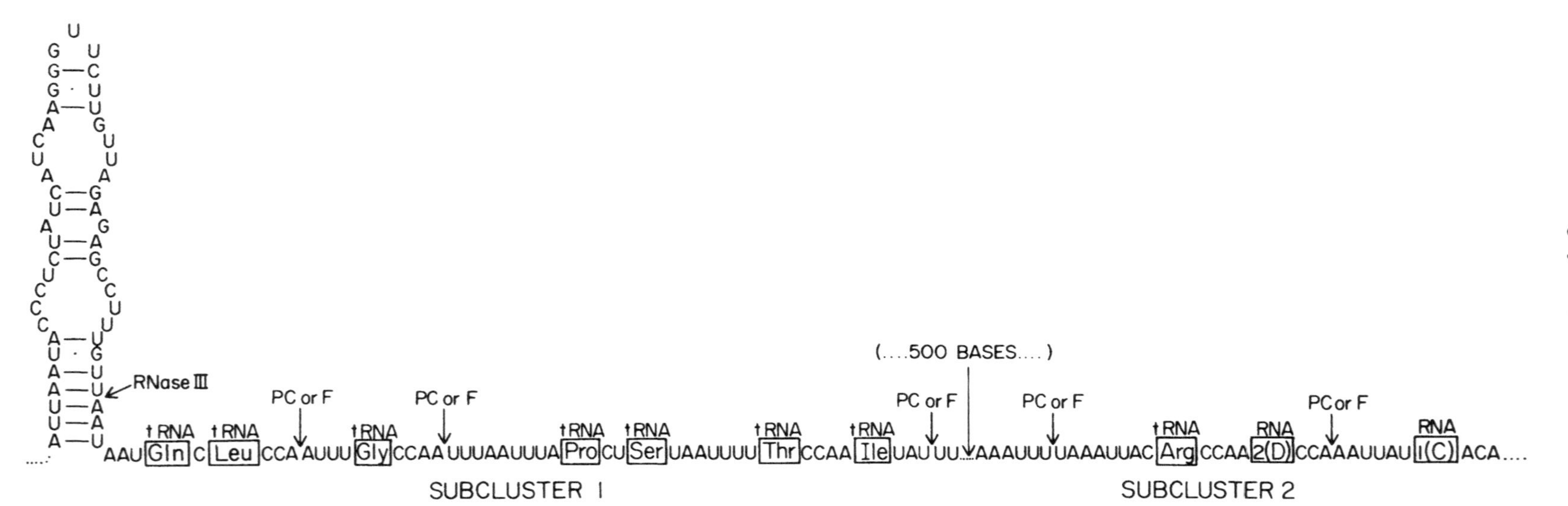


FIGURE 2. Nucleotide sequence arrangement of T4 rRNA region. Transcription is from left to right. Only the sequences between the RNA species are shown. Processing sites in the primary transcript are shown.

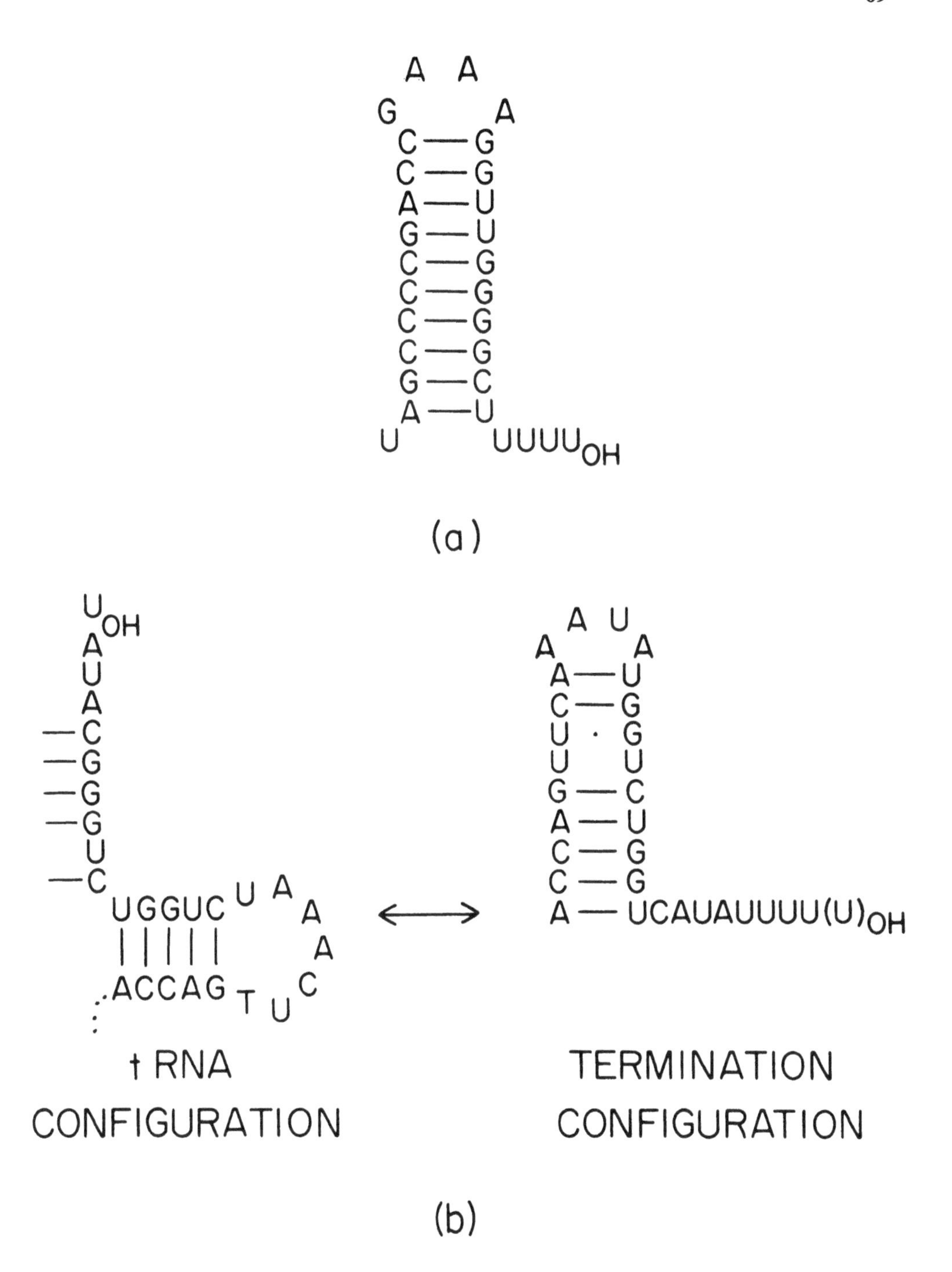


FIGURE 3. Putative transcription termination sequences for the T4 tRNA cluster. (a) Terminator distal to the entire tRNA gene cluster. The sequence of the RNA transcript is shown. (b) Stem and loop which can be formed from part of the tRNA$^{Ile}$ sequence. This weak potential hairpin precedes a U-rich sequence in the RNA transcript. The alternative pairings are shown.[17]

### D. From Transcript to Precursor

Analysis of the processing of the T4 tRNA species and many other large RNAs is complicated by the involvement of multiple enzymes acting on a multidomain structure. As a result, processing blocks (especially in vitro) are not always absolute, even in the complete absence of the relevant activity. Nevertheless, some conclusions can be drawn.

1. Synthesis of the immature 5′ end of the $tRNA^{Gln}$-$tRNA^{Leu}$ precursor RNA is accomplished by RNase III. A long, hairpin double-stranded region of the transcript can be formed. Sequence analysis of precursor forms synthesized in an *rnc* host shows that, in common with many other RNase III sites, cleavage occurs within this hairpin distal to a pyrimidine-rich sequence (see Figure 2).[35] In the presence of RNase III, about 50 nucleotides are removed in vitro or in vivo to give a precursor whose further maturation is independent of RNase III. The RNA on which RNase III acts is itself derived by an undetermined enzymatic step. In the absence of RNase III, $tRNA^{Leu}$ is still made, even though its normal mode of synthesis involves a dimeric precursor containing $tRNA^{Gln}$ and $tRNA^{Leu}$. The $tRNA^{Leu}$ apparently arises after cleavage with the $tRNA^{Gln}$ sequence to yield a maturation-competent molecule. No other tRNAs seem to require RNase III for their synthesis.[35]
2. The processing to yield the 3′ end of the $tRNA^{Gln}$-$tRNA^{Leu}$ precursor is accomplished by another enzyme. This enzyme is apparently the PC nuclease described by Goldfarb and Daniel,[34] and which may be related to the RNase F found by Apirion and Watson.[36]

When an *rne* mutant deficient in RNase E is infected with T4, a tRNA precursor accumulates whose 5′ end is identical to that produced by RNase III.[33] In itself this might be taken to indicate that RNase E cleaves the RNase III cleavage product but $rne^-$ cell extract was still capable of correct processing to give the $tRNA^{Gln}$-$tRNA^{Leu}$ dimer. Two interpretations of this result are possible: (1) RNase E is normally required for correct processing of the RNase III cleavage product but other enzymes(s) can do this in vitro, or (2) RNase E is an accessory enzyme in this pathway. The possibilities cannot be distinguished with certainty but RNase E absence appears to affect RNase P activity.[33] In vitro data alone cannot be used to predict processing pathways, however,[37] and the question is still unresolved.

The generation of other tRNA precursors, $tRNA^{Pro}$-$tRNA^{Ser}$, pre-$tRNA^{Arg}$ and precursors to stable RNAs 1(C) and 2(D), and probably pre-$tRNA^{Gly}$ can be accomplished by nuclease PC in vitro, although these product RNAs have not been characterized by sequence analysis.[34] The in vitro synthesis of $tRNA^{Thr}$-$tRNA^{Ile}$ dimeric precursor by a purified enzyme has not been reported.

### E. From Precursor to tRNA

When wild-type *E. coli* cells are infected with T4, several tRNA precursors in addition to the mature RNA species can be detected. Historically these precursors served as the first substrates for the elucidation of the tRNA biosynthetic reactions and many of the concepts which arose have guided further analysis of tRNA biosynthetic reactions.

Structurally, the T4 tRNA precursors can be envisioned as a series of tRNA-like domains joined by linker regions. Evidence for this comes from the inability of base-change mutations to interfere with base-modification of more than one tRNA in a dimer[38] and from in vitro solution studies using nuclease $S_1$[39] as a structural probe. Furthermore, processing enzymes fail to recognize oligonucleotides spanning the cleavage site with a precursor.[40]

Each of the similar T4 tRNA domains, however, possesses a unique maturation pathway, determined by combinations of requirements for removal of 3′ nucleotides either distal to or in place of the C-C-$A_{OH}$ sequence, for synthesis of the C-C-$A_{OH}$ sequence in whole or in part, and for RNase P to act on a C-C-A-containing precursor molecule. As the gene sequences are different, so too are the maturation pathways.

*1. The $tRNA^{Pro}$-$tRNA^{Ser}$ Precursor RNA*

This species was the first tRNA dimer whose nucleotide sequence and maturation pathway were known in detail; many aspects of this pathway have been reviewed by McClain.[41]

The $tRNA^{Pro}$-$tRNA^{Ser}$ precursor is a relatively abundant RNA species in normal labeling of T4-infected *E. coli*, being about one third as abundant as a mature tRNA by 15 min after infection. Other T4 tRNA precursors are present in significantly lower amounts. This must mean that the metabolic step immediately following $tRNA^{Pro}$-$tRNA^{Ser}$ precursor RNA synthesis must be rate-limiting in vivo. A clue to the nature of this limiting event is provided by nucleotide sequence analysis of this precursor RNA: consistent with the gene sequence, the 3′ end of this precursor does not contain a C-C-$A_{OH}$ terminus. Instead U-A-$A_{OH}$ and shorter 3′ ends are found. The C-C-$A_{OH}$ sequence must be derived by processing. The maturation pathway (Figure 4) was deduced from an analysis of precursor RNA forms which accumulate in various mutant cells. First, in RNase P-deficient cells, precursor RNA ending in the mature C-C-$A_{OH}$ sequence is found; this observation shows that the C-C-$A_{OH}$ sequence can be added prior to RNase P cleavage.[42] In cells deficient in tRNA nucleotidyltransferase, a shortened 3′ sequence ending in $G_{OH}$, i.e., with neither U-A-$A_{OH}$ or C-C-$A_{OH}$, is found and further precursor maturation is blocked.[42,43] In *E. coli* strain BN, the U-A-$A_{OH}$ form of precursor RNA predominates and maturation is blocked. Strain BN was shown to be lacking a specific nuclease capable of removing extra 3′ nucleotides from immature tRNA species.[44]

This scheme (Figure 4) makes several predictions about the enzymology of tRNA processing which were confirmed in vitro and in vivo: first, tRNA nucleotidyltransferase is capable of accurate synthesis of C-C-$A_{OH}$ on the precursor RNA.[43] Secondly, since the G-U-A-$A_{OH}$ and $G_{OH}$ but not G-C-A-$A_{OH}$ forms of precursor RNA can be seen in labeled cells, the latter precursor should be a better substrate for RNase P than the former two species; this was shown by kinetic experiments to be the case.[40] Finally, a precursor whose C-C-A sequence is encoded in the genome should not require tRNA nucleotidyltransferase or BN RNase for maturation. Fortuitously, such a species exists: phage T2 codes for an identical $tRNA^{Pro}$-$tRNA^{Ser}$ whose C-C-A sequence is genomically encoded; its cleavage in vivo does not depend on the presence of cellular BN RNase or tRNA nucleotidyltransferase.[45]

Once the 3′ end of $tRNA^{Ser}$ has matured to C-C-$A_{OH}$, RNase P acts to cleave the precursor RNA. In vitro, cleavage at the interstitial site is about 2.5 times faster than at the 5′ site.[37] This difference is magnified in vivo, so that an extensive search for $tRNA^{Pro}$-$tRNA^{Ser}$ dimers with cleaved 5′ ends was not successful.[37] On the other hand, RNase P cleavage of the monomeric tRNA precursor to generate the mature 5′ end of $tRNA^{Pro}$ occurs without the necessity for 3′ maturation, as evidenced by the presence of T2 $tRNA^{Pro}$ with a mature 5′ but not 3′ end in *cca* or BN hosts (T2 $tRNA^{Ser}$ is synthesized accurately in these mutant hosts).[45] This facile cleavage of a 3′ immature monomeric precursor contrasts sharply with the restricted cleavage of a 3′ immature dimer; the structural or enzymatic basis for this discrimination by RNase P is not shown. (It is also possible that the $tRNA^{Pro}$ cleavage can be accomplished by another cellular or phage-coded nuclease.)

To summarize: both T4 $tRNA^{Pro}$ and $tRNA^{Ser}$ maturations from their common precursor require the participation of three host enzymes. Two of these enzymes, BN RNase and tRNA nucleotidyltransferase, are required for 3′ C-C-$A_{OH}$ synthesis while RNase P is solely responsible for 5′ terminal maturation.

*2. The $tRNA^{Gln}$-$tRNA^{Leu}$ Precursor RNA*

This precursor RNA is present only in transient amounts; unlike the $tRNA^{Pro}$-$tRNA^{Ser}$ precursor, it does not accumulate in hosts lacking either RNase BN or tRNA nucleotidyltransferase.

Both of these tRNAs are available in suppressor form which permits in vivo estimation of the enzymes required for tRNA synthesis. As noted above, $tRNA^{Gln}_{Su+}$ but not $tRNA^{Leu}_{Su+}$

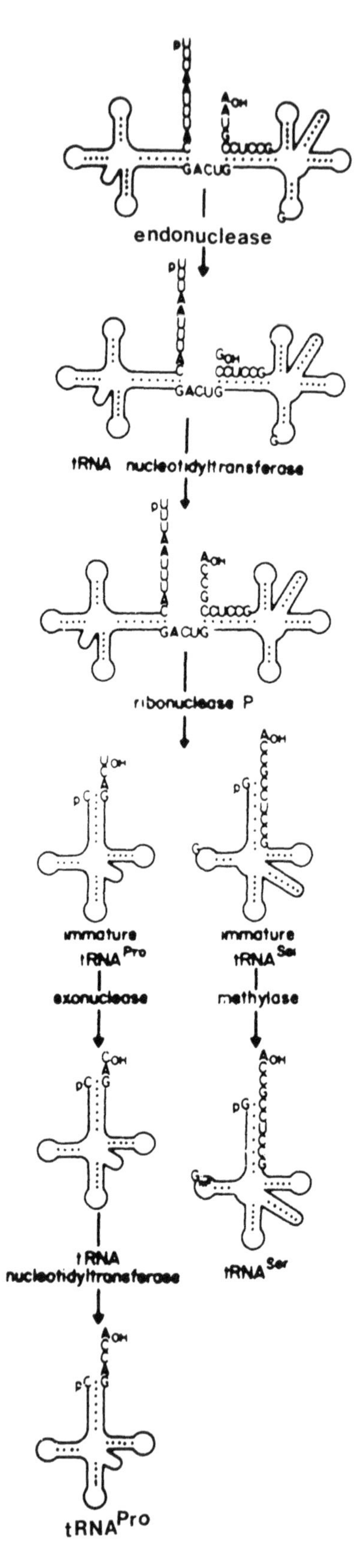
endonuclease
tRNA nucleotidyltransferase
ribonuclease P
immature
tRNA^Pro
exonuclease
immature
tRNA^Ser
methylase
tRNA
nucleotidyltransferase
tRNA^Ser
tRNA^Pro

FIGURE 4

requires RNase III activity for its accurate synthesis; this enzyme is required to generate the 5′ end of the dimeric precursor from the primary transcript. Further, $tRNA^{Gln}_{Su+}$ expression is restricted by *cca* hosts[26] but not those lacking RNase BN, while $tRNA^{Leu}_{Su+}$ requires neither of these activities for its expression.[46] These observations permit the following conclusions about these tRNA species: (1) neither tRNA has extra nucleotides in place of the C-C-A sequence, and (2) the C-C-A sequence of $tRNA^{Leu}$ is genomically encoded but that of $tRNA^{Gln}$ is synthesized by tRNA nucleotidyltransferase. These conclusions were borne out by the DNA and RNA sequence analysis (see Figure 2, above).

The $tRNA^{Gln}$-$tRNA^{Leu}$ precursor RNA is present only in very small amounts in wild-type *E. coli,* consistent with the notion that synthesis of the C-C-$A_{OH}$ sequence is the rate-limiting step in processing the $tRNA^{pro}$-$tRNA^{Ser}$ precursor RNA. (Since the 3′ C-C-A sequence of $tRNA^{Leu}$ is genomically encoded, its cleavage is faster than that of the other tRNA dimers.) RNase P is required for the cleavage of this precursor RNA, however; a fact which permitted Guthrie[47] to determine its sequence and in vitro processing pathway. Purified (but not homogeneous) RNase P cleaved the dimer accurately to generate mature 5′ termini of both $tRNA^{Leu}$ and $tRNA^{Glu}$. Further, the interstitial $tRNA^{Leu}$ cleavage occurred faster than the $tRNA^{Gln}$ cleavage; whether this is due to the state of the respective 3′ terminal sequences (C-C-$A_{OH}$ vs. $C_{OH}$) has not been determined. This feature of RNase P cleavage has also been noted in the in vitro processing of the $tRNA^{Pro}$-$tRNA^{Ser}$ precursor RNA.[37]

The T4 $tRNA^{Gln}$ requires tRNA nuceotidyltransferase to synthesize its 3′ end. Like the 3′ terminal maturation of $tRNA^{Pro}$, this can occur *after* RNase P cleavage, since McClain et al.[48] found that $tRNA^{Gln}$ with a mature 5′ and immature 3′ end was synthesized in *cca* hosts.

### 3. *The $tRNA^{Thr}$-$tRNA^{Ile}$ Precursor RNA*

The T4 $tRNA^{Ile}$ is unusual in several aspects: its coding specificity is such that it recognizes A-U-A exclusively.[49] Since T4 DNA is more A-T rich than *E. coli,* it has been hypothesized that this tRNA replaces the major $tRNA^{Ile}_{AUC}$ of the host.[10] The T4 $tRNA^{Ile}$, moreover, is required for phage growth in some strains of *E. coli* isolated from the human gut although these hosts do not seem to be deficient in the capacity to read A-U-A.[49] Most surprisingly, the gene sequence of the T4 tRNA gene cluster shows that the ability to read A-U-A arises from modification of the C in the anticodon sequence C-A-U which would normally read the methionine condon A-U-G. In contrast to other modified bases, this nucleotide is always found in full molar yield, presumably ensuring that missense reading of methionine codons cannot occur.

Furthermore, $tRNA^{Ile}$ is normally underproduced in phage-infected cells, even compared to $tRNA^{Thr}$ which is in the same dimeric precursor.[50] Two explanations of this phenomenon have been proposed which invoke either processing or transcriptional events to account for the underproduction of $tRNA^{Ile}$. Both explanations have features which commend them.

An explanation involving aberrant processing of the $tRNA^{Thr}$-$tRNA^{Ile}$ dimer has been proposed by Guthrie and Scholla.[50] They observed that this precursor does not accumulate in *E. coli A49,* a thermosensitive RNase P mutant, at high temperature. This is despite the fact that the $tRNA^{Thr}$-$tRNA^{Ile}$ precursor accumulates in wild-type,[51] BN,[44] and *cca*[105] hosts. The immature 3′ ends of $tRNA^{Ile}$ precursors in mutant hosts indicate that the pathway for synthesizing these tRNAs can be arranged in the same form as that for $tRNA^{Pro}$ and $tRNA^{Ser}$. However, these blockages are not complete: in both BN and *cca* hosts, 3′ immature, 5′ mature monomeric $tRNA^{Ile}$ species are synthesized, albeit inefficiently, and mature $tRNA^{Thr}$ is made as well.[28,48,50] The immature $tRNA^{Ile}$ species arise by cleavage of the dimeric precursor at the correct interstitial site prior to 3′ maturation. Since the $tRNA^{Thr}$ requires only RNase P for its 5′ maturation and RNase D or its homologue for its 3′ maturation, it

FIGURE 4. Processing pathway for the T4 $tRNA^{Pro}$-$tRNA^{Ser}$ precursor RNA. (Reprinted with permission from McClain, W. H., *Acc. Chem. Res.*, 10, 418, 1977. Copyright 1977 American Chemical Society.)

is synthesized normally after RNase P cleavage. In *rnp* cells at the restrictive temperature, a monomeric precursor to $tRNA^{Thr}$ can be observed; it contains a 5′ immature, 3′ mature tRNA which can be accurately processed in vitro. The synthesis of $tRNA^{Thr}$, therefore, seems to require RNase P for 5′ cleavage and either RNase D or some other enzyme to generate its 3′ terminus.

This explanation still leaves the question: how is the $tRNA^{Ile}$ synthesized? Extracts deficient in RNase P activity were used to cleave the $tRNA^{Thr}$-$tRNA^{Ile}$ precursor RNA. While some of the dimer was cleaved to tRNA-sized product, this product was deficient in $tRNA^{Ile}$, i.e., $tRNA^{Thr}$ sequences appeared in about a 2:1 molar ratio. Guthrie and Scholla[50] concluded that another RNase activity besides RNase P could cleave the precursor with the concomitant degradation of $tRNA^{Ile}$. It has been noted that mature $tRNA^{Ile}$ contains a fully modified wobble base; perhaps this modification protects $tRNA^{Ile}$ from degradation.[20]

A second explanation, not necessarily incompatible with the one above, invokes a transcriptional component to account for the underproduction of $tRNA^{Ile}$. Goldfarb and Daniel[23] observed that transcription of the T4 subcluster I of tRNA genes terminated in a *rho*-dependent reaction after (or in) $tRNA^{Ile}$. They also noted that $tRNA^{Ile}$ was not synthesized from this terminated transcript. It is possible that this underproduction of $tRNA^{Ile}$ is due at least in part to its preferential degradation, but it is also possible that the terminated in vitro transcript did not contain a complete $tRNA^{Ile}$ sequence at all. In support of this possibility, Mazzara et al.[19] found in a computerized but not exhaustive search of the gene sequence that a potential terminator could be formed within $tRNA^{Ile}$. This terminator structure has been shown in Figure 3b, and it is apparent on examination that it is mutually incompatible with the normal cloverleaf folding of $tRNA^{Ile}$. Attenuation (mechanisms of gene regulation involving mutually incompatible RNA secondary structures, one of which is a transcription terminator), has been proposed to account for the regulation of other operons besides the classic *trp* case.[52,53] The present mechanism, if true, would be an example of an attenuative mechanism in a tRNA operon. In other attenuation mechanisms translation apparently discriminates between the two structures. Since this region of the tRNA operon is not known to be translated, the extent of tRNA base modification can perhaps discriminate between these alternate foldings.

### 4. *Cluster II: $tRNA^{Arg}$ and Stable RNAs 1(C) and 2(D)*

Precursors to these adjacent RNAs can be generated by RNase PC cleavage of an in vitro transcript and subsequently matured by a crude S100 extract to mature RNA species.[34] Examination of the DNA sequence indicates that the 3′ C-C-A ends of $tRNA^{Arg}$ and Band 2(D) are genomically encoded, but that of Band 1(C) is derived by processing (Figure 2). The maturation of Species 1(C) probably follows a pathway similar to a tRNA whose C-C-$A_{OH}$ sequence is not genomically encoded; after RNase PC cleaves the primary transcript, the 3′ C-C-$A_{OH}$ is synthesized by the sequential actions of RNase BN and tRNA nucleotidyltransferase. The 3′ terminal C-C-A sequences of $tRNA^{Arg}$ and Species 2(D) are genomically encoded; maturation of these RNAs does not, therefore, require tRNA nucleotidyltransferase or RNase BN activities.[44,48] The question of RNase P activity is unresolved since precursors to these RNAs accumulate only minimally in *rnpA49* cells under restrictive conditions.[31] The pathways have not been elucidated, but may be similar in some respects to the unusual metabolism of the pre-$tRNA^{Thr}$-$tRNA^{Ile}$ dimer.

### 5. *Monomeric Precursor to $tRNA^{Gly}$*

This tRNA is unique among those of cluster I in that its cleavage from the primary transcript is as a monomer. An inspection of the gene sequence shows that this tRNA is flanked by A-T-rich sequences on both sides, while other tRNAs are either preceded or followed by these sequences.[16,17] Presumably the enzyme(s) required for cleavage of the transcript must recognize these sequences (by base composition or structure). Since $tRNA^{Gly}$

**Table 1**
**HOST PROCESSING FUNCTIONS INVOLVED IN T4 tRNA SYNTHESIS**

| Enzyme | RNAs affected | Function |
|---|---|---|
| RNase P | All | 5′ Cleavage to generate mature tRNAs |
| RNase BN | $tRNA^{Pro}$, $tRNA^{Ser}$, $tRNA^{Ile}$, Species 1(C) | Removal of extra nucleotides from 3′ end of RNAs without encoded C-C-$A_{OH}$ |
| RNase PC (or F?) | All | Cleavage of primary transcript |
| RNase III | $tRNA^{Gln}$ | Cleavage of primary transcript at 5′ end |
| RNase D or homologue | $tRNA^{Leu}$, $tRNA^{Gly}$, $tRNA^{Thr}$, $tRNA^{Arg}$, Species 2(D), T2 $tRNA^{Ser}$ | Removal of 3′ nucleotides from 3′ end of RNAs of encoded C-C-A sequence |
| RNase E | Species 1(C) | Accessory cleavage factor in processing large precursor molecule |
| tRNA nucleotidyltransferase | $tRNA^{Pro}$, $tRNA^{Ser}$, $tRNA^{Gln}$, $tRNA^{Ile}$ | Synthesis of 3′ C-C-$A_{OH}$ sequence |

has a C-C-A sequence encoded, precursor maturation requires only the removal of an extra nucleotide by RNase D or a homologue before or after RNase P cleavage.

### F. Enzymatic Features of T4 tRNA Biosynthesis

A summary of the host enzymes involved in T4 tRNA synthesis is shown in Table 1. All of the characterized enzymes are host-encoded, even though one, the BN RNase, is apparently dispensable in normal laboratory growth. Two phage functions are known which have an effect on processing. One gene near rII, called *M1*[54] or *mb*[55] is required for the synthesis of Band 2(D), $tRNA^{Pro}$, $tRNA^{Ser}$, and $tRNA^{Ile}$. These tRNA species also require host RNase BN to remove extra nucleotides from their 3′ ends. If *M1 (mb)* gene product were required for the efficient action of RNase BN, infection of wild-type cells with *m1 (mb)* phage should show the same labeling pattern as infection of BN by wild-type phage. This, however, is not the case: all LMW RNAs are reduced in amount, and pre-$tRNA^{Pro}$-$tRNA^{Ser}$ does not accumulate.[54] It is known, however, that the *M1 (mb)* gene product is *trans*-active in mixed infections and requires phage protein synthesis for activity.[31,106] Fukada and Abelson[18] have hypothesized that *M1 (mb)* protein is a nuclease inhibitor.

A second phage-coded function which may affect tRNA processing is the T4-induced modification of RNase D which was described by Deutscher and co-workers.[56] The modification results in a larger molecular weight RNase D, but detailed kinetic studies of the modified enzyme have not been reported. Modification of RNase D is not the product of the *M1 (mb)* gene.[107]

### G. Other Phages Encoding tRNA Species

The synthesis of tRNAs by naturally occurring phage species is a phenomenon so far limited to the T-phages and their close relatives. tRNAs and low molecular weight RNAs are encoded by T-even phage T2, T4, T6, and RB69. Although the tRNA-coding capacities of these species vary widely, it is of interest that stable RNA Species 2(D) is conserved; the functional significance of this is unknown.[57] Some inferences about the evolutionary plasticity of this region of the genome as it affects processing can be drawn from comparative patterns of RNA synthesis. First, the C-C-$A_{OH}$ sequence may have evolved independently from the rest of the tRNA genes. T2 and T4 $tRNA^{Ser}$ species are identical, except that the C-C-A sequence in T2 $tRNA^{Ser}$ is genomically encoded but that of T4 must be derived by processing.[48] Secondly, the $tRNA^{Gln}$ sequence of T2(H), but not T2(L), contains a mutation in the anticodon stem. This altered sequence results in an unstable $tRNA^{Gln}$, but dimeric $tRNA^{Gln}$-$tRNA^{Leu}$ precursor RNA can be observed in T2(H)-infected *rnpA49* cells.[57] This means the degradative mechanism that edits defective $tRNA^{Gln}$ must act primarily on monomer-sized RNAs; this may be relevant to the explanation of the unusual metabolism of T4 $tRNA^{Ile}$ (see above). A further interspecies difference is found in the $tRNA^{Ile}$ from T6; like the case of T2 $tRNA^{Ser}$, the 3′ C-C-A sequence of this tRNA is derived transcriptionally.[48] This property may also be useful in experiments regarding T4 $tRNA^{Ile}$ processing (see above).

T5 and its close relative BF23 synthesize a large enough variety of tRNA species to account for nearly the entire genetic code.[58,59] Whether this capacity is related to phage virulence is unknown. DNA sequence analyses of these tRNA genes are underway which may define the enzymatic requirements for biosynthesis of these tRNAs more precisely.

## III. PROCESSING OF BACTERIOPHAGE MESSENGER RNA AND ITS ROLE IN CONTROL OF GENE EXPRESSION

Consider a lytic bacteriophage on encountering a host. The problem of phage survival and propagation is essentially one of timing: the phage genome must be expressed with enough efficiency to appropriate the cellular machinery before the host restriction system can act in defense of the cell (or its siblings). Since bacterial gene expression is often regulated at the level of transcription initiation, the phage must be able to do this with dispatch. In this view, the function of messenger processing, then, would be to allow differential expression of the cistrons on a single transcription product, that is, to separate the rate-limiting step in gene expression from transcription. The paradigm of such expression is provided by the RNase III-mediated processing of T7 early messenger RNAs.

### A. Bacteriophage T7 Biology

Bacteriophage T7 DNA is a linear molecule, terminally redundant but not circularly permuted, of about 44,000 bp. T7 DNA is injected in a single direction, apparently in two stages: transcription of the first 20% of the DNA (the early region) is apparently required for injection of the later region.[60]

Early transcription after T7 infection uses host RNA polymerase and transcribes the leftward 20% of the genome exclusively. When transcripts are made in vivo and analyzed by gel electrophoresis, five RNA species are observed, which can be mapped to genetic loci by the use of phage deletion mutants.[61] However, in vitro transcription of T7 DNA by *E. coli* RNA polymerase holoenzyme results in RNA species which cover the entire early region;[62] these RNA transcripts originate at three closely related promoters at positions 496, 626, and 750 of the T7 genome.[63] The existence of in vivo RNAs which are derived from longer transcription products is prima facie evidence for RNA processing and a search was initiated for the ''sizing factor'' present in *E. coli* extracts. Since the smaller mRNAs can be observed before T7-specific translation occurs, this processing must be due to a host enzyme. The ''sizing factor'' was shown to be identical to host RNase III by two criteria. First, T7-infected *E. coli* mutant strain AB105-301, which is deficient in RNase III, does not process T7 early messenger RNA in vivo.[64,65] This *rnc* mutation has widespread effects on cellular RNA synthesis (see Chapter 2, this volume) and also affects T4 tRNA processing (see above). A second criterion was the demonstration that highly purified RNase III accurately cleaved the T7 primary transcript in vitro.[66]

The DNA sequences around the T7 cleavage sites are shown in Figure 5. All of the DNA sequences contain regions of dyad symmetry which result in a potential hairpin structure in the RNA transcript. A contrast between the pre-23S and pre-16S rRNA precursors and T7 mRNA sites is the existence in the hairpin of an unpaired ''bubble'' around the RNase III cut site. It is possible that this unpaired region is responsible for some difference between T7 mRNA and rRNA processing by the same enzyme, but it can also be argued[67] that the unpaired bases would nevertheless stack into a helix under physiological conditions. The detailed enzymology of RNase III recognition and cleavage of T7 mRNAs awaits further study, but some potential advances aided by genetic studies are discussed below.

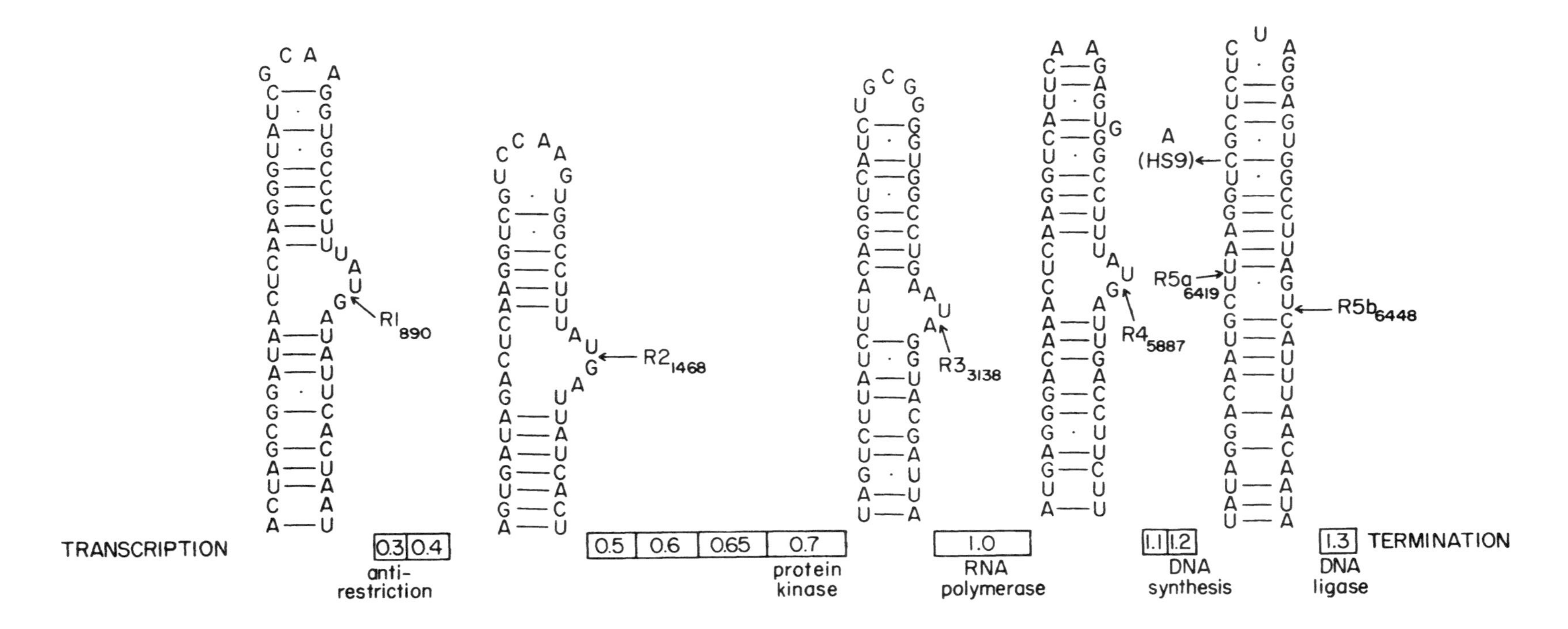


FIGURE 5. Sequences of the RNase III cleavage sites in T7 early mRNAs and gene organization of the T7 DNA. Transcription is from left to right. RNase III cleavage sites are given by R1, R2, R3, R4, and R5; the site of cleavage is indicated by the arrows. Numbering of the residues is taken from Reference 63. The site of the HS9 mutation[76] ($C_{6426} \rightarrow A$) which disrupts cleavage at R5a is shown. The R5 hairpin is the only one known to be cleaved twice by RNase III.

*1. Processing Required for the Synthesis of T7 0.3 Product*

If the enzymatic mechanism of T7 mRNA processing is obscure, the physiological function of these cleavages is less so. Infection of an *rnc* host (deficient in RNase III) does decrease the burst size of the infected cells severalfold, but progeny phage are nonetheless produced.[68,69] Further, in vitro protein synthesis using cleaved and uncleaved polycistronic mRNA showed that most proteins were made from uncleaved RNA. The sole exception to this rule was the gene *0.3* product, a protein which binds to the *Eco B* restriction endonuclease and therefore overcomes the host restriction system.[70] It is not known why RNA processing is required for the synthesis of this protein. Several possibilities exist. First, the unprocessed hairpin may mask the *0.3* ribosome binding site in some way. A second possibility, raised by studies of abnormal metabolism of transfer and ribosomal RNA precursors in mutant *E. coli* (see, e.g., Chapter 2, this volume, and above) is that failure to correctly process the gene *0.3* mRNA renders it sensitive to other cellular nucleases. Current evidence seems to favor the former hypothesis. The binding of gene *0.3* mRNA to ribosomes in vitro is known to be weak[71] and a search for *0.3* mutants yielded alterations in the 4-base Shine-Dalgarno sequence complementary to the 3′ end of 16S rRNA.[72] The biology of *0.3* mRNA reflects this phenomenon: the *0.3* mRNA produced in vivo is stable enough to persist throughout the infectious cycle, but the *0.3* gene product is not synthesized during late infection because it competes poorly for ribosomes. If late transcription is prevented, for example by a mutation in gene *1* (T7 RNA polymerase), then *0.3* protein is synthesized persistently.[73,74]

*2. Gene Control by Processing of T7 1.1 and 1.2 Messages*

RNase III cleaves the T7 early transcript between genes *1* and *1.1* (site R4) and again before gene *1.3* (site R5; see Figure 5). The message so generated codes for two products: the *1.1* DNA codes for a basic protein of 42 amino acids and the *1.2* protein is similarly predicted to be 85 amino acids long.[63] As yet no function has been assigned to the *1.1* gene and neither the *1.1* or *1.2* gene product is required for growth on laboratory strains of *E. coli*. A function has been found, however, for the *1.2* gene product. Saito and Richardson[75] isolated a mutation *(opt A)* in *E. coli* which is nonpermissive for *1.2* strains of T7. The mutant phage fail to replicate their DNA in an *opt A* host. This accomplishment, reminiscent of examples of suppressor-restricting hosts of T4 (see above), illustrates the power of phage genetics when applied to a biologically subtle system: bacteriophage are uniquely suited for the isolation of conditional mutants since two genetic systems can be independently manipulated. In the present cases, the *opt A* mutation appears to have no deleterious effect for laboratory growth of the bacterial host while the *1.2* gene product is not required for ordinary phage propagation either. Only when the two mutations are combined is phage growth inhibited. Since RNA processing events are often manifested only by subtle effects, the ability to amplify these effects in certain genetic backgrounds is crucial to gathering in vivo information.

Saito and Richardson[76] mutagenized wild-type T7 and found strains which could not grow on *opt A* hosts. The mutants fell into three classes. Class I mutations were simply base-change or frame-shift mutations within the gene *1.2* sequence and Class II mutations were polar mutations in the *1.1* cistron. Class III contained only a single isolate, mutant strain HS9; in this mutant the altered nucleotide sequence was in the R5 RNase III processing site distal to the *1.2* coding sequence (see Figure 5).

Revertants of the HS9 mutant were readily isolated by selecting phage able to plaque on an *opt A* host; sequencing showed that most of these were second-site mutations in the hairpin stem of the RNase III cleavage site. The HS9 mutation presumably destabilizes the hairpin sequence, interfering with RNase III recognition at the R5 sites; revertants either can restabilize this structure, or more commonly and unexpectedly, destabilize it further. A

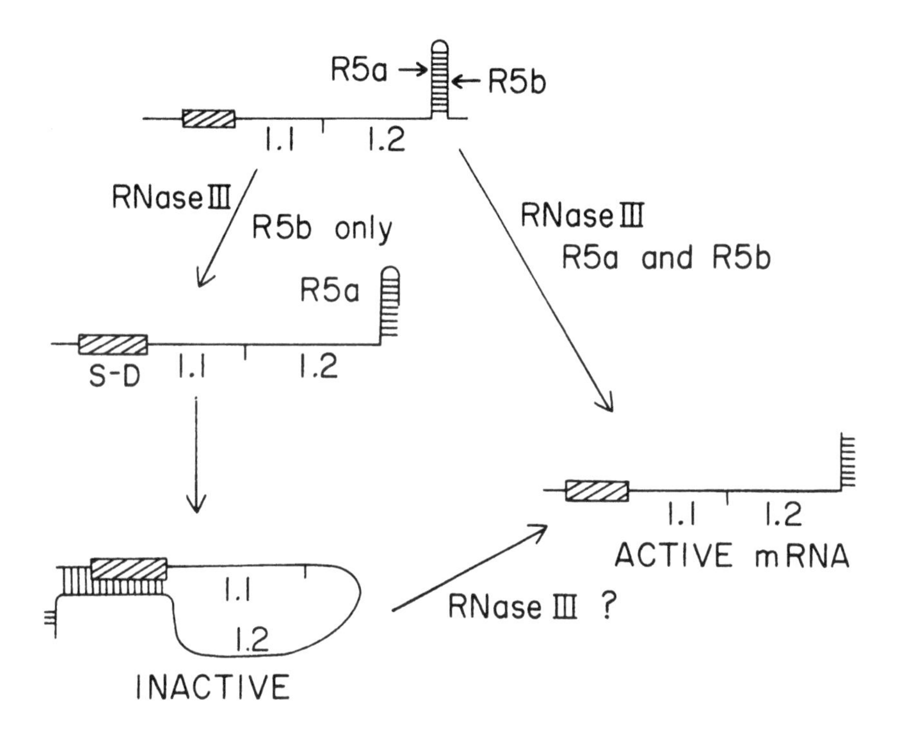


FIGURE 6. Model for negative control of T7 gene *1.1* gene expression by mRNA processing at R5b only or, alternatively, at both R5 sites. The Shine-Dalgarno ribosome binding sequence (S-D) is indicated by the box and the R5 cleavage stem is indicated by the stem-and-loop structure. The RNase III cleavage indicated by the question mark has not been shown to occur in vivo.

proposed explanation[76] starts from the observation that the H59 mutation interferes with processing at the R5a site only. T7 *HS9* plates on an *rnc opt A* host with nearly the efficiency of wild-type phage, indicating that the *1.2* gene product is made in the absence of any processing at either the R5a or R5b sites. Taken together, these data must mean that message that is cleaved only at the stronger site (R5b) is capable of interfering with the synthesis of gene 1.2 product, but message that is cleaved twice, or not at all, is not. Note that this is opposite to the normal direction of genetic polarity since gene *1.2* lies upstream (promoter-proximal) from the R5 site in the mRNA. A search for upstream sequences complementary to the R5 hairpin revealed strong homology between the hairpin and the ribosome binding site of gene *1.1*; if message were cleaved at the R5a site this homology no longer could exist. Saito and Richardson proposed a model (Figure 6) whereby a polycistronic message cleaved first at site R5b could undergo one of two fates: it could bind back on itself and sequester the *1.1* ribosome binding site thereby blocking both *1.1* and *1.2* expression or it could be further cleaved at site R5b removing the homology and permitting *1.1* and *1.2* synthesis. This model makes some testable predictions: (1) second site revertants of HS9 permitting both *1.1* and *1.2* synthesis should be found which only disrupt the foldback of once-cleaved RNA to the Shine-Dalgarno sequence preceding gene *1.1*; these could occur either around the ribosome binding site or in the unpaired region of R5, and (2) *opt A* hosts containing a leaky *rnc* mutation should restrict somewhat the growth of wild-type phage since the stronger R5b cut would be made preferentially.

### *3. Oligoadenylation of the T7 mRNAs*

The RNase III-cleaved T7 early mRNAs are also adenylated at the 3′ end. In their analysis of the sequence of T7 mRNAs produced in vivo, Kramer et al.[77] found oligoadenyl sequences present at the 3′ end. The DNA and RNA sequence data indicate unequivocally that these residues must be added posttranscriptionally. Whether, as is the general model in eukaryotic systems, these oligo A sequences contribute to mRNA stability is unknown, but Strome and Young[74] have shown that the *0.3* mRNA is unusually stable in vivo.

**B. Processing of the Leftward Transcript of λ and Its Consequences for the Lysis-Lysogeny Decision**

When λ infects a permissive host, the phage genome is expressed in such a way that either (1) λ DNA is replicated and packaged and the host is lysed; or (2) the phage genome is integrated into the host chromosome, where it is replicated passively with host DNA. The choice of replication modes depends on the physiological state of the host: rapid growth favors lysis while cells in stationary phase are more easily lysogenized. A detailed exegesis of the λ control circuits is beyond the scope of this article and the reader is referred to the eminently readable review by Herskowitz and Hagen[78] for an account of recent research in this area.

*1. The λ N Transcript*

The leftward control region of the phage λ genome is shown schematically in Figure 7. Immediately after infection, host RNA polymerase initiates transcription at the leftward promoter, *pL*, which is terminated at the rho-dependent termination site *tL1*, located between the *N* and *xis* genes (Figure 7). Transcription is also initiated at three other promoters on the rightward arm; these are not discussed here. The 12*S* transcript initiated from *pL* codes for protein N. The *N* gene product is required for lysogenic phage growth[79] and is an antiterminator at both rho-dependent and independent termination sites.[80] Thus, after sufficient N protein is made, *pL*-initiated transcription continues past the *tL1* terminator site to produce a long transcript continuing into the distal genes in the operon. Among the products of this region is cIII protein which is essential for the lysogenic response. A site *(nut L)* of N action in its own transcript has been identified. *Nut L* is defined as a 15-bp segment containing two copies of a 5-bp G-C rich (4/5) repeat; presumably the transcript of this region is capable of forming an RNA polymerase pause site which is overcome by N action.[81] Lozeron et al.[82,83] showed that in an RNase III-deficient host the N message: (1) was attached to a longer message which extended through the entire leftward region, whereas in *rnc*$^+$ hosts it was not part of this long message, and (2) was significantly more stable than in an *rnc*$^+$ host. Part of the basis for these observations may be seen in an examination of the *N* DNA sequence as determined by Franklin and Bennett[81] and the corresponding RNA sequence determined by Lozeron et al.[82] RNA processing can occur at two sites within or near N gene sequence: at positions 72 and 88 from the mRNA start. This processing removes the *nut L* site from the mRNA; the functional significance of this cannot be assessed but it is known that *nut L* does not function by a simple antagonism of termination factor rho. Rather, the *nut L* site in some way must "signal" the RNA polymerase at the *tL1* termination sequence, perhaps through rho. Processing at these sites would remove *nut L* from the nascent mRNA and thereby promote transcription termination at *tL1*.

If transcription does not terminate at *tL1*, a further processing event occurs at the 3′ side of N, so that the *N* message is separated from the longer leftward transcript.[84,85] This processed mRNA is believed to be less stable than the terminated mRNA. Thus, processing negatively controls *N* expression, but only in the presence of N protein itself; if there is no antitermination the message is more stable. This is the opposite of the effects of processing on T7 *0.3* and *1.2* mRNAs, where processing promotes gene expression.

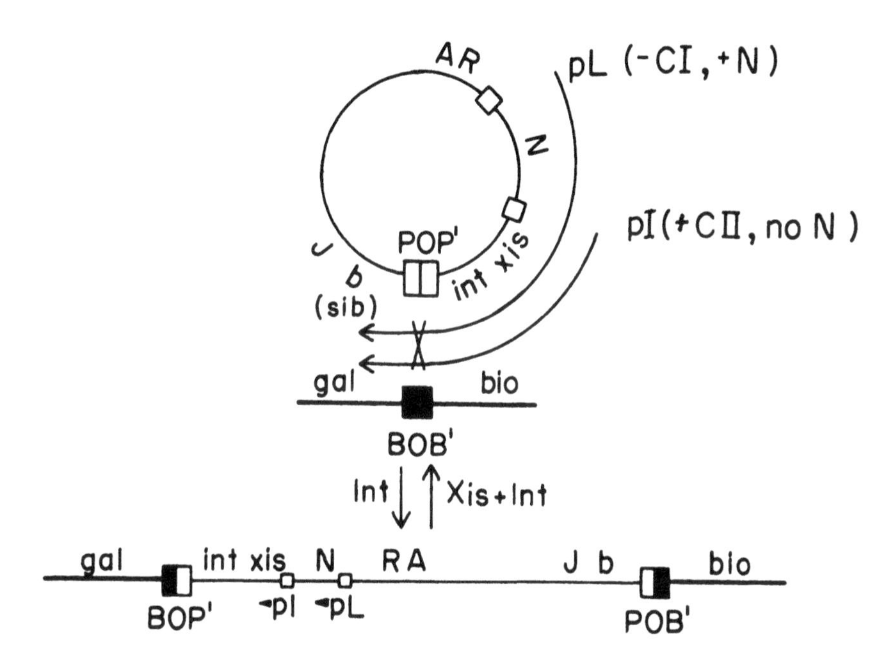


FIGURE 7. Configuration of the λ genome in the unintegrated (top) and integrated (bottom) states. The crossover between the host and phage genomes is indicated. This reversible process required Int and Xis proteins; the former for both integration and excision and the latter for excision only. In the circular phage, genome transcription from pL is repressed by cI repressor and antiterminated by N protein. This message continues into the *b* region (across POP′) and is degraded as a result of the action of RNase III at the *sib* site. In the absence of N and presence of cII, *int* transcription is initiated at pI; since it is at or near *sib* the transcript is not subject to RNase III-initiated degradation. (Adapted, with permission, from Herskowitz, I. and Hagen, D., *Annu. Rev. Genet.*, 14, 399, 1980. © 1980 by Annual Reviews Inc.)

*2. The Functional Control of* λ Int *Gene Expression by RNA Processing*

N-dependent transcription of the leftward λ operon of a circular phage genome continues through the integration *(POP′)* sequence and beyond. This transcript includes the integrase *(int)* gene but Int protein is not expressed from this message. Instead, the productive transcription of the *int* gene initiates from its own promoter, *pI,* and is dependent on cII protein. The cII protein is the end product of a process which senses the metabolic state of a host cell; if cII is high, integration and lysogenic replication occur.[78] Teleologically, it makes sense for the phage to avoid making Int protein and thereby committing itself to integration until it tests the cellular waters. However, Int protein in addition to Xis protein is also required for prophage excision on induction. In this case, Int protein is made from *pL;* the wild-type signal is the destruction of cI repressor by host RecA protein.[86] The synthesis of Int protein must, therefore, respond to two signals which arise from different proteins. Both cI at *pL* (negatively), and cII at *pI* (positively) affect *int* expression. What feature of the leftward transcript makes it act differently when λ is circular (immediately after infection or during lytic growth), or linear (during vegetative growth or at induction)? Campbell's model[86a] provides an answer: in the former state transcription continues across the integration site, while in the latter it stops making λ transcript and continues into the *E. coli gal* operon (see Figure 7).

The region clockwise from *int* in Figure 7 (top) is the λ *b* region: it can be deleted without affecting lytic growth and its function has remained obscure. Guarneros and Galindo[87] found that deletions in the *b* region allow synthesis of integrase from *pL*. The function of *b* termed *sib* acts in a negative fashion, since the mutant allows *int* expression. Furthermore, the *sib* mutation acts only in *cis* in a complementation test;[88] this behavior classically defines a site of regulation rather than a diffusible regulator (compare with *o* mutations in *lac*[89]).

The molecular basis of *sib* action has recently been shown to involve RNA processing.[90,92,108,109] Consider two modes of Int protein synthesis in a circular phage genome: one from *pI* (activated by cII protein) and one from *pL* (antiterminated by N protein). Transcription from *pI* starts within the *xis* structural gene and continues through the phage attachment site into a terminator, *tI*, located 190 bp into the *b* region (see Figure 8). The *pL* transcript, on the other hand, continues through this terminator with the help of N protein. This longer transcript is now a substrate for RNase III which cleaves on either side of the terminator stem-and-loop sequence in a manner very like the RNase III cleavage of pre-16S ribosomal RNA[93] (see also Chapter 2, this volume). RNase III cleavage destroys the stem-and-loop part of the terminator; this leads to message degradation by other cellular nucleases (RNase II?) and the Int protein is not synthesized. This model is supported on several grounds. First, Schindler and Echols[91] have shown that if *int* is deleted, the upstream gene product *Ea 22* is not made in an *rnc*$^+$ host. This can be explained as a result of processive cleavage on the shorter *pL* transcript which reaches to *Ea 22* rather than *int* mRNA sequences. Second, and in a most convincing fashion, Rosenberg and co-workers[110] have shown that this regulation can act on other genes as well. They constructed a plasmid with the following organization:

| ⟶ | transcription | | |
|---|---|---|---|
| λ | *E. coli* | λ | λ |
| pL | gal K | tL | sib |

In this chimera, galactokinase was under the control of the same elements as *int* and behaved similarly: in the absence of N gene product, transcription continued from *pL* through *gal K* and stopped at *tL*. This message was stable enough to support the synthesis of galactokinase. On the other hand, if N was present (analogous to the uncommitted leftward transcription of λ), it acted as an antiterminator at *tL*, transcription continued through *tL* into *sib*, RNase III processed the mRNA, cellular nucleases degraded the *gal K* message and no galactokinase was made. This elegant demonstration of both the power of recombinant DNA techniques and the relationship between RNA processing and gene control may have implications for cellular control as well, since the expression of many *E. coli* proteins is altered in an *rnc* background (see Chapter 2).

The *sib* regulation of lambda Int synthesis and the processing of the T7 *1.2* message (see above) can be generalized to predict genetic and biochemical features of regulatory RNA processing sites (see below). We can confidently expect many more examples of this phenomenon in the near future.

### C. Other Processed Phage Messages

A variety of phage mRNAs are derived by processing but the molecular biology and enzymology of these systems is less advanced than in T7 and λ. T3 early mRNAs are derived by processing in much the same manner as T7 early mRNA.[94] This is consistent with previous observations that these two phage are closely related. One point is of interest: the T3 *0.3* gene, like its counterpart in T7, interferes with host restriction and modification but the enzyme function is quite different. T3 *0.3* protein hydrolyzes *S*-adenosylmethionine while T7 *0.3* protein binds Type I restriction enzymes. The T3 *0.3* gene product also functions

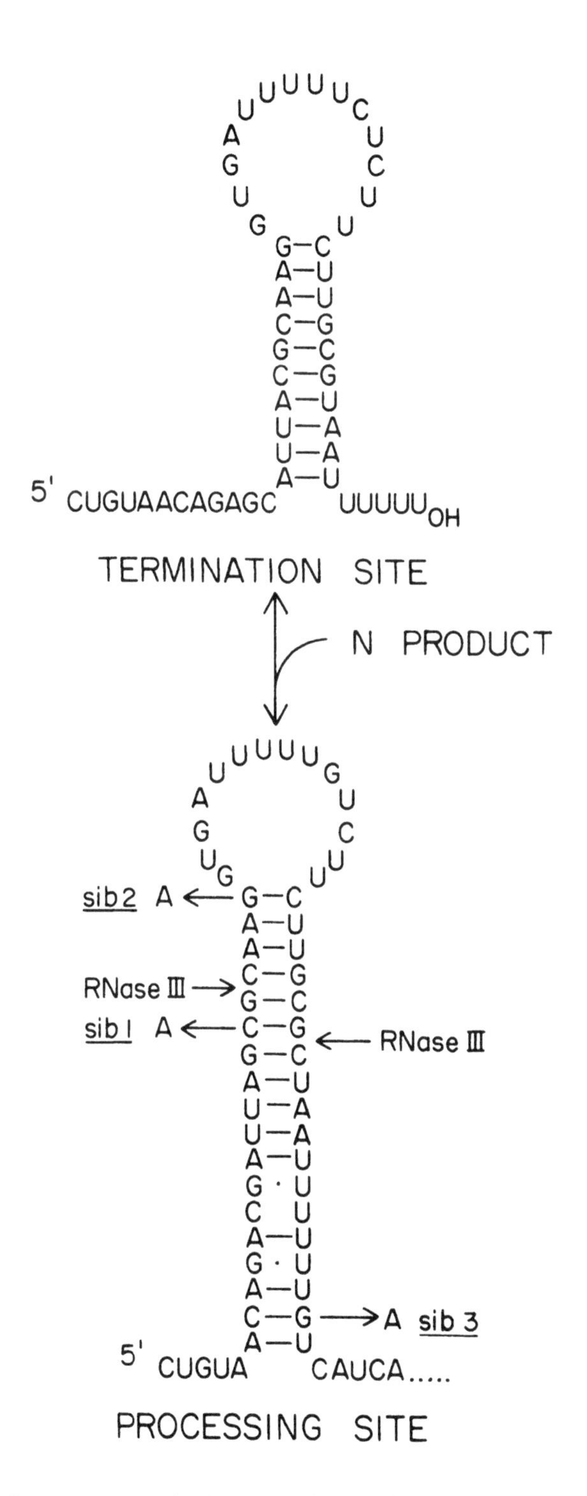


FIGURE 8. Termination (top) and processing site distal to the leftward λ transcript.[91,92]

in a reaction which T7 does not accomplish: in a starved host, T3 is able to act as a facultative lysogen. Although not integrated, its DNA is not replicated until the host starts to grow. This ability depends on the *S*-adenosylmethionine hydrolase, since *0.3* mutants of T3 are immediately lytic in starved cells.[95] Since the expression of T7 *0.3* is known to depend on messenger processing and processing sequences are conserved, one might expect that starved *rnc* hosts would be immediately lysed by a wild-type T3 since its *0.3* gene would not be expressed. Such a system could be useful in examining T3 processing signals genetically.

Both T7[96] and T3[97] late mRNA classes are transcribed by phage-coded RNA polymerases. Some, but not all, of these late mRNAs are processed by RNase III.[98] No function has yet been ascribed to these processing reactions and no sites have been sequenced directly. The known T7 DNA sequence does contain a region to the right of gene *3.5* (lysozyme), which is likely to be a processing site in vivo.[73] This site is capable of forming a stem and loop similar to the T7 early mRNA sites cleaved by RNase III.[63]

Messenger RNAs from the filamentous, male-specific phage fd are known to be processed as well. In contrast to all the other known cases of phage mRNA processing, however, processing of fd messages takes place in an *rnc* host. Webster and co-workers[99,100] have argued that this implies another enzymatic system for mRNA processing; its identity is not known at present.

Processed phage RNAs have also been reported from *B. subtilis* phage SP*82*[101] and *Caulobacter crescentus* phage $\phi$cdl.[102]

## IV. PROSPECTS AND CONCLUSIONS

We now have sufficient information from several systems to infer some general rules about RNA processing and its function as a control mechanism in the cell. First, since processing occurs at a site in the RNA transcript, processing mutants in a target RNA sequence will behave genetically as *cis*-dominant. Thus, for example, Guarneros and Galindo[87] showed that *sib* mutants in $\lambda$ could not be complemented by a wild-type *sib*$^+$ function on another phage.[87] The tools of phage genetics make screening for such *cis*-dominant mutations out of a large collection relatively easy.

Further insight can be derived from an examination of the behavior of wild-type and mutant phage on a bacterial host deficient in a specific processing enzyme. If the effect of a cellular enzyme deficiency is to mimic a *cis*-dominant processing mutation, then processing exerts a positive effect on gene expression. Many examples of this are known: thus *E. coli* deficient in RNase III, RNase P, RNase BN, or tRNA nucleotidyltransferase restrict the synthesis of various phage tRNAs and an RNase III-deficient host restricts the synthesis of the T7 *0.3* gene product. The converse can also hold: a deficient host can phenotypically cure a particular mutation in a processing site. In this case, processing exerts a negative control on gene expression. An example of this phenomenon is the ability of a cellular *rnc* mutation to permit synthesis of $\lambda$ integrase in a *sib*$^+$ phage. Processing of the leftward transcript leads to destruction of the mRNA by nucleolytic degradation. Both situations hold in the processing of the T7 *1.1* mRNA: the initial processing event (R5b) results in masking of the ribosome binding site at the start of the *1.1* mRNA sequence while the R5a cut releases the complementary piece of RNA and frees the *1.1* ribosome binding site. Since the R5b site is cut more rapidly than R5a, this is also a temporal control of gene expression. Initially, uncleaved mRNA would be active in protein synthesis while R5b-cleaved RNA is inactive. Later, when the R5a site was also cleaved, the mRNA would again be active in translation.

An important point about these RNA processing control mechanisms is that, like the translational control of ribosomal protein synthesis,[103] these regulatory mechanisms act on operator-proximal genes. Schindler and Echols[91] have termed this phenomenon "retro-regulation". A regulatory element acting on an upstream gene can be predicted to act differently than an element whose target is downstream. Since the initiation of protein synthesis closely follows transcription, in some cases at least, synthesis of the target protein could begin before the processing site was transcribed. Only later, when RNA cleavage initiated the inactivation of the mRNA, would synthesis of the target protein cease. An example of this might occur in the T7 early message where the single processing event at site R5b inactivates gene *1.1* mRNA. If the folded-back *1.1* and *1.2* mRNA were cleaved again by RNase III, synthesis of these proteins could then proceed. The net effect of such a sequence of events would be a small amount of gene product synthesized initially, a cessation of synthesis from once-cleaved message, and finally, a large amount of gene product made from mature mRNA. It remains to be seen whether such a regulatory sequence occurs or is biologically relevant (see the question mark in Figure 6), but the speculation is intriguing.

A striking result of the data on two disparate systems, T4 $tRNA^{Ile}$ and the λ leftward transcript, is that processing is the first step on the road to ruin. We may expect that other examples of control by RNA degradation will arise in the future; but currently, the mechanisms, enzymes and pathways that carry out this degradation are completely unknown. In their initial purification of *E. coli* RNase P, Robertson et al.[104] described a ribosome-bound "obliterase" fraction with all the activity the name implies. If specific RNA degradative mechanisms exist, the tools of phage genetics will be most useful in elucidating them.

## ACKNOWLEDGMENTS

I thank those who provided information before publication, Jeff Robbins for a critical reading of the manuscript, and David Apirion for helpful discussions. Work in my laboratory is sponsored by the National Institutes of Health.

## REFERENCES

1. **Mills, D. R., Kramer, F. R., and Spiegelman, S.,** Complete nucleotide sequence of a replicating RNA molecule, *Science,* 180, 916, 1973.
2. **Guthrie, C.,** Folding up a transfer RNA molecule is not simple, *Quart. Rev. Biol.,* 55, 335, 1981.
3. **Abelson, J.,** RNA processing and the intervening sequence problem, *Annu. Rev. Biochem.,* 48, 1035, 1979.
4. **Gegenheimer, P. and Apirion, D.,** Processing of procaryotic ribonucleic acid, *Microbiol. Rev.,* 45, 502, 1981.
5. **Mazzara, G. P. and McClain, W. H.,** tRNA synthesis, in *Transfer RNA: Biological Aspects,* Soll, D., Abelson, J., and Schimmel, P., Eds., Cold Spring Harbor Laboratory, Cold Spring Harbor, N.Y., 1980, 3.
6. **Daniel, V.,** Biosynthesis of transfer RNA, *CRC Crit. Rev. Biochem.,* 11, 253, 1981.
7. **Dunn, J. J., Ed.,** Processing of RNA, *Brookhaven Symp. Biol. 26,* Brookhaven National Laboratory, Upton, N.Y., 1975.
8. **Weiss, S. B., Hsu, W.-T., Froft, J. W., and Scherberg, N. H.,** Transfer RNA coded by the T4 bacteriophage genome, *Proc. Natl. Acad. Sci. U.S.A.,* 61, 114, 1968.
9. **Daniel, V., Sarid, S., and Littauer, U. Z.,** Coding by T4 phage DNA of soluble RNA containing pseudouridylic acid, *Proc. Natl. Acad. Sci. U.S.A.,* 60, 1403, 1968.
10. **Scherberg, N. M. and Weiss, S. B.,** T4 transfer RNAs: Recognition and coding properties, *Proc. Natl. Acad. Sci. U.S.A.,* 69, 1114, 1972.
11. **Wilson, J. H. and Kells, S.,** Bacteriophage T4 transfer RNAs: Isolation and characterization of two phage-coded nonsense suppressors, *J. Mol. Biol.,* 69, 39, 1972.
12. **McClain, W. H.,** UAG suppressor coded by bacteriophage T4, *FEBS Lett.,* 6, 99, 1970.

13. **Kao, S.-H. and McClain, W. H.,** U-G-A suppressor of bacteriophage T4 associated with tRNA$^{Arg}$, *J. Biol. Chem.*, 252, 8254, 1977.
14. **Seidman, J. G., Comer, M. M., and McClain, W. H.,** Nucleotide alterations in the bacteriophage T4 glutamine transfer RNA that affect ochre suppressor activity, *J. Mol. Biol.*, 90, 677, 1974.
16. **Wilson, J. H., Kim, J. S., and Abelson, J. N.,** Clustering of the genes for the T4 transfer RNAs, *J. Mol. Biol.*, 71, 547, 1972.
15. **Comer, M. M., Guthrie, C., and McClain, W. H.,** An ochre suppressor of bacteriophage T4 that is associated with a transfer RNA, *J. Mol. Biol.*, 90, 665, 1974.
17. **McClain, W. H., Guthrie, C., and Barrell, B. G.** Eight transfer RNAs induced by infection of *Escherichia coli* with bacteriophage T4, *Proc. Natl. Acad. Sci. U.S.A.*, 69, 3703, 1972.
18. **Fukada, K. and Abelson, J.,** DNA sequence of a transfer RNA gene cluster, *J. Mol. Biol.*, 139, 377, 1980.
19. **Mazzara, G. P., Plunkett, G., and McClain, W. H.,** DNA sequence of the transfer RNA region of bacteriophage T4: Implications for transfer RNA synthesis, *Proc. Natl. Acad. Sci. U.S.A.*, 78, 889, 1981.
20. **Schimmel, P., Söll, D., and Abelson, J., Eds.,** *Transfer RNA: Structure, Properties and Recognition*, Cold Spring Harbor Laboratory, Cold Spring Harbor, N.Y., 1980, 522.
21. **Fukada, K., Gossens, L., and Abelson, J.,** The cloning of a T4 tRNA gene cluster, *J. Mol. Biol.*, 137, 213, 1980.
22. **Kaplan, D. A. and Nierlich, D. P.,** Initiation and transcription of a set of transfer ribonucleic acid genes *in vitro*, *J. Biol. Chem.*, 250, 934, 1974.
23. **Goldfarb, A. and Daniel, V.,** Mapping of transcription units in the bacteriophage T4 tRNA gene cluster, *J. Mol. Biol.*, 146, 393, 1981.
24. **Brody, E. N., Black, L. W., and Gold, L. M.,** Transcription and translation of sheared bacteriophage T4 DNA in vitro, *J. Mol. Biol.*, 60, 389, 1971.
25. **Goldfarb, A. and Daniel, V.,** Transcriptional control of two gene subclusters in the tRNA operon of bacteriophage T4, *Nature*, 286, 418, 1980.
26. **Deutscher, M. P., Foulds, J., and McClain, W. H.,** Transfer ribonucleic acid nucleotidyltransferase plays an essential role in the normal growth of *Escherichia coli* and in the biosynthesis of some bacteriophage T4 transfer ribonucleic acids, *J. Biol. Chem.*, 249, 6696, 1974.
27. **Maisurian, A. N. and Buyanovskaya, E. A.,** Isolation of an *Escherichia coli* strain restricting bacteriophage suppressor, *Mol. Gen. Genet.*, 120, 227, 1973.
28. **Seidman, J. G., Schmidt, F. J., Foss, K., and McClain, W. H.,** A mutant of *Escherichia coli* defective in removing 3′ terminal nucleotides from some transfer RNA precursor molecules, *Cell*, 5, 389, 1975.
29. **Barrell, B. G., Seidman, J. G., Guthrie, C., and McClain, W. H.,** Transfer RNA biosynthesis: The nucleotide sequence of a precursor to serine and proline transfer RNAs, *Proc. Natl. Acad. Sci. U.S.A.*, 71, 413, 1974.
30. **McClain, W. H.,** A role for ribonuclease III in synthesis of bacteriphage T4 transfer RNAs, *Biochem. Biophys. Res. Commun.*, 86, 718, 1979.
31. **Abelson, J., Fukada, K., Johnson, P., Lamfrom, H., Nierlich, D. P., Otsuka, A., Paddock, G. V., Pinkerton, T. C., Sarabhai, A., Stahl, S., Wilson, J. H., and Yesian, H.,** Bacteriophage T4 tRNAs: Structure, genetics, biosynthesis, *Brookhaven Symp. Biol.*, 26, 77, 1975.
32. **Guthrie, C., Seidman, J. G., Comer, M. M., Bock, R. M., Schmidt, F. J., Barrell, B. G., and McClain, W. H.,** The biology of bacteriophage T4 transfer RNAs, *Brookhaven Symp. Biol.*, 26, 106, 1975.
33. **Pragai, B. and Apirion, A.,** Processing of bacteriophage T4 tRNAs: Structural analysis and *in vitro* processing of precursors which accumulate in RNAase E$^-$ strains, *J. Mol. Biol.*, 154, 465, 1982.
34. **Goldfarb, A. and Daniel, V.,** An *Escherichia coli* endonuclease responsible for primary cleavage of *in vitro* transcripts of bacteriophage T4 tRNA gene cluster, *Nucleic Acids Res.*, 8, 4501, 1981.
35. **Pragai, B. and Apirion, D.,** Processing of bacteriophage T4 tRNAs: The role of RNAase III, *J. Mol. Biol.*, 153, 619, 1981.
36. **Watson, N. and Apirion, D.,** Ribonuclease F: A putative processing endoribonuclease from *Escherichia coli*, *Biochem. Biophys. Res. Commun.*, 103, 543, 1981.
37. **Schmidt, F. J. and McClain, W. H.,** Transfer RNA biosynthesis. Alternate orders of ribonuclease P cleavage occur *in vitro* but not *in vivo*, *J. Biol. Chem.*, 253, 4730, 1978.
38. **McClain, W. H. and Seidman, J. G.,** Genetic perturbations that reveal tertiary conformation of tRNA precursor molecules, *Nature (London)*, 257, 106, 1975.
39. **Manale, A., Guthrie, C., and Colby, D.,** S1 nuclease as a probe for the conformation of a dimeric tRNA precursor, *Biochemistry*, 18, 77, 1979.
40. **Schmidt, F. J., Seidman, J. G., and Bock, R. M.,** Transfer ribonucleic acid biosynthesis. Substrate specitificity of ribonuclease P, *J. Biol. Chem.*, 251, 2440, 1976.

41. **McClain, W. H.**, Seven terminal steps in a biosynthetic pathway leading from DNA to transfer RNA, *Acc. Chem. Res.*, 10, 418, 1977.
42. **Seidman, J. G. and McClain, W. H.**, Three steps in conversion of large precursor RNA into serine and proline transfer RNAs, *Proc. Natl. Acad. Sci. U.S.A.*, 72, 1491, 1975.
43. **Schmidt, F. J.**, A novel function of *Escherichia coli* transfer RNA nucleotidyltransferase: Biosynthesis of the C-C-A sequence in a phage T4 transfer RNA precursor, *J. Biol. Chem.*, 250, 8399, 1975.
44. **Schmidt, F. J. and McClain, W. H.**, An *Escherichia coli* ribonuclease which removes an extra nucleotide from a biosynthetic intermediate of bacteriophage T4 proline transfer RNA, *Nucleic Acids Res.*, 5, 4129, 1978.
45. **Seidman, J. G., Barrell, B. G., and McClain, W. H.**, Five steps in the conversion of a large precursor RNA into bacteriophage proline and serine transfer RNAs, *J. Mol. Biol.*, 99, 733, 1975.
46. **Foss, K., Kao, S.-H., and McClain, W. H.**, Three suppressor forms of bacteriophage T4 leucine transfer RNA, *J. Mol. Biol.*, 135, 1013, 1979.
47. **Guthrie, C.**, The nucleotide sequence of the dimeric precursor to glutamine and leucine transfer RNAs coded by bacteriophage T4, *J. Mol. Biol.*, 95, 529, 1975.
48. **McClain, W. H., Seidman, J. G., and Schmidt, F. J.**, Evolution of the biosynthesis of 3′ terminal C-C-A residues in T-even bacteriophage transfer RNAs, *J. Mol. Biol.*, 119, 519, 1978.
49. **Guthrie, C. and McClain, W. H.**, Conditionally lethal mutants of bacteriophage T4 defective in production of a transfer RNA, *J. Mol. Biol.*, 81, 137, 1973.
50. **Guthrie, C. and Scholla, C. A.**, Asymmetric maturation of a dimeric transfer RNA precursor, *J. Mol. Biol.*, 139, 349, 1980.
51. **Guthrie, C., Seidman, J. G., Altman, S., Barrell, B. G., Smith, J. D., and McClain, W. H.**, Identification of tRNA precursor molecules made by phage T4, *Nature New Biol.*, 246, 6, 1973.
52. **Gardner, J. F.**, Regulation of the threonine operon: Tandem threonine and isoleucine codons in the control region and translational control of transcription termination, *Proc. Natl. Acad. Sci. U.S.A.*, 76, 1706, 1979.
53. **Horinouchi, S. and Weisblum, B.**, Posttranscriptional modification of mRNA conformation: Mechanism that regulates erythromycin-induced resistance, *Proc. Natl. Acad. Sci. U.S.A.*, 77, 7079, 1980.
54. **Wilson, J. H. and Abelson, J. N.**, Bacteriophage T4 transfer RNA. II. Mutants of T4 defective in the formation of functional suppressor transfer RNA, *J. Mol. Biol.*, 69, 57, 1972.
55. **McClain, W. H., Guthrie, C., and Barrell, B. G.**, The *psu*$^+$ amber suppressor gene of bacteriophage T4: Identification of its amino acid and transfer RNA, *J. Mol. Biol.*, 81, 157, 1973.
56. **Cudny, M., Roy, P., and Deutscher, M. P.**, Alteration of *Escherichia coli* RNase D by infection with bacteriophage T4, *Biochem. Biophys. Res. Commun.*, 98, 337, 1980.
57. **Likover-Moen, T., Seidman, J. G., and McClain, W. H.**, A catalogue of transfer RNA-like molecules synthesized following infection of *Escherichia coli* by T-even bacteriophage, *J. Biol. Chem.*, 253, 7910, 1978.
58. **Hunt, C., Desai, S. M., Vaughan, J., and Weiss, S. B.**, Bacteriophage T5 transfer RNA: Isolation and characterization of tRNA species and refinement of the tRNA gene maps, *J. Biol. Chem.*, 255, 3164, 1980.
59. **Ozeki, H., Sakano, H., Yamada, S., Ikemura, T., and Shimura, Y.**, Temperature sensitive mutants of *Escherichia coli* defective in tRNA biosynthesis, *Brookhaven Symp. Biol.*, 26, 89, 1975.
60. **Zabriev, S. K. and Shemyakin, M. F.**, Blockage of phage T7 injection into *E. coli* cells by bacterial RNA polymerase inhibitors (in Russian), *Dokl. Akad. Nauk. S.S.S.R.*, 246, 475, 1979.
61. **Studier, F. W.**, Analysis of bacteriophage T7 early RNAs and proteins on slab gels, *J. Mol. Biol.*, 79, 237, 1973.
62. **Dunn, J. J. and Studier, F. W.**, T7 early RNAs are generated by site-specific cleavages, *Proc. Natl. Acad. Sci. U.S.A.*, 70, 1559, 1973.
63. **Dunn, J. J. and Studier, F. W.**, Nucleotide sequence from the genetic left end of bacteriophage T7 DNA to the beginning of gene 4, *J. Mol. Biol.*, 148, 303, 1981.
64. **Dunn, J. J. and Studier, F. W.**, T7 early RNAs and *Escherichia coli* ribosomal RNAs are cut from large precursor RNAs *in vivo* by ribonuclease III, *Proc. Natl. Acad. Sci. U.S.A.*, 70, 3296, 1973.
65. **Nikolaev, N., Silengo, L., and Schlessinger, D.**, A role for ribonuclease III in processing of ribosomal riboucleic acid and messenger ribonucleic acid precursors in *Escherichia coli, J. Biol. Chem.*, 248, 7967, 1973.
66. **Rosenberg, M., Kramer, R. A., and Steitz, J. A.**, T7 early messenger RNAs are the direct products of ribonuclease III cleavage, *J. Mol. Biol.*, 89, 777, 1974.
67. **Robertson, H. D. and Barany, F.**, Enzymes and mechanisms in RNA processing, *FEBS Proc. Meet.*, 51, 285, 1979.
68. **Apirion, D.**, The fate of mRNA and rRNA in *Escherichia coli, Brookhaven Symp. Biol.*, 26, 286, 1975.
69. **Dunn, J. J. and Studier, F. W.**, Effect of RNAase III cleavage on translation of bacteriophage T7 messenger RNAs, *J. Mol. Biol.*, 99, 487, 1975.

70. **Mark, K. K. and Studier, F. W.**, Purification of the gene 0.3 protein of bacteriophage T7, an inhibitor of the DNA restriction system of *Escherichia coli, J. Biol. Chem.*, 256, 2573, 1981.
71. **Rosa, M. D.**, Structure analysis of three T7 late mRNA ribosome binding sites, *J. Mol. Biol.*, 147, 55, 1981.
72. **Dunn, J. J., Buzash-Pollert, E., and Studier, F. W.**, Mutations of bacteriophage T7 that affect initiation of synthesis of the gene 0.3 protein, *Proc. Natl. Acad. Sci. U.S.A.*, 75, 2741, 1978.
73. **Strome, S. and Young, E. T.**, Translational control of the expression of bacteriophage T7 gene *0.3, J. Mol. Biol.*, 136, 417, 1980.
74. **Strome, S. and Young, E. T.**, Translational discrimination against bacteriophage T7 gene *0.3* messenger RNA, *J. Mol. Biol.*, 136, 433, 1980.
75. **Saito, H. and Richardson, C. C.**, Genetic analysis of gene 1.2 of bacteriophage T7: Isolation of a mutant of *Escherichia coli* unable to support the growth of T7 gene *1.2* mutants, *J. Virol.*, 37, 343, 1981.
76. **Saito, H. and Richardson, C. C.**, Processing of mRNA by ribonuclease III regulates expression of gene *1.2* of bacteriophage T7, *Cell*, 27, 533, 1981.
77. **Kramer, R. A., Rosenberg, M., and Steitz, J. A.**, Nucleotide sequences of the 5′ and 3′ termini of bacteriophage T7 mRNA synthesized *in vivo:* Evidence for sequence specificity of mRNA processing, *J. Mol. Biol.*, 89, 767, 1974.
78. **Herskowitz, I. and Hagen, D.**, The lysis-lysogeny decision of phage λ: Explicit programming and responsiveness, *Annu. Rev. Genet.*, 14, 399, 1980.
79. **Kleckner, N.**, Amber mutants in the *O* region of bacteriophage λ are not efficiently complemented in the absence of *N* gene function, *Virology*, 79, 174, 1977.
80. **Rosenberg, M. and Court, D.**, Regulatory sequences involved in the promotion and termination of RNA transcription, *Annu. Rev. Genet.*, 13, 319, 1979.
81. **Franklin, N. C. and Bennett, G. N.**, The N protein of bacteriophage lambda, defined by its DNA sequence, is highly basic, *Gene*, 8, 107, 1979.
82. **Lozeron, H. A., Dahlberg, J. E., and Szybalski, W.**, Processing of the major leftward mRNA of coliphage lambda, *Virology*, 71, 262, 1976.
83. **Lozeron, H. A., Anevski, P. J., and Apirion, D.**, Antitermination and absence of processing of the leftward transcript of coliphage lambda in the RNAase III-deficient host, *J. Mol. Biol.*, 109, 359, 1977.
84. **Wilder, D. A. and Lozeron, H. A.**, Differential modes of processing and decay for the major *N*-dependent RNA transcript of coliphage λ, *Virology*, 99, 241, 1979.
85. **Anevski, P. and Lozeron, H. A.**, Multiple pathways of RNA processing and decay for the major leftward N-independent RNA transcript of coliphage lambda, *Virology*, 113, 39, 1981.
86. **Roberts, J. W., Roberts, C. W., and Craig, N.**, *Escherichia coli rec A* gene product inactivates phage λ repressor, *Proc. Natl. Acad. Sci. U.S.A.*, 75, 4714, 1978.
86a. **Campbell A.**, Episomes, *Adv. Genet.*, 11, 101, 1962.
87. **Guarneros, G. and Galindo, J. M.**, The regulation of integrative recombination by the *b2* region and the *cII* gene of bacteriophage λ, *Virology*, 95, 119, 1979.
88. **Epp, C., Pearson, M., and Engquist, L.**, Downstream regulation of *int* gene expression by the *b2* region of phage lambda, *Gene*, 13, 327, 1981.
89. **Jacob, F. and Monod, J.**, Genetic regulatory mechanisms in the synthesis of proteins, *J. Mol. Biol.*, 3, 318, 1961.
90. **Belfort, M.**, The cII-independent expression of the phage λ *int* gene in RNase III-defective *E. coli, Gene*, 11, 149, 1980.
91. **Schindler, D. and Echols, H.**, Retroregulation of the *int* gene of bacteriophage λ: Control of translation completion, *Proc. Natl. Acad. Sci. U.S.A.*, 78, 4475, 1981.
92. **Guarneros, G., Montañez, C., Hernandez, T., and Court, D.**, Posttranscriptional control of bacteriophage λ *int* gene expression from a site distal to the gene, *Proc. Natl. Acad. Sci. U.S.A.*, 79, 238, 1982.
93. **Young, R. A. and Steitz, J. A.**, Complementary sequences 1700 nucleotides apart form a ribonuclease III cleavage site in *Escherichia coli* ribosomal precursor RNA, *Proc. Natl. Acad. Sci. U.S.A.*, 75, 3593, 1978.
94. **Dunn, J. J., Anderson, J. W., Atkins, J. F., Bartlett, D. C., and Crockett, W. C.**, Bacteriophage T7 and T3 as model systems for RNA synthesis and processing, in *Progress in Nucleic Acid Research and Molecular Biology*, Vol. 19, Cohn, W. E. and Volkin, E., eds., Academic Press, New York, 1976, 263.
95. **Kruger, D. H. and Schroeder, C.**, Bacteriophage T3 and bacteriophage T7 virus-host cell interactions, *Microbiol. Rev.*, 45, 9, 1981.
96. **Golomb, M. and Chamberlin, M. J.**, T7- and T3-specific RNA polymerases: Characterization and mapping of the *in vitro* transcripts read from T3 DNA, *J. Virol.*, 21, 743, 1977.
97. **Majunder, H. K., Bishayee, S., Chakraborty, P. R, and Maitra, U.**, RNase III cleavage of bacteriophage T3 RNA polymerase transcripts to late T3 mRNAs, *Proc. Natl. Acad. Sci. U.S.A.*, 74, 4891, 1977.

98. **Pachl, C. A. and Young, E. T.,** The size and messenger RNA activity of bacteriophage T7 late transcripts synthesized *in vivo, J. Mol. Biol.,* 122, 6, 1978.
99. **Cashman, J. S. and Webster, R. E.,** Bacteriophage f1 infection of *Escherichia coli:* Identification and possible processing of f1-specific mRNA, *in vivo, Proc. Natl. Acad. Sci. U.S.A.,* 76, 1169, 1979.
100. **Cashman, J. S., Webster, R. E., and Steege, D. A.,** Transcription of bacteriophage f1: The major *in vivo* RNAs, *J. Biol. Chem.,* 255, 2554, 1980.
101. **Panganiban, A. T. and Whitely, H. R.,** Analysis of bacteriophage SP82 major ''early'' *in vitro* transcripts, *J Virol.,* 37, 372, 1981.
102. **Raboy, B., Shapiro, L., and Ameniya, K.,** Physical map of *Caulobacter crescentus bacteriophage* ϕCd 1 DNA, *J. Virol.,* 34, 542, 1980.
103. **Nomura, M., Yates, J. L., Dean, D., and Post, L. E.,** Feedback regulation of ribosomal protein gene expression in *Escherichia coli:* Structural homology of ribosomal RNA and ribosomal protein mRNA, *Proc. Natl. Acad. Sci. U.S.A.,* 77, 7084, 1980.
104. **Robertson, H. D., Altman, S., and Smith, J. D.,** Purification and properties of a specific *Escherichia coli* ribonuclease which cleaves a tyrosine transfer ribonucleic acid precursor, *J. Biol. Chem.,* 247, 5243, 1972.
105. **Schmidt, F. J., Seidman, J. G., and McClain, W. H.,** unpublished data.
106. **Schmidt, F. S. and McClain, W. H.,** unpublished data.
107. **Deutscher, M.,** personal communication.
108. **Schmiessner, U., McKenney, K., Court, D., and Rosenberg, M.,** Manuscript in preparation.
109. **Court, D., Huang, T. F., and Oppenhein, A.,** Manuscript in preparation.
110. **Rosenberg, M.,** personal communication.

Chapter 4

# GENETIC AND BIOCHEMICAL STUDIES OF RNA PROCESSING IN YEAST

**Anita K. Hopper**

## TABLE OF CONTENTS

# I. INTRODUCTION

This chapter describes aspects of processing of precursors to tRNA, rRNA, and mRNA in yeast. In general, RNA processing in yeast appears to be similar to that of higher eukaryotes. Why, then, a separate chapter for yeast? The answer regards the distinct manner and ease in which much of the information regarding RNA processing in yeast has been obtained. This difference is due to the well-defined genetics which has allowed identification and characterization of many mutations which alter RNA substrates or products essential for the production of mature RNAs. Additionally, the well-defined genetics, in conjunction with the development of cloning technology and the yeast transformation system[1] and surrogate and in vitro transcription-processing systems, have aided the analysis of individual genes and their intermediate and mature RNA products. Hence, the intention of this review is to emphasize how the combination of genetic, recombinant DNA, and biochemical approaches have been used to study aspects of RNA processing in yeast. Only the processing of nuclear-encoded RNA species is considered. The description of steps involved in the production of mature nuclear-encoded RNAs has been restricted to those steps which occur after an initial precursor transcript has been generated and does not include any discussion of gene regulation, promotion or termination of transcription, or RNA turnover. Although most of the work described here is from studies of *Saccharomyces cerevisiae,* some studies of *Schizosaccharomyces pombe* and *Saccharomyces carlsbergensis* have been included where these contribute important additional or corroborative information. The reader is also directed to recent and excellent reviews by others discussing tRNA processing[2,3] and rRNA processing[4] in yeast.

# II. PROCESSING OF YEAST TRANSFER RNA PRECURSORS

The processing of yeast tRNAs has been especially amenable to combined biochemical and genetic studies for several reasons. First, tRNAs are small, stable molecules. Second, the primary and secondary structures have been solved for many of them,[5] and the tertiary structure is known for some tRNAs.[6] Third, in several cases the tRNA genes have been cloned and sequenced.[7-20] Furthermore, specific tRNA-encoding genes can be mutated to produce tRNA species capable of recognizing (''suppressing'') nonsense or frameshift codons, whose presence gives the cell a readily identifiable phenotype.[8,17,18,21-23] Additionally, it has been possible to identify mutants which are altered in tRNA synthetic steps.[24-33] The wealth of structural and genetic data provide valuable tools for the in vivo and in vitro studies of tRNA processing.

A variety of biochemical techniques have been used to study the steps of tRNA production. Initially tRNA precursor molecules were detected on polyacrylamide gels as 4.5S, transiently labeled species possessing modified bases.[34] However, when recombinant DNA clones encoding tRNAs[7-20] and mutants affecting tRNA processing[25,27,28,30] became available, it was possible to identify, purify, and sequence particular tRNA precursors.[3,30,35-39] Furthermore, the development of *Xenopus* oocytes as a surrogate in vivo transcription and processing system[40] and the development of in vitro transcript-processing systems from *Xenopus,*[41,42] *HeLa,*[43] and yeast[44,45] have allowed the characterization of transcripts and processing intermediates from a variety of wild-type and mutant tRNA cloned genes.

## A. Mutations Affecting tRNA Production

There are two general types of mutations which affect tRNA processing. The first are mutations of a tRNA gene which alter the precursor tRNA substrate such that it is not properly recognized by the processing and modification enzymes; these mutations will be referred to as ''intragenic'' lesions. The second type are lesions of the genes encoding the

## Table 1
## GENETIC SCREENS FOR DETECTING LOSS OF SUPPRESSION LESIONS

"ade-can screen"[a]

| Genotype | Color on rich medium | Growth on medium lacking adenine | Growth on canavanine-containing medium |
|---|---|---|---|
| *ADE2⁺ CAN1⁺* | White | +, white | − |
| *ade2-1 can1-100* | Red | −, red[c] | + |
| *ade2-1 can1-100 SUP4* | White | +, white | − |
| *ade2-1 can1-100 SUP4 losx* | Red | −, red | + |

"ade-gal screen"[b]

| | Color on rich glucose medium | Growth on glucose-containing medium lacking adenine | Growth and color on galactose-indicator medium |
|---|---|---|---|
| *ADE2⁺ GAL1⁺ GAL7⁺* | White | +, white | +, white |
| *ade2-1 gal1-o gal7* | Red | −, red | −, gray-white[c] |
| *ade2-1 gal1-o gal7 SUP4* | White | +, white | −, yellow[c] |
| *ade2-1 gal1-o gal7 SUP4 losx* | Red | −, red | −, gray-white |

[a] The "ade-can" procedure[50] scores the ability of cells to suppress *ade2-1* and *can1-100*. Cells which are defective in suppression are red on rich medium and are able to grow in the presence of canavanine. Since growth is required to monitor the loss of ability to suppress, only "leaky" mutations in essential genes or mutations in genes which are unessential for growth will be detected by this procedure.

[b] The "ade-gal" procedure does not require that colonies be able to grow more than a few generations at the nonpermissive temperature in order to monitor loss of suppressor activity. On rich medium containing the pH indicator bromthymol-blue and the carbon source galactose, replicas of *gal1* colonies are gray-white whereas replicas of *gal7* mutant colonies are yellow. Since *GAL1* codes for the first enzyme and *GAL7* for the second enzyme in the Leloir pathway, *gal1 gal7* double mutants have a "*gal1*" phenotype. Strains with a *gal1* ochre (*gal1-o*) mutation and a *gal7* missense mutation produce colonies which are gray-white in the absence of an ochre suppressor and are yellow in the presence of the suppressor. Strains containing *gal1-o, gal7, ade2-1,* and *SUP4* alleles therefore provide a double color assay for loss of suppression.[51]

[c] Although replicas of colonies do not grow significantly on these media, the color changes can be readily detected.

processing, modification, or other products necessary for production of mature tRNAs; these will be referred to as "extragenic" lesions.

Loss of ability to suppress nonsense mutations has been used as a means to identify mutations affecting tRNA synthesis.[46-49] The rationale to this approach is that since nonsense suppression requires tRNA function, lesions which result in the loss of ability to suppress nonsense mutations could be localized to the suppressing tRNA locus itself or to loci whose products are involved in the production of the functional suppressor tRNA. There are two convenient assays which have proved successful in identifying such lesions of yeast.[50,51] (Table 1). All of the intragenic mutations[28,30] and a few of the extragenic mutations (e.g., *los1*[27] and *mod5*[26]) affecting tRNA production have been identified by these assays. Other extragenic mutations affecting tRNA production have been identified by screening conditionally lethal yeast mutants for defects in tRNA processing (e.g., *rna1*[25]), or screening mutagenized cells for defects in tRNA modification (e.g., *trm1*[24]). A few mutants have been found serendipitously (e.g., *trm2*[32] and *mia*[31]).

The 69 intragenic second-site loss of suppression mutations of the *SUP4* $tRNA^{Tyr}$ ("*sup4⁻*"

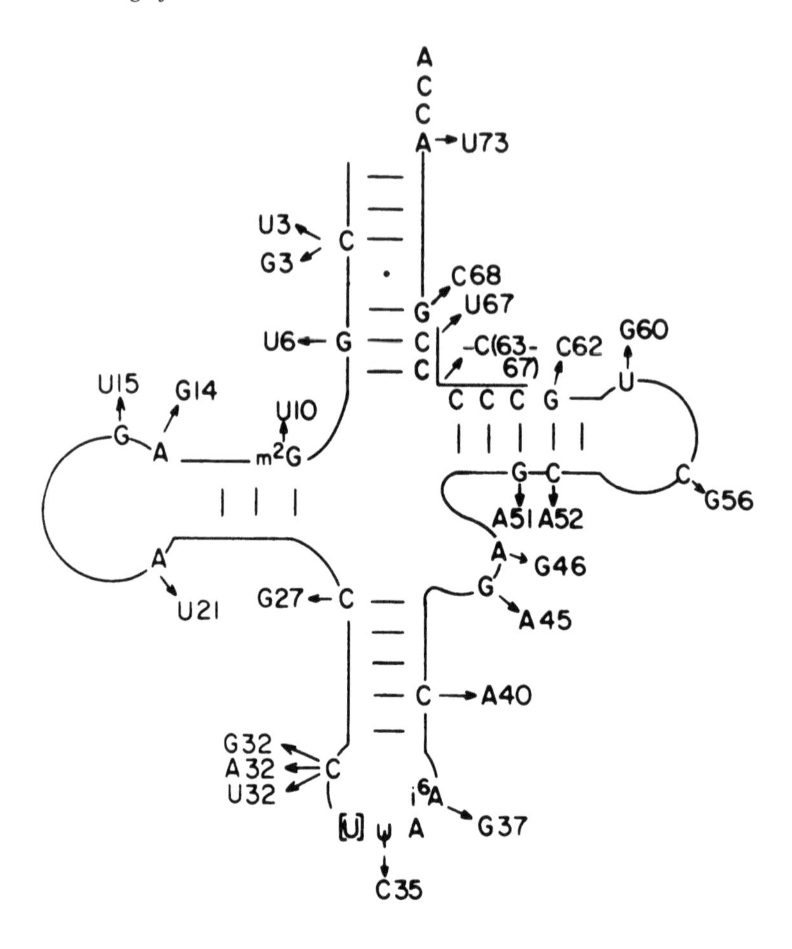


FIGURE 1. Mutations of the yeast *SUP4* tRNA$^{Tyr}$. The suppressor tRNA$^{Tyr}$ structure is shown with the position and alterations found in the various mutants indicated. The numbering system is that of Sprinzl and Gauss.[5] (From Kurjan, J., Hall, B. D., Gillam, S., and Smith, M., *Cell*, 20, 701, 1980. Copyright by MIT Press. With permission.)

mutations) locus which have been obtained by Kurjan et al.[28,52] have been genetically mapped, and 26 of these mutant tRNA$^{Tyr}$ genes have been cloned and sequenced (Figure 1). Most surprisingly, none of the lesions maps outside the 5′ or 3′ boundaries of the mature tRNA$^{Tyr}$ sequence. Recombinant DNAs containing the *sup4*$^-$ mutant sequences have been transcribed and processed in the *Xenopus* oocyte system.[40,53,53a] Independently, Colby et al.[30] studied the effect of a *sup4*$^-$ lesion upon in vivo processing of pre-tRNA$^{Tyr}$. All together, these studies (example results are given in Table 2) have shown that lesions mapping to various locations of the tRNA gene affect processing of precursor tRNA$^{Tyr}$ species.

Mutations in two extragenic loci (*rna1*[25] and *los1*[27]) have been identified which cause the accumulation of a particular subset of precursor tRNAs. At the time that the accumulation of pre-tRNAs in the *rna1* mutant was discovered, it was already known that some yeast tRNA genes contained intervening sequences.[8,9] By oligonucleotide mapping it was shown that the species which accumulated in *rna1* cells were mature at the 5′ and 3′ termini, but contained intervening sequences.[35-38] Further studies showed that the *los1* mutants accumulated the same precursor tRNAs as *rna1* cells.[27] Even though both *rna1* and *los1* cells are defective in removing intervening sequences, both of these mutants appear to retain tRNA-splicing enzyme activity.[54] The *rna1* mutation, which purportedly affects the transport

**Table 2**
**EFFECT OF *sup4*⁻ MUTATIONS UPON PRECURSOR tRNA PRODUCTION**

| Position of lesion | Nucleotide change | Effect on processing[a] |
|---|---|---|
| U6 (aa stem) | G → U | No processing[62] |
| G37 (AC loop) | $i^6A$ → G | 92-Nucleotide species poorly processed[30,53] |
| G32 (AC loop) | C → G | 92-Nucleotide species not spliced[53a] |
| A40 (AC stem) | C → A | 92-Nucleotide species not spliced[53a] |
| G60 (TΨ loop) | U → G | No processing[53a] |
| C62 (TΨ stem) | G → C | Shortened transcript,[42] shortened mature species[39] |
| C68 (aa stem) | G → C | Poor processing of the 108 species[53a] |

[a] Superscript numerals refer to references in the bibliography.

**Table 3**
**MUTATIONS AFFECTING THE PRODUCTION OR MODIFICATION OF MATURE tRNA SPECIES**

| Locus | tRNA defect[a] | Phenotype[a] |
|---|---|---|
| *rna1*[b] | Accumulates pre-tRNAs containing intervening sequences[25,35-38] | Conditional lethal[55] |
| *los1*[b] | Accumulates pre-tRNAs containing intervening sequences[27] | Loss of suppression[27] |
| *mod5*[b] | No $i^6A$[26] | Loss of suppression[26] |
| *sin1*[c] | No $i^6A$[29] | Loss of suppression[29] |
| *trm1*[b] | No $m^2_2G$[24] | None[51,72] |
| *trm2*[b] | No $m^5U$[32] | None[32] |
| *mia*[b] | Decreased hU[31] | None[71] |
| *sin3*[c] | No $mcm^5S^2U$[33] | Loss of suppression[33] |

[a] Numbered superscripts refer to references in the bibliography.
[b] Lesions of *Saccharomyces cerevisiae.*
[c] Lesions of *Schizosaccharomyces pombe.*

of RNA from the nucleus to the cytoplasm,[55,56] also affects mRNA[55,56] and rRNA.[25,57] The *los1* locus segregates independently from *rna1* and appears to affect only tRNA.[27]

Four *Saccharomyces cerevisiae (mod5-1, mia, trm1,* and *trm2)* and two *Schizosaccharomyces pombe,*[29,33] mutants defective in tRNA nucleoside modification have been characterized. In *mod5-1* cells the amount of isopentenyladenosine ($i^6A$) is reduced to approximately 1% of wild-type levels.[26] An $i^6A$-deficient mutant of *S. pombe* has also been found.[29] The *mia* strain has lowered levels of dihydrouridine (hU).[31] The *trm1* mutant has little or no $m^2_2G^{24}$ although tRNA from wild-type cells contains 0.6 to 0.8% $m^2_2G^{58}$. Finally, the *trm2* mutant has little or no $m^5U^{32}$ in tRNA although almost all known tRNAs have $m^5U$ in the TΨC loop.[5]

A summary of the known extragenic mutations and their phenotypes is given in Table 3.

### B. Production of Mature tRNAs

Genetic and physical studies have shown that, in general, the yeast tRNA genes are scattered as singlets throughout the yeast genome,[7,59-61] and consequently the initial tRNA transcripts can contain only a single tRNA sequence. So far, only two exceptions have been found where recombinant DNAs encoding information for two different tRNA species have been transcribed into dimeric precursors and processed to mature-sized tRNAs in *Xenopus* germinal vesicle extracts.[11,12] Therefore yeast, most likely, encodes enzymes capable of processing dimeric pre-tRNAs to monomeric tRNA species. As the initial tRNA transcripts may also contain extra nucleotides at both the 5′ and 3′ termini,[11,12,39-43] these posttranscriptional nucleolytic processing steps are needed to remove these sequences. Some, but not all, yeast tRNA genes contain intervening sequences which are removed posttranscriptionally from the precursor tRNAs.[3,8-10,16,35-38] Since all tRNAs possess a 3′ CCA sequence which is not encoded[8-20] this sequence must be added posttranscriptionally by nucleotidyltransferase after 3′ end trimming.[172] For both $tRNA^{His}$ of *Drosophila* and *S. pombe* an entirely novel mechanism for generating mature 5′ termini has been recently described. In these cases, the 5′ terminal moiety is not encoded, but, rather, it is added in an ATP-dependent step which occurs after removal of the encoded extra 5′ nucleotides.[61a] Finally, since a significant fraction of the nucleotides of a given tRNA molecule are modified to nonstandard nucleosides,[5] many other posttranscriptional addition steps also occur before a mature functional tRNA is generated.

*1. Order of the Nucleolytic Processing Events*

When the cloned *SUP8*$^+$ wild type $tRNA^{Tyr}$ yeast gene was injected into *Xenopus* oocytes, three precursor RNA species of 108, 104, and 92 nucleotides were generated in addition to the 78 nucleotide mature $tRNA^{Tyr}$.[40] The largest (108-nucleotide) species contained a 5′ triphosphate and extra 5′ leader, the 14-nucleotide intervening sequence, and a 3′ trailer sequence.[40] Recombinant DNA clones encoding different $tRNA^{Tyr}$ loci generated transcripts with 5′ leaders of different lengths.[40,42,45] The second largest (104-nucleotide) species synthesized from *SUP8*$^+$ DNA in *Xenopus* oocytes contained a shortened 5′ leader, 3′ trailer, and an intervening sequence.[40] The 92-nucleotide species possessed mature 5′ and 3′ termini and the posttranscriptionally added CCA, but contained the intervening sequence.[40] The same 92-nucleotide species accumulates in vivo in the *rna1*[25] and *los1*[27] mutants.

It has also been possible to detect precursor tRNA species of $tRNA^{Tyr}$ and $tRNAs^{Ser}$ by the Northern hybridization technique.[39] Three species homologous to a $tRNA^{Tyr}$ gene were detected in these studies — a large ~108-nucleotide presumptive initial transcript, a 92-nucleotide presumptive intervening sequence-containing species, and the 78-nucleotide mature tRNA. In general there is very good agreement between the species detected in vivo in yeast and those detected when cloned yeast $tRNA^{Tyr}$ genes are transcribed and processed by *Xenopus* oocytes or by extracts from *Xenopus* or yeast cells. However no heterogeneity in the largest species (arising from different initiation sites on the various $tRNA^{Tyr}$ genes) as would be expected from the *Xenopus* studies was detected by the Northern hybridization. No precursor $tRNA^{Tyr}$ species from *Saccharomyces cerevisiae* which lack the intervening sequence and possess unprocessed 5′ and 3′ termini have been detected in *Xenopus* or in wild-type or mutant yeast cells.

Removal of the intervening sequences was also found to follow 5′ and 3′ termini maturation when the *S. cerevisiae* $tRNA^{Trp}$ gene was transcribed and processed by a nuclear extract from *Xenopus* oocytes.[41] When a recombinant DNA encoding *Schizosaccharomyces pombe* $tRNA^{Ser}_{UCG}$ and $tRNA^{Met}$ was transcribed in the *Xenopus* germinal vesicle system, the initial transcript was a dimeric precursor molecule. This molecule was first cleaved to two monomeric species which were further processed at the 5′ and 3′ termini. The last step of the processing was the removal of the intervening sequence from the pre-$tRNA^{Ser}_{UCG}$.[11]

Thus, studies of the *rna1* and *los1* mutants which show that 5′ and 3′ termini can be removed prior to excision of intervening sequences and studies of the processing of pre-tRNA$^{Tyr}$, pre-tRNA$^{Trp}$, and pre-tRNA$^{Ser}_{UCG}$ which provide no evidence for removal of intervening sequences prior to 5′ and 3′ maturation all indicate that 5′ and 3′ termini processing is generally followed by removal of intervening sequences. However, this need not be a mandatory order since HeLa in vitro transcription-processing extracts fail to remove 5′ and 3′ termini but are able to remove intervening sequences from a yeast tRNA$^{Leu}_{3}$ transcript.[43]

Products with mature 5′ and unprocessed 3′ termini or vice versa or products which have been cleaved at the 3′ terminus but lack the CCA addition usually are not detected when *Saccharomyces cerevisiae* tRNA$^{Tyr}$ or *Schizosaccharomyces pombe*$^{Ser}_{UCG}$ sequences are transcribed and processed by *Xenopus* oocytes.[40,11] However, small quantities of such sequences have been detected in vivo.[62] Presumably, then, 5′ and 3′ termini processing and CCA addition are approximately concurrent. Figure 2 summarizes the pathway of processing events for *SUP8*$^{+}$ tRNA$^{Tyr}$ of *S. cerevisiae* based on the above data.

*2. tRNA Nucleolytic Processing Enzymes*

The 5′ termini of precursor tRNAs of *Escherichia coli* are processed by RNase P.[48] This enzyme has been purified and has been shown to contain a protein component and an RNA component both of which are essential for activity.[63] A RNase P-like enzyme has been partially purified from *Schizosaccharomyces pombe*.[64] Although the *S. pombe* enzyme was isolated using a synthetic substrate, it has been shown to properly process the 5′ termini of the pre-tRNA$^{Tyr}$ of *E. coli*[64] and a variety of other tRNA precursors.[65] Like RNase P from *E. coli* and other eukaryotic cells,[48] the 5′ processing enzyme from *S. pombe* requires an RNA component for activity[64,65] and makes a single endonucleolytic cleavage in pre-tRNA to generate a mature 5′ terminus.[64] This activity is apparently also responsible for generating monomeric pre-tRNAs from the *S. pombe* dimeric tRNA$^{Met}$-tRNA$^{Ser}_{UCG}$ species since the initial cleavage event in *Xenopus* germinal vesicles generates a pre-tRNA$^{Met}$ with a mature 5′ terminus.[11] Although there have been few studies of the enzyme activity which generates mature 3′ tRNA termini, recently Tekamp[66] has identified an exonucleolytic enzyme from yeast which accurately processes the 3′ terminus of tRNA$^{Leu}_{3}$ precursor generated by HeLa cell extracts.

Studies of the *sup4*$^{-}$ mutants indicate that the enzymes which process the 5′ and 3′ termini may recognize similar structures of the tRNA precursor molecules. The tRNA transcripts from several of these mutant genes generated in the *Xenopus* surrogate system[53,53a] are not completely processed and therefore the lesions define sites recognized by, or associated with, the 5′ and 3′ nucleases and splicing enzyme. However, so far no intermediates which lack only 5′ or 3′ extra sequences have been detected (Table 2). Perhaps disruption of 5′ processing recognition sites can affect 3′ processing and/or vice versa.

Studies of the enzymes responsible for removal of intervening sequences from precursor tRNAs have utilized the precursor species which accumulate in *rna1* cells as substrates. Knapp et al.[35] and O'Farrell et al.[36] were able to describe an enzyme activity which splices the intervening sequences from these precursor molecules. Initially it was thought that tRNA splicing was comprised of two steps: (1) an endonucleolytic step ("cleaving") removing the intervening sequences and producing a 5′ tRNA half-fragment bearing a 3′ phosphate and a 3′ half-tRNA fragment bearing a 5′ hydroxyl termini; (2) an ATP-dependent ligation step joining the resulting 5′ and 3′ halves.[3,67] Subsequently, Konarska et al.[68a] described a novel RNA ligase in wheat germ which utilizes 2′,3′ cyclic phosphate and 5′ hydroxyl nucleotides as substrates and generates a 2′-phosphomonoester, 3′,5′-phosphodiester linkage. Further studies showed that the enzyme mechanism of the yeast tRNA ligase is similar to the wheat germ ligase.[68a] The current proposed mechanism for tRNA splicing is described below and summarized in Figure 3.[69]

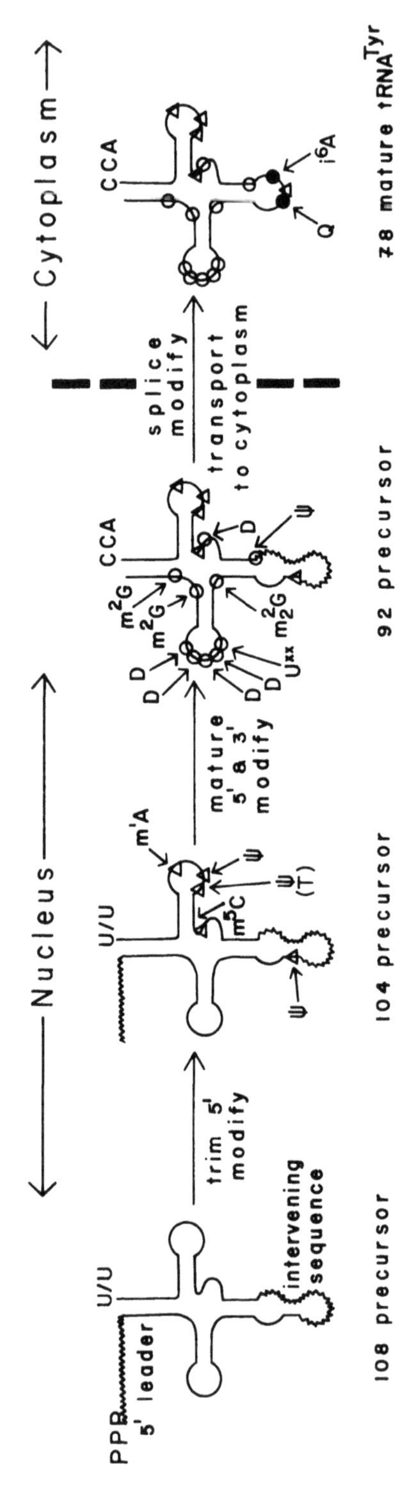


FIGURE 2. The pathway of events leading to the production of mature $tRNA^{Tyr}$.[40,74] (Adapted with permission from Nishikura, K. and DeRobertis, E. M., *J. Mol. Biol.*, 145, 405, 1981. Copyright: Academic Press Inc. (London) Ltd.)

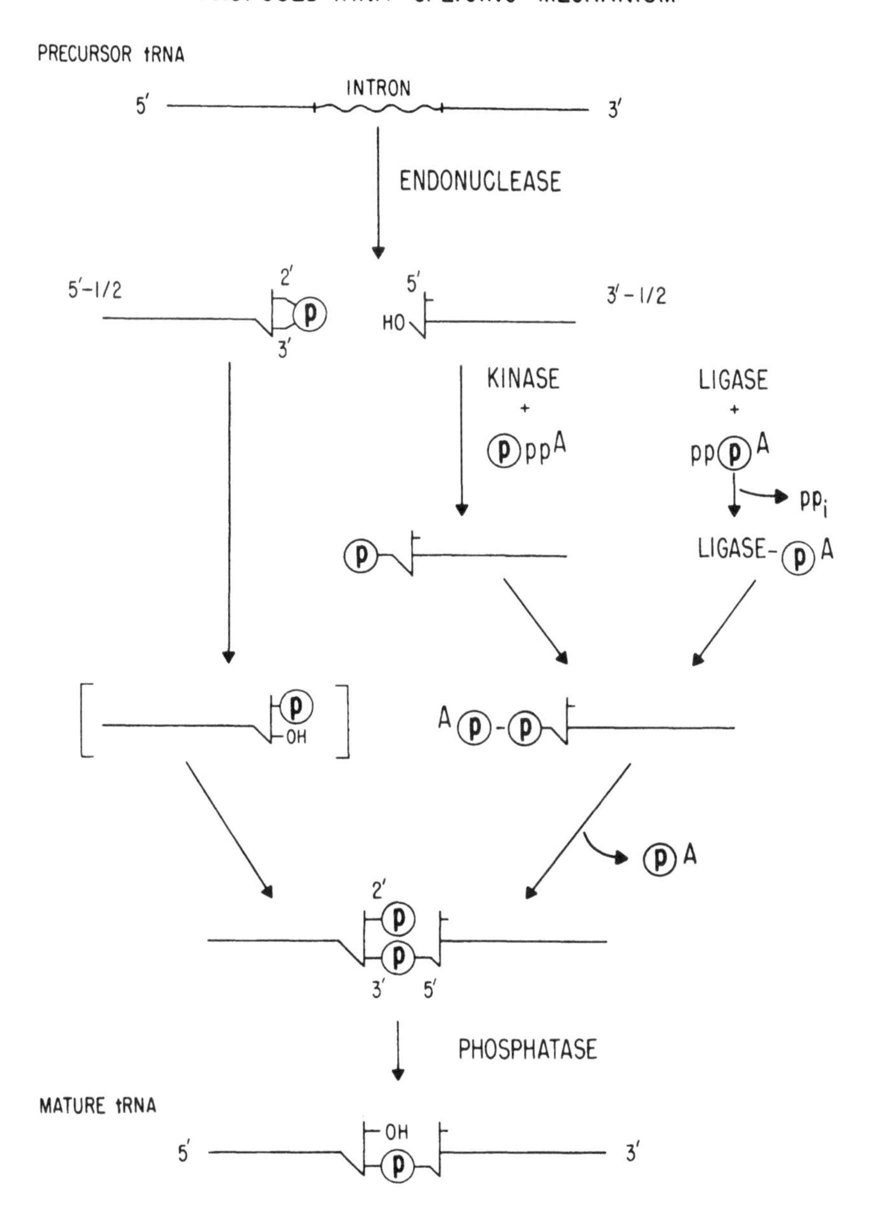


FIGURE 3. Proposed mechanism for tRNA splicing. Details of the mechanism are described in the text. (From Greer, C. L., Peebles, C. L., Gegenheimer, P., and Abelson, J., *Cell*, 32, 537, 1983. Copyright by MIT Press. With permission.)

1. Endonuclease step — the initial step of splicing is the endonucleolytic cleavage of pre-tRNAs which produces 5′ half molecules with 2′,3′ cyclic phosphoryl termini, 3′ half molecules with 5′ hydroxyl termini, and excised linear intervening sequences.
2. Cyclic 2′,3′ phosphate opening — the phosphate in the cyclic linkage is eventually localilzed to the 2′ position of the ligated half molecule.
3. Kinase activity — gamma phosphate of ATP is transferred to the 5′ position of the 3′ half molecule.

4. Adenylation — AMP is transferred from ATP to a single yeast polypeptide (presumably the tRNA splicing ligase or a component of this enzyme).
5. RNA activation — transfer of an AMP moiety, presumably from the adenylated ligase, to the 3′ half molecule produces a 5′,5′ pyrophosphate linkage.
6. Joining — the 5′ half molecule (either 2′,3′ cyclic or 2′ phosphate) is joined to the activated 5′,5′ pyrophosphate-linked 3′ half molecule and produces a 3′,5′ monophosphodiester bond at the junction and releases AMP.
7. Phosphatase — it is postulated that after joining, the 2′ phosphate is removed enzymatically.

The splicing endonuclease and ligase activities purify separately. The nuclease appears to be membrane bound whereas the ligase does not copurify with membrane fractions. It is not yet clear whether all the activities associated with RNA ligation (2′,3′ phosphate opening, RNA kinase, protein adenylation, RNA activation, and 2′ phosphatase) are due to a single multifunctional or to individual enzymes.[69] Partially purified preparations of the endonuclease cleave many intervening sequence-containing pre-tRNAs with similar efficiency. Likewise, partially purified ligase preparations join various cognate pairs of cleaved tRNA half molecules with similar efficiency.[68] It therefore appears likely that one enzyme system is able to splice all intervening sequence-containing pre-tRNAs which accumulate in *rna1* cells.[3,67,68]

The tRNA splicing activity is unlikely to be responsible for processing precursor mRNA since the intervening sequences of tRNA genes are unlike those of mRNA genes. The intervening sequences of tRNA genes are short (14 to 60 bases) sequences located one nucleotide removed from the 3′ end of the anticodon and contain sequences complementary to the anticodon.[3,8-10,16,35-38] The intervening sequences do not contain the GT and AG conserved junction sequences typical of mRNA genes.[3,8-10,16,35-38]

There is some information regarding what features of the pre-tRNAs are recognized by the splicing activity. Base pairing of the anticodon to the intervening sequence is apparently not required since suppressor $tRNAs^{Tyr}$, which are efficiently expressed,[21,22] are altered in one base of the anticodon. In fact, in *Xenopus SUP4* precursor tRNA was spliced more efficiently than *wild-type SUP4*$^{+}$ precursor tRNA.[53,53a] Of the 69 mutations generated by Kurjan et al.,[28] only one lesion which was isolated three independent times (A → T at the 4th nucleotide of the intron) of the intervening sequence has been found to inactivate *SUP4* activity. The suppression defect caused by this lesion is due to the creation of a transcription termination signal in the intron.[42,45,52] These results imply that none of the remaining 13 nucleotides of the intron can be altered individually to cause defects in the removal of the intervening sequence. In in vitro studies Johnson et al.[69a] inserted a heterologous sequence into the intron of a yeast $tRNA^{Leu}_{3}$ gene. The altered gene was transcribed efficiently and the product processed by *Xenopus* germinal vesicle extracts. The "precursor" RNA containing the 21-nucleotide insert in the putatively unpaired intron loop region could also be processed properly in vitro in reactions containing crude yeast cleavage-ligating activity.[69a] Therefore both in vivo and in vitro data indicate that the nucleotide sequence of the intron is not recognized by the splicing activity. Somewhat different results were obtained when a plasmid (PYTC) encoding a yeast $tRNA^{Tyr}$ with a natural polymorphism at the 5th nucleotide of the intron was studied in *Xenopus* oocytes. In this case the $tRNA^{Tyr}$ was processed more poorly than products from other yeast $tRNA^{Tyr}$ genes.[53]

At least one of the nucleotides (#37) at the exon splice junction of *SUP4* $tRNA^{Tyr}$ is important for splicing since a mutation (A → G) at this site results in inefficient removal of the intron and a sevenfold increase in the steady-state level of the 92-nucleotide precursor species.[30,53,53a] In addition, mutations at positions #32 of the anticodon loop and #40 of the anticodon stem of the *sup4*$^{-}$ genes prevent splicing of pre-tRNA by *Xenopus* oocytes.[53a]

Finally, it has not been possible to ligate in vitro half molecules of different tRNAs using the yeast tRNA splicing enzymes.[3] Thus, it appears that both the nucleolytic and ligating activities recognize elements of tRNA secondary or tertiary structure. Interestingly, the RNA ligase from wheat germ can form a 3′,5′ phosphodiester linkage from RNAs of any sequence[68a,69] and hence this enzyme may serve additional or different functions in wheat germ.

*3. Modification of tRNA Species*

Yeast tRNAs contain many different types of modified bases and a significant percent of the nucleosides in any given tRNA may be altered.[5] The functions of these modifications and their relationship to other tRNA processing steps are poorly understood.[70]

Studies of the yeast mutants defective in tRNA modifications have led to the general conclusion that the $i^6A$, hU, $m^2_2G$, and $m^5U$ modifications are not essential for tRNA processing events. Specifically, the *mia* (hU deficient) and *trm2* ($m^5U$ negative) strains are indistinguishable from wild-type under various growth conditions and suppress nonsense mutations as well as wild-type cells.[32,71] Thus far only deficiencies in $i^6A$ *(mod5-1, sin1)* and $mcm^5S^2U$ *(sin3)*, both modifications of the anticodon loop, appear to cause phenotypic differences.[26,29,33] Although *mod5-1* cells have no growth defect, *mod5-1 SUP7-1* haploid cells fail to suppress some nonsense mutations and *mod5-1/mod5-1* diploid cells sporulate poorly.[26,73] The $i^6A$-deficient mutant of *S. pombe* is also defective in suppression.[29] No secondary tRNA processing defects have been evidenced in the *mod5-1*,[73] *mia*,[31,71] *trm1* ($m^2_2G$ negative),[72] and *trm2*[32] mutants.

Studies of the yeast $tRNA^{Tyr}$ synthesis in *Xenopus* oocytes indicate that nucleoside modification takes place in the nucleus during tRNA processing in a well-defined order[74] (see Figure 2). In yeast, however, a particular modification step may not be a prerequisite for another. For example, *trm1* mutants lack $m^2_2G$, but have normal levels of $m^5U$, and *trm2* mutants lack $m^5U$ but have normal levels of $m^2_2G$.[32] On the other hand, recent studies have shown that deletion of the intron of a tRNA gene results in the failure to produce a pseudouridine modification in the anticodon.[74a]

*4. Cellular Localization of tRNA Synthetic Steps*

The precursor tRNAs which accumulate in the *rna1* mutant have processed termini, contain 3′ CCA, and many modifications.[35-38] If this mutant is defective in transporting RNA from the nucleus to the cytoplasm as claimed,[55,56] then most tRNA processing steps must occur in the nucleus. Consistent with this conclusion, yeast nuclear preparations have been shown to be a source of splicing activity.[36] Elegant experiments in *Xenopus,* wherein processing in nuclear and cytoplasmic compartments for individual precursor $tRNA^{Tyr}$ species was studied, have contributed additional strong support for this conclusion. In these experiments precursors containing 5′ leader and 3′ trailer sequences were localized only in the nucleus, while intervening sequence-containing pre-tRNA (92 nucleotides) was found in both nuclear and cytoplasmic compartments. Upon reinjection of the latter species into nuclei, enucleated oocytes, or oocyte cytoplasm, splicing occurred only in the nucleus[75] (Figure 2). Thus the localization of the splicing enzyme to the nucleus appears to be efficient. Further studies in *Xenopus* have indicated that the splicing activity is not bound to the nuclear membrane, but rather is nucleoplasmic.[76]

Studies utilizing *Xenopus* oocytes and studies of the *rna1* yeast mutant both show that most nucleoside modifications occur in the nucleus. It is thought, however, that the same genes which encode nuclearly located modification enzymes also encode the enzymes which reside in the mitochondria.[32,77] Thus one gene may encode tRNA synthetic enzymes for multiple cellular compartments.

### C. Speculations and Summary

It is not known if processing of tRNAs takes place on ribonucleoprotein particles, as has been shown for pre-rRNA. Nor is it known if tRNA is transported from the nucleus to the cytoplasm on such particles. Until recently, tRNA was not thought to be associated with protein. However recent experiments utilizing sera containing antibodies to RNPs have provided evidence that mammalian tRNAs may exist in ribonucleoprotein particles both in the nucleus and the cytoplasm.[78] As yet, there is no evidence for yeast tRNA-protein particles. Since the *los1* and *rna1* mutants are both defective in tRNA splicing, but both contain pre-tRNA substrate and splicing enzyme activity, these mutants could be altered in genes encoding RNP particles or components of nucleoplasm-cytoplasm transport. Alternatively, they could be altered in some control pathway.

To process tRNA molecules, then, the yeast cell must encode 5′ and 3′ nucleases, splicing nucleolytic and ligation enzymes, nucleotidyltransferase, 5′ G addition activity, a multitude of nucleoside modification enzymes and, perhaps, ribonucleoproteins. All of these proteins must be synthesized on cytoplasmic ribosomes and then be shuttled to the nucleus. The data of Melton et al.[75] regarding the cellular location of the tRNA-splicing enzyme suggests that such partitioning is efficient. It is anticipated that further combined genetic and biochemical studies will reveal not only the details and complexities of the nucleolytic and modification events involved in generating mature functional tRNAs, but also yield information on the role and mechanisms of cellular compartmentalization.

## III. PROCESSING OF YEAST RIBOSOMAL RNA PRECURSORS

The processing of yeast precursor rRNAs has been studied for many years. Early studies of *Saccharomyces cerevisiae* and *S. carlsbergensis* determined the pathway of the nucleolytic events and nucleoside modifications and the role of ribosomal and other nuclear proteins in this pathway.[79-84] More recently, rRNA genes from these organisms have been cloned[85-90] and the exact locations of the termini of the initial rRNA transcript and mature rRNA species have been determined.[90-97] Furthermore, sequences surrounding some processing sites have now been determined.[94,97,97a]

Studies of the yeast ribosomal DNAs and processing of precursor ribosomal RNAs have been the subject of previous recent reviews, and in many respects the conclusions of these studies mirror the conclusions of studies of rRNA synthesis for other eukaryotes.[4,82,84,98] In this review, kinetic studies of the synthesis of yeast rRNAs will be described only briefly and recent studies utilizing cloned rDNA sequences and studies of yeast mutants which are altered in rRNA production will be described in greater detail. The reader is referred to the chapter on rRNA processing (Chapter 9 of this volume), which also describes the production of yeast rRNAs.

### A. Nucleolytic Steps of Precursor rRNA Processing

There are approximately 140 tandemly repeated copies of 9060 ± 100 nucleotides of the ribosomal RNA genes in the yeast genome.[91,99-101] The 18S*, 5.8S, and 25S rDNA sequences

* The sedimentation coefficients derived for *S. cerevisiae* mature and precursor rRNA species[79] will be used here. Although these are somewhat different than the values derived for *S. carlsbergensis*, this probably does not reflect major differences in the sizes of the various RNA species. It is therefore assumed that 25S and 18S mature rRNAs of *S. cerevisiae* are the same as 26S and 17S species of *S. carlsbergensis;* likewise it is assumed that the 35S, 27S, and 20S pre-rRNAs of *S. cerevisiae* are equivalent to the 37S, 29S, and 18S species, respectively, of *S. carlsbergensis*.

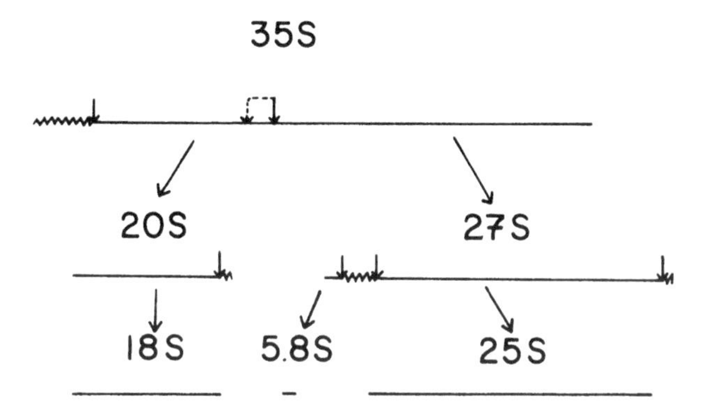


FIGURE 4. Pathway of processing steps for yeast rRNAs: ~~ sequences eliminated during processing; arrows indicate cut sites.

are arranged (Figure 5) similarly to those in other eukaryotes (i.e., 5′-sp*-18S-sp-5.8S-sp-25S-sp-3′).[86-91] Unlike other eukaryotes, the 5S rDNA sequences of yeast are interspersed within the rDNA repeats and are transcribed in the opposite direction from the other rDNAs.[100,102]

Kinetic studies employing in vivo labeling of rRNA species showed that the 18S, 25S, and 5.8S ribosomal genes are transcribed into a large 35S precursor[79,80,82] (~6657 nucleotides[92-94,96]). This precursor is cleaved, perhaps via another intermediate,[83] to the 27S and 20S species.[79,80,82] The 27S species is processed to the 25S (3392 nucleotides[103]) and 5.8S (158 to 164 nucleotides[104]) mature rRNAs.[79,80,82] In *S. carlsbergensis* 5.8S is derived via another, 7S, precursor.[105] No 7S rRNA has been detected as yet in *S. cerevisiae*. The 20S species is processed to the 18S (1789 nucleotides[91,95,97]) mature species.[79,80,82] All of the processing steps with the exception of the 20S to 18S cleavage occur in the nucleus; 20S pre-rRNA is processed in the cytoplasm.[81,106] The processing pathway for *S. cerevisiae* is summarized in Figure 4.

The 5S rRNA species contains a 5′ terminal triphosphate[107] and hence is thought to be the initial transcript. No species larger than 5S has been detected in vivo although in vitro transcription of the 5S gene can give rise to a larger transcript which is extended at the 3′ terminus.[108]

Since the ribosomal RNA genes have been cloned, characterized, and partially sequenced, it has been possible to localize the termini of the various rRNA species to the rDNA using the techniques of R-looping, S1-protection mapping, and reverse transcriptase-mediated primer extension and direct sequence comparison; since the 35S initial transcript contains a 5′ triphosphate[91a,91b] it has been considered to be the primary rRNA transcript. However, others[91c] have found evidence for transcription inition at a site 2.2 kb upstream from the 35S 5′ terminus. The 35S species contains a 5′ sequence of ~704 nucleotides prior to the 18S rRNA sequence,[92,93] ~363 nucleotides between the 18S and the 5.8S RNAs[97] and ~235 nucleotides between the 5.8S and the mature 25S rRNA sequence,[91,94,95,104] and has been reported to extend 7 to 15 nucleotides beyond the 3′ terminus of 25S rRNA[94,96] (see Figure 5). It is unknown whether the 35S species is generated by transcription termination or undergoes 3′ processing.[94] Of the ~6657 nucleotides of the initial transcript, ~5348 nucleotides encode the rRNAs and ~1309 nucleotides are transcribed spacer sequences.

The 5′ and 3′ termini of the 20S precursor and 18S mature *S. carlsbergensis* rRNAs have

* sp = spacer DNA.

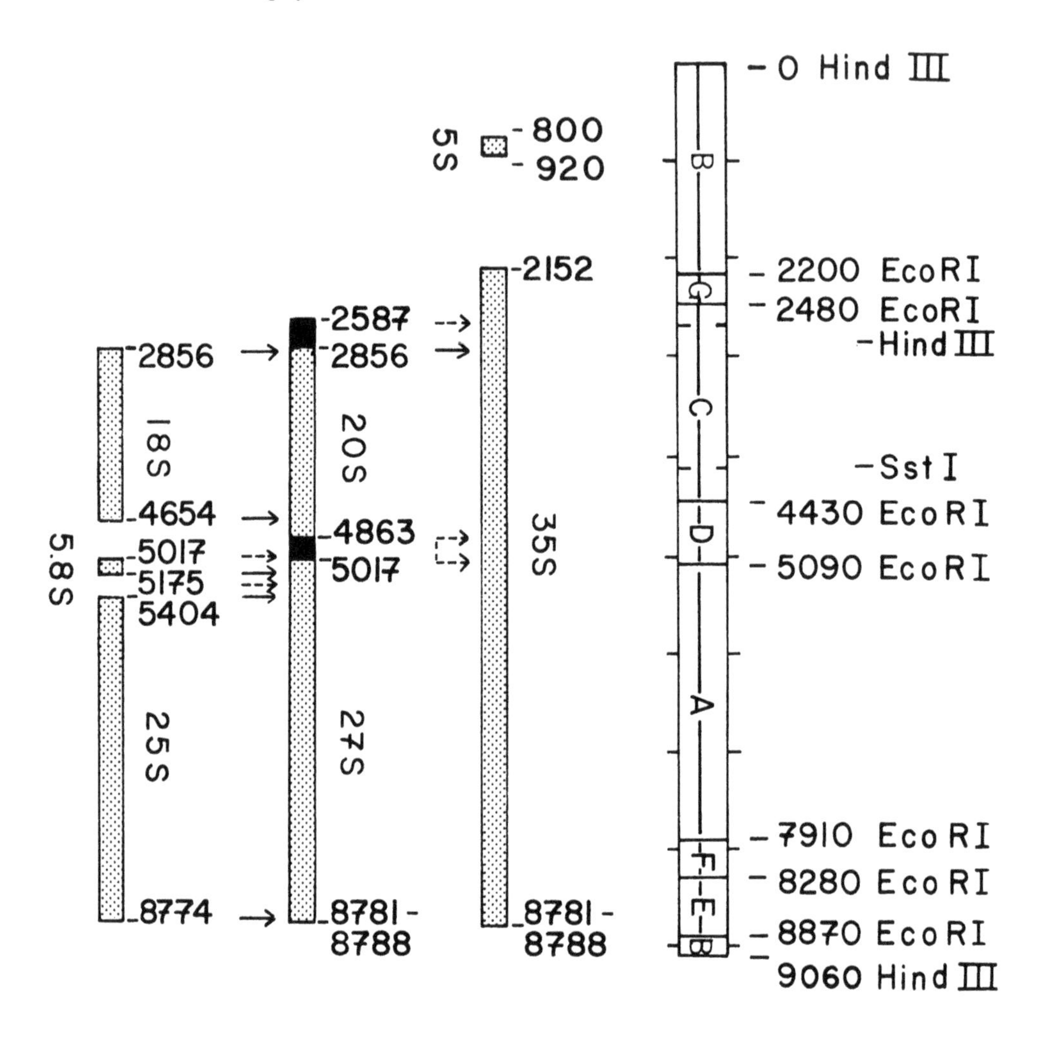


FIGURE 5. Transcripts and processing products of the yeast rRNA genes. Top row—restriction map of a rDNA repeat (after Philippsen et al.[91]) Rows 2, 3, and 4 — location of various rRNA species with reference to the restriction map. Stippled areas indicate regions which have been shown to be included in the respective RNAs; filled areas indicate regions which have not been precisely determined. Arrows: ( ↑ ) probable endonucleolytic cut sites; ( ⇡ ) uncertain cut sites. The references to the studies determining the exact location of the various RNA termini are given in the text and in Table 4.

been studied. It has been found that the 18S and 20S rRNAs possess the same hexanucleotide sequence at the 5′ end but have different sequences at the 3′ termini.[109,110] Sequences of the rDNA corresponding to the 3′ terminus of 20S rRNA were found 109 nucleotides downstream from the 3′ terminus of 18S rRNA coding region.[97] Since the differences in the oligonucleotide fingerprint pattern of the 18S and 20S rRNAs can be accommodated by this 109-nucleotide sequence, 18S rRNA is generated, most likely, from 20S precursor by a single endonucleolytic cleavage 109 nucleotides upstream from the 3′ terminus of the 20S species.[97]

The 5′ and 3′ termini of the *S. cerevisiae* 27S precursor have not yet been localized to the rDNA sequence, but it is likely that this species is generated from 35S rRNA by a single endonucleolytic cleavage either at the 5.8S 5′ end or within the 3′ 18S to 5′ 5.8S spacer region.[97] In *S. carlsbergensis* it is thought that 5.8S rRNA is produced via a 7S RNA

## Table 4
## TERMINI OF rRNA SPECIES

| Species | Location of 5′ terminus | Location of 3′ terminus |
|---|---|---|
| 5S | 800[a91] | 920[91] |
| 35S | 2152[92,93] | 8781— |
| | (2122)[b] | 8788[94,96] |
| 20S | 2856[97] | 4863[97] |
| | (2587)[c] | |
| 18S | 2856[91,94,95] | 4654[95,97] |
| 27S | 4863—5017[d] | 8781—8788[94] |
| 5.8S | 5017[91,97] | 5175 (5181)[e] |
| 25S | 5404[91,94] | 8774[91,94f] |

[a] Position numbers are in reference to the restriction map (Figure 4) with the leftmost Hind III site defined as 0 (Philippsen et al.[91]); superscript numerals refer to references in bibliography.

[b] A second possible initiation site described by Bayev et al.[93]

[c] A sequence which could encode the known 5′ nucleotides of 20S is repeated 269 nucleotides 5′ to the 18S 5′ terminus,[94] however, since the differences between 18 and 20S rRNAs can be accommodated by 3′ sequences[97] it is not likely that 2587 is the 5′ 20S terminus.

[d] The exact location of the 5′ terminus of 27S has not been determined.

[e] The major species of 5.8S is 158 nucleotides; however, a minor species 6 nucleotides longer also is found on ribosomes.[104]

[f] DNA sequencing has determined that 25S is 3392 nucleotides[103] whereas the transcript mapping studies indicate that 25S is 3370;[91] probably slight errors in the size of the restriction fragments account for the difference. Similarly, the DNA sequence analysis of 18S has shown this species to be slightly (1789 vs. 1798) different than the value determined by restriction site analyses.

intermediate[105] and 5.8S and 7S RNAs differ at the 3′ termini.[97] Hence, in *S. cerevisiae*, generation of the 5.8S and 25S mature rRNAs from 27S precursor requires a minimum of 3 nucleolytic cleavages if 27S and 5.8S share the 5′ terminus (at the 3′ end of 5.8S and 5′ and 3′* ends of 25S sequences), but could require as many as 5 cleavages if the 27S and 5.8S do not share a common 5′ terminus and if generation of 5.8S proceeds via a 7S intermediate. The various cleavage sites are given in Figures 4 and 5 and Table 4.

The sequences of DNA surrounding the termini of the various precursor and mature rRNA species have been determined.[94,96,97,97a] In contrast to *E. coli* rDNA sequences, no long, extended inverted repeats surrounding the rRNA sequences in either *S. cerevisiae* or *S. carlsbergensis*, have been found. However, short stretches of complimentary bases surrounding the processing sites have been found in *S. carlsbergensis*.[97a] Veldman et al.[97a] have proposed a model in which these short sequences are held in base pairing on the ribonucleoprotein particle and processed by an enzyme activity like RNase III which is responsible for rRNA processing in *E. coli*.[111] It has been found that some of the 3′ terminal yeast rRNA sequences are flanked by AT-rich regions. Additionally, short, conserved, direct repeats are

* It is not known whether the 3′ 25S terminus is generated by endo- or exonucleolytic activity.

found at sites around rRNA termini.[94,96,97] Although the short complementary bases and direct repeated sequences are candidates for rRNA processing enzyme recognition sites, this has not been proven. To date there have been no reports of in vivo- or in vitro-generated altered ribosomal genes which might be used to define further the processing enzyme(s) recognition sites.

Very little is known about the enzyme(s) which processes the ribosomal precursors to mature species. Tekamp[108,112] has identified and partially purified an endonucleolytic activity which matures the 3′ terminus of a 5S RNA species which is generated in in vitro transcription systems. Although several *S. cerevisiae* mutants exist which fail to process some of the rRNA precursors, none of these have been shown to be defective in RNA processing activities (see below). rRNA processing in mutants known to have RNase deficiencies has been investigated; neither the *pep4-3*[113] nor the *rnase3*[114] lesions noticeably affect rRNA processing.[115]

### B. Nucleoside Modification

In yeast the pattern and timing of nucleoside modification of ribosomal RNA is quite similar to that of other eukaryotes. Most of the 43 methylations of 25S and 26 methylations of 18S (predominantly at 2′-O ribose moieties) are introduced on to the 35S precursor rRNA.[83,116] Nearly all of these modifications occur at sites maintained in the final mature rRNAs.[83,116] At later stages during rRNA processing, base modifications are introduced on to the 27S and 20S precursor species or the 18S mature species.[117] As in other eukaryotes,[118] methylation appears to be a prerequisite for the production of at least 25S mature rRNA.[119,120]

### C. The Necessity of Ribosomal Structural Proteins for rRNA Processing

In mammalian cells, precursor rRNA processing and ribosomal protein synthesis are obligately linked. The initial precursor rRNA is assembled into a particle which contains both ribosomal structural proteins and other nuclear proteins not found on the mature ribosome. All subsequent RNA processing steps occur on pre-ribosomal-protein particles.[82,121]

Studies of *S. carlsbergensis* indicate that yeast ribosomal processing also follows a similar course.[81,105] The 35S transcript and 5S rRNA species are assembled into a 90S particle which has a protein/RNA ratio greater than mature ribosomes. From kinetic studies it would appear that the 90S particle is converted to a 66S and a 43S particle. The 66S particle contains the 27S precursor rRNA, 5S rRNA, and proteins, and the 43S particle contains the 20S pre-rRNA and proteins. Further rRNA processing and loss of proteins result in the formation of mature 60S ribosomal particles from the 66S particle and the 40S from the 43S particle. With the exception of the last step, all of the ribosomal particle formations and conversions occur in the nucleus. Therefore, the ribosomal and nonstructural proteins of the ribosomal precursor particles must be synthesized on cytoplasmic ribosome and be localized to the nucleus, where assembly occurs.

### D. Mutations Affecting Precursor rRNA Processing

Mutations of several yeast loci which affect the production of rRNA have been described. Some of these lesions affect 25S, 18S, and 5.8S rRNA synthesis whereas other lesions affect only 25S and 5.8S or 18S RNAs.

#### *1. Mutations Affecting the Production of All rRNAs*

Ten temperature-sensitive lethal mutations *(rna2-rna11)* were found to be defective in the synthesis 25S, 18S, and 5.8S rRNAs at the nonpermissive temperature.[57,122] Further studies showed that the rRNA genes are transcribed in the mutants, but that rRNA processing is defective, and unprocessed rRNA intermediates are rapidly degraded.[57,123] The primary defect in these mutants is at the level of ribosomal protein synthesis, although general protein synthesis does not seem to be defective.[123] More recent studies have shown that the *rna2* mutant fails to properly process many precursor mRNAs encoding the ribosomal proteins[124,125,125a] (see Section IV for details). Although it had been established previously that protein synthesis is required for the processing of rRNA precursors,[79,82] these studies show that the synthesis of ribosomal proteins, in particular, is required for rRNA processing. These results support the notion that not only does ribosomal RNA processing occur upon the preribosomal protein particles, but that such particle formation is essential for proper processing.

The *rna1* mutant is also defective in rRNA production. At the nonpermissive temperature there is accumulation of a large, presumably 35S pre-rRNA species and little or no mature rRNA can be detected.[25,57] The molecular basis of the defect in rRNA processing is not understood but may be an indirect consequence of the deficiency of mRNA encoding ribosomal proteins. Other cold-sensitive and slow-growing yeast mutants which are defective in the production of rRNAs have been reported;[126,127] it is not known whether these are allelic to the *rna1-rna11* lesions.

*2. Mutations Affecting the Production of 25S and 5.8S rRNAs*

Andrew et al.[128] searched for mutants defective in RNA processing by selecting strains which were conditionally lethal, but which were able to grow at least 1 to 2 doublings at the nonpermissive temperature and showed defects in incorporation of radioactive precursor into protein as well as RNA. The rationale of this approach was that cells defective in the processing of particular stable RNA species would not show a preferential shut-off of RNA accumulation relative to protein accumulation and would continue to divide until the concentration of the stable RNA became limiting. One mutant, ts 351 (which defines the locus *rrp1* — "*r*ibosomal *R*NA *p*rocessing"[129]) has been shown to be defective in the accumulation of 25S and 5.8S rRNAs, but was able to synthesize 18S rRNA. At the nonpermissive temperature this strain fails to process 27S rRNA to 25S and 5.8S mature RNAs. The precursor 27S species does not accumulate and is rapidly degraded.[128] Unlike *rna2-rna11*, ribosomal proteins are synthesized at near normal rates in the *rrp1* mutant; however, these proteins fail to accumulate and are rapidly degraded.[130] Other *S. cerevisiae*[127] and *Neurospora crassa*[131] mutants with similar phenotypes have been reported. The molecular nature of the *rrp1* and similar lesions is unknown. The mutants could be defective in an endonucleolytic activity which is necessary solely for the processing of 27S pre-rRNA or in ribosomal and/or nuclear proteins which are components of the precursor ribosomal protein particle.

*3. Mutations Affecting the Production of 18S rRNA*

Yeast mutants resistant to trichodermin, an antibiotic which affects the peptidyltransferase center of the 60S ribosomal subunit, have lesions in the genes encoding the L3 protein of the 60S subunit.[132-134] A trichodermin-resistant yeast strain was found which also demonstrated a defect in the production of 40S subunits.[135] The lesions responsible for trichodermin resistance and 40S ribosomal subunit deficiency segregate independently, but the 40S deficiency lesion is only expressed in progeny which also are trichodermin resistant.[135] Thus the 40S defect is dependent upon an altered protein of the 60S ribosomal subunit. Further studies of the double mutant showed that the deficiency in 40S ribosomal subunits is a result of slowed transport of the 20S pre-rRNA from the nucleus to the cytoplasm and a slow conversion of the 20S species to 18S rRNA.[136] Unlike the *rrp1* mutant, the double mutant demonstrates a slow, but not conditionally lethal, growth and the 20S pre-RNA is not rapidly degraded.[136]

The molecular basis of the transport/processing defect in the mutant strain is not yet understood. It is not likely that this strain possesses a defective endonuclease uniquely responsible for conversion of 20S precursor to 18S rRNA, since this processing step occurs independently of the 60S subunit, and it therefore would be difficult to explain the dependency of the endonuclease-defective phenotype on the trichodermin resistance mutation. The strain could be defective in a protein component of the precursor and/or mature ribosome particle.[136] This possibility can accommodate the trichodermin resistance dependency since at the 90S pre-particles stage the particle contains both 40S and 60S proteins; it can be imagined that an altered 60S protein could affect such assembly. Two-dimensional gel electrophoretic analysis failed to identify any aberrant proteins from the 40S mature ribosomal subunit of the mutant.[136]

### E. Summary

Identification and in vitro manipulation of sites surrounding the termini of the various precursor rRNA species should define the nature of the substrate(s) of the rRNA processing enzyme(s). To date little is known about the enzymatic machinery responsible for the processing events and although several yeast mutants with rRNA processing defects exist none of these have been shown to be altered in a nuclease activity. The notion that some of these mutants will be found to be defective in preribosomal particle or proteins comprising the nuclear architecture seems appealing. If true, such mutants may help reveal the complexities of the role of RNPs in rRNA processing and transport to the cytoplasm.

## IV. PROCESSING OF YEAST MESSENGER RNA PRECURSORS

### A. Generation of Mature 5′ Termini

As in all eukaryotes,[137] yeast messenger RNAs are capped at the 5′ terminus. Unlike higher eukaryotes, the cap structure of yeast mRNAs possesses only one methyl group, 7-methylguanine, at the 5′ terminus (Cap 0 structure) and lacks further methylation at the 1st and 2nd internal nucleotides (Cap 1 and Cap 2 structures).[138,139] It is not known whether the 5′ termini of yeast mRNAs are generated by nucleolytic processing of larger precursors or whether, as in other eukaryotes, capping occurs at the 5′ terminus of the initial transcript.[137,140]

### B. Generation of Mature 3′ Termini

The 3′ termini of yeast mRNAs, including even histone mRNA,[141] possess poly(A) sequences. These sequences are somewhat shorter, about 50 residues on average, than the poly(A) tracts (~200 residues) of higher eukaryotes.[142,143] Both the yeast nuclear and cytoplasmic compartments contain enzymatic activities which catalyze addition of A residues onto the 3′ terminus of mRNA.[144] Although there has been a report of encoded poly(T) tracts in yeast DNA,[145] those mRNA genes which have been cloned and sequenced do not encode the poly(A) tract.[140,146-155]

The 3′ termini of several yeast mRNAs have been mapped to their respective gene sequences. Of these, the genes encoding actin[149] and *MATa1*[150] contain an AATAAA putative polyadenylation recognition site[156,157] in the mRNA trailer region. Additionally, AATAAA sequences have been found beyond the protein coding region for two yeast glyceraldehyde-3-phosphate dehydrogenases and two enolase genes[151-153] although the 3′ termini of the mRNA for these genes have not been mapped. On the other hand, the putative polyadenylation site is absent from the yeast genes encoding iso-1-cytochrome *c*,[146-148] *matα1*,[150]

*MATα2,*[150] and *ADHI.*[140] Other 3′ consensus sequences for the yeast mRNA 3′ regions have been found.[140,148] Hence it seems that the poly(A) polymerases of yeast may recognize some sequence other than AAUAAA.

It is not yet known if the 3′ termini are generated by nucleolytic processing or by transcription termination. Studies of the *CYC1* gene suggest that the 3′ termini of some yeast genes may be generated by transcription termination. The gene encoding iso-1-cytochrome *c (CYC1)* has been found to encode a ~630-nucleotide transcript which extends at the 3′ end ~180 nucleotides beyond the TAA protein termination signal.[147] A 38-bp deletion extending from 130 to 167 bp beyond the TAA codon results in the production of several additional transcripts (850, 1350, 1450, 1650, and 2000 nucleotides) which are elongated at the 3′ terminus. Since none of the larger transcripts have ever been detected in wild-type cells, it has been concluded that the 3′ end of *CYC1* mRNA is generated by transcription termination and that the termination signal or a component of this signal is located in the 130 to 167 bp region.[148] These results further suggest that there is a tight coupling of transcription termination and polyadenylation since each of the aberrant transcripts bears a poly(A) sequence.[148] Alternatively, one could postulate that the *cyc1* mRNA initial transcript contains extra 3′ sequences which are very rapidly degraded and that the *cyc1* deletion mutation disrupts sequences necessary for recognition of a 3′ processing enzyme.

Studies of transcripts produced by the *rna1* mutant also provide evidence for RNA transcripts with elongated 3′ termini. The *rna1* mutant affects mRNA production[55,56] as well as tRNA and rRNA production. Two-dimensional gel electrophoretic analysis of the proteins synthesized at the nonpermissive temperature by *rna1* cells showed that most mRNAs are affected by this lesion.[158,159] It has also been reported that poly(A) RNA which accumulates in the nucleus of *rna1* cells is larger than cytoplasmic poly(A) RNA.[56] Recent experiments studying the appearance of transcripts homologous to cloned yeast sequences confirm this result. For example, when wild-type cells are induced for the Leloir pathway, three transcripts (1.65 kb, 1.75 kb, and 4.5 kb) homologous to the *GAL1* gene are produced.[160] RNA from *rna1* cells at the nonpermissive temperature show, in addition, two novel (5.5 kb and 6.5 kb) transcripts which have the same 5′ terminus as the other *GAL1* transcripts but are extended at the 3′ terminus.[160] It is not known if these novel transcripts are normal products which are stabilized in *rna1* cells at the nonpermissive temperature or whether these result from aberrant transcription termination at the nonpermissive conditions. The latter interpretation may be supported by the fact that deletions of the *GAL1* gene 3′ to the protein coding sequence but including regions covered by the *rna1* novel transcripts do not affect the ability of the altered gene to complement *gal1* mutations.[160] The *rna1* lesion does not affect the *HIS3* transcripts[161] or the transcripts of two ribosomal protein genes which are affected by the *rna2-rna11* lesions.[124] There are no reports concerning the effect of the *rna1* lesion upon the *CYC1* transcript.

### C. Removal of Intervening Sequences

Until recently it was thought that yeast mRNA genes did not contain intervening sequences. Indeed, studies of the yeast genes which were first cloned and characterized provided no evidence for intervening sequences (e.g., *HIS3,*[161] *CYC1,*[146,147,162] *CYC7,*[154] glyceraldehyde-3-phosphate dehydrogenase,[151,152] enolase,[153] *TRP1,*[155] histones H2B1 and H2B2,[163] and alcohol dehydrogenase I[140]). However, it was later shown that the yeast actin gene does contain a 304 bp intervening sequence at the triplet encoding amino acid 4.[164,165] Since there is only one copy of the actin gene in the *S. cerevisiae* genome,[169] precursor actin mRNA must contain an intervening sequence. This intervening sequence contains the conserved GT, AG nucleotides at the splice junction.[164,165] Gallwitz[164a] studied the RNA processing of transcripts from actin genes altered in vitro and transcribed in vivo by transformed yeast cells. These studies showed that significant parts of the intron can be removed without affecting splicing; however, removal of the conserved T residue of the 5′ splice junction caused accumulation of unspliced actin precursor RNA. Thus at least one nucleotide of the splice junction must be recognized by the splicing machinery.

Studies of the *rna2* yeast mutant have provided further evidence for intervening sequences in other precursor mRNAs. The *rna2* mutation, as well as the *rna3-rna11* mutations, affects the production of rRNA by coordinately affecting the production of mRNAs encoding most of the ribosomal proteins.[57,123] However, as assayed by pulse-labeling of proteins or in vitro translation, these lesions do not affect the production of most nonribosomal protein mRNAs.

Rosbash et al.[124] showed that RNA isolated from *rna2* cells at the nonpermissive temperature contains less RNA homologous to specific probes encoding ribosomal proteins than does RNA isolated from cells grown at permissive temperature, thereby confirming earlier results.[166,167] RNA from cells grown at permissive conditions contains a major, presumably mature mRNA species, and barely detectable larger species homologous to the probes. RNA from cells grown at the nonpermissive conditions has little mature mRNA but contains disproportionate amounts of the larger species. Further S1 mapping studies showed that for the ribosomal protein-51 gene the larger RNA species contains an intervening sequence, several hundred nucleotides in length, which is lacking in the lower molecular weight species; in other words, the *rna2* mutation results in an increased proportion, at the nonpermissive temperature, of a precursor RNA which contains an intervening sequence.[124] These studies have been extended to other cloned ribosomal protein genes. Fried et al.[125] have found that for each of 8 of 11 additional ribosomal protein genes a larger, presumably intervening sequence containing RNA accumulates in the mutant at nonpermissive conditions. It has also been found that the *rna2* lesion inhibits the processing of actin mRNA.[168] Where investigated, the effect of the other, *rna3-rna11*, mutations upon removal of intervening sequences from precursor mRNAs is similar or identical to the effect of the *rna2* lesion.[124,169,171]

The precursor RNAs which accumulate in *rna2* strains may serve as suitable substrates to aid isolation of the mRNA cleaving-ligating activities and to identify and quantitate the number of yeast genes which possess intervening sequences. As for all the other yeast lesions which affect RNA processing in yeast, the molecular basis of the defect in the *rna2-rna11* strains is not understood; however, it seems most unlikely that all of these loci could encode mRNA splicing activity. Recently a dominant mutation, *SRN1*, was isolated which is capable of suppressing the temperature-sensitive phenotype of many of the *rna* mutants including *rna2-rna6, rna8,* and *rna1*.[170] Further studies of *SRN1* may lead to a better understanding of the *RNA* genes.

### D. Summary

It is clear that generation of mature yeast mRNAs requires 5′ capping enzymes, 3′ poly(A) polymerases, and mRNA splicing enzymes. It is not yet clear if, in addition, nucleases capable of processing 5′ and/or 3′ termini of initial transcripts are required. Further studies of the mutants which affect mRNA splicing and 3′ termination/processing should help the efforts to resolve the steps involved in mRNA synthesis and to isolate and characterize the enzyme activities essential for the production of mature yeast mRNAs.

## V. EPILOGUE

Within the last five years considerable advances have been made in describing the steps of RNA processing in yeast. It is clear, however, that much remains to be learned. The enzymology of processing of yeast RNAs (and, for that matter, all eukaryotic RNAs) is in a very early stage as only the nucleotidyltransferase,[172] RNase P-like, and tRNA splicing activities are well characterized. Additionally, while there are several interesting yeast mutants which are defective in a variety of steps of tRNA, rRNA, and mRNA precursor processing which have proved most useful for identifying RNA precursors and purifying processing enzymes, the molecular nature of the defects in these mutants has not been elucidated. Finally the synthesis of mature RNAs requires well-controlled cytoplasmic and nuclear interactions since all the processing steps take place in the nucleus but all of the processing enzymes and ribonucleoproteins are first synthesized in the cytoplasm. Available data indicate that compartmentalization of precursor RNAs and processing enzymes to the nucleus and mature RNAs to the cytoplasm is efficient. However, the mechanism(s) for shuttling of proteins and RNA molecules across compartmental barriers remains obscure. One hopes that mutants will again be useful to unravel this fascinating and important area of molecular biology.

## ACKNOWLEDGMENTS

I thank E. DeRobertis, C. Greer, B. D. Hall, C. Peebles, M. Rosbash, D. Standring, D. Soll, P. Tekamp, and A. Weil for providing unpublished data, and G. Fabian, D. Hurt, S. Johnston, J. Hopper, and C. Peebles for comments on drafts of this review.

## REFERENCES

1. **Hinnen, A., Hicks, J. B., and Fink, G. R.,** Transformation of yeast, *Proc. Natl. Acad. Sci. U.S.A.*, 75, 1929, 1978.
2. **Abelson, J.,** RNA processing and the intervening sequence problem, *Annu. Rev. Biochem.*, 48, 1035, 1979.
3. **Guthrie, C. and Abelson, J.,** Organization and expression of tRNA genes in *S. cerevisiae*, in *The Molecular Biology of the Yeast Saccharomyces*, Strathern, J. N., Jones, E. W., and Broach, J. R., Eds., Cold Spring Harbor Laboratory, Cold Spring Harbor, N.Y., 1982, 487.
4. **Warner, J. R.,** Yeast ribosome: structure, function, and synthesis, in *The Molecular Biology of the Yeast Saccharomyces*, Strathern, J. N., Jones, E. W., and Broach, J. R., Eds., Cold Spring Harbor Laboratory, Cold Spring Harbor, N.Y., 1982, 529.
5. **Sprinzl, M. and Gauss, D. H.,** Compilation of tRNA sequences, *Nucleic Acids Res.*, 10, r1, 1982.
6. **Rich, A. and RajBhandary, U. L.,** Transfer RNA: molecular structure, sequence, and properties, *Annu. Rev. Biochem.*, 45, 805, 1976.
7. **Beckmann, J. S., Johnson, P. F., and Abelson, J.,** Cloning of the yeast transfer RNA genes in *Escherichia coli*, *Science*, 196, 205, 1977.
8. **Goodman, H. M., Olson, M. V., and Hall, B. D.,** Nucleotide sequences of a mutant eukaryotic gene; the yeast tyrosine inserting suppressor, *SUP4-0*, *Proc. Natl. Acad. Sci. U.S.A.*, 74, 5453, 1977.
9. **Valenzuela, P., Venegas, A., Weinberg, F., Bishop, R., and Rutter, W. J.,** Structure of yeast phenylalanine-tRNA genes: an intervening DNA segment within the region coding for the tRNA, *Proc. Natl. Acad. Sci. U.S.A.*, 75, 190, 1978.
10. **Vengas, A., Quiroga, M., Zaldivar, J., Rutter, W. J., and Valenzuela, P.,** Isolation of yeast $tRNA_3^{Leu}$ genes: DNA sequence of a cloned $tRNA^{Leu}$ gene, *J. Biol. Chem.*, 254, 12306, 1979.
11. **Mao, J., Schmidt, O., and Söll, D.,** Dimeric transfer RNA precursors in *S. pombe*, *Cell*, 21, 509, 1980.
12. **Schmidt, O., Mao, J., Ogden, R., Beckmann, J., Sakano, H., Abelson, J., and Söll, D.,** Dimeric tRNA precursors in yeast, *Nature (London)*, 287, 750, 1980.
13. **Olah, J. and Feldmann, H.,** Structure of a yeast non-initiating methionine-tRNA gene, *Nucleic Acids Res.*, 8, 1975, 1980.
14. **Page, G. S. and Hall, B. D.,** Characterization of the yeast $tRNA^{Ser}$ gene family: genomic organization and DNA sequence, *Nucleic Acids Res.*, 9, 921, 1981.

15. **Eigel, A., Olah, J., and Feldmann, H.,** Structural comparison of two yeast $tRNA^{Glu}_3$ genes, *Nucleic Acids Res.*, 9, 2961, 1981.
16. **Olson, M. V., Page, G. S., Sentenac, A., Piper, P. W., Worthington, M., Weiss, R., and Hall, B. D.,** Only one of two closely related yeast suppressor tRNA genes contains an intervening sequence, *Nature (London)*, 291, 464, 1981.
17. **Broach, J.. R., Friedman, L., and Sherman, F.,** Correspondence of yeast UAA suppressors to cloned $tRNA^{Ser}_{UCA}$ genes, *J. Mol. Biol.*, 150, 375, 1981.
18. **Cummins, C. M. and Culbertson, M. K.,** Molecular cloning of the *SUF2* frameshift suppressor gene from *Saccharomyces cerevisiae, Gene*, 14, 263, 1981.
19. **Venegas, A., Gonzalez, E., Bull, P., and Valenzuela, P.,** Isolation and structure of a yeast initiation $tRNA^{Met}$ gene, *Nucleic Acids Res.*, 10, 1093, 1982.
20. **Sprinzl, M. and Gauss, D. H.,** Compilation of sequences of tRNA genes, *Nucleic Acids Res.*, 10, r57, 1982.
21. **Stewart, J. W. and Sherman, F.,** Demonstration of UAG as a nonsense codon in bakers yeast by amino acid replacements in iso-1-cytochrome c, *J. Mol. Biol.*, 68, 429, 1972.
22. **Hawthorne D. C. and Leupold, U.,** Suppressors in yeast, *Curr. Topics Microbiol. Immunol.*, 64, 1, 1974.
23. **Waldron, C., Cox, B., Willis, N., Gesteland, R., Piper, P., Colby, D., and Guthrie, C.,** Yeast ochre suppressor *SUQ5-o1* is an altered $tRNA^{Ser}$ gene, *Nucleic Acids Res.*, 9, 3077, 1981.
24. **Phillips, J. H. and Kjellin-Straby, K.,** Studies on microbial ribonucleic acid. IV. Two mutants of *Saccharomyces cerevisiae* lacking $N^2$-dimethylguanine in soluble ribonucleic acid, *J. Mol. Biol.*, 26, 509, 1967.
25. **Hopper, A. K., Banks, F., and Evangelidis, V.,** Yeast mutant which accumulates precursor tRNAs, *Cell*, 14, 211, 1978.
26. **Laten, H., Gorman, J., and Bock, R. M.,** Isopentenyladenosine deficient tRNA from an anti-suppressor mutant of *Saccharomyces cerevisiae, Nucleic Acids Res.*, 5, 4329, 1978.
27. **Hopper, A. K., Schultz, L. D., and Shapiro, R. A.,** Processing of intervening sequences: a new yeast mutant which fails to excise intervening sequences from precursor tRNAs, *Cell*, 19, 741, 1980.
28. **Kurjan, J., Hall, B. D., Gillam, S., and Smith, M.,** Mutations at the *SUP4* $tRNA^{Tyr}$ locus: DNA sequence changes in mutants lacking suppressor activity, *Cell*, 20, 701, 1980.
29. **Janner, F., Vogeli, G., and Fluri, R.,** The antisuppressor strain of *Schizosaccharomyces pombe* lacks the modification isopentyladenosine in transfer RNA, *J. Mol. Biol.*, 139, 207, 1980.
30. **Colby, D., Leboy, P., and Guthrie, C.,** Yeast tRNA precursor mutated at a splice junction is correctly processed *in vivo, Proc. Natl. Acad. Sci.*, 78, 415, 1981.
31. **Lo, R. Y. C., Bell, J. B., and Roy, K. L.,** Dihydrouridine deficient tRNAs in *Saccharomyces cerevisiae, Nucleic Acids Res.*, 10, 889, 1982.
32. **Hopper, A. K., Furukawa, A. H., Pham, H. D., and Martin, N. C.,** Defects in modification of cytoplasmic and mitochondrial transfer RNAs are caused by single nuclear mutations, *Cell*, 28, 543, 1982.
33. **Kohli, J.,** personal communication, 1982.
34. **Blatt, B. and Feldmann, H.,** Characterization of precursors to tRNA in yeast, *FEBS Lett.*, 37, 129, 1973.
35. **Knapp, G., Beckmann, J. S., Johnson, P. F., Fuhrman, S. A., and Abelson, J.,** Transcription and processing of intervening sequences in yeast tRNA genes, *Cell*, 14, 221, 1978.
36. **O'Farrell, P. Z., Cordell, B., Valenzuela, P., Rutter, W. J., and Goodman, H. M.,** Structure and processing of yeast precursor tRNAs containing intervening sequences, *Nature (London)*, 274, 438, 1978.
37. **Etcheverry, T., Colby, D., and Guthrie, C.,** A precursor to a minor species of $tRNA^{Ser}$ contains an intervening sequence, *Cell*, 18, 11, 1979.
38. **Knapp, G., Ogden, R. C., Peebles, C. L., and Abelson, J.,** Splicing of yeast tRNA precursors: structure of the reaction intermediates, *Cell*, 18, 37, 1979.
39. **Hopper, A. K. and Kurjan, J.,** tRNA synthesis: identification of *in vivo* precursor tRNAs from parental and mutant yeast strains, *Nucleic Acids Res.*, 9, 1019, 1981.
40. **DeRobertis, E. M. and Olson, M. V.,** Transcription and processing of yeast tyrosine tRNA genes microinjected into frog oocytes, *Nature (London)*, 278, 137, 1979.
41. **Ogden, R. C., Beckman, J. S., Abelson, J., Kang, H. S., Soll, D., and Schmidt, O.,** *In vitro* transcription and processing of a yeast tRNA gene containing an intervening sequence, *Cell*, 17, 399, 1979.
42. **Koski, R. A., Clarkson, S. G., Kurjan, J., Hall, B. D., and Smith, M.,** Mutations of the yeast *SUP4* $tRNA^{Tyr}$ locus: transcription of the mutant genes *in vitro, Cell*, 22, 415, 1980.
43. **Standring, D. N., Venegas, A., and Rutter, W. J.,** Yeast $tRNA^{Leu}_3$ gene transcribed and spliced in a *HeLa* cell extract, *Proc. Natl. Acad. Sci.*, 78, 5963, 1981.
44. **Klekamp, M. S. and Weil, P. A.,** Specific transcription of class III genes in soluble yeast cell-free extracts, *J. Biol. Chem.*, 257, 8432, 1982.

45. **Koski, R. A., Worthington, M., Allison, D. S., and Hall, B. D.,** An *in vitro* RNA polymerase III system from *S. cerevisiae:* Effects of deletions and point mutations upon *SUP4* gene transcription, *Nucleic Acids Res.*, 10, 8127, 1983.
46. **Schedl, P. and Primakoff, P.,** Mutations of *Escherichia coli* thermosensitive for the synthesis of transfer RNA, *Proc. Natl. Acad. Sci. U.S.A.*, 70, 2091, 1973.
47. **Sakano, H., Yamada, S., Ikemura, T., Shimura, Y. and Ozeki, H.,** Temperature sensitive mutants of *Escherichia coli* for tRNA synthesis, *Nucleic Acids Res.*, 1, 355, 1974.
48. **Altman, S.,** Transfer RNA biosynthesis, *Int. Rev. Biochem.*, 17, 19, 1978.
49. **McCready, S. J. and Cox, B. S.,** Anti-suppressors in yeast, *olec. Gen. Genet.*, 124, 305, 1973.
50. **Rasse-Messenguy, F. and Fink, G. R.,** Temperature-sensitive nonsense suppressors in yeast, *Genetics*, 75, 459, 1973.
51. **Hopper, A. K., Nolan, S. L., Kurjan, J., and Hama-Furukawa, A.,** Genetic and biochemical approaches to studying *in vivo* intermediates in tRNA biosynthesis, in *Molecular Genetics in Yeast*, von Wettstein, D., Friis, J., Kielland-Brandt, M., and Stenderup, A., Eds., Munksgaard, Copenhagen, 1981, 302.
52. **Kurjan, J., Hall, B. D., Koski, R. A., Clarkson, S. G., and Smith, M.,** Mutations at the *SUP4* $tRNA^{Tyr}$ locus: DNA sequence changes which alter transcription,, in *Molecular Genetics in Yeast*, von Wettstein, D., Friis, J., Kielland-Brandt, M., Stenderup, A., Eds., Munksgaard, Copenhagen, 1981, 245.
53. **Nishikura, K. and DeRobertis, E. M.,** Processing of yeast $tRNA^{Tyr}$ in *Xenopus oocytes* microinjected with cloned genes, in *Developmental Biology Using Purified Genes*, Vol. 23, ICN-UCLA Symp. Molec. Cell. Biol., Brown, D. D. and Fox, C. F., Eds., Academic Press, N.Y., 1981, 483.
53a. **Nishikura, K., Kurjan, J., Hall, B. D., and DeRobertis, E. M.,** Genetic analysis of the processing of a spliced tRNA, *EMBO J.*, 1, 263, 1982.
54. **Schultz, L. S. and Hopper, A. K.,** unpublished results, 1980.
55. **Hutchinson, H. T., Hartwell, L. H., and McLaughlin, C. S.,** Temperature sensitive yeast mutant defective in ribonucleic acid production, *J. Bacteriol.*, 99, 807, 1969.
56. **Shiokawa, K. and Pogo, A. O.,** The role of cytoplasmic membranes in controlling the transport of nuclear messenger RNA and initiation of protein synthesis, *Proc. Natl. Acad. Sci. U.S.A.*, 71, 2658, 1974.
57. **Warner, J. R. and Udem, S. A.,** Temperature sensitive mutations affeccting ribosome synthesis in *Saccharomyces cerevisiae*, *J. Mol. Biol.*, 65, 243, 1972.
58. **Martin, R., Schneller, J. M., Stahl, A. J. C., and Dirheimer, G.,** Studies of odd bases in yeast mitochondrial tRNA. II. Characterization of rare nucleosides, *Biochim. Biophys. Res. Commun.*, 70, 997, 1976.
59. **Hawthorne, D. C. and Mortimer, R. K.,** Genetic mapping of nonsense suppressors in yeast, *Genetics*, 60, 735, 1968.
60. **Mortimer, R. K. and Schild, D.,** The genetic map of *Saccharomyces cerevisiae*, *Microbiol. Rev.*, 44, 519, 1980.
61. **Olson, M. V., Hall, B. D., Cameron, J. R., and Davis, R. W.,** Cloning of the yeast tyrosine transfer RNA genes in bacteriophage Lambda, *J. Mol. Biol.*, 127, 285, 1979.
61a. **Cooley, L., Appel, B., and Soll, D.,** Post-transcriptional nucleotide addition is responsible for the formation of the 5′ terminus of histidine tRNA, *Proc. Natl. Acad. Sci.*, 79, 6479, 1982.
62. **Colby, D. and Guthrie, C.,** personal communication, 1982.
63. **Kole, R., Baer, M. F., Stark, B. C., and Altman, S.,** *E. coli* RNAase P has a required RNA component *in vivo*, *Cell*, 19, 881, 1980.
64. **Kline, L., Nishikawa, S., and Soll, D.,** Partial purification of RNase P from *Schizosaccharomyces pombe*, *J. Biol. Chem.*, 256, 5058, 1981.
65. **Soll, D.,** personal communication, 1981.
66. **Tekamp, P.,** personal communication, 1981.
67. **Peebles, C. L., Ogden, R. C., Knapp, G., and Abelson, J.,** Splicing of yeast tRNA precursors: a two-stage reaction, *Cell*, 18, 27, 1979.
68. **Peebles, C. L., Gegenheimer, P., and Abelson, J.,** Precise excision of intervening sequences from precursor tRNAs by a membrane-associated yeast endonuclease, *Cell*, 32, 525, 1983.
68a. **Konarska, M., Filipowicz, W., Domdey, H., and Gross, H. J.,** Formation of a 2′-phosphomonester, 3′5′-phosphodiester linkage by a novel RNA ligase in wheat germ, *Nature (London)*, 293, 112, 1981.
69. **Greer, C. L., Peebles, C. L., Gegenheimer, P., and Abelson, J.,** Mechanism of action of a yeast RNA ligase in tRNA splicing, *Cell*, 32, 537, 1983.
69a. **Johnson, J. D., Ogden, R., Johnson, P., Abelson, J., Dembeck, P., and Itakura, K.,** Transcription and processing of a yeast tRNA gene containing a modified intervening sequence, *Proc. Natl. Acad. Sci. U.S.A.*, 77, 2564, 1980.
70. **Nau, F.,** The methylation of tRNA, *Biochimie*, 58, 629, 1976.
71. **Lo, R. Y. C.,** A Mutation in *Saccharomyces cerevisiae* Defective in the Production of Dihydrouridine in tRNA, Ph.D. thesis, Department of Genetics, University of Alberta, Edmonton, 1981.

72. **Hopper, A. K. and Furukawa, A. H.,** unpublished data, 1981.
73. **Laten, H., Gorman, J., and Bock, R. M.,** $i^6A$-deficient tRNA from an antisuppressor mutant of *Saccharomyces cerevisiae*, in *Transfer RNA: Biological Aspects*, Soll, D., Abelson, J. N., and Schimmel, P. R., Eds., Cold Spring Harbor Laboratory, Cold Spring Harbor, N.Y., 1980, 395.
74. **Nishikura, K. and DeRobertis, E. M.,** RNA processing in microinjected *Xenopus* oocytes: sequential addition of base modifications in a spliced transfer RNA, *J. Mol. Biol.*, 145, 405, 1981.
74a. **Johnson, P. F. and Abelson, J.,** The yeast $tRNA^{Tyr}$ gene intron is essential for correct modification of its tRNA product, *Nature*, 302, 681, 1983.
75. **Melton, D. A., DeRobertis, E. M., and Cortese, R.,** Order and intracellular location of the events involved in the maturation of a spliced tRNA, *Nature (London)*, 284, 143, 1980.
76. **DeRobertis, E. M., Black, P., and Nishikura, K.,** Intranuclear location of the tRNA splicing enzymes, *Cell*, 23, 89, 1981.
77. **Martin, N. C. and Hopper, A. K.,** Isopentenylation of both cytoplasmic and mitochondrial tRNA is affected by a single nuclear mutation, *J. Biol. Chem.*, 257, 10562, 1982.
78. **Hendrick, J. P., and Wolin, S. L., Rinker, J., Lerner, M. R., and Steitz, J. A.,** Ro small cytoplasmic ribonucleoproteins are a subclass of La ribonucleoproteins: further characterization of the Ro and La small ribonucleoproteins from uninfected mammalian cells, *Molec. Cell. Biol.*, 1, 1138, 1981.
79. **Udem, S. A. and Warner, J. R.,** Ribosomal RNA synthesis in *Saccharomyces cerevisiae, J. Mol. Biol.*, 65, 227, 1972.
80. **Trapman, J. and Planta, R. J.,** Maturation of ribosomes in yeast. I. Kinetic analysis by labelling of high molecular weight rRNA species, *Biochim. Biophys. Acta*, 442, 265, 1976.
81. **Trapman, J., Retel, J., and Planta, R. J.,** Ribosomal precursor particles from yeast, *Exptl. Cell Res.*, 90, 95, 1975.
82. **Hadjiolov, A. A. and Nikolaev, N.,** Maturation of ribonucleic acids and the biogenesis of ribosomes, *Prog. Biophys. Molec. Biol.*, 31, 95, 1976.
83. **Planta, R. J., Retel, J., Klootwijk, J., Meyerink, J. H., deJonge, P., Van Keulen, H., and Brand, R. C.,** Synthesis and processing of ribosomal ribonucleic acid in eukaryotes, *Biochem. Soc. Trans.*, 5, 462, 1977.
84. **Warner, J. R., Tushinski, R. J., and Wejksnora, P. J.,** Coordination of RNA and proteins in eukaryotic ribosome production, in *Ribosomes: Structure, Function, and Genetics*, Chambliss, G., Craven, G. R., Davies, J., Davis, K., Kahan, L., and Nomura, M., Eds., University Park Press, Baltimore, 1979, 889.
85. **Kramer, R. A., Cameron, J. R., and Davis, R. W.,** Isolation of bacteriophage containing yeast ribosomal RNA genes: screening by *in situ* RNA hybridization to plaques, *Cell*, 8, 227, 1976.
86. **Cramer, J. H., Farrelly, F. W., Barnitz, J. T., and Round, R. H.,** Construction and restriction mapping of hybrid plasmids containing *Saccharomyces cerevisiae* ribosomal DNA, *Molec. Gen. Genet.*, 151, 229, 1977.
87. **Nath, K. and Bollon, A. P.,** Organization of the yeast ribosomal RNA gene cluster via cloning and restriction analysis, *J. Biol. Chem.*, 252, 6562, 1977.
88. **Bell, G. I., DeGennaro, L. J., Gelfand, D. H., Bishop, R. J., Valenzuela, P., and Rutter, W. J.,** Ribosomal RNA genes of *Saccharomyces cerevisiae*. I. Physical map of the repeating unit and location of the regions coding for 5S, 5.8S, and 25S ribosomal RNAs, *J. Biol. Chem.*, 252, 8118, 1977.
89. **Valenzuela, P., Bell, G. I., Venegas, A., Sewell, E. T., Masiarz, F. R., De Gennaro, L., Weinberg, F., and Rutter, W. J.,** Ribosomal RNA genes of *Saccharomyces cerevisiae*. II. Physical map and nucleotide sequence of the 5S ribosomal RNA gene and adjacent intergenic regions, *J. Biol. Chem.*, 252, 8126, 1977.
90. **Planta, R. J. and Meyerink, J. H.,** Organization of the ribosomal RNA genes in eukaryotes, in *Ribosomes: Structure, Function, and Genetics*, Chambliss, G., Craven, G.. R., Davies, J., Davis, K., Kahan, L. and Nomura, M., Eds., University Park Press, Baltimore, 1979, 871.
91. **Philippsen, P., Thomas, M., Kramer, R. A., and Davis, R. W.,** Unique arrangement of coding sequences for 5S, 5.8S, 18S and 25S ribosomal RNA in *Saccharomyces cerevisiae* as determined by R-loop and hybridization analysis, *J. Mol. Biol.*, 123, 387, 1978.
91a. **Klootwijk, J., de Jonge, P., and Planta, R. J.,** The primary transcript of the ribosomal repeating unit in yeast, *Nucleic Acids Res.*, 6, 27, 1979.
91b. **Nikolaev, Georgiev, O. I., Venkov, P. V., and Hadjiolov, A. A.,** The 37S precursor to ribosomal RNA is the primary transcript of ribosomal genes in *Saccharomyces cerevisiae, J. Mol. Biol.*, 127, 297, 1979.
91c. **Swanson, M. E. and Holland, M. J.,** RNA polymerase I-dependent selective transcription of yeast ribosomal DNA, *J. Biol. Chem.*, 258, 3252, 1982.
92. **Klemenz, R. and Geiduschek, E. P.,** The 5′ terminus of the precursor ribosomal RNA of *Saccharomyces cerevisiae, Nucleic Acids Res.*, 8, 2679, 1980.
93. **Bayev, A. A., Georgiev, O. I., Hadjiolov, A. A., Kermekchiev, M. B., Nikolaev, N., Skryabin, K. G., and Zakharyev, V. M.,** The structure of the yeast ribosomal RNA genes. II. The nucleotide sequence of the initiation site for ribosomal RNA transcription, *Nucleic Acids Res.*, 8, 4919, 1980.

94. **Bayev, A. A., Georgiev, O. I., Hadjiolov, A. A., Nikolaev, N., Skryabin, K. G., and Zakharyev, V. M.,** The structure of the yeast ribosomal RNA genes. III. Precise mapping of the 18S and 25S rRNA genes and structure of the adjacent regions, *Nucleic Acids Res.*, 9, 789, 1981.
95. **Rubtsov, P. M., Musakhanov, M. M., Zakharyev, V. M., Krayev, A. S., Skryabin, K. G., and Bayev, A. A.,** The structure of the yeast ribosomal RNA genes. I. The complete nucleotide sequence of the 18S ribosomal RNA gene from *Saccharomyces cerevisiae, Nucleic Acids Res.*, 8, 5779, 1980.
96. **Veldman, G. M., Klootwijk, J., deJonge, P., Leer, R. J., and Planta, R. J.,** The transcription termination site of the ribosomal RNA operon in yeast, *Nucleic Acids Res.*, 8, 5179, 1980.
97. **Veldman, G. M., Brand, R. C., Klootwijk, J., and Planta, R. J.,** Some characteristics of processing sites in ribosomal precursor RNA of yeast, *Nucleic Acids Res.*, 8, 2907, 1980.
97a. **Veldman, G. M., Klootwijk, J., van Heerikhuizen, H., and Planta, R. J.,** The nucleotide sequence of the integenic region between the 5.8S and 26S rRNA genes of the yeast ribosomal RNA operon. Possible implications for the interaction between 5.8S and 26S rRNA and the processing of the primary transcript, *Nucleic Acids Res.*, 9, 4847, 1981.
98. **Perry, R. P.,** Processing of RNA, *Annu. Rev. Biochem.*, 45, 605, 1976.
98a. **Crouch, R.,** Ribosomal RNA processing in eukaryotes, in *RNA Processing*, Apirion, D., Ed., CRC Press, Boca Raton, Fla., 1983, chap. 9.
99. **Schweiger, E., MacKechnie, C., and Halvorson, H. O.,** The redundancy of ribosomal and transfer RNA genes in *Saccharomyces cerevisiae, J. Mol. Biol.*, 40, 261, 1969.
100. **Rubin, G. M. and Sulston, J. E.,** Physical linkage of the 5S cistrons to the 18S and 28S ribosomal RNA cistrons in *Saccharomyces cerevisiae, J. Mol. Biol.*, 79, 521, 1973.
101. **Petes, T. D.,** Molecular genetics of yeast, *Annu. Rev. Biochem.*, 49, 1980.
102. **Kramer, R. A., Philippsen, P., and Davis, R. W.,** Divergent transcription in the yeast ribosomal RNA coding region as shown by hybridization to separate strands and sequence analysis of cloned DNA, *J. Mol. Biol.*, 123, 405, 1978.
103. **Georgiev, O. I., Nikolaev, N., and Hadjiolov, A. A.,** The structure of the yeast ribosomal RNA genes. IV. Complete sequence of the 25S rRNA gene from *Saccharomyces cerevisiae, Nucleic Acids Res.*, 9, 6953, 1981.
104. **Rubin, G. M.,** The nucleotide sequences of *Saccharomyces cerevisiae* 5.8S ribosomal ribonucleic acids, *J. Biol. Chem.*, 248, 3860, 1973.
105. **Trapman, J., Planta, R. J., and Raué, H. A.,** Maturation of ribosomes in yeast. II. Position of the low molecular weight rRNA species in the maturation process, *Biochim. Biophys. Acta*, 442, 275, 1976.
106. **Udem, S. A. and Warner, J. R.,** The cytoplasmic maturation of a ribosomal precursor ribonucleic acid in yeast, *J. Biol. Chem.*, 248, 1412, 1973.
107. **Miyazaki, M.,** Studies on the nucleotide sequence of pseudouridine-containing 5S RNA from *Saccharomyces cerevisiae, J. Biochem.*, 75, 1407, 1974.
108. **Tekamp, P. A., Garcea, R. L., and Rutter, W. J.,** Transcription and *in vitro* processing of yeast 5S rRNA, *J. Biol. Chem.*, 255, 9501, 1980.
109. **DeJonge, P., Klootwijk, J., and Planta, R. J.,** Terminal nucleotide sequence of 17-S ribosomal RNA and its immediate precursor 18-S RNA in yeast, *Eur. J. Biochem.*, 72, 361, 1977.
110. **DeJonge, P., Klootwijk, J., and Planta, R. J.,** Sequence of the 3'-terminal 21 nucleotides of yeast 17S ribosomal RNA, *Nucleic Acids Res.*, 4, 3655, 1977.
111. **Gegenheimer, P. and Apirion, D.,** Processing of procaryotic ribonucleic acid, *Microbiol. Rev.*, 45, 502, 1981.
112. **Tekamp, P. A.,** personal communication, 1981.
113. **Jones, E. W.,** Proteinase mutants of *Saccharomyces cerevisiae, Genetics*, 85, 29, 1977.
114. **Littlewood, R., Shaffer, B., and Davies, J. E.,** Mutants of *S. cerevisiae* with reduced levels of ribonuclease activity, *Genetics*, 68, s39, 1971.
115. **Fabian, G. R. and Hopper, A. K.,** unpublished data, 1982.
116. **Klootwijk, J. and Planta, R. J.,** Analysis of methylation sites in yeast ribosomal RNA, *Eur. J. Biochem.*, 39, 325, 1973.
117. **Brand, R., Klootwijk, J., Van Steenbergen, T. J. M., DeKok, A. J., and Planta, R. J.,** Secondary methylation of yeast ribosomal precursor RNA, *Eur. J. Biochem.*, 75, 311, 1977.
118. **Vaughan, M. H., Jr., Soeiro, R., Warner, J. R., and Darnell, J. E., Jr.,** Effects of methionine deprivation on ribosome synthesis in *HeLa* cells, *Proc. Natl. Acad. Sci. U.S.A.*, 58, 1527, 1967.
119. **Retel, J. and Planta, R. J.,** Ribosomal precursor RNA in *Saccharomyces carlsbergensis, Eur. J. Biochem.*, 3, 248, 1967.
120. **Wejksnora, P. and Haber, J. E.,** Methionine dependent synthesis of ribosomal ribonucleic acid during sporulation and vegetative growth of *Saccharomyces cerevisiae, J. Bacteriol.*, 102, 1344, 1974.

121. **Warner, J. R. and Soeiro, R.,** Nascent ribosomes from *HeLa* cells, *Proc. Natl. Acad. Sci. U.S.A.*, 58, 1984, 1967.
122. **Hartwell, L. H., McLaughlin, C. S., and Warner, J. R.,** Identification of ten genes that control ribosome formation in yeast, *Molec. Gen. Genet.*, 109, 42, 1970.
123. **Gorenstein, C. and Warner, J. R.,** Coordinate regulation of the synthesis of eukaryotic ribosomal proteins, *Proc. Natl. Acad. Sci. U.S.A.*, 73, 1547, 1976.
124. **Rosbash, M., Harris, P. K. W., Woolford, J. L., Jr., and Teem, J. L.,** The effect of temperature sensitive RNA mutants on the transcription products from cloned ribosomal protein genes of yeast, *Cell*, 24, 679, 1981.
125. **Fried, H. M., Pearson, N. J., Kim, C. H., and Warner, J. R.,** The genes for fifteeen ribosomal proteins of *Saccharomyces cerevisiae*, *J. Biol. Chem.*, 256, 10176, 1981.
125a. **Bromley, S., Hereford, L., and Roshbash, M.,** Further evidence that the *rna2* mutation of *Saccharomyces cerevisiae* affects mRNA processing, *Molec. Cell. Biol.*, 2, 1205, 1982.
126. **Venkov, P. V. and Vasileva, A. P.,** *Saccharomyces cerevisiae* mutants defective in the maturation of ribosomal RNA, *Molec. Gen. Genet.*, 173, 203, 1979.
127. **Ursie, D. and Davies, J.,** Cold sensitive mutant of *Saccharomyces cerevisiae* defective in ribosome processing, *Molec. Gen. Genet.*, 175, 313, 1979.
128. **Andrew, C., Hopper, A. K., and Hall, B. D.,** A yeast mutant defective in the processing of 27S r-RNA precursor, *Molec. Gen. Genet.*, 144, 29, 1976.
129. **Fabian, G. R. and Hopper, A. K.,** unpublished data.
130. **Gorenstein, C. and Warner, J. R.,** Synthesis and turnover of ribosomal proteins in the absence of 60S subunit assembly in *Saccharomyces cerevisiae*, *Molec. Gen. Genet.*, 157, 327, 1977.
131. **Loo, M. W., Schricker, N. S., and Russell, P. J.,** Heat-sensitive mutant strain of *Neurospora crassa*, 4M(t), conditionally defective in 25S ribosomal ribonucleic acid production, *Molec. Cell. Biol.*, 1, 199, 1981.
132. **Schindler, D., Grant, P., and Davies, J.,** Trichodermin resistance-mutation affecting eukaryotic ribosomes, *Nature (London)*, 248, 535, 1974.
133. **Jimenez, A. and Vazquez, D.,** Quantitative binding of antibiotics to ribosomes from a yeast mutant altered on the peptidyl-transferase center, *Eur. J. Biochem.*, 54, 483, 1975.
134. **Fried, H. M. and Warner, J. R.,** Cloning of yeast gene for trichodermin resistance and ribosomal protein L3, *Proc. Natl. Acad. Sci. U.S.A.*, 78, 238, 1981.
135. **Carter, C. J., Cannon, M., and Jimenez, A.,** A trichodermin-resistant mutant of *Saccharomyces cerevisiae* with an abnormal distribution of native ribosomal subunits, *Eur. J. Biochem.*, 107, 173, 1980.
136. **Carter, C. J. and Cannon, M.,** Maturation of ribosomal precursor RNA in *Saccharomyces cerevisiae*. A mutant with a defect in both the transport and terminal processing of the 20S species, *J. Mol. Biol.*, 143, 179, 1980.
137. **Banerjee, A. K.,** 5′-Terminal cap structure in eukaryotic messenger ribonucleic acids, *Microbiol. Rev.*, 44, 175, 1980.
138. **DeKloet, S. R. and Andrean, B. A. G.,** Methylated nucleosides in polyadenylate-containing yeast messenger ribonucleic acid, *Biochim. Biophys. Acta*, 425, 401, 1975.
139. **Sripati, C. E., Groner, Y., and Warner, J. R.,** Methylated, blocked 5′ termini of yeast mRNA, *J. Biol. Chem.*, 251, 2898, 1976.
140. **Bennetzen, J. L. and Hall, B. D.,** The primary structure of the *Saccharomyces cerevisiae* gene for alcohol dehydrogenase I, *J. Biol. Chem.*, 257, 3018, 1982.
141. **Fahrner, K., Yarger, J., and Hereford, L.,** Yeast histone mRNA is polyadenylated, *Nucleic Acids Res.*, 8, 5725, 1980.
142. **McLaughlin, C. S., Warner, J. R., Edmonds, M., Nakazato, H., and Vaughan, M. H.,** Polyadenylic acid sequences in yeast messenger ribonucleic acid, *J. Biol. Chem.*, 248, 1466, 1973.
143. **Groner, B., Hynes, N., and Phillips, S.,** Length heterogeneity in the polyadenylic acid region of yeast messenger RNA, *Biochemistry*, 13, 5378, 1971.
144. **Haff, L. A. and Keller, E. B.,** The polyadenylate polymerases from yeast, *J. Biol. Chem.*, 250, 1838, 1975.
145. **Phillips, S. L., Tse, C., Serventi, T. I., and Hynes, N.,** Structure of polyadenylic acid in the ribonucleic acid of *Saccharomyces cerevisiae*, *J. Bacteriol.*, 138, 542, 1979.
146. **Smith, M., Leung, D. W., Gillam, S., Astell, C., Montgomery, D. L., and Hall, B. D.,** Sequence of the gene for iso-1-cytochrome c in *Saccharomyces cerevisiae*, *Cell*, 16, 753, 1979.
147. **Boss, J. M., Darrow, M. D., and Zitomer, R. S.,** Characterization of yeast iso-1-cyctochrome c mRNA, *J. Biol. Chem.*, 255, 8023, 1980.
148. **Zaret, K. S. and Sherman, F.,** DNA sequence required for efficient transcription termination in yeast, *Cell*, 28, 563, 1982.

149. **Gallwitz, D., Perrin, F., and Seidel, R.,** The actin gene in yeast *Saccharomyces cerevisiae:* 5′ and 3′ end mapping, flanking and putative regulatory sequences, *Nucleic Acids Res.*, 9, 6339, 1981.
150. **Astell, C. R., Ahlstrom-Jonasson, L., Smith, M., Tatchell, K., Nasmyth, K. A., and Hall, B. D.,** The sequence of the DNAs coding for the mating-type genes of *Saccharomyces cerevisiae, Cell,* 27, 15, 1981.
151. **Holland, J. P. and Holland, M. J.,** The primary structure of a glyceraldehyde-3-phosphate dehydrogenase gene from *Saccharomyces cerevisiae, J. Biol. Chem.*, 254, 9839, 1979.
152. **Holland, J. P. and Holland, M. J.,** Structural comparison of two nontandemly repeated yeast glyceraldehyde-3-phosphate dehydrogenase genes, *J. Biol. Chem.*, 255, 2596, 1980.
153. **Holland, M. J., Holland, J. P., Thill, G. P., and Jackson, K. A.,** The primary structures of two yeast enolase genes. Homology between the 5′ noncoding regions of yeast enolase and glyceraldehyde-3-phosphate dehydrogenase, *J. Biol. Chem.*, 256, 1385, 1981.
154. **Montgomery, D. L., Leung, D. W., Smith, M., Shalit, P., Faye, G., and Hall, B. D.,** Isolation and sequence of the gene for iso-2-cytochrome c in *Saccharomyces cerevisiae, Proc. Natl. Acad. Sci. U.S.A.*, 77, 541, 1980.
155. **Tschumper, G. and Carbon, J.,** Sequence of a yeast DNA fragment containing a chromosomal replicator and the *TRP1* gene, *Gene,* 10, 157, 1980.
156. **Proudfoot, N. J. and Brownlee, G. G.,** 3′ Noncoding region sequences in eukaryotic messenger RNA, *Nature (London)*, 263, 211, 1976.
157. **Fitzgerald, M. and Shenk, T.,** The sequence 5′-AAUAAA-3′ forms part of the recognition site for polyadenylation of late SV40 mRNAs, *Cell,* 24, 251, 1981.
158. **Koch, H. and Friesen, J. D.,** Individual messenger RNA half-lives in *Saccharomyces cerevisiae, Molec. Gen. Genet.*, 170, 129, 1979.
159. **Chia, L.-L. and McLaughlin, C.,** The half-life of mRNA in *Saccharomyces cerevisiae, Molec. Gen. Genet.*, 170, 137, 1979.
160. **St. John, T. and Davis, R. W.,** The organization and transcription of the galactose gene cluster of *Saccharomyces, J. Mol. Biol.*, 152, 285, 1981.
161. **Struhl, K. and Davis, R. W.,** Transcription of the *his3* gene region in *Saccharomyces cerevisiae, J. Mol. Biol.*, 152, 535, 1981.
162. **Boss, J. M., Gillam, S., Zitomer, R. S., and Smith, M.,** Sequence of the yeast iso-1-cytochrome c mRNA, *J. Biol. Chem.*, 256, 12958, 1981.
163. **Wallis, J. W., Hereford, L., and Grunstein, M.,** Histone H2B genes of yeast encode two different proteins, *Cell,* 22, 799, 1980.
164. **Gallwitz, D. and Sures, I.,** Structure of a split yeast gene: complete nucleotide sequence of the actin gene in *Saccharomyces cerevisiae, Proc. Natl. Acad. Sci. U.S.A.*, 77, 2546, 1980.
164a. **Gallwitz, D.,** Construction of a yeast actin gene intron deletion mutant that is defective in splicing and leads to the accumulation of precursor RNA in transformed yeast cells, *Proc. Natl. Acad. Sci.*, 79, 3493, 1982.
165. **Ng, R. and Abelson, J.,** Isolation and sequence of the gene for actin in *Saccharomyces cerevisiae, Proc. Natl. Acad. Sci. U.S.A.*, 77, 3912, 1980.
166. **Warner, J. R. and Gorenstein, C.,** The synthesis of eukaryotic ribosomal proteins, *in vitro, Cell,* 11, 201, 1977.
167. **Hereford, L. M. and Rosbash, M.,** Regulation of a set of abundant mRNA sequences, *Cell,* 10, 463, 1977.
168. **Rosbash, M.,** personal communication, 1981.
169. **Woolford, J. L.,** personal communication, 1982.
170. **Pearson, N. J., Thornburn, P. C., and Haber, J. E.,** A suppressor of temperature-sensitive rna mutations that affect mRNA metabolism in *Saccharomyces cerevisiae, Molec. Cell. Biol.*, 2, 571, 1982.
171. **Pearson, N. J.,** personal communication, 1983.
172. **Deutscher, M. P.,** tRNA nucleotidyl transferase, *The Enzymes,* 15, 183, 1982.

Chapter 5

# 5′ TERMINAL CAP STRUCTURES OF EUKARYOTIC AND VIRAL mRNAs

**Bernard Moss**

## TABLE OF CONTENTS

# I. OCCURRENCE OF CAP STRUCTURES IN EUKARYOTIC mRNAs

Messenger RNAs of eukaryotic cells and viruses contain a small number of methylated nucleotides.[1,2,3] Some are present at the 5′ end of the RNA in a cap structure (Figure 1) consisting of 7-methylguanosine ($m^7G$) connected via a triphosphate bridge to the 5′ position of the adjacent nucleotide.[4-6] The occurrence of capped mRNAs in a variety of organisms and viruses has been the subject of a number of reviews, the most recent and comprehensive of which is by Banerjee.[7] Similar cap structures are not present in prokaryotic RNAs or eukaryotic tRNAs, rRNAs, or mitochondrial mRNAs.[8,9] However, a related structure containing a trimethylated guanosine is found in certain low-molecular-weight nuclear RNAs.[10,11]

The simplest 5′ terminal modification of mRNA, $m^7G(5')pppN$-, is referred to as cap 0. When the penultimate nucleotide, that is the one adjacent to $m^7G$, is methylated at the 2′ position of the ribose moiety, the structure $m^7G(5')pppN^m$- is called cap 1. A cap 2 structure $m^7G(5')pppN^m$-$N^m$- contains two consecutive nucleotides methylated at the 2′ position. In some cap 1 and cap 2 structures, the penultimate nucleoside is a dimethylated adenosine ($m^6A^m$) containing a methyl group at the $N^6$ position of the base as well as on the 2′ position of ribose.

The extent of 2′-methylation of cap structures varies among phylogenetic groups. Cap 0 is found in mRNAs of yeast[12-14] and higher plants.[15-17] Slime mold mRNA contains predominantly cap 0 as well as some cap 1,[18] while sea urchin embryos have cap 1 mRNAs exclusively.[19,20] Mammals[21-25] and some insects[26-28] contain both cap 1 and 2 mRNAs. Penultimate $m^6A^m$ has been described in mammalian mRNAs only.[25,29]

Viral mRNAs synthesized in eukaryotic cells generally have cap structures similar to those of their hosts. Thus, all known DNA viruses that replicate in mammalian cells have cap 1 and 2 mRNAs.[30-33] The situation with RNA viruses is more complex. Viral mRNAs isolated from mammalian cells infected with vesicular stomatitis virus,[34] reovirus,[35] and influenza[36] have cap 1 or 2. However, Sindbis virus[37] has cap 0, and picornaviruses such as poliovirus[38] have uncapped mRNAs with a single terminal phosphate. Similarly, many plant viruses have RNAs with cap 0 structures[39-42] but at least three, including tobacco necrosis virus, satellite tobacco necrosis virus, and cowpea mosaic virus[43-48] appear to be uncapped. With regard to the latter, however, the polyribosomal mRNAs have not been examined.

Eukaryotic and some viral mRNAs also contain internally located nucleotides that are methylated. Most frequently, this is $N^6$-methyladenosine ($m^6A$);[2] the presence of 5-methylcytosine has also been reported.[2,25,37] In all cases examined, $m^6A$ occurs in the sequence A-$m^6A$-C or G-$m^6A$-C[49-52] which may be in the coding region of the message.[53,54] On average, there is one $m^6A$ residue per 800 to 1000 nucleotides.[21-24,55] However, some mRNAs have more $m^6A$[53] and others such as histone[56] and globin[57,58] mRNA have none.

# II. FUNCTION OF CAP STRUCTURES IN EUKARYOTIC mRNAs

Possible functions of cap structures have been considered in several excellent reviews[59-62] and will not be considered here in any detail. Accumulated reports cited above, however, clearly demonstrate that the $m^7G$ portion of the cap facilitates ribosome binding and translation of mRNA in cell-free systems. This effect may be mediated through a specific cap binding protein.[63,64] Additional experiments suggest that uncapped mRNAs are less stable than capped mRNAs when injected into *Xenopus laevis* oocytes or in cell-free extracts of wheat germ or L-cells.[65,66] Virtually nothing is known regarding the function of the 2′-*O*-methylated residues of cap 1 and 2 structures.

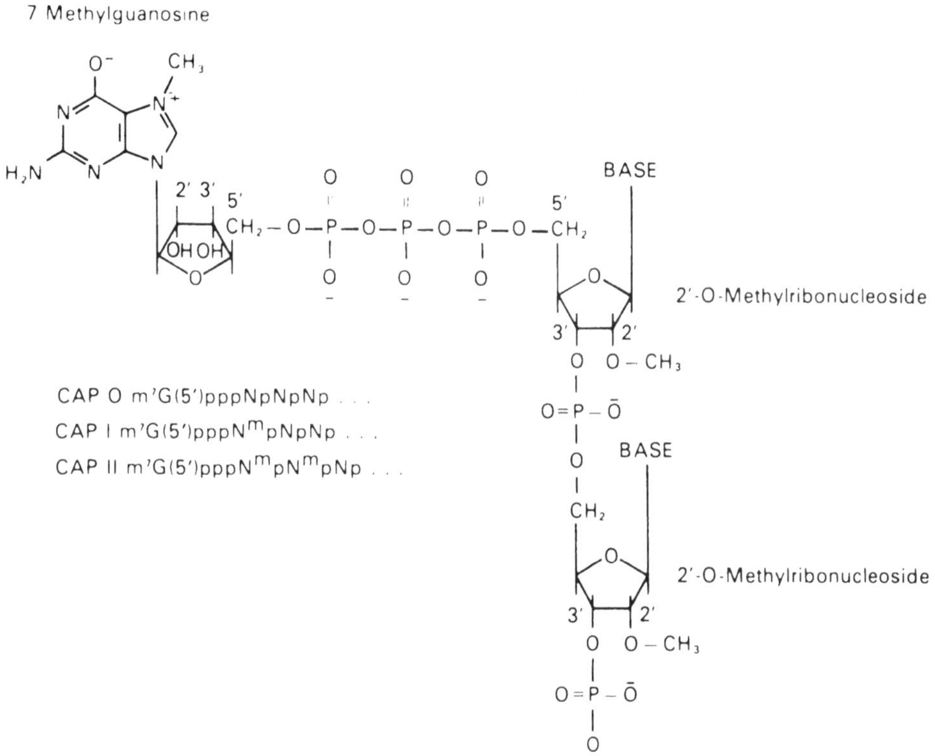

7 Methylguanosine

2'-O-Methylribonucleoside

2'-O-Methylribonucleoside

CAP 0 $m^7G(5')pppNpNpNp$ . . .

CAP I $m^7G(5')pppN^mpNpNp$ . . .

CAP II $m^7G(5')pppN^mpN^mpNp$ . . .

FIGURE 1. 5' Terminal cap structures of eukaryotic mRNAs.

# III. MECHANISMS OF CAP FORMATION

## A. Viruses

### *1. Vaccinia Virus*

Viruses, particularly those that package enzymes within the infectious particle for the synthesis and modification of mRNA, have provided particularly useful systems for studying transcription.[67] Vaccinia, a member of the poxvirus family, has a DNA genome of about 180,000 base pairs. When purified vaccinia virus is incubated with a nonionic detergent and a reducing agent, the envelope is disrupted and the core becomes permeable to ribonucleoside triphosphates and *S*-adenosylmethionine (AdoMet). Under these conditions, translatable mRNA containing $m^7G(5')pppA^m$ or $m^7G(5')pppG^m$ ends are formed.[68-70] By using ribonucleoside triphosphates labeled with $^{32}P$ in the $\alpha$, $\beta$, or $\gamma$ position it was possible to demonstrate that the basic capping reaction consists of transfer of a GMP residue from GTP to the diphosphate end of nascent RNA.[71] This was confirmed by disrupting the virus core and using solubilized enzymes to cap exogenously added synthetic polyribonucleotides.[72]

A multifunctional enzyme complex with a molecular weight of 127,000 and containing two dissimilar subunits of about 95,000 and 31,500 has been isolated from disrupted vaccinia particles.[73] The purified enzyme catalyzes the following reactions.[74-76]

$$pppN_1\text{-}N_2\text{-} \rightarrow ppN_1\text{-}N_2\text{-} + Pi \qquad (i)$$

$$GTP + ppN_1\text{-}N_2\text{-} \rightleftharpoons G(5')pppN_1\text{-}N_2\text{-} + PPi \qquad (ii)$$

$$AdoMet + G(5')pppN_1\text{-}N_2\text{-} \rightarrow m^7G(5')pppN_1\text{-}N_2\text{-} + AdoHcy \qquad (iii)$$

AdoMet is *S*-adenosylmethionine; AdoHcy is *S*-adenosylhomocysteine.

In reaction (i), $pppN_1$-$N_2$- represents the 5′ end of a nascent RNA chain. Although the effect of RNA chain length on this reaction has not been studied in detail, simple ribonucleoside triphosphates are not good substrates compared to oligomers or polymers.[77] Only the terminal phosphate is removed in step (i) forming a diphosphate-terminated RNA as acceptor for reaction (ii). In the latter, a GMP residue is transferred from GTP to the diphosphate end of the nascent RNA. Significantly, the 7-methyl derivative of GTP cannot serve as a donor, thus establishing the order of reactions (ii) and (iii). An alternative scheme in which reactions (i) and (ii) are coupled obligatorily[78,79] has not been confirmed.[76,80]

Recently, reaction (ii) has been shown to occur via a covalent enzyme guanylate (E-GMP) intermediate:[80,81]

$$\text{GTP} + \text{E} \rightarrow \text{E-GMP} + \text{PPi} \qquad \text{(iia)}$$

$$\text{E-GMP} + \text{ppN}_1\text{-N}_2\text{-} \rightarrow \text{G(5')pppN}_1\text{-N}_2\text{-} + \text{E} \qquad \text{(iib)}$$

The E-GMP complex can transfer GMP either to pyrophosphate, thereby regenerating GTP, or to the end of the RNA forming an unmethylated cap. The GMP residue is joined to the 95,000-dalton polypeptide, apparently through a phosphoamide bond.

The unmethylated cap then serves as a methyl acceptor for reaction (iii) resulting in the formation of a cap 0 structure. Although the unmethylated cap can undergo rapid enzyme-catalyzed phosphorylysis to reverse reaction (ii), methylation at the 7-position of guanine prevents this.[74]

Methylation of the penultimate nucleotide is catalyzed by a separable viral enzyme that has a molecular weight of 55,000:[82,83]

$$\text{AdoMet} + \text{m}^7\text{G(5')pppN}_1\text{-N}_2\text{-} \rightarrow \text{m}^7\text{G(5')pppN}_1{}^{\text{m}}\text{-N}_2\text{-} + \text{AdoHcy} \qquad \text{(iv)}$$

The order of reaction (iii) and (iv) was suggested first by finding increased ratios of $m^7G/A^m + G^m$ when suboptimal levels of AdoMet were used for RNA synthesis by vaccinia virus particles.[71] Furthermore, RNAs containing cap 0 structures were the best substrates for the purified methyltransferase.[83] Nevertheless, some investigators isolated oligonucleotides containing $G(5')pppA^m$ and $G(5')pppG^m$ from virus particles and proposed that reaction (iv) precedes (iii).[84]

The possibility that some 5′ termini are generated by endonucleolytic cleavage of RNA precursors, followed by addition of phosphates and capping, has been considered. Circumstantial evidence for this has included the formation of large RNAs that can be chased into smaller ones under certain in vitro conditions[85-87] and the isolation of endoribonuclease[88] and 5′-phosphate polyribonucleotide kinase[89] activities from vaccinia virus. Nevertheless, this hypothesis remains speculative since UV target size experiments do not support such a processing model.[90-92] Moreover, the caps of those mRNAs examined appear to contain the β-phosphate of the initial ribonucleoside triphosphate.[71,93]

Although vaccinia virus contains enzymes for the synthesis of cap 1 structures only, viral specific mRNA isolated from infected cells contains cap 2 structures as well as $m^6A^m$.[32] It is believed that these additional methylations are catalyzed by host enzymes.

*2. Reovirus*

The genome of reovirus consists of ten double-stranded RNA segments packaged within each virus particle. When the latter are treated with chymotrypsin, encapsidated enzymes can use ribonucleoside triphosphates and AdoMet to form mRNAs ending in $m^7G(5')pppGpC$.[5,94] Kinetic experiments suggest that cap formation is an early step in transcription.[95] Although the capping and methylating enzymes have not been solubilized from reovirus, the dinucleotide ppGpC can penetrate the core and is converted to G(5′)pppGpC in the presence of GTP and AdoHcy and to $m^7$G(5′)pppG$^m$pC in the presence of GTP and AdoMet.[95] Significantly, pGpC is not a substrate; however, pppGpC is used presumably after removal of the terminal phosphate. Furuichi et al.[95] have proposed a scheme for reovirus cap formation that is similar to the one discussed for vaccinia virus.

Reovirus mRNA isolated from infected cells contains cap 1 and cap 2 structures.[35] Presumably, the additional methylation step needed to form the latter is catalyzed by a host enzyme. Interestingly, recent studies indicate that reovirus mRNAs synthesized late in infection[96] or in vitro by progeny subviral particles[97,98] lack cap structures and this is correlated with changes in the translation system of infected cells.[99]

*3. Cytoplasmic Polyhedrosis Virus*

Cytoplasmic polyhedrosis virus, like reovirus, contains ten double-stranded RNA genome segments and can synthesize mRNAs with $m^7$G(5′)pppA$^m$pGp ends in vitro.[6] Unlike reovirus or vaccinia virus, however, transcription is greatly stimulated by AdoMet or AdoHcy.[100,101] These results, as well as the inhibition of transcription by the $\beta,\gamma$-imido analogue of ATP, led Furuichi[101] to propose that capping is an obligatory pretranscriptional event. Nevertheless, the capping reaction, as with vaccinia virus and reovirus, involves the transfer of GMP from GTP to an RNA acceptor.

*4. Vesicular Stomatitis Virus*

Vesicular stomatitis virus has a single-stranded RNA genome that can be transcribed in vitro by enzymes present in the virus to form five mRNAs, all ending in $m^7$G(5′)pppA$^m$pApCpApGp.[102] The basic capping mechanism, however, differs from that observed with the previous virus systems in that the net reaction involves transfer of a GDP residue from GTP to the 5′ end of mRNA. Evidently, the tight coupling of transcription and cap formation has made it impossible to cap or methylate exogenously added RNA or oligonucleotides. This feature has precluded a detailed analysis of vesicular stomatitis capping reactions. Since the predominant cap structure formed at low AdoMet concentrations is G(5′)pppA$^m$-, it was suggested that 2-*O*-methylation precedes guanine-7-methylation.[103] However, the opposite order was proposed from an analysis of capped structures formed in vivo in the presence of cycloleucine.[104] Vesicular stomatitis virus mRNAs synthesized in vivo have cap 2 structures as well as dimethylated adenosine.[105,106] As with other viral mRNAs, we presume that these additional methylations are catalyzed by host enzymes.[105]

**B. Eukaryotic Cells**

*1. In Vivo Studies*

In mammalian cells, the initial capping events take place in the nucleus as indicated by the presence of cap 1 structures and $m^6$A$^m$ in heterogeneous nuclear RNA (hnRNA).[107,108] Since nuclear transcripts only several hundred nucleotides in length are capped, these modifications probably occur relatively early in hnRNA biosynthesis.[109-111] In fact one of the latter studies,[109] in contrast to some earlier ones,[112-114] suggests that large hnRNA molecules rarely contain di- or triphosphate ends.

Partly because of a preconception that eukaryotic mRNAs are initiated with purines exclusively, the presence of pyrimidines in the penultimate position of cap structures[21-24] initially led to speculation that some caps are added at sites of endonucleolytic processing. However, there are no convincing data to support such a mechanism and careful studies have failed to detect RNA polymerase II transcription of cellular or viral DNA sequences upstream of mRNA cap sites.[115,116]

Experiments with methylation inhibitors indicated the presence of increased levels of mRNA with cap 0 structures which normally do not accumulate in mammalian cells.[117]

Hence, guanine-7-methylation can precede 2′-*O*-methylation of the penultimate nucleotide, at least under the above conditions. The absence of cap 2 in hnRNA[107,108] and the results of pulse-chase experiments indicate that this final methylation step occurs in the cytoplasm.[118,119]

*2. In Vitro Studies*

Studies with isolated nuclei or nuclear homogenates from uninfected or DNA virus infected cells indicate that RNA polymerase II transcripts are capped and methylated.[120,121] The incorporation of the β-phosphate from ATP or CTP into cap structures implies that the initiating ribonucleotide, whether a purine or pyrimidine triphosphate, is capped.[122-124] Further evidence for the transfer of a GMP residue from GTP to the diphosphate end of RNA was obtained using a nuclear extract that had been passed through DEAE-cellulose to make it dependent on exogenous substrates.[125] Thus, the basic mechanism of capping in mammalian cells appears to be similar to that found for vaccinia and reovirus. The efficiency of capping transcripts in template-dependent homogenates suggests that capping might be closely linked with initiation of RNA synthesis.[126,127]

*3. Enzymes*

***a. RNA Guanylyltransferase***

RNA guanylyltransferase has been partially purified from several sources including calf thymus,[128] rat liver nuclei,[129] HeLa cell nuclei,[130,131] and wheat germ.[132] Enzymes from the latter three sources are of similar size and appear to have similar catalytic properties. Unlike the situation with vaccinia virus, the purified eukaryotic enzymes have no RNA (guanine-7-)methyltransferase and little or no RNA triphosphatase activities. Whether the three activities are more intimately associated prior to cell disruption or purification, however, has not been ascertained.

Because of the absence of RNA triphosphatase from the most highly purified eukaryotic RNA guanylyltransferase preparations, diphosphate-ended polyribonucleotides are better acceptors than those with triphosphate ends. Significantly, polyribonucleotides with a single terminal phosphate were not capped at all. The inability of these enzymes to cap a mononucleotide, such as ADP, indicates that pretranscriptional capping is an unlikely event. Since diphosphate-ended dinucleotides[131] and trinucleotides[129] are capped, the minimal requirement is one phosphodiester bond. The purified HeLa cell and wheat germ RNA guanylyltransferases were found to cap both ppC- and ppA-ended polyribonucleotides, suggesting that one enzyme is responsible for forming pyrimidine and purine caps.

Of a variety of donors tested, the HeLa cell RNA guanylyltransferase could only transfer a mononucleotide from GTP or ITP to a suitable acceptor polyribonucleotide.[131] The inability to use $m^7$GTP as a donor indicated that methylation must follow transfer of the GMP residue.

Both HeLa cell[133] and wheat germ[134] RNA guanylyltransferases were shown to form covalent GMP intermediates that have molecular weights of approximately 65,000 and 85,000, respectively. The chemical reactivity of the covalent bonds suggest a phosphoamide linkage. In the case of the HeLa cell enzyme, a single tryptic peptide with bound GMP was resolved by two-dimensional techniques. Significantly, the purified enzyme-GMP complexes catalyze the transfer of the GMP moiety to either PPi, regenerating GTP, or to the 5′ diphosphate end of poly(A), forming the cap structure. These results explain the pyrophosphate exchange reactions of the rat liver enzyme[129] and indicate that the eukaryotic capping mechanism is similar or identical to that of vaccinia virus.

A detailed comparison of the properties of viral and eukaryotic capping and methylating enzymes has been published elsewhere.[135]

***b. RNA (Guanine-7-)methyltransferase***

RNA (guanine-7-)methyltransferase has been isolated from HeLa cells,[136] rat liver,[129]

*Neurospora crassa,*[137] and wheat germ,[134,138] and detected in chick lens extracts.[139] The enzymes appear to be entirely specific for cap structures either free or at the end of RNA molecules. Either the enzyme can exist in various multimeric forms or there is a large variation in size of these enzymes since the values reported for HeLa cell and rat liver are 56,000 and 130,000, respectively.

### c. *RNA (Nucleoside-2′-)methyltransferases*

Thus far, RNA (nucleoside-2′-)methyltransferases have been partially purified from only one eukaryotic source: HeLa cells. Two separate enzymes were isolated.[140] The cap 1 methyltransferase was specific for the penultimate nucleotide of capped RNA and was assayed by conversion of cap 0 structures to cap 1 form. The cap 2 methyltransferase modified the adjacent nucleotide to form cap 2 structures. Cap 2 methyltransferase was found almost exclusively in cytoplasmic fractions while some cap 1 methyltransferase also was found in the nucleus — its apparent biological site of action. The finding that a capped terminus with at least two additional nucleotides, e.g., G(5′)pppNpNp was required for cap 1 methyltransferase activity indicated that initiation of transcription, minimal chain extension, and capping must precede 2′-*O*-methylation. However, since the activity of the cap 1 methyltransferase for polyribonucleotides ending in G(5′)pppN- and $m^7$G(5′)pppN- were virtually identical, no conclusion regarding the order of guanine-7-methylation and 2′-*O*-methylation could be drawn from these data. Both purine and pyrimidine nucleosides in the N position of $m^7$G(5′)pppN- were methylated, suggesting that the same enzyme modifies any of the four nucleosides when present in the penultimate position. Incomplete information regarding the substrate specificity of cap 2 methyltransferase was obtained, partly because of some contamination with RNA (guanine-7-)methyltransferase activity.

### d. *RNA (2′-O-methyladenosine-$N^6$-)methyltransferase*

An enzyme responsible for methylating the $N^6$ position of 2′-*O*-methyladenosine located in the penultimate position of the capped 5′ end of mRNA has been purified from HeLa cells.[141] This methyltransferase, which has a molecular weight of approximately 65,000, exhibited greatest activity with RNA acceptors ending in $m^7$G(5′)pppA$^m$-. Less activity was obtained with RNAs ending in $m^7$G(5′)pppA-; activity was barely detected with RNA ending in G(5′)pppA- and not detected at all with RNA ending in ppA-. Further, no activity was found with oligonucleotides such as $m^7$G(5′)pppA$^m$ and $m^7$G(5′)pppA$^m$pN, indicating that a longer oligomer or polymer is required. It can be concluded from the substrate specificity of the enzyme that formation of $m^6$A$^m$ follows the biosynthesis of molecules containing $m^7$G(5′)pppA$^m$pNpN.

### e. *Additional Enzymes*

The putative RNA triphosphatase, necessary for removal of the terminal phosphate at the end of a nascent RNA chain to provide a diphosphate suitable for capping, has not yet been characterized. However, during the purification of the RNA guanylyltransferase from HeLa cells[130] and wheat germ[132] such an activity was removed during specific column chromatography steps.

The enzyme responsible for methylating internal adenosine residues at the $N^6$ position has not been identified. It is important to point out that the RNA (2′-*O*-methyladenosine-$N^6$-)methyltransferase is not a candidate since it is specific for cap structures.[141] An RNA (cytosine-5-)methyltransferase that has specificity for tRNA[142] might also be responsible for the occasional methylation of internal cytosine residues of mRNA and viral RNA. However, no definite information regarding this possibility has been obtained.

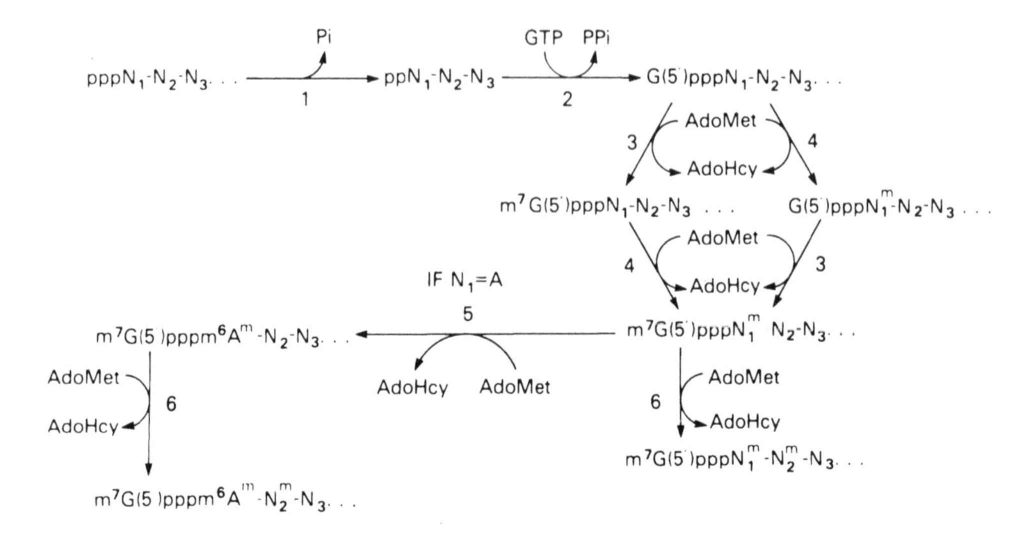


FIGURE 2. Enzymatic steps in the modification of the 5′ end of eukaryotic mRNA. The enzymes that catalyze the indicated reactions are 1, RNA triphosphatase; 2, RNA guanylyltransferase; 3, RNA (guanine-7-)methyltransferase; 4 cap 1 RNA (nucleoside-2′-)methyltransferase; 5, RNA (2′-*O*-methyladenosine-$N^6$-) methyltransferase; and 6, cap 2 RNA (nucleoside-2′-)methyltransferase. (From Langberg, S. R. and Moss, B., *J. Biol. Chem.*, 256, 10054, 1981. With permission.)

## IV. CONCLUSIONS

The formation of biologically active mRNA in eukaryotic organisms involves a complex series of events including initiation and termination of transcription, capping, internal methylation, polyadenylation, splicing, and cytoplasmic transport. Of these processes, probably capping is understood best largely because the individual enzymes involved have been isolated and characterized in vitro; the following scheme is compatible with available information (Figure 2). Removal of the terminal phosphate from nascent RNA is followed by transfer of GMP from GTP through a covalent enzyme intermediate to the 5′ diphosphate end of the transcript. The next two steps are methylation of the terminal guanosine at the $N^7$ position and the penultimate nucleoside at the 2′ position. However, which of the two methylation steps occurs first in vivo cannot be inferred from the substrate specificities of the enzymes. When the cap 1 structure contains penultimate adenosine, methylation of the $N^6$ position proceeds. The final step, 2′-*O*-methylation of the adjacent nucleotide to form the cap 2 structure, takes place in the cytoplasm. Since oligonucleotides can be used as substrates for the initial capping and methylation reactions, it is possible that these events occur very soon after initiation of the polyribonucleotide. Such an early event would be compatible with in vivo data. However, pretranscriptional capping, that is, transfer of a GMP residue to a mononucleoside diphosphate, is not consistent with the properties of the purified HeLa cell capping enzyme.

A number of viruses that replicate within the cytoplasm of infected cells specify their own enzymes for the formation of cap structures and have proven to be particularly useful model systems. For vaccinia virus and reovirus, the capping mechanism is very similar to that employed by their host cells. However, both cytoplasmic polyhedrosis viruses and vesicular stomatitis virus have unique features that distinguish their capping mechanisms which are still incompletely understood.

## ACKNOWLEDGMENTS

I am indebted to past and present members of my laboratory including Cha Mer Wei, Scott Martin, Enzo Paoletti, Marcia Ensinger, Jerry Keith, Sundararajan Venkatesan, and Steven Langberg for their collaboration in elucidating the mechanism of cap formation.

## REFERENCES

1. **Perry, R. P. and Kelly, D. E.,** Existence of methylated messenger RNA in mouse L cells, *Cell,* 1, 37, 1974.
2. **Desrosiers, R., Frederici, K., and Rottman, F.,** Identification of methylated nucleosides in messenger RNA from Novikoff hepatoma cells, *Proc. Natl. Acad. Sci. U.S.A.,* 71, 3971, 1974.
3. **Furuichi, Y.,** Methylation-coupled transcription by virus-associated transcriptase of cytoplasmic polyhedrosis virus containing double-stranded RNA, *Nucleic Acids Res.,* 1, 802, 1974.
4. **Wei, C. M. and Moss, B.,** Methylated nucleotides block 5'-terminus of vaccinia virus messenger RNA, *Proc. Natl. Acad. Sci. U.S.A.,* 72, 318, 1975.
5. **Furuichi, Y., Morgan, M., Muthukrishnan, S., and Shatkin, A. J.,** Reovirus messenger RNA contains a methylated, blocked 5'-terminal structure, $m^7G(5')ppp(5')GpCp$, *Proc. Natl. Acad. Sci. U.S.A.,* 72, 362, 1975.
6. **Furuichi, Y. and Miura, K.-I.,** A blocked structure at the 5'-terminus of messenger RNA from cytoplasmic polyhedrosis virus, *Nature (London),* 253, 374, 1975.
7. **Banerjee, A. K.,** 5'-Terminal cap structure in eucaryotic messenger ribonucleic acids, *Microbiol. Rev.,* 44, 175, 1980.
8. **Taylor, R. H. and Dubin, D. T.,** The methylation status of mammalian mitochondrial messenger RNA, *J. Cell Biol.,* 67, 4282A, 1975.
9. **Grohmann, K., Amalric, F., Crews, S., and Attardi, G.,** Failure to detect "cap" structures in mitochondrial DNA-coded poly(A)-containing RNA from HeLa cells, *Nucleic Acids Res.,* 5, 637, 1978.
10. **Ro-Choi, T. S., Yong, C. C., Henning, D., McCloskey, J., and Busch, H.,** Nucleotide sequence of U-2 ribonucleic acid, *J. Biol. Chem.,* 250, 3921, 1975.
11. **Cory, S. and Adams, J. M.,** Modified 5'-termini in small nuclear RNAs of mouse myeloma cells, *Mol. Biol. Rep.,* 2, 287, 1975.
12. **DeKloet, S. R. and Andrean, B. A. G.,** Methylated nucleosides in polyadenylate-containing yeast messenger ribonucleic acid, *Biochim. Biophys. Acta,* 425, 401, 1976.
13. **Mager, W. H., Klootwijk, J., and Klein, I.,** Minimal methylation of yeast messenger RNA, *Mol. Biol. Rep.,* 3, 9, 1976.
14. **Sripati, C. E., Groner, Y., and Warner, J. R.,** Methylated, blocked 5' termini of yeast mRNA, *J. Biol. Chem.,* 251, 2898, 1976.
15. **Haffner, M. H., Chin, M. B., and Lane, B. G.,** Wheat embryo ribonucleates. XII. Formal characterization of terminal and penultimate nucleoside residues at the 5'-ends of "capped" RNA from imbibing wheat embryos, *Can. J. Biochem.,* 56, 729, 1978.
16. **Nichols, J. L.,** "Cap" structures in maize poly(A)-containing RNA, *Biochim. Biophys. Acta,* 563, 490, 1979.
17. **Takagi, S. and Mori, T.,** The 5'-terminal structure of poly(A)-containing RNA of soybean seeds, *J. Biochem.,* 86, 231, 1979.
18. **Dottin, R. P., Weiner, A. M., and Lodish, H. F.,** 5'-Terminal nucleotide sequences of the messenger RNAs of *Dictyostellium discoideum, Cell,* 8, 233, 1976.
19. **Surrey, S. and Nemer, M.,** Methylated blocked 5' terminal sequences of sea urchin embryo messenger RNA classes containing and lacking poly(A), *Cell,* 9, 589, 1976.
20. **Faust, M., Millward, S., Duchastel, A., and Fromson, D.,** Methylated constituents of poly$(A)^-$ and poly$(A)^+$ polyribosomal RNA of sea urchin embryos, *Cell,* 9, 597, 1976.
21. **Wei, C. M., Gershowitz, A., and Moss, B.,** Methylated nucleotides block 5' terminus of HeLa cell messenger RNA, *Cell,* 4, 379, 1975.
22. **Perry, R. P., Kelley, D. E., Frederici, K., and Rottman, F.,** The methylated constituents of L cell messenger RNA: evidence for an unusual cluster at the 5' terminus, *Cell,* 4, 387, 1975.

23. **Adams, J. M. and Cory, S.,** Modified nucleosides and bizarre 5′-termini in mouse myeloma mRNA, *Nature (London)*, 255, 28, 1975.
24. **Furuichi, Y., Morgan, M., Shatkin, A. J., Jelinek, W., Salditt-Georgieff, M., and Darnell, J. E.,** Methylated, blocked 5′ termini in HeLa cell mRNA, *Proc. Natl. Acad. Sci. U.S.A.*, 72, 1904, 1975.
25. **Dubin, D. T. and Taylor, R. H.,** The methylation state of poly A-containing messenger RNA from cultured hamster cells, *Nucleic Acids Res.*, 2, 1653, 1975.
26. **Levis, R. and Penman, S.,** 5′-Terminal structures of poly(A)$^-$ cytoplasmic messenger RNA and of poly(A)$^+$ and poly(A)$^-$ heterogeneous nuclear RNA of cells of the dipteran *Drosophila melanogaster, J. Mol. Biol.*, 120, 487, 1978.
27. **Yang, N. S., Manning, R. F., and Gage, L. P.,** The blocked and methylated 5′ terminal sequence of a specific cellular messenger: the mRNA for silk fibroin of *Bombyx mori, Cell*, 7, 339, 1976.
28. **Hsuchen, C.-C. H. and Dubin, D. T.,** Methylated constituents of *Aedes Albopictus* poly(A)-containing messenger RNA, *Nucleic Acids Res.*, 4, 2671, 1977.
29. **Wei, C. M., Gershowitz, A., and Moss, B.,** $N^6$ 2′-0-dimethyladenosine, a novel methylated ribonucleoside next to the 5′-terminal of animal cell and virus mRNAs, *Nature (London)*, 257, 251, 1975.
30. **Lavi, S. and Shatkin, A. J.,** Methylated simian virus 40-specific RNA from nuclei and cytoplasm of infected BSC-1 cells, *Proc. Natl. Acad. Sci. U.S.A.*, 72, 2012, 1975.
31. **Moss, B. and Koczot, F.,** Sequence of methylated nucleotides at the 5′ terminus of adenovirus-specific RNA, *J. Virol.*, 17, 385, 1976.
32. **Boone, R. F. and Moss, B.,** Methylated 5′-terminal sequences of vaccinia virus mRNA species made *in vivo* at early and late times after infection, *Virology*, 79, 67, 1977.
33. **Moss, B., Gershowitz, A., Stringer, J. R., Holland, L. E., and Wagner, E. K.,** 5′-Terminal and internal methylated nucleosides in herpes simplex virus type 1 messenger RNA, *J. Virol.*, 23, 234, 1977.
34. **Moyer, S. A., Abraham, G., Adler, R., and Banerjee, A. K.,** Methylated and blocked 5′ termini in vesicular stomatitis virus *in vivo* mRNAs, *Cell*, 5, 59, 1975.
35. **Desrosiers, R. C., Sen, G. C., and Lengyel, P.,** Difference in 5′ terminal structure between the mRNA and the double-stranded virion RNA of reovirus, *Biochem. Biophys. Res. Commun.*, 73, 32, 1976.
36. **Krug, R. M., Morgan, M. A., and Shatkin, A. J.,** Influenza viral mRNA contains internal $N^6$-methyladenosine and 5′-terminal 7-methylguanosine in cap structures, *J. Virol.*, 20, 45, 1976.
37. **Dubin, D. T., Stollar, V., Hsu-Chen, C.-C., Timko, K., and Guild, G. M.,** Sindbis virus messenger RNA: the 5-′-termini and methylated residues of 26 and 42S RNA, *Virology*, 77, 457, 1977.
38. **Hewlett, M. J., Rose, J. K., and Baltimore, D.,** 5′-Terminal structure of poliovirus polyribosomal RNA is pUp, *Proc. Natl. Acad. Sci. U.S.A.*, 73, 327, 1976.
39. **DasGupta, R., Harada, F., and Kaesberg, P.,** Blocked 5′ termini in brome mosaic virus RNA, *J. Virol.*, 18, 260, 1976.
40. **Keith, J. and Fraenkel-Conrat, H.,** Tobacco mosaic virus RNA carries 5′ terminal triphosphorylated guanosine blocked by 5′ linked 7-methylguanosine, *FEBS Lett.*, 57, 31, 1975.
41. **Zimmern, D.,** The 5′ end group of tobacco mosaic virus RNA is $m^7G(5')ppp(5')Gp$, *Nucleic Acids Res.*, 2, 1189, 1975.
42. **Pinck, L.,** The 5′-end groups of alfalfa mosaic virus RNA are $^7mG(5')ppp(5')Gp$, *FEBS Lett.*, 59, 24, 1975.
43. **Wimmer, E., Chang, A. Y., Clark, J. M., Jr., and Reichmann, M. E.,** Sequence studies of satellite tobacco necrosis virus RNA. Isolation and characterization of a 5′-terminal trinucleotide, *J. Mol. Biol.*, 38, 59, 1968.
44. **Lesnaw, J. A. and Reichmann, M. E.,** Identity of the 5′-terminal RNA nucleotide sequence of the satellite tobacco necrosis virus and its helper virus: possible role of the 5′ terminus in the recognition by virus-specific RNA replicase, *Proc. Natl. Acad. Sci. U.S.A.*, 66, 140, 1970.
45. **Leung, D. W., Browning, K. S., Heckman, J. E., RajBhandary, U. L., and Clark, J. M.,** Nucleotide sequence of the 5′-terminus of satellite tobacco necrosis virus ribonucleic acid, *Biochemistry*, 18, 1361, 1979.
46. **Klootwijk, J., Klein, I., Zabel, P., and VanKammen, A.,** Cowpea mosaic virus RNAs have neither $^7mGpppN$.. nor mono-, di-, or triphosphates at their 5′ ends, *Cell*, 11, 73, 1977.
47. **Daubert, S. D., Bruening, G., and Najarian, R. C.,** Protein bound to the genome RNAs of cowpea mosaic virus, *Eur. J. Biochem.*, 92, 45, 1978.
48. **Stanley, J., Rottler, P., Davies, J. W., Zabel, P., and VanKammen, A. V.,** A protein linked to the 5′ termini of both RNA components of the cowpea mosaic virus genome, *Nucleic Acids Res.*, 5, 4505, 1978.
49. **Wei, C. M., Gershowitz, A., and Moss, B.,** 5′-Terminal and internal methylated nucleotide sequences in HeLa cell mRNA, *Biochemistry*, 15, 397, 1976.
50. **Wei, C. M. and Moss, B.,** Nucleotide sequences at the $N^6$-methyladenosine sites of HeLa cell messenger ribonucleic acid, *Biochemistry*, 16, 1672, 1977.

51. **Schibler, U., Kelley, D. E., and Perry, R. P.,** Comparison of methylated sequences in messenger RNA and heterogeneous nuclear RNA from mouse L cells, *J. Mol. Biol.*, 115, 695, 1977.
52. **Dimock, K. and Stoltzfus, C. M.,** Sequence specificity of internal methylation in B77 avian sarcoma virus RNA subunits, *Biochemistry*, 16, 471, 1977.
53. **Canaani, D., Kahana, C., Lavi, S., and Groner, Y.,** Identification and mapping of $N^6$-methyladenosine sequences in simian virus 40 RNA, *Nucleic Acids Res.*, 6, 2879, 1979.
54. **Beemon, K. and Keith, J.,** Localization of $N^6$-methyladenosine in the Rous sarcoma virus genome, *J. Mol. Biol.*, 113, 165, 1977.
55. **Lavi, U., Fernandez-Munoz, R., and Darnell, J. E., Jr.,** Content of N-6 methyladenylic acid in heterogeneous nuclear and messenger RNA of HeLa cell, *Nucleic Acids Res.*, 4, 63, 1977.
56. **Moss, B., Gershowitz, A., Weber, L. A., and Baglioni, C.,** Histone mRNAs contain blocked and methylated 5′ terminal sequences but lack methylated nucleosides at internal positions, *Cell*, 10, 113, 1977.
57. **Cheng, T.-C. and Kazazian, H. H., Jr.,** The 5′-terminal structures of murine alpha globin and beta globin messenger RNA, *J. Biol. Chem.*, 252, 1758, 1977.
58. **Cho, S., Cheng, T.-C., and Kazazian, H. H., Jr.,** Characterization of 5′-terminal methylation of human alpha and beta globin mRNAs, *J. Biol. Chem.*, 253, 5558, 1978.
59. **Shatkin, A. J.,** Capping of eukaryotic mRNAs, *Cell*, 9, 645, 1976.
60. **Filipowicz, W.,** Function of the 5′-terminal $^7$mG cap in eukaryotic mRNA, *FEBS Lett.*, 96, 1, 1978.
61. **Kozak, M.,** How do eukaryotic ribosomes select initiation regions in messenger RNA?, *Cell*, 15, 1109, 1978.
62. **Revel, M. and Groner, Y.,** Post-transcriptional and translational controls of gene expression in eucaryotes, *Annu. Rev. Biochem.*, 47, 1079, 1977.
63. **Sonenberg, N., Morgan, M. A., Merrick, W. C., and Shatkin, A. J.,** A polypeptide in eucaryotic initiation factors that crosslinks specifically to the 5′-terminal cap in mRNA, *Proc. Natl. Acad. Sci. U.S.A.*, 75, 4843, 1978.
64. **Sonenberg, N., Rupprecht, K. M., Hecht, S. M., and Shatkin, A. J.,** Eucaryotic mRNA cap binding protein: purification by affinity chromatography on Sepharose-coupled $m^7$-GDP, *Proc. Natl. Acad. Sci. U.S.A.*, 76, 4345, 1979.
65. **Furuichi, Y., LaFiandra, A., and Shatkin, A. J.,** 5′-Terminal structure and mRNA stability, *Nature (London)*, 266, 235, 1977.
66. **Lockard, R. E. and Lane, C.,** Requirement for 7-methylguanosine in translation of globin mRNA *in vivo*, *Nucleic Acids Res.*, 5, 3237, 1978.
67. **Rhagow, R. and Kingsbury, D. W.,** Endogenous viral enzymes involved in messenger RNA production, *Annu. Rev. Microbiol.*, 30, 21, 1976.
68. **Wei, C. M. and Moss, B.,** 5′-Terminal capping of RNA by guanylyltransferase from HeLa cell nuclei, *Proc. Natl. Acad. Sci. U.S.A.*, 74, 3758, 1977.
69. **Wei, C. M. and Moss, B.,** Methylation of newly synthesized viral messenger RNA by an enzyme in vaccinia virus, *Proc. Natl. Acad. Sci. U.S.A.*, 71, 3014, 1974.
70. **Urushibara, T., Furuichi, Y., Nishimura, C., and Miura, K.,** A modified structure at the 5′-terminus of mRNA of vaccinia virus, *FEBS Lett.*, 49, 385, 1975.
71. **Moss, B., Gershowitz, A., Wei, C.-M., and Boone, R.,** Formation of the guanylylated and methylated 5′-terminus of vaccinia virus mRNA, *Virology*, 72, 341, 1976.
72. **Ensinger, M. J., Martin, S. A., Paoletti, E., and Moss, B.,** Modification of the 5′-terminus of mRNA by soluble guanylyl and methyl transferases from vaccinia virus, *Proc. Natl. Acad. Sci. U.S.A.*, 72, 2525, 1975.
73. **Martin, S. A., Paoletti, E., and Moss, B.,** Purification of mRNA guanylyltransferase and mRNA (guanine-7-)-methyltransferase from vaccinia virions, *J. Biol. Chem.*, 250, 9322, 1975.
74. **Martin, S. A. and Moss, B.,** Modification of RNA by mRNA guanylyltransferase and mRNA (guanine-7-)-methyltransferase from vaccinia virions, *J. Biol. Chem.*, 250, 9330, 1975.
75. **Martin, S. A. and Moss, B.,** mRNA guanylyltransferase and mRNA (guanine-7-)-methyltransferase from vaccinia virions, *J. Biol. Chem.*, 251, 7313, 1976.
76. **Venkatesan, S., Gershowitz, A., and Moss, B.,** Modification of the 5′ end of mRNA: association of RNA triphosphatase with the RNA guanyltransferase-RNA (guanine-7-)-methyltransferase complex from vaccinia virus, *J. Biol. Chem.*, 255, 903, 1980.
77. **Tutas, D. J. and Paoletti, E.,** Purification and characterization of core-associated polynucleotide 5′-triphosphatase from vaccinia virus, *J. Biol. Chem.*, 252, 3092, 1977.
78. **Monroy, G., Spencer, E., and Hurwitz, J.,** Purification of mRNA guanylyltransferase from vaccinia virions, *J. Biol. Chem.*, 253, 4481, 1978.
79. **Monroy, G., Spencer, E., and Hurwitz, J.,** Characteristics of reaction catalyzed by purified guanylyltransferase from vaccinia virus, *J. Biol. Chem.*, 253, 4490, 1978.

80. **Shuman, S., Surks, M., Furneaux, H., and Hurwitz, J.**, Purification and characterization of a GTP-pyrophosphate exchange activity from vaccinia virions, *J. Biol. Chem.*, 255, 11588, 1980.
81. **Shuman, S. and Hurwitz, J.**, Mechanism of mRNA capping by vaccinia virus guanylyltransferase: characterization of an enzyme-guanylate intermediate, *Proc. Natl. Acad. Sci. U.S.A.*, 78, 187, 1981.
82. **Barbosa, E. and Moss, B.**, mRNA (nucleoside-2′-)-methyltransferase from vaccinia virus: purification and physical properties, *J. Biol. Chem.*, 253, 7692, 1978.
83. **Barbosa, E. and Moss, B.**, mRNA (nucleoside-2′-)-methyltransferase from vaccinia virus: characteristics and substrate specificity, *J. Biol. Chem.*, 253, 7698, 1978.
84. **Urushibara, T., Nishimura, C., and Miura, K. I.**, Process of cap formation of messenger RNA by vaccinia virus particles carrying an organized enzyme system, *J. Gen. Virol.*, 52, 49, 1981.
85. **Paoletti, E.**, The high molecular weight virion-associated RNA of vaccinia: a possible precursor to 8-12S mRNA, *J. Biol. Chem.*, 252, 872, 1977.
86. **Paoletti, E. and Lipinskas, B. R.**, The role of ATP in the biogenesis of vaccinia virus mRNA *in vitro*, *Virology*, 87, 317, 1978.
87. **Gershowitz, A., Boone, R. F., and Moss, B.**, Multiple roles for ATP in the synthesis and processing of mRNA by vaccinia virus: specific inhibitory effects of adenosine (β,γ-imido)triphosphate, *J. Virol.*, 27, 339, 1978.
88. **Paoletti, E. and Lipinskas, B. R.**, Soluble endoribonuclease activity from vaccinia virus: specific cleavage of virion-associated high-molecular-weight RNA, *J. Virol.*, 26, 822, 1978.
89. **Spencer, E., Loring, D., Hurwitz, J., and Monroy, G.**, Enzymatic conversion of 5′-phosphate-terminated RNA to 5′-di- and triphosphate-terminated RNA, *Proc. Natl. Acad. Sci. U.S.A.*, 10, 4793, 1978.
90. **Pelham, H. R. B.**, Use of coupled transcription and translation to study mRNA production by vaccinia cores, *Nature (London)*, 269, 532, 1977.
91. **Bossart, W., Paoletti, E., and Nuss, D. L.**, Cell-free translation of purified virion-associated high-molecular-weight RNA synthesized *in vitro* by vaccinia virus, *J. Virol.*, 28, 905, 1978.
92. **Cooper, J. A., Wittek, R., and Moss, B.**, Hybridization selection and cell-free translation of mRNAs encoded within the inverted terminal repetition of the vaccinia virus genome, *J. Virol.*, 37, 284, 1981.
93. **Venkatesan, S. and Moss, B.**, *In vitro* transcription of the inverted terminal repetition of the vaccinia virus genome: correspondence of initiation and cap sites, *J. Virol.*, 37, 738, 1981.
94. **Faust, M., Hastings, K. E. M., and Millward, S.**, $^{7}$mG(5′)ppp(5′)G$^{m}$pCpUp at the 5′-terminus of reovirus messenger RNA, *Nucleic Acids Res.*, 2, 1329, 1975.
95. **Furuichi, Y., Muthukrishnan, S., Tomasz, J., and Shatkin, A. J.**, Mechanism of formation of reovirus mRNA 5′-terminal blocked and methylated sequence, $^{7}$mGpppGmC, *J. Biol. Chem.*, 251, 5043, 1976.
96. **Skup, D., Zarbl, H., and Millward, S.**, Regulation of translation in L-cells infected with reovirus, *J. Mol. Biol.*, 151, 35, 1981.
97. **Zarbl, H., Skup, D., and Millward, S.**, Reovirus progeny subviral particles synthesize uncapped mRNA, *J. Virol.*, 34, 497, 1980.
98. **Skup, D. and Millward, S.**, mRNA capping enzymes are masked in reovirus progeny subviral particles, *J. Virol.*, 34, 490, 1980.
99. **Skup, D. and Millward, S.**, Reovirus-induced modification of cap-dependent translation in infected L cells, *Proc. Natl. Acad. Sci. U.S.A.*, 77, 152, 1980.
100. **Furuichi, Y.**, Methylation coupled transcription by virus-associated transcriptase of cytoplasmic polyhedrosis virus containing double-stranded RNA, *Nucleic Acids Res.*, 1, 802, 1974.
101. **Furuichi, Y.**, "Pretranscriptional capping" in the biosynthesis of cytoplasmic polyhedrosis virus mRNA, *Proc. Natl. Acad. Sci. U.S.A.*, 75, 1086, 1978.
102. **Abraham, G., Rhodes, D. P., and Banerjee, A. K.**, The 5′-terminal structure of the methylated messenger RNA synthesized *in vitro* by vesicular stomatitis virus, *Cell*, 5, 51, 1975.
103. **Testa, D. and Banerjee, A. K.**, Two methyltransferase activities in the purified virions of vesicular stomatitis virus, *J. Virol.*, 24, 786, 1977.
104. **Moyer, S. A.**, Alteration of the 5′ terminal caps of the mRNAs of vesicular stomatitis virus by cycloleucine *in vivo*, *Virology*, 112, 157, 1981.
105. **Moyer, S. A. and Banerjee, A. K.**, *In vivo* methylation of vesicular stomatitis virus and its host cell messenger RNA species, *Virology*, 70, 339, 1976.
106. **Rose, J. K.**, Heterogeneous 5′ terminal structures occur on vesicular stomatitis virus mRNA's, *J. Biol. Chem.*, 250, 8098, 1975.
107. **Perry, R. P., Kelley, D. E., Frederici, K. H., and Rottman, F. M.**, Methylated constituents of heterogeneous nuclear RNA: presence in blocked 5′ terminal structures, *Cell*, 6, 13, 1975.
108. **Salditt-Georgieff, M., Jelinek, W., Darnell, J. E., Furuichi, Y., Morgan, M., and Shatkin, A.**, Methyl labeling of HeLa cell HnRNA: a comparison with mRNA, *Cell*, 7, 227, 1976.
109. **Salditt-Georgieff, M., Harpold, M., Chen-Kiang, S., and Darnell, J. E.**, The addition of 5′-cap structures occurs early in HnRNA synthesis and prematurely terminated molecules are capped, *Cell*, 19, 69, 1980.

110. **Tamm, I., Kikuchi, T., Darnell, J. E., Jr., and Salditt-Georgieff, M.,** Short capped hnRNA precursor chains in HeLa cells: continued synthesis in the presence of 5,6-dichloro-1-β-D-ribofuranosylbenzimidazole, *Biochemistry,* 19, 2743, 1980.
111. **Babich, A., Nevins, J. R., and Darnell, J. E., Jr.,** Early capping of transcripts from the adenovirus major late transcription unit, *Nature (London),* 287, 246, 1980.
112. **Schibler, U. and Perry, R. P.,** Characterization of the 5′-termini of HnRNA in mouse L cells: implications for processing and cap formation, *Cell,* 9, 121, 1976.
113. **Schibler, U. and Perry, R. P.,** The 5′-termini of heterogeneous nuclear RNA: a comparison among molecules of different sizes and ages, *Nucleic Acids Res.,* 4, 4133, 1977.
114. **Bajszar, G., Samarina, O. P., and Georgiev, G. P.,** On the nature of 5′ termini in nuclear pre-mRNA of Ehrlich carcinoma cells, *Cell,* 9, 323, 1975.
115. **Ziff, E. B. and Evans, R. M.,** Coincidence of the promoter and capped 5′ terminus of RNA from the adenovirus 2 major late transcription unit, *Cell,* 15, 1463, 1978.
116. **Weaver, R. F. and Weissman, C.,** Mapping of RNA by a modification of the Berk-Sharp procedure: the 5′ termini of 15 S β-globin mRNA precursor and mature 10 S β-globin mRNA have identical map coordinates, *Nucleic Acids Res.,* 7, 1175, 1979.
117. **Kaehler, M., Coward, J., and Rottman, F.,** *In vivo* inhibition of Novikoff cytoplasmic messenger RNA methylation by S-tubercidinyl-homocysteine, *Biochemistry,* 16, 5770, 1977.
118. **Perry, R. P. and Kelley, D. E.,** Kinetics of formation of 5′ terminal caps in mRNA, *Cell,* 8, 433, 1976.
119. **Frederici, K., Kaehler, M., and Rottman, F.,** Kinetics of Novikoff cytoplasmic messenger RNA methylation, *Biochemistry,* 15, 5234, 1976.
120. **Groner, Y. and Hurwitz, J.,** Synthesis of RNA containing a methylated blocked 5′ terminus by HeLa nuclear homogenates, *Proc. Natl. Acad. Sci. U.S.A.,* 72, 2930, 1975.
121. **Winicov, I. and Perry, R. P.,** Synthesis, methylation and capping of nuclear RNA by a subcellular system, *Biochemistry,* 15, 5039, 1976.
122. **Groner, Y., Gilboa, E., and Aviv, H.,** Methylation and capping of RNA polymerase II primary transcripts by HeLa nuclear homogenates, *Biochemistry,* 17, 977, 1978.
123. **Gidoni, D., Kahana, C., Canaani, D., and Groner, Y.,** Specific *in vitro* initiation of transcription of simian virus 40 early and late genes occurs at the various cap nucleotides including cytidine, *Proc. Natl. Acad. Sci. U.S.A.,* 78, 2174, 1981.
124. **Hagenbuchle, O. and Schibler, U.,** Mouse β-globin and adenovirus-2 major late transcripts are initiated at the cap site *in vitro, Proc. Natl. Acad. Sci. U.S.A.,* 78, 2283, 1981.
125. **Wei, C.-M. and Moss, B.,** 5′-Terminal capping of RNA by guanylyltransferase from HeLa cell nuclei, *Proc. Natl. Acad. Sci. U.S.A.,* 74, 3758, 1977.
126. **Weil, P. A., Luse, D. S., Segall, J., and Roeder, R. G.,** Selective and accurate initiation of transcription at the Ad 2 major late promoter in a soluble system dependent on purified RNA polymerase II and DNA, *Cell,* 18, 469, 1979.
127. **Manley, J. L., Fire, A., Cano, A., Sharp, P. A., and Gefter, M. L.,** DNA-dependent transcription of adenovirus genes in a soluble whole cell extract, *Proc. Natl. Acad. Sci. U.S.A.,* 77, 3855, 1980.
128. **Laycock, D. E.,** Purification of eucaryoticguanylyltransferase, *Fed. Proc.,* 36, 770, 1977.
129. **Mizumoto, K. and Lipmann, F.,** Transmethylation and transguanylylation in 5′-RNA capping system ioslated from rat liver nuclei, *Proc. Natl. Acad. Sci. U.S.A.,* 76, 4961, 1979.
130. **Venkatesan, S., Gershowitz, A., and Moss, B.,** Purification and characterization of mRNA guanylyltransferase from HeLa cell nuclei, *J. Biol. Chem.,* 255, 2829, 1980.
131. **Venkatesan, S. and Moss, B.,** Donor and acceptor specificities of HeLa cell mRNA guanylyltransferase, *J. Biol. Chem.,* 255, 2835, 1980.
132. **Keith, J. M., Venkatesan, S., Gershowitz, A., and Moss, B.,** Purification and characterization of the mRNA capping enzyme GTP: RNA guanylyltransferase from wheat germ, *Biochemistry,* 21, 327, 1982.
133. **Venkatesan, S. and Moss, B.,** Eucaryotic mRNA capping enzyme-guanylate covalent intermediate, *Proc. Natl. Acad. Sci. U.S.A.,* 79, 340, 1982.
134. **Keith, J. M., Galer, D., and Westreich, L.,** Wheat germ mRNA capping and methylating enzymes, in *Biochemistry of S-Adenosylmethionine and Related Compounds,* Usdin, E., Borchardt, R. T., and Creveling, C. R., Eds., MacMillan, New York, 1982.
135. **Keith, J. M.,** 5′-Terminal modification of mRNAs by viral and cellular enzymes, in *Enzymes of Nucleic Acid Synthesis and Processing,* Jacobs, S. T., Ed., CRC Press, Boca Raton, Fla., 1983, 111.
136. **Ensinger, M. J. and Moss, B.,** Modification of the 5′ terminus of mRNA by an RNA (guanine-7-)-methyltransferase from HeLa cells, *J. Biol. Chem.,* 251, 5283, 1976.
137. **Germerhausen, J., Goodman, D., and Somberg, E. W.,** 5′-Cap methylation of homologous poly A(+) RNA by a RNA (guanine-7-)methyltransferase from *Neurospora crassa, Biochem. Biophys. Res. Commun.,* 82, 871, 1978.

138. **Locht, C., Bouchet, H., and Delcour, J.,** Partial purification of the mRNA(guanine-7-)methyltransferase from wheat germ, in *Biochemistry of S-Adenosylmethionine and Related Compounds*, Usdin, E., Borchardt, R. T , and Creveling, C. R., Eds., MacMillan, New York, 1982.
139. **Lavers, G. C.,** Detection of methyltransferase activities which modify GpppG to m[7]GpppGm in embryonic chick lens, *Mol. Biol. Rep.*, 3, 275, 1977.
140. **Langberg, S. R. and Moss, B.,** Post-transcriptional modification of mRNA. Purification and characterization of cap I and cap II RNA (nucleoside-2′-)-methyltransferase from HeLa cells, *J. Biol. Chem.*, 256, 10054, 1981.
141. **Keith, J. M., Ensinger, M. J., and Moss, B.,** HeLa cell RNA (2′-0-methyladenosine-N[6]-)-methyltransferase specific for the capped 5′-end of messenger RNA, *J. Biol. Chem.*, 253, 5033, 1978.
142. **Keith, J. M., Winters, E. M., and Moss, B.,** Purification and characterization of a HeLa cell transfer RNA(cytosine-5-)-methyltransferase, *J. Biol. Chem.*, 255, 4646, 1980.

Chapter 6

# POLY(A) IN EUKARYOTIC mRNA

**Joseph R. Nevins**

## TABLE OF CONTENTS

# I. INTRODUCTION

The formation of a eukaryotic mRNA involves a complex series of biochemical events. The primary transcript is not the messenger RNA, but instead it must first undergo modifications such as capping of the 5′ terminus, methylation of certain internal adenylate residues, the addition of poly(A) at specific 3′ termini, and the splicing together of sequences that are noncontiguous in the genome.[1,2,3] Finally, the fully processed mRNA must then be actively transported from the nucleus to the cytoplasm in order to function in directing the synthesis of some particular protein.

Probably the earliest known modification to mRNA is the segment of polyadenylic acid — poly(A) — found at the 3′ terminus of most eukaryotic mRNAs. The discovery of the association of poly(A) with mRNA was made some 12 years ago.[4-6] Although a considerable amount of information has accumulated pertaining to the structure and properties of poly(A), the determination of the function of the poly(A) sequence in mRNA has been elusive. Recently, however, experiments have been performed that suggest physiological roles for the poly(A). In this presentation, no attempt will be made to reiterate all of the facts concerning the structure, occurrence, and enzymatic biosynthesis of poly(A). These subjects have, for the most part, been thoroughly reviewed (for instance, see References 7 and 8), and little in the way of new information is now available. Rather, I will attempt to cover the various aspects of poly(A) as it relates to the biogenesis and metabolism of eukaryotic mRNA. As will become obvious, the material that is covered draws heavily on experiments utilizing adenovirus and to a lesser degree SV40 to study poly(A) function. This bias in material stems not from a prejudice but rather a reflection of the workability of the virus systems rendering them particularly useful for such studies.

# II. ADDITION OF POLY(A) TO RNA

## A. Generation of Poly(A) Addition Sites

The addition of poly(A) to nuclear RNA molecules that are the precursors to messenger RNA, the so-called heterogeneous nuclear RNA (hnRNA), occurs by the sequential polymerization of approximately 200 to 250 adenylate residues onto the proper 3′ terminus of the RNA. The size of the newly synthesized poly(A) in the nucleus has been determined both by gel electrophoresis of RNase-liberated poly(A) as well as by a determination of the ratio of AMP to $A_{OH}$ residues after labeling.[8a,8b] The mechanism by which a rather precise number of adenylate residues are polymerized into a poly(A) tract is not at all clear at this point. The fact that this reaction is template independent is well established.[9-11] But how is a 3′ terminus generated and what mechanism determines the proper 3′ terminus? Two distinct mechanisms would produce a free 3′ terminus capable of serving as a primer for poly(A) formation. As depicted in Figure 1, RNA transcripts might terminate precisely at the appropriate poly(A) addition site leaving a free 3′ OH for priming. Alternatively, endonucleolytic cleavage of a larger precursor molecule might also produce a proper 3′ terminus. In the more complex situation of multiple poly(A) sites within a single transcription unit, again either termination of transcription or RNA chain cleavage could generate poly(A) addition sites. In this complex instance, if the two poly(A) sites were utilized with equal frequencies, then termination of transcription vs. RNA chain cleavage can be distinguished. A termination mechanism would result in more transcription near the first poly(A) site than near the second. That is, for every two polymerases transcribing the first poly(A) site, only one polymerase would transcribe the second poly(A) site. Therefore, if transcription rates were found to be equal at the first and second poly(A) sites, then this would indicate that an RNA chain cleavage was necessary to produce the first poly(A) site, although not necessarily the second poly(A) site. The late adenovirus transcription unit is an example of such

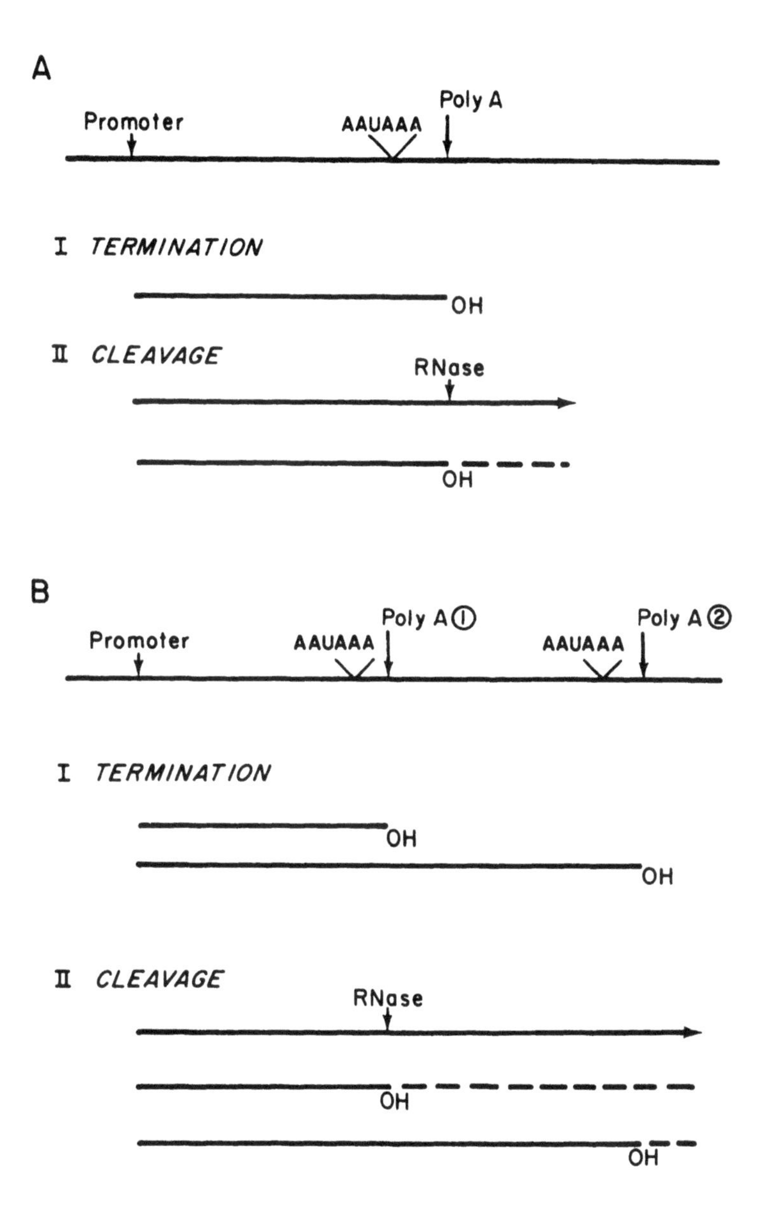


FIGURE 1. Possible mechanisms for generation of poly(A) addition sites. A. Example of a simple transcription unit containing a single site for poly(A) addition. *Termination* of transcription at the poly(A) site generates a 3′ OH end suitable for poly(A) addition. *Cleavage* of an RNA transcript to generate the poly(A) site would be required if transcription continued beyond the poly(A) site. B. Example of a complex transcription unit containing two sites, (1) and (2), for poly(A) addition. Again, as described in part A, either transcription termination or RNA chain cleavage could generate the poly(A) addition sites. Furthermore, a combination of termination and cleavage could operate. Thus, termination might prevent the use of the second poly(A) site, but cleavage would still be the mechanism for the generation of the first poly(A) site, that is, if termination occurred before site (2) but after site (1).

a complex transcription unit (Figure 2). In this case, five poly(A) sites are contained within this single large transcription unit.[12-15] Since the addition of poly(A) at the five sites occurs with approximately equal frequency,[16] then transcriptional termination as a mechanism for

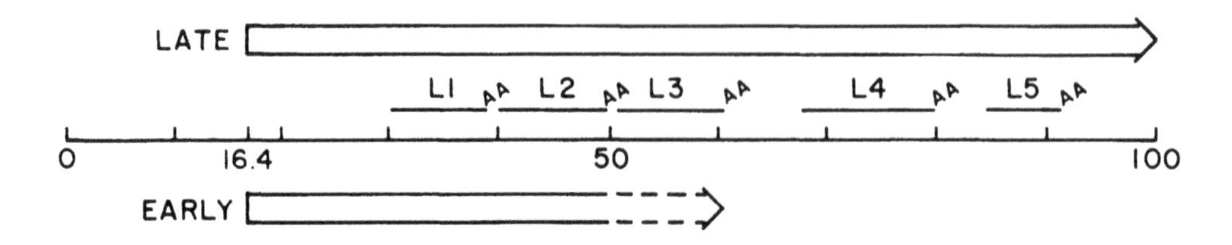


FIGURE 2. Poly(A) addition sites in the late adenovirus transcription unit. The late adenovirus transcription unit is defined by the promoter at map coordinate 16.4 and the termination site at ~99. Five families of 3′ co-terminal mRNAs, L1 through L5, are encoded within this transcription unit. The mRNAs of a particular family share a poly(A) site. Certain sequences between mRNA families (such as between L3 and L4) do not appear in mRNA. Poly(A) addition sites are located at map coordinates 38.5 (L1), 50.0 (L2), 61.5 (L3), 79.5 (L4), and 91.5 (L5). For a discussion of the details of this transcription unit see References 90 and 91. Transcription from the same promoter during early infection (prior to viral DNA replication) terminates between 50 and ~65. Therefore, as referred to in Figure 1, termination prevents the utilization of the L4 and L5 sites but the generation of the L1 and the L2 sites, and possibly the L3 site, still requires an RNA chain cleavage.

generating poly(A) sites should yield transcription rates that are five times greater at the first poly(A) site than at the fifth poly(A) site. Such an experiment performed with the late adenovirus transcription unit has in fact demonstrated that the first four poly(A) sites are created by an RNA chain cleavage.[16] That is, transcription across the entire transcription unit was found to be equal. The last poly(A) site is also generated by cleavage since it was demonstrated that transcription continued beyond the final poly(A) site and at a rate equal to the transcription preceding the poly(A) site.[17]

The finding that even the last poly(A) site in the late adenovirus transcription unit is generated by cleavage presents the possibility that a simple transcription unit, one with only a single poly(A) site, might also produce the poly(A) site by an RNA chain cleavage. This possibility was confirmed by analyzing the transcription from the late region of SV40.[18] It was found that transcription could be detected extending beyond the single poly(A) addition site at a rate indicating that a chain cleavage was required virtually every time to generate the poly(A) site. Furthermore, when the nuclear RNA in SV40-infected cells was mapped by the S1 technique, it was found that the poly(A)$^-$ RNAs possessed 5′ termini in common with the poly(A)$^+$ RNAs, but whereas the poly(A)$^+$ RNAs had 3′ termini at the poly(A) site, the poly(A)$^-$ molecules extended beyond this site with heterogeneous 3′ termini.[19] Although there is no proof in this case that these poly(A)$^-$ molecules are precursors to the poly(A)$^-$ molecules, when taken together with the labeling studies it seems certain that endonucleolytic cleavage must generate the poly(A) site. Transcription beyond poly(A) sites has also been deduced from the nature of the formation of late polyoma virus nuclear RNA[20,21] and two early adenovirus transcription units have been shown to generate poly(A) addition sites through an RNA chain cleavage.[22] Thus, even in the cases where no choice must be made between multiple sites, a chain cleavage is utilized to produce the poly(A) site. And very recently, for a cellular transcription unit, the mouse beta globin gene,[23] it has been shown that the poly(A) site must be generated by an RNA chain cleavage rather than by transcription termination. Therefore, in every case where pulse-labeled RNA has been examined, a poly(A) site is produced through an RNA chain cleavage rather than by termination of transcription at the poly(A) addition site. There is at least one instance where it has been suggested that the poly(A) site is produced by transcription termination. Analysis of the nuclear precursors to the ovalbumin mRNA revealed that the 3′ ends were the same as in the mature mRNA.[24] Furthermore, no nuclear RNA sequences could be detected that were encoded in the DNA distal to the poly(A) site. However, pulse-labeled RNA was not measured, making the interpretation difficult if potential transcripts beyond the poly(A) site were rapidly destroyed. Nevertheless, it does remain a possibility that both mechanisms, transcription termination or RNA chain cleavage, exist for producing poly(A) sites. However, the demonstration of termination at a specific site is oviously much more difficult to prove than to simply demonstrate transcription beyond a poly(A) site that then requires an RNA chain cleavage.

### B. Role of the AAUAAA Sequence

If the general mechanism for generating poly(A) sites is by an endonucleolytic cleavage, then what determines the cleavage site? A common feature of most eukaryotic mRNAs is the occurrence of the sequence AAUAAA preceding the poly(A) by 11 to 30 nucleotides.[25] The sequence appears to be very highly conserved, implying that it may have an important role in either recognition of the site to produce the cleavage or in recognition by the poly(A) polymerase so as to achieve poly(A) addition. (There is now at least one example of a variation on the sequence in two separate mRNAs; AUUAAA occurs in both the chicken lysozyme mRNA[26] and in one of the adenovirus E3 mRNAs.[26a] In addition, the pancreatic alpha amylase mRNA has a sequence of AAUAUA.[27]) The role of the AAUAAA sequence in determining the poly(A) site has been directly investigated through the use of deletion mutants of SV40.[28] The SV40 late mRNAs, although heterogeneous at their 5′ termini, are quite homogeneous at their 3′ termini. Poly(A) is added at the same site for each mRNA, and preceding this site by 12 nucleotides is the sequence AAUAAA. That this AAUAAA sequence is actually involved in poly(A) addition is shown by the fact that the deletion of the sequence eliminates poly(A) addition. Viable SV40 variants were constructed so as to possess a late region containing two tandem poly(A) sites (and thus two AAUAAA sequences). Deletions in one of the AAUAAA sequence were then constructed and the RNAs resulting from these mutants analyzed. Virus containing both poly(A) sites produced mRNAs that utilized both sites. Deletion of one of the AAUAAA sequences, however, resulted in the loss of the mRNA that normally utilized this site while the other mRNA continued to be produced.

Deletions were also generated around the AAUAAA, but not affecting the hexanucleotide sequence itself. Deletions 3′ to the AAUAAA, that is between the AAUAAA and the poly(A), produced RNAs in which poly(A) addition occurred downstream from the normal poly(A) addition site. In each case, poly(A) addition took place at a site within 11 to 19 nucleotides from the AAUAAA. Thus it would appear that there is no sequence requirement between the AAUAAA and the poly(A) site but rather just a spatial requirement. Finally, deletions placed 5′ to the AAUAAA also altered the normal site of poly(A) addition, but not in a predictable manner.

Variability in the site of poly(A) addition has also been observed in an in vivo situation. An examination of the sequences adjacent to bovine prolactin mRNA revealed that the poly(A) was added at several sites within an 11 base-pair region, although this presumably still represents a single "poly(A) site" as defined by a single AAUAAA.[29] The same situation appears also to be the case for one of the adenovirus E3 mRNAs.[26a] Therefore, the formation of the poly(A) site, whether this occurs through transcription termination or by endonuclease cleavage, may not always be an exact event. An alternative explanation could be that exonuclease degradation before poly(A) addition occurs results in heterogeneity of the nucleotide sequence adjacent to the poly(A).

Clearly then, the AAUAAA sequence appears to be essential for the formation of a poly(A) containing RNA. However, the precise function of the sequence in the poly(A) addition event still remains uncertain. Does the AAUAAA specify the cleavage of the RNA chain at the proper site or does it serve as a recognition site for the poly(A) polymerase? In other words, are correct 3′ ends generated in the absence of the AAUAAA sequence with the subsequent failure of poly(A) addition, or are the RNA chains simply not cleaved and thus do not serve as substrates for poly(A) addition? This question has been dealt with in recent experiments utilizing a mutant in the adenovirus EIA transcription unit. Site specific mutagenesis was employed to convert the AAUAAA to AAGAAA (29a). Analysis of the nuclear RNAs produced by this mutant revealed that the normal EIA 3′ terminus produced by wild type virus was found in only about 10% of the transcripts in the mutant-infected cells. The majority of the transcripts apparently proceeded beyond into the adjacent transcription unit. The small percentage of the RNAs that were cleaved at the normal site were in fact polyadenylated. Thus, it would appear that the AAUAAA sequence was required for the proper cleavage reaction but not for the subsequent poly(A) polymerization reaction.

### C. In Vitro Addition of Poly(A) to RNA

From the preceding discussions it can be seen that certain aspects of the details of poly(A) addition are becoming clear. However, what is totally obscure at the moment are the protein factors that carry out the poly(A) addition, and more important, those that impart the specificity in poly(A) site selection. Possibly the best way to answer questions of how poly(A) is added and what determines the specificity is through the use of in vitro systems capable of faithfully reproducing the in vivo situation. Once in hand, an in vitro system can then be manipulated so as to define essential components including, hopefully, those imparting specificity. Numerous poly(A) polymerase activities have been detected in the nucleus and the cytoplasm of a wide variety of cell types and tissues (for a review, see Reference 30). However, there has yet to be a clear demonstration that any of these enzymes are responsible for the *de novo* synthesis of the poly(A) segment that appears on nuclear RNA. A possible exception is a poly(A) synthesizing activity that has been detected in the chromatin fraction of cells. This activity is sensitive to cordycepin triphosphate, suggesting that it may be the activity that normally adds the initial poly(A) segment.[31] However, progress toward the goal of in vitro synthesis of poly(A) and its addition to RNA principally involves the use of isolated nucleus transcription systems or crude extracts capable of carrying out transcription of specific genes. It is most important to assess the fidelity of the system, that is, whether poly(A) *addition* occurs at the correct sites rather than just the polymerization of ATP.

Isolated nuclei from adenovirus-infected cells have been shown by several groups to synthesize viral RNA containing poly(A).[32,33] The poly(A) sequence found in the in vitro synthesized RNA appears to be of the proper size and to be added at correct sites, indicating that the isolated nuclei retain the in vivo specificity of poly(A) addition. Furthermore, the relative selection of poly(A) sites in vitro appeared to reflect the in vivo situation.[16,32]

Recently, a system prepared from late adenovirus-infected cells that is template independent has been described.[34] Infected cell extracts utilizing the endogenous viral DNA were shown to synthesize viral RNA in vitro with poly(A) of appropriate size added at the correct sites. Of interest was the added finding that rather large complexes of ~70S could simultaneously synthesize RNA and polyadenylate, suggesting a physical association of transcriptional activity and poly(A) polymerase activity. This is of relevance in light of the manner in which these sites become polyadenylated in vivo; the two events of transcription and poly(A) addition appear to be tightly linked (see below).

## III. ROLE OF POLY(A) IN RNA PROCESSING

### A. Temporal Relationship of Poly(A) Addition and RNA Processing

In order to assess the role of poly(A) in mRNA biogenesis, an important first consideration must be the temporal relationship of poly(A) addition and the various aspects of RNA processing. The available evidence indicates that the addition of poly(A) to nuclear RNAs is a very early event in the biogenesis of an mRNA. This has been determined in two ways. First, kinetic measurements of incorporation of label into the body of the RNA and the appearance of the sequences as poly(A)$^+$ indicate that there is very little lag time between the labeling of the RNA and the addition of the poly(A) segment.[16,35] In fact, it was concluded that RNA chain cleavage of the late adenovirus transcripts to produce poly(A) addition sites took place while transcripts were still nascent, with the subsequent polyadenylation occurring in a rapid fashion.[16]

However, since it is possible that splicing of the RNA could also occur while the RNA is nascent, then, in fact, poly(A) addition could actually follow splicing even though poly(A) addition occurred quite rapidly. To answer such a question, a second approach regarding the timing of poly(A) addition and RNA processing has been to analyze the size of poly(A) containing nuclear RNAs. In each case examined, the newly synthesized nuclear RNAs that contained poly(A) were, as best as could be judged, unspliced molecules that were still collinear with their template.[16,18,36-39] Analysis of pulse-labeled adenovirus nuclear RNAs that contained poly(A) demonstrated that only with longer labeling times did spliced molecules begin to appear (see Figure 3). The initially labeled nuclear RNA that is the precursor to the fiber mRNA is of a size suggesting an unspliced molecule. Only after a much longer period of labeling does the mature mRNA-sized molecule appear in the poly(A)$^+$ fraction. In addition, fingerprint analysis of poly(A)$^+$ nuclear RNA from late adenovirus-infected cells indicated that the RNAs still retained the sequences that eventually are spliced out to form the mRNA.[40] Thus it appears that splicing of the 5′ sequences occurs subsequent to poly(A) addition. However, this is not always the case, since some adenovirus transcripts that lacked poly(A) were found to have undergone partial splicing of the 5′ segments.[41]

The demonstration of pulse-labeled poly(A)$^+$ nuclear precursors for cellular gene transcripts has, for the most part, been hampered by the relatively low levels of transcription of most cellular genes. There are reports of newly synthesized precursor molecules that contain poly(A),[42-45] although labeling times have not been short enough such that only the initial precursor was labeled. Furthermore, poly(A) containing nuclear transcripts in the steady-state population that are collinear with the gene (unspliced) have been shown for a variety of transcripts.[3] These experiments thus demonstrate that poly(A) *can* be added prior to splicing but they do not say whether this is the obligatory pathway. For instance, even the pulse-labeling experiments cannot rule out the possibility that poly(A) is also added to older, already spliced molecules. However, analysis of the size of specific viral RNA molecules to which labeled poly(A) was added (that is, the substrates for poly [A] addition) indicated that mainly large molecules greater than the size of the mature mRNA served as the poly(A) acceptor.[16,39] Thus it seems likely that poly(A) addition takes place prior to splicing of transcripts.

### B. Involvement of Poly(A) Addition in Selective RNA Processing

Since poly(A) addition appears to precede the splicing event, then one might argue that the selection of a poly(A) site is a crucial first event in mRNA biogenesis. What must be emphasized at this point is that the event of 3′ end formation and poly(A) addition is indeed a *selection*. For example, in each instance in which a transcription unit is composed of multiple poly(A) sites, the mature mRNAs contain common sequences that were derived from the 5′ end of the transcription unit. If it were true that the AAUAAA sequence is always a signal for RNA chain cleavage, then cleavage at the first site would eliminate the ability to form the mRNA using the second poly(A) site. The presence of multiple AAUAAA sequences within a complex transcription unit that are coincident with sites of poly(A) addition has been verified for the adenovirus major late transcription unit.[47a] In other words

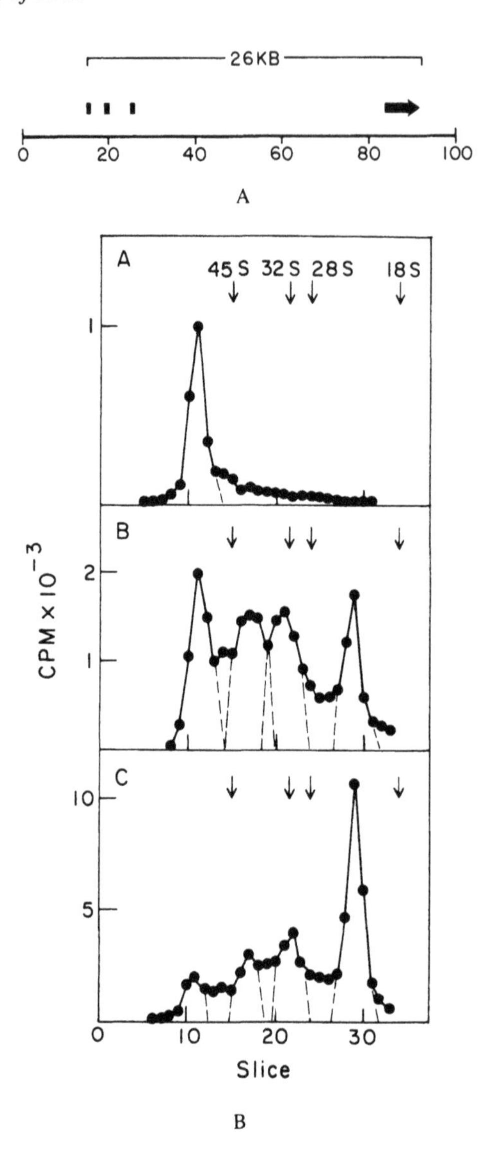


FIGURE 3. Poly(A) containing nuclear RNA that is the precursor to the adenovirus fiber mRNA. A. The adenovirus mRNA encoding the fiber protein derives from the L5 region of the late transcription unit (see Figure 2) and is constructed of four exons as depicted. The distance from the promoter at 16.4 to the poly(A) site at 91.5 is approximately 26 kb. The mature mRNA is about 3 kb in length. Therefore, if poly(A) addition *preceded* splicing, the initial poly(A)$^+$ fiber mRNA would be about 26 kb. However, if poly(A) was added *after* splicing, then the initial poly(A)$^+$ molecule would be about 3 kb in length. B. Poly(A)$^+$ nuclear RNA labeled for 5 min (A), 25 min (B), or 80 min (C) was fractionated in polyacrylamide agarose gels and hybridized with a fiber mRNA-specific DNA probe. The mature mRNA migrates in gel slice 29. The two peaks of RNA in slice 13 and slice 17 are intermediates in RNA splicing. The size markers are ribosomal RNA molecules of 14 kb (45S), 6.7 kb (32S), 5.1 kb (28S), and 2.1 kb (18S). (Figure 3B reproduced from Nevins, J. R., *J. Mol. Biol.*, 130, 493, 1979. Copyright: Academic Press, Inc. [London] Ltd. With permission.)

where more than one poly(A) site exists in a transcriptional unit, then a *choice* must be made to utilize one and only one of the sites. Conceivably, some mechanism might insure that only one selection be made per transcription event, with the selection itself being a random event. Alternatively, the selection could be active and predetermined, a possibility suggested by the manner in which the adenovirus poly(A) sites are selected. As mentioned previously, the late adenovirus transcription unit is complex in that more than one poly(A) addition site is utilized (see Figure 2). It was found that poly(A) site selection and poly(A) addition occurred *during* transcription well before the polymerase terminated.[16] For example, the Ll poly(A) site is selected and adenylated before the L5 site is even transcribed. Therefore, if the selection were purely random one would predict a polar effect whereby the first poly(A) site was selected most frequently. This was not the case. Rather, the selection was roughly equal among the five sites implying an active selection of a particular site instead of a chance selection.

This situation that requires selection of a poly(A) site among several possibilities is now not unique in that the transcription unit encoding the immunoglobulin μ heavy chain mRNAs also contain multiple poly(A) addition sites (see Figure 4).[47-49] Since in each case only a single mRNA results from a transcriptional event, then a choice must be made between the two poly(A) sites.

Furthermore, there is now good evidence that poly(A) site selection can be a regulatory event. A portion of the late adenovirus transcription unit is expressed early in infection prior to viral DNA replication.[50-55] However, in contrast to the expression of this transcription unit late in infection, only a single mRNA is produced early in infection.[50,51,54] Poly(A) site selection operating at two levels is largely responsible for this control. First, termination of transcription prior to the L4 poly(A) site prevents expression of the L4 and L5 regions[51,53,54] and thus selection of the L4 and L5 poly(A) sites. Second, although the L1, L2, and possibly L3 poly(A) sites are transcribed early in infection, selection of the L1 site predominates.[51] This is in contrast to late in infection when the L1 site is selected less frequently than the L2 site.[16]

By analogy the same scenario could hold true for the expression of the immunoglobulin heavy-chain mRNAs. Initially, the μ membrane mRNA is produced, which must involve a preferential utilization of the μ membrane poly(A) site over the μ-secreted poly(A) site since both sites must be transcribed if the μ membrane site is to be used (see Figure 4). The change, then, to production of the μ-secreted mRNA could be mediated either by transcriptional termination prior to μ membrane sequences or by a change in posttranscriptional poly(A) site selection now favoring the μ-secreted poly(A) site (see Figure 4). The situation now appears even more complex in that the delta heavy-chain mRNAs can be encoded in the same transcription unit as described above.[56] In this case, two additional poly(A) sites encoding the delta mRNAs that are 3′ to the μ membrane site are now available. Once again, control of the expression of the mRNAs would take place through poly(A) site selection, either by transcriptional termination or by 3′ end selection.

Finally, examples of multiple poly(A) sites for a single mRNA can be found for the alpha amylase mRNA[57] as well as the mRNA for dihydrofolate reductase.[58] However, it is yet to be shown whether these variations in poly(A) site, that still produce apparently the same mRNA with respect to protein product, have any biological significance.

### C. Poly(A) and RNA Splicing

Because poly(A) is added prior to splicing, a natural question arises as to the requirement for poly(A) in splicing. Obviously, the way to answer such a question would be to prevent poly(A) addition to nuclear RNA without disturbing normal RNA synthesis. One could then follow the fate of such RNA. Ideally, a viral mutant conditionally deficient in poly(A) formation would provide the best experimental circumstances for such an analysis. However, since the nuclear DNA viruses utilize host enzymes, no such mutants are available and one must then turn to the use of inhibitors. Cordycepin (3′-deoxyadenosine) has been used over the years to selectively prevent poly(A) synthesis.[9,59] Employing short exposures to the drug, there is little effect on RNA polymerase II activity (although RNA polymerase I is severely inhibited).[60] Therefore, under appropriate conditions, poly(A) addition can be prevented without significantly altering RNA synthesis. Although there has been a report indicating

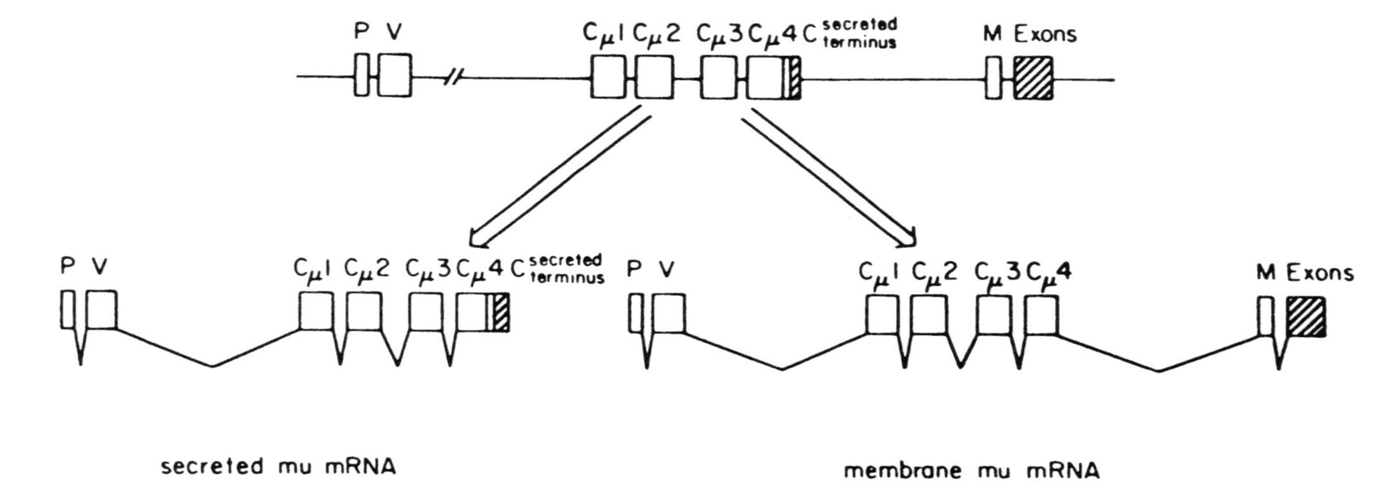


FIGURE 4. Structure of the transcription unit encoding the mouse μ membrane and secreted immunoglobulin heavy-chain mRNAs. The hatched boxes represent the exons to which poly(A) is added. (From Early, P., Rogers, J., Davis, M., Calame, K., Bond, M., Wall, R., and Hood, L., *Cell*, 20, 313, 1980. Copyright by M.I.T., Cambridge, Mass. With permission.)

that globin transcription is severely affected by cordycepin,[61] other experiments measuring transcription of various specific viral and cellular genes have shown little or no effect on transcription.[62] That poly(A) has a crucial role in mRNA biogenesis has been known for many years, based on just such an experiment. That is, there is no accumulation of mRNA in the cytoplasm in the presence of cordycepin.[9,11,59]

The question of poly(A) involvement in splicing has been investigated using adenovirus transcription as the system. Adenovirus-infected cells were exposed to cordycepin and then labeled for 30 min with $^3$H-uridine. The labeled nuclear RNA specific to the viral early region 2 (a simple transcription unit containing one poly(A) site) was then analyzed and compared to RNA labeled in untreated cells.[62] In untreated cells, the region 2 poly(A)-containing RNA consists of a 5-kb unspliced molecule, a 4-kb splicing intermediate and a 2.1-kb spliced product,[37,52] The 2.1-kb spliced product that is the size of the mature mRNA is the major species in the nucleus after such a labeling period. In the cordycepin-treated cells, the major species was a 1.9-kb RNA, the size expected for the fully spliced product lacking the 200-nucleotide poly(A) segment (Figure 5). It was further shown that this RNA did, in fact, lack a poly(A) segment and contained the same sequences that are noncontiguous in the genome as did the control RNA, and therefore indicated that splicing had occurred in the absence of poly(A) addition. Therefore, these results indicate that, in the absence of poly(A) synthesis, a nuclear precursor RNA can still be correctly processed to yield a mature mRNA and at about the same efficiency as occurs when poly(A) has been added to the RNA.

### D. Involvement of Poly(A) in Nuclear-Cytoplasmic Transport of mRNA

The role of poly(A) in transport of mRNA sequences from the nucleus to the cytoplasm has been investigated primarily in two ways. The first approach seeks to measure the efficiency by which mRNA sequences are transported to the cytoplasm with respect to the presence of poly(A). Although these experiments cannot establish a requirement for poly(A) in transport, they do provide a picture of the involvement of poly(A) in mRNA biogenesis. The earliest experiments measuring conservation of poly(A) during nuclear-cytoplasmic transport yielded conflicting results, suggesting either near-complete transfer of poly(A) from the nucleus to the cytoplasm[10] or only a partial transfer with a large fraction turning over within the nucleus.[63] These experiments are complicated by the fact that the total populations of poly(A) were assayed and also by the fact that in short labeling times the terminal addition labeling of poly(A) in the cytoplasm is significant.[8a,8b] However, it does appear safe to conclude that the RNA that is transported to the cytoplasm, by and large, contains poly(A), and that a significant portion of the transcripts formed in the nucleus does not acquire a poly(A) and does not appear in the cytoplasm.[64]

Of course, at least one group of mRNAs, the histone mRNAs, are not adenylated and certainly are transported.[65-68] (whether in fact this is the only group of poly(A)$^-$ RNAs is still in doubt — for a discussion see References 8, 69 and 70) Thus it is clear that at least in the case of these particular mRNAs there is not an obligatory requirement for poly(A) in transport.

The analysis of specific sequences has provided more detailed and conclusive information. Late adenovirus mRNA,[16,71] early adenovirus mRNA,[51] and mRNA specific to certain CHO cell cDNAs[72] as well as the rat growth hormone RNA[45] all show efficient transport of the poly(A) containing RNA. There appears to be no significant transport of specific sequences that do not contain poly(A) (with the exception of the histone RNAs), and the specific sequences that do become polyadenylated are efficiently transported. This conclusion is perhaps not an inflexible rule, however, since certain early transcripts from the adenovirus late transcription unit are poorly transported despite the fact that they contain poly(A).[51] This finding, therefore, raises the possibility that transport may be regulated as well as suggesting that the addition of poly(A) to an RNA does not insure successful transport.

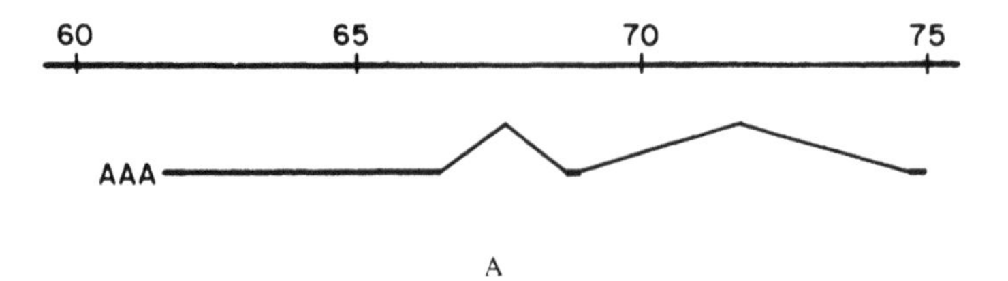


A

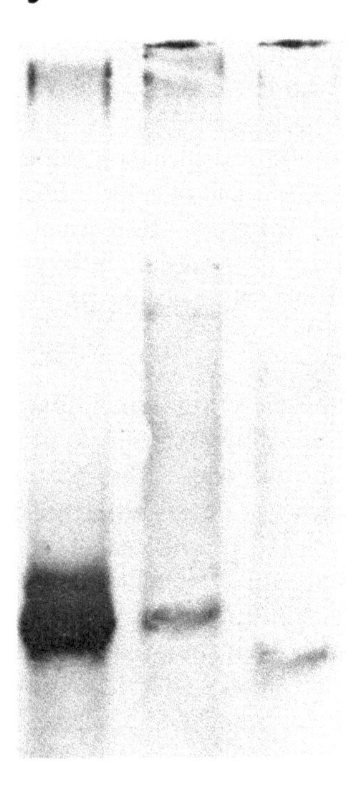


B

FIGURE 5. RNA splicing in the absence of poly(A) addition. A. The E2 transcription unit of adenovirus has a promoter at map coordinate 75.1 and a poly(A) site at 61.5. The major mRNA product is 2.1 kb in length, including the poly(A), and is constructed of 3 exons as shown. B. E2-selected nuclear RNA, pulse-labeled with $^3$H-uridine in the absence (control) or the presence of cordycepin (3′ dA), was analyzed by gel electrophoresis followed by fluorography. The $A^+$ cyto RNA was run as a marker for the fully spliced E2 RNA which is the predominant nuclear species. The band of E2 nuclear RNA from control cells is 2.1 kb in length. The band of E2 nuclear RNA from cordycepin-treated cells is 1.9 kb in length and lacks a poly(A) segment. (Adapted from Zeevi, M., Nevins, J. R., and Darnell, J. E., Jr., *Cell,* 26, 39, 1981. Copyright M.I.T., Cambridge, Mass. With permission.)

The second approach involves the prevention of poly(A) synthesis using cordycepin with the subsequent analysis of the fate of the nuclear RNA minus poly(A). From the earliest experiments performed with cordycepin, it has been concluded that poly(A) was essential for mRNA transport. That is, no labeled mRNA appeared in the cytoplasm in the absence of poly(A) synthesis.[9,11,59] However, these experiments examined the accumulation of $^3$H-labeled RNA after 30 to 60 min of labeling, leaving open the possibility that the mRNA minus the poly(A) was in fact transported by was then rapidly degraded once in the cytoplasm. That this latter possibility might be the case has been suggested by recent experiments employing cordycepin to prevent poly(A) formation in early adenovirus-infected cells. As mentioned before, spliced nuclear RNA lacking a poly(A) segment was found to accumulate in the presence of cordycepin in amounts comparable to the spliced poly(A)-containing form.[62] Furthermore, transcription rate measurements indicated near-normal levels of synthesis of viral RNAs in the presence of cordycepin. The amount of 45-min-labeled RNA in the cytoplasm of cordycepin-treated cells was only 10% of the control when the labeling time was 45 min, implying a failure to transport the properly spliced nuclear RNAs that lacked the poly(A) segment. However, when the very initial appearance of mRNA in the cytoplasm was measured, it was found that nearly normal amounts of RNA exited from the nucleus in the cordycepin-treated cells.[73] Furthermore, the RNA that could be found in the cytoplasm in cordycepin-treated cells was present in polysomes indicating that the cytoplasmic appearance was not simply due to nonspecific nuclear leakage. Therefore, it appears that mRNA lacking a poly(A) segment can still be transported from the nucleus to the cytoplasm with a near-normal efficiency. However, the mRNA fails to accumulate in the cytoplasm to the levels normally found when it contains poly(A). Thus, although mRNA that is transported to the cytoplasm normally contains poly(A), the presence of a poly(A) segment is not an obligatory requirement.

## IV. ROLE OF POLY(A) IN mRNA STABILITY

It has been suggested for quite a long while that the poly(A) segment may play a role in determining the stability of mRNA once in the cytoplasm. One reason for suggesting a cytoplasmic role for poly(A) is the presence of poly(A) in mRNAs that only reside in the cytoplasm (for instance, the mRNA of vaccinia virus, poliovirus, and vesicular stomatitis virus).[7] Many experiments have been performed demonstrating that the presence or absence of the poly(A) segment does not influence the in vitro translation of mRNA.[7,8] However, in vitro translation probably will not demonstrate subtle differences in mRNAs and certainly are not likely to reflect differences in the stabilities of mRNAs. Another approach that also measures the translation of functional mRNA, but in an in vivo situation, has provided evidence linking the stability of the mRNA to the presence of a poly(A) segment. Hemoglobin mRNA, when injected into *Xenopus* oocytes, is translated into globin for long periods of time (several days). However, if the poly(A) segment is removed from the globin mRNA prior to injection, globin is produced, but only for a few hours.[74] This loss of ability to synthesize globin was found to be due to a rapid degradation of the mRNA, in contrast to the poly(A)-containing globin mRNA which appeared to be quite stable upon injection.[75] That this increased lability of the globin mRNA was, in fact, due to the removal of the poly(A) was clearly demonstrated by the finding that the stability could be recovered by the in vitro addition of poly(A) to the deadenylated mRNA.[76] Therefore, there is a direct correlation between the presence of poly(A) and the stability of the mRNA. Furthermore, the addition of poly(A) to histone mRNA resulted in a similar phenomenon. Histones were translated for only a short time after injection of the native, poly(A)$^-$ mRNA, whereas the in vitro adenylated RNA was active for much longer periods.[77] The measurement of stability for an RNA with or without its poly(A) segment has now also been made by injection into HeLa cells. Once again, globin mRNA was found to be stable after injection whereas removal of the poly(A) resulted in an increased lability of the mRNA.[78]

This analysis was further refined to define a length requirement for the poly(A) segment in order for the stability of the mRNA to be maintained. Globin mRNA containing varying lengths of poly(A) was prepared and assayed for activity by oocyte injection.[79] Globin mRNAs containing poly(A) segments of 30 residues or greater were stable upon injection, whereas RNA with a poly(A) segment of 16 residues in length was unstable, behaving as if it had no poly(A). Whether this defines a functional domain for the poly(A) is not yet clear.

There is, however, at least one example of mRNA stability in the oocyte system apparently not linked to poly(A). Interferon mRNA, when injected into oocytes, decays with the same rate whether in a native form, containing poly(A), or in a poly(A)$^-$ form.[80] However, in this instance there was no demonstration of the efficiency of the removal of the poly(A), which would appear to be critical since as discussed before, the presence of only 16 residues could stabilize globin mRNA.

That the poly(A) segment contributes to mRNA stability has also been suggested by the experiments with cordycepin utilizing adenovirus infection. As described in the preceding sections, adenovirus RNA produced in the presence of cordycepin is transported to the cytoplasm at nearly the same rate as in control cells.[73] However, after 45 min of labeling there is only 10% as much in the cordycepin-treated cells as in the control. Thus, transcription continues and transport continues but the RNA does not accumulate upon entering the cytoplasm, indicating that degradation rapidly takes over in the cytoplasm.

What then might be the mechanism whereby poly(A) imparts stability to mRNA? If a major mechanism for RNA decay were by 3′ exonucleolytic action, then the poly(A) might provide a buffer against this attack. For example, it may be that 5′ end capping and 3′ end poly(A) addition are the means by which a mRNA is protected. One possibility is that the poly(A) provides material for nuclease attack that is not essential to mRNA function, since it is known that poly(A) undergoes a shortening process while it is in the cytoplasm.[81,82] Another possibility stems from the oocyte injection experiments that suggested a minimal length of poly(A) was necessary for mRNA stability. As the poly(A) segment shortens to a critical minimal length, a site for protein binding might be lost. If the presence of such a protein were responsible for maintaining stability of the mRNA, then this would signal the destruction of the mRNA. Such a candidate protein has been identified as a protein of 75 to 78 kd that specifically interacts with the poly(A).[83-85] Of added interest is the finding that this protein may also be a poly(A) polymerase.[86-87] It is known that poly(A) undergoes a terminal adenylate addition reaction while in the cytoplasm.[8a,8b] Thus, once the protein can no longer bind to the poly(A) there would be no way to prevent its destruction (by adding additional adenylate residues). An alternative explanation for a critical minimal length of poly(A) might be that a certain length of poly(A) may be necessary to maintain a secondary structure, involving either the poly(A) alone or the poly(A) with the mRNA body, critical for protection of the mRNA.

## V. CONCLUSIONS

It would appear that the primary role for poly(A), and perhaps the sole function, would be in conveying cytoplasmic stability to mRNA; the absence of poly(A) on an mRNA normally possessing poly(A) results in a failure of the accumulation of the mRNA in the cytoplasm. However, although the addition of poly(A) to a poly(A) site does not appear to be essential for mRNA biogenesis, the correct formation of this site is probably a critical event. Certainly, in the case of a transcription unit containing multiple poly(A) sites, the selection of a site is an important event.

It is possible that the addition of poly(A) in the nucleus also plays a role in stabilization of nuclear transcripts while they are further processed. However, the time required for splicing and transport out of the nucleus may be short relative to degradation, thus making this role less evident. Furthermore, it is also conceivable that the presence of poly(A) facilitates mRNA processing and/or transport, but that the effect is subtle, such that it is not obvious in the measurements that have been made.

Finally, if poly(A) is indeed involved in imparting stability to mRNA then how is it that different poly(A)$^+$ mRNAs have different half-lives? And how to do some poly(A)$^+$ mRNAs that are normally stable, become unstable as a result of a changed environment? An example of such a phenomenon can be found in developing *Dictyostelium*. Specific mRNAs that accumulate as a result of cellular aggregation are specifically degraded upon dispersal of the aggregates.[88] Why do these poly(A)$^+$ mRNAs decay rapidly while the remaining *Dictyostelium* poly(A)$^+$ mRNAs remain stable, despite the fact that both groups possess poly(A)? Obviously, the mechanism by which a mRNA turns over in the cytoplasm is complex, involving more than just the presence of a poly(A) segment.

One mechanism postulated for mRNA decay involves the shortening of the poly(A) to a critical size that then renders the mRNA unstable. It has been shown that the poly(A) of certain unstable adenovirus RNAs appears to shorten more rapidly than does the poly(A) of total cellular RNA.[89] Whether this is also the case for the *Dictyostelium* aggregation stage mRNAs when they become short-lived is not known. However, even if this correlation holds true, one still must explain what causes the poly(A) to shorten faster, either for some particular group of RNAs or for a single RNA under different circumstances. The answers to these remaining questions will surely require the identification and isolation of the proteins involved in messenger RNA decay.

## REFERENCES


1. **Darnell, J. E., Jr.,** Transcription units for mRNA production in eukaryotic cells and their DNA viruses, *Prog. Nucleic Acid. Res. Mol. Biol.*, 22, 327, 1979.
2. **Abelson, J.,** RNA processing and the intervening sequence problem, *Ann. Rev. Biochem.*, 48, 1035, 1979.
3. **Breathnach, R. and Chambon, P.,** Organization and expression of eucaryotic split genes coding for proteins, *Ann. Rev. Biochem.*, 50, 349, 1981.

3a. **Shatkin, A. J.,** Capping of eukaryotic mRNAs, *Cell*, 9, 645, 1976.

4. **Kates, J. R.,** Transcription of the vaccinia virus genome and the occurrence of polyriboadenylic acid sequences in messenger, RNA, *Cold Spring Harbor Symp. Quant. Biol.*, 35, 743, 1970.
5. **Edmonds, M., Vaughan, M. H., and Nakazato, H.,** Polyadenylic acid sequences in the heterogeneous nuclear RNA and rapidly-labeled polyribosomal RNA of HeLa cells; possible evidence for a precursor relationship, *Proc. Natl. Acad. Sci. U.S.A.*, 68, 1336, 1971.
6. **Lee, S. Y., Mendecki, J., and Brawerman, G.,** A polynucleotide segment rich in adenylic acid in the rapidly-labeled polyribosomal RNA component of mouse sarcoma 180 ascites cells, *Proc. Natl. Acad. Sci. U.S.A.*, 68, 1331, 1971.
7. **Brawerman, G.,** Characteristics and significance of the polyadenylate sequence in mammalian messenger RNA, *Prog. Nucleic Acid Res. Mol. Biol.*, 17, 117, 1976.
8. **Brawerman, G.,** The role of the poly(A) sequence in mammalian messenger RNA, *CRC Crit. Rev. Biochem.*, 10, 1, 1981.

8a. **Brawerman, G. and Diez, J.,** Metabolism of the polyadenylate sequence of nuclear RNA and messenger RNA in mammalian cells, *Cell*, 5, 271, 1975.

8b. **Sawicki, S., Jelinek, W., and Darnell, J. E.,** 3′ Terminal addition to HeLa cell nuclear and cytoplasmic poly(A), *J. Mol. Biol.*, 113, 219, 1977.

9. **Darnell, J. E., Jr., Philipson, L., Wall, R., and Adesnik, M.,** Polyadenylic acid sequences: Role in conversion of nuclear RNA into messenger RNA, *Science*, 174, 507, 1971.
10. **Jelinek, W., Adesnik, M., Salditt, M., Sheiness, D., Wall, R., Molloy, G., Philipson, L., and Darnell, J. E., Jr.,** Further evidence on the nuclear origin and transfer to the cytoplasm of polyadenylic acid sequences in mammalian cell RNA, *J. Mol. Biol.*, 75, 515, 1973.

11. **Philipson, L., Wall, R., Glickman, G., and Darnell, J. E., Jr.,** Addition of polyadenylate sequences to virus-specific RNA during adenovirus replication, *Proc. Natl. Acad. Sci. U.S.A.*, 68, 2806, 1971.
12. **Chow, L. T., Roberts, J., Lewis, J., and Broker, T.,** A map of cytoplasmic RNA transcripts from lytic adenovirus type 2, determined by electron microscopy of RNA:DNA hybrids, *Cell*, 11, 819, 1977.
13. **McGrogan, M. and Raskas, H. J.,** Species identification and genome mapping of cytoplasmic adenovirus type 2 RNAs synthesized late in infection, *J. Virol.*, 23, 240, 1977.
14. **Nevins, J. R., and Darnell, J. E.,** Groups of adenovirus type 2 mRNAs derived from a large primary transcript: Probable nuclear origin and possible common 3′ ends, *J. Virol.*, 25, 811, 1978.
15. **Fraser, N. F. and Ziff, E. B.,** RNA structures near poly(A) of adenovirus-2 late messenger RNAs, *J. Mol. Biol.*, 124, 27, 1978.
16. **Nevins, J. R. and Darnell, J. E., Jr.,** Steps in the processing of Ad2 mRNA: Poly(A)$^+$ nuclear sequences are conserved and poly(A) addition precedes splicing, *Cell*, 15, 1477, 1978.
17. **Fraser, N. W., Nevins, J. R., Ziff, E., Jr., and Darnell, J. E.,** The major late adenovirus type-2 transcription unit: termination is downstream from the last poly(A) site, *J. Mol. Biol.*, 129, 643, 1979.
18. **Ford, J. and Hsu, M.-T.,** Transcription pattern of in vivo labeled late SV40 RNA, *J. Virol.*, 28, 795, 1978.
19. **Lai, C.-J., Dhar, R., and Khoury, G.,** Mapping the spliced and unspliced late lytic SV40 RNAs, *Cell*, 14, 971, 1978.
20. **Acheson, H. N.,** Polyoma virus giant RNAs contain tandem repeats of the nucleotide sequence of the entire viral genome, *Proc. Natl. Acad. Sci. U.S.A.*, 75, 4754, 1978.
21. **Legon, S., Flavell, A. J., Cowie, A., and Kamen, R.,** Amplification in the leader sequence of late polyoma virus mRNAs, *Cell*, 16, 373, 1979.
22. **Nevins, J. R., Blanchard, J.-M., and Darnell, J. E., Jr.,** Transcription units of adenovirus type 2: termination of transcription beyond the poly(A) addition site in early regions 2 and 4, *J. Mol. Biol.*, 144, 377, 1980.
23. **Hofer, E. and Darnell, J. E., Jr.,** The primary transcription unit of the mouse β-major globin gene, *Cell*, 23, 585, 1981.
24. **Roop, D. R., Tsai, M.-J., and O'Malley, B. W.,** Definition of the 5′ and 3′ ends of transcripts of the ovalbumin gene, *Cell*, 19, 63, 1980.
25. **Proudfoot, N. J. and Brownelee, G. G.,** 3′ Non-coding region sequences in eukaryotic messenger RNA, *Nature (London)*, 263, 211, 1976.
26. **Jung, A., Sippel, A. E., Grez, M., and Schutz, G.,** Exons encode functional and structural units of chicken lysosyme, *Proc. Nat. Acad. Sci. U.S.A.*, 77, 5759, 1980.
26a. **Ahmed, C. M. I., Chanda, R., Stow, N., and Zain, B. S.,** personal communication, 1982.
27. **Hagenbuchle, O., Bovey, R., and Young, R. A.,** Tissue-specific expression of mouse α-amylase genes: nucleotide sequence of isoenzyme mRNAs from pancrease and salivary gland, *Cell*, 21, 179, 1980.
28. **Fitzgerald, M. and Shenk, T.,** The sequence 5′-AAUAAA-3′ forms part of the recognition site for polyadenylation of late SV40 RNAs, *Cell*, 24, 251, 1981.
29. **Sasavage, N., Smith, M., Gillam, S., Woychik, R. P., and Rottman, F. M.,** Variation in the polyadenylation site of bovine prolactin messenger RNA, *Proc. Natl. Acad. Sci. U.S.A.*, 79, 223, 1982.
29a. **Montell, C., Fisher, E., Caruthers, M., and Berk, A. J.,** personal communication, 1982.
30. **Edmonds, M. and Winters, M. A.,** Polyadenylate polymerases, *Prog. Nucleic Acid Res. Mol. Biol.*, 17, 149, 1976.
31. **Rose, K. M., Bell, L. E., and Jacob, S. T.,** Specific inhibition of chromatin-associated poly(A) synthesis in vitro by cordycepin 5′-triphosphate, *Nature (London)*, 267, 178, 1977.
32. **Manley, J. L., Sharp, P. A., and Gefter, M. L.,** RNA synthesis in isolated nuclei: identification and comparison of adenovirus-2 encoded transcripts synthesized in vitro and in vivo, *J. Mol. Biol.*, 135, 171, 197.
33. **Yang, V. W., Binger, M.-H., and Flint, S. J.,** Transcription of adenoviral genetic information in isolated nuclei, *J. Biol. Chem.*, 255, 2097, 1980.
34. **Chen-Kiang, S., Wolgemuth, D., Hsu, .-T., and Darnell, J. E.,** Transcription and accurate polyadenylation in vitro of RNA from the major late adenovirus 2 transcription unit, *Cell*, 28, 575, 1982.
35. **Salditt-Georgieff, M., Harpold, M., Sawicki, S., Nevins, J., and Darnell, J. E., Jr.,** The addition of poly(A) to nuclear RNA occurs soon after RNA synthesis, *J. Cell Biol.*, 86, 844, 1980.
36. **Craig, E. A. and Raskas, H. J.,** Nuclear transcripts larger than the cytoplasmic mRNAs are specified by segments of the adenovirus genome coding for early functions, *Cell*, 8, 205, 1976.
37. **Goldenberg, C. and Raskas, H. J.,** Splicing patterns of nuclear precursors to the mRNA for adenovirus 2 DNA binding protein, *Cell*, 16, 131, 1979.
38. **Nevins, J. R.,** Processing of adenovirus nuclear RNA to mRNA: kinetics of formation of intermediates and demonstration that all events are nuclear, *J. Mol. Biol.*, 130, 493, 1979.
39. **Weber, J., Blanchard, J.-M., Ginsberg, H. S., and Darnell, J. E., Jr.,** Order of polyadenylic acid addition and splicing events in early adenovirus mRNA formation, *J. Virol.*, 33, 286, 1980.

40. **Ziff, E. B. and Evans, R. M.,** Coincidence of the promoter and capped 5′ terminus of RNA from the adenovirus 2 major late transcription unit, *Cell,* 15, 1463, 1978.
41. **Berget, S. M. and Sharp, P. A.,** Structure of late adenovirus 2 heterogeneous nuclear RNA, *J. Mol. Biol.,* 129, 547, 1979.
42. **Ross, J. and Knecht, D. A.,** Precursors of alpha and beta globin messenger RNAs, *J. Mol. Biol.,* 119, 1, 1978.
43. **Schibler, U., Marcu, K. B., and Perry, R. P.,** The synthesis and processing of the messenger RNAs specifying heavy and light chain immunoglobulins in MPC-11 cells, *Cell,* 15, 1495, 1978.
44. **Gilmore-Hebert, M. and Wall, R.,** Nuclear RNA precursors in the processing pathway to MOPC 21 kappa light chain messenger RNA, *J. Mol. Biol.,* 135, 879, 1979.
45. **Harpold, M. M., Dobner, P. R., Evans, R., Bancroft, F. C., and Darnell, J. E.,** The synthesis and processing of a nuclear RNA precursor to rat pregrowth hormone messenger RNA, *Nucleic Acids Res.,* 6, 3133, 1979.
46. **Fraser, N. W., Baker, C. C., Moore, M. A., and Ziff, E. B.,** Poly(A) sites of adenovirus serotype 2 transcription units, *J. Mol. Biol.,* 155, 207, 1982.
47. **Alt, F. W., Bothwell, A. L. M., Knapp, M., Siden, E., Mather, E., Koshland, M., and Baltimore, D.,** Synthesis of secreted and membrane-bound immunoglobulin mu heavy chain is directed by mRNAs that differ at their 3′ ends, *Cell,* 20, 293, 1980.
48. **Early, P., Rogers, J., Davis, M., Calame, K., Bond, M., Wall, R., and Hood, L.,** Two mRNAs can be produced from a single immunoglobulin μ gene by alternative RNA processing pathways, *Cell,* 20, 313, 1980.
49. **Rogers, J., Early, P., Carter, C., Calame, K., Bond, M., Hood, L., and Wall, R.,** Two mRNAs with different 3′ ends encode membrane-bound and secreted forms of immunoglobulin μ chain, *Cell,* 20, 303, 1980.
50. **Chow, L. T., Broker, T. R., and Lewis, J. B.,** Complex splicing patterns of RNA from the early regions of Ad2, *J. Mol. Biol.,* 134, 265, 1979.
51. **Nevins, J. R. and Wilson, M. C.,** Regulation of adenovirus-2 gene expression at the level of transcriptional termination and RNA processing, *Nature (London),* 290, 113, 1981.
52. **Kitchingman, G. R. and Westphal, H.,** The structure of adenovirus 2 early nuclear and cytoplasmic RNAs, *J. Mol. Biol.,* 137, 23, 1980.
53. **Shaw, A. R. and Ziff, E. B.,** Transcripts from the adenovirus-2 major late promoter yield a single early family of 3′ coterminal mRNAs and five late families, *Cell,* 22, 905, 1980.
54. **Akusjarvi, G. and Persson, H.,** Control of RNA splicing and termination in the major late adenovirus transcription unit, *Nature (London),* 290, 420, 1981.
55. **Lewis, J. B. and Mathews, M. B.,** Control of adenovirus early gene expression: a class of immediate early products, *Cell,* 21, 303, 1980.
56. **Maki, R., Roeder, W., Traunecker, A., Sidman, C., Wabl, M., Raschke, W., and Tonegawa, S.,** The role of DNA rearrangement and alternative RNA processing in the expression of immunoglobulin delta genes, *Cell,* 24, 353, 1981.
57. **Tosi, M., Young, R. A., Hagenbuchle, O., and Schibler, U.,** Multiple polyadenylation sites in a mouse α-amylase gene, *Nucleic Acids Res.,* 9, 2313, 1981.
58. **Setzer, P. R., McGrogan, M., Nunberg, J. H., and Schimke, R. T.,** Size heterogeneity in the 3′ end of dihydrofolate reductase messenger RNAs in mouse cells, *Cell,* 22, 361, 1980.
59. **Penman, S., Rosbash, M., and Penman, M.,** Messenger and heterogeneous nuclear RNA in HeLa cells: different inhibition by cordycepin, *Proc. Natl. Acad. Sci. U.S.A.,* 67, 1878, 1970.
60. **Siev, M., Weinberg, R., and Penman, S.,** The selective interruption of nucleolar RNA synthesis in HeLa cells by cordycepin, *J. Cell Biol.,* 41, 510, 1969.
61. **Beach, L. R. and Ross, J.,** Cordycepin, an inhibitor of newly synthesized globin messenger RNA, *J. Biol. Chem.,* 253, 2628, 1978.
62. **Zeevi, M., Nevins, J. R., and Darnell, J. E., Jr.,** Nuclear RNA is spliced in the absence of poly(A) addition, *Cell,* 26, 39, 1981.
63. **Perry, R. P., Kelley, D. E., and Latorre, J.,** Synthesis and turnover of nuclear and cytoplasmic poly A in mouse L cells, *J. Mol. Biol.,* 82, 315, 1974.
64. **Salditt-Georgieff, M., Harpold, M., Wilson, M., and Darnell, J. E., Jr.,** Large hnRNA has three times as many 5′ caps as poly(A) segments and most caps do not enter polyribosomes, *Mol. Cell. Biol.,* 1, 179, 1981.
65. **Adesnik, M. and Darnell, J. E., Jr.,** Biogenesis and characterization of histone messenger RNA in HeLa cells, *J. Mol. Biol.,* 67, 397, 1972.
66. **Adesnik, M., Salditt, M., Thomas, W., and Darnell, J. E., Jr.,** Evidence that all messenger RNA molecules (except histone messenger RNA) contain poly(A) sequences and that the poly(A) has a nuclear function, *J. Mol. Biol.,* 71, 21, 1972.

67. **Grunstein, M. and Schedl, P.,** Isolation and sequence analysis of sea urchin *(Lytechinus pictus)* histone H4 messenger RNA, *J. Mol. Biol.*, 104, 323, 1976.
68. **Perry, R. P. and Kelley, D. E.,** Messenger RNA turnover in mouse L cells, *J. Mol. Biol.*, 79, 681, 1973.
69. **Lewin, B.,** *Gene Expression*, Vol. 2, John Wiley & Sons, New York, 1980.
70. **Darnell, J. E.,** Variety in the mechanisms of gene control, *Nature (London)*, 297, 365, 1982.
71. **Chen-Kiang, S., Nevins, J. R., and Darnell, J. E.,** N-6-methyl adenosine in adenovirus type 2 nuclear RNA is conserved in the formation of messenger RNA, *J. Mol. Biol.*, 135, 733, 1979.
72. **Harpold, M. W., Wilson, M. C., and Darnell, J. E., Jr.,** Chinese hamster polyadenylated mRNA: relationship to nonpolyadenylated sequences and relative conservation during mRNA processing, *Mol. Cell. Biol.*, 1, 188, 1981.
73. **Zeevi, M., Nevins, J. R., and Darnell, J. E., Jr.,** Newly formed mRNA lacking poly(A) enters the cytoplasm and the polyribosomes but has a shorter half-life in the absence of poly(A), *Mol. Cell. Biol.*, 2, 517, 1982.
74. **Huez, G., Marbaix, G., Hubert, E., Leclercq, M., Nudel, U., Soreq, H., Salomon, R., Lebleu, B., Revel, M., and Littauer, U. Z.,** Role of polyadenylate segment in the translation of globin messenger RNA in Xenopus oocytes, *Proc. Natl. Acad. Sci. U.S.A.*, 71, 3143, 1974.
75. **Marbaix, G., Huez, G., Burny, A., Cleuter, Y., Hubert, E., Leclercq, M., Chantrenne, H., Soreq, H., Nudel, U., and Littauer, U. Z.,** Absence of polyadenylate segment in globin messenger RNA accelerates its degradation in *Xenopus* oocytes, *Proc. Natl. Acad. Sci. U.S.A.*, 72, 3065, 1975.
76. **Huez, G., Marbaix, G., Hubert, E., Cleuter, Y., Leclercq, M., Chantrenne, H., Devos, R., Soreq, H., Nudel, U., and Littauer, U. Z.,** Readenylation of polyadenylate-free globin messenger RNA restores its stability *in vivo*, *Eur. J. Biochem.*, 59, 589, 1975.
77. **Huez, G., Marbaix, G., Gallwitz, D., Weinberg, E., Devos, R., Hubert, E., and Cleuter, Y.,** Functional stabilization of HeLa cell histone messenger RNAs injected into *Xenopus* oocytes by 3'-OH polyadenylation, *Nature (London)*, 271, 572, 1978.
78. **Huez, G., Bruck, C., and Clueter, Y.,** Translational stability of native and deadenylated rabbit globin mRNA injected into HeLa cells, *Proc. Natl. Acad. Sci. U.S.A.*, 78, 908, 1981.
79. **Nudel, U., Soreq, H., Littauer, U. Z., Marbaix, G., Huez, G., Leclercq, M., Hubert, E., and Chantrenne, H.,** Globin mRNA species containing poly(A) segments of different lengths, their functional stability in *Xenopus* oocytes, *Eur. J. Biochem.*, 64, 115, 1976.
80. **Sehgal, P. B., Soreq, H., and Tamm, I.,** Dose 3'-terminal poly(A) stabilize human fibroblast interferon mRNA in oocytes of *Xenopus laevis?*, *Proc. Natl. Acad. Sci. U.S.A.*, 75, 5030, 1978.
81. **Sheiness, D. and Darnell, J. E., Jr.,** Polyadenylic acid segment in mRNA becomes shorter with age, *Nature (London)*, 241, 265, 1973.
82. **Sheiness, D., Puckett, L., and Darnell, J. E., Jr.,** Possible relationship of poly(A) shortening to mRNA turnover, *Proc. Natl. Acad. Sci. U.S.A.*, 72, 1077, 1975.
83. **Blobel, G.,** A protein of molecular weight 78,000 bound to the polyadenylate region of eukaruotic messenger RNAs, *Proc. Natl. Acad. Sci. U.S.A.*, 70, 924, 1973.
84. **Schwartz, H. and Darnell, J. E., Jr.,** The association of protein with the polyadenylic acid of HeLa cell messenger RNA: evidence for a "transport" role of a 75,000 molecular weight polypeptide, *J. Mol. Biol.*, 104, 833, 1974.
85. **Setyono, B. and Greenberg, J. R.,** Protein associated with poly(A) and other regions of mRNA and hnRNA molecules as investigated by crosslinking, *Cell*, 24, 775, 1981.
86. **Rose, K. M., Jacob, S. T., and Kumar, A.,** Poly(A) polymerase and poly(A)-specific mRNA binding protein are antigenically related, *Nature (London)*, 279, 260, 1979.
87. **Haynes, S. R. and Jelinek, W. R.,** personal communication, 1982.
88. **Chung, S., Landfear, S. M., Blumberg, D. D., Cohen, N. S., and Lodish, H. F.,** Synthesis and stability of developmentally regulated Dictyostelium mRNAs are affected by cell-cell contact and cAMP, *Cell*, 24, 785, 1981.
89. **Wilson, M. C., Sawicki, S., White, P. A., and Darnell, J. E., Jr.,** A correlation between the rate of poly(A) shortening and half-life of messenger RNA in adenovirus transformed cells, *J. Mol. Biol.*, 126, 23, 1978.
90. **Nevins, J. R. and Chen-Kiang, S.,** Processing of adenovirus nuclear RNA to mRNA, *Adv. Virus Res.*, 26, 1, 1981.
91. **Ziff, E. B.,** Transcription and RNA processing by the DNA tumor viruses, *Nature (London)*, 287, 491, 1981.

Chapter 7

# PROCESSING OF mRNA PRECURSORS IN EUKARYOTIC CELLS

**S. J. Flint**

## TABLE OF CONTENTS

## I. INTRODUCTION

Our understanding of the mechanisms whereby mature mRNA species are made in eukaryotic cells has improved dramatically since the development of molecular cloning methods that permit detailed examination of the products of transcription of individual genes. This technical advance and the consequent recognition that eukaryotic cellular and viral mRNA species may be constructed from coding sequences that are not contiguous in the genome have focused the picture of RNA processing in eukaryotic cells from a generalized blur to specific images.

The properties and metabolism of nonribosomal, nuclear RNA in eukaryotic cells, so-called hnRNA, have been examined since the initial discovery of this RNA class and have produced a large body of evidence consistent with the notion that hnRNA populations indeed contain precursors to mature mRNA species. Such evidence (reviewed by Perry[1] and Lewin[2]) includes comparison of the sizes of nuclear and mRNA populations, as well as of specific RNA species therein; comparison of the sequence complexities of hnRNA and mRNA populations; analysis of the kinetics of hnRNA and mRNA synthesis; and comparison of the sizes of transcriptional units to the those of mRNA. The discovery of polyadenylation, capping, and methylation of newly synthesized RNA provided direct demonstration of specific posttranscriptional processing reactions. These alone, however, could not fully account for the differences between hnRNA and mRNA populations, particularly those in their size and complexity.

Controversy, or at least confusion, about which regions and sequences of an hnRNA molecule were actually conserved during the cleavages that appeared necessary to create functional mRNA reigned until the characterization of specific genes and their transcripts led to the realization that many genes, especially those of higher eukaryotes, include noncoding (intervening) sequences interspersed among regions that contribute to mature mRNA species. It was rapidly established that both coding and intervening sequences are transcribed, but that the latter are removed posttranscriptionally in a reaction that must comprise precise endonucleolytic cleavages at the junctions between coding and intervening sequences and the subsequent, or concomitant, ligation of the ends of the coding segments thus liberated.

The discovery of such splicing reactions has clarified considerably our conception of posttranscriptional processing of mRNA precursors, not least because many questions can now be framed in specific terms and be addressed by direct experimentation. This is not, however, to imply that RNA processing is fully appreciated: a more precise outline can certainly be drawn, but the discovery of splicing reactions, like all seminal discoveries, has raised more questions than it has answered. And the answers to these questions may well lie in realms even more remote than, for example, the enzymatic reactions that fashion a mature mRNA species from its primary transcript.

## II. ENZYMATIC PROCESSING REACTIONS

The biogenesis of mature mRNA species from primary products of transcription in eukaryotic cells can be described as an ordered sequence of discrete enzymatic reactions that alter the chemical properties of the RNA: novel 5′ and 3′ termini are made and internal sequences may be modified or removed. These reactions can be distinguished from a second class of processing events during which the chemical nature of sequence of the RNA is not altered, examples of which include association of mRNA precursors with proteins to form ribonucleoproteins, and movement of the RNA from its sites of synthesis, cellular chromatin, to the cytoplasm where it can be translated. Such a distinction, made here for descriptive convenience, is of course artificial and it is important to keep in mind that the substrates of the various processing reactions are not naked RNA molecules but rather ribonucleoproteins.

Possible implications of this frequently ignored fact are discussed in Section III; the processing reactions to which an eukaryotic mRNA precursor may be subjected are described in subsequent parts of this section.

### A. Capping, Internal Methylation, and Polyadenylation of mRNA Precursors

The posttranscriptional modification of presumed precursors to mRNA by addition of 5′ blocking groups (capping), methylation of 5′ terminal and internal residues, and addition of poly(A) to 3′ termini have been recognized for a decade or more. Indeed, capping and polyadenylation have been well studied and are the subjects of Chapters 5 and 6 in this volume. In the general context of synthesis of mature mRNAs, it is, however, worth recapitulating that capping is both ubiquitous among cellular mRNA species and the first of the enzymatic processing reactions to which pre-mRNA species are subject. Addition of poly(A) is a later event and some mature mRNA species are devoid of poly(A) tracts. By contrast, relatively little is known of the mechanism, or significance, of internal methylation.

In addition to the ribose and base methylations characteristic of mature, capped 5′-termini of mRNA (see Chapter 5), internal adenosine residues may be methylated at the $N^6$ position.[3-7] Internal methylation has been less intensively studied than other RNA processing reactions, but initial studies established several general properties. The total mRNA population of higher eukaryotic cells contains, on average, 1 to 2 $^{m6}$Ap residues per chain, whereas nuclear RNA has been estimated to contain 2 to 6 such modified bases per molecule, the lower and higher estimates being made for poly(A)-containing and total nuclear RNA preparations, respectively.[3,5-11] These estimates translate to values of about 0.5 to 1.0 and 0.1 to 0.4 $^{m6}$A residues per 1000 nucleotides in mammalian cell mRNA and hnRNA, respectively. At face value, this difference suggests that methylated internal residues are conserved during processing, especially as the content of $^{m6}$A in poly(A)-containing hnRNA approaches that of mRNA.

Although our understanding of internal methylation is generally quite sketchy, it does seem clear that recognition by the relevant methylase(s) includes sequence specificity: within RNA, $^{m6}$Ap residues are present exclusively at the 5′ side of cytidine residues and the sequences A$^{m6}$AC G$^{m6}$AC account for essentially all internal methylated sequences.[7,12,13]

Nonpolyadenylated histone mRNA lacks internal $^{m6}$A residues.[14,15] Nevertheless, the presence of $^{m6}$A in the nonhistone poly(A)-lacking cytoplasmic RNA class and in nonpolyadenylated RNA[6] indicates that methylation of internal A residues is not dependent on prior polyadenylation. Rather, the occurrence of $^{m6}$A in the largest, nuclear RNA molecules examined, including those complementary to a specific transcriptional unit,[6,16] argues that methylation is an early processing event, preceding polyadenylation. This conclusion remains to be demonstrated kinetically. Analysis of the size of nuclear RNA molecules complementary to the adenovirus type 2 late transcriptional unit (see Reference 17 and Chapter 8) made in virus-infected HeLa cells does, however, indicate that unspliced RNA sequences can be methylated, that is, methylation can take place before splicing.[16]

The distribution of methylated nucleotides along a pre-mRNA molecule is not clearly established. The results of early experiments appeared to indicate that $^{m6}$A residues, while conserved during processing, were not located in the 3′ terminal portion of mRNA.[6,8,18] Similarly, Chen-Kiang et al.[16] have concluded, from a comparison of the kinetics of labeling of $^{m6}$A residues to that of total RNA transcribed from the Ad2 late transcriptional unit, that $^{m6}$A residues are conserved. These authors have therefore suggested that methylated residues might in some way serve as signals to identify those regions of an mRNA precursor that are retained during processing. Although detailed analysis of the methylation of precursor species transcribed from a typical cellular gene have not been reported, several mRNA species whose biogenesis requires an otherwise typical repertoire of processing reactions do not appear to be methylated at internal positions. Such examples are provided by the rabbit α- and β-globin mRNAs[19-21] and *Bombyx mori* fibroin mRNA.[22] If these mRNA species were fashioned from transcripts that never become methylated, then it must be that methylation plays no mandatory role in subsequent processing reactions. If, on the other hand, these primary transcripts were methylated, then there can be no conservation of methylated residues in these cases. Either way, the existence of mRNA species that carry no $^{m6}$A residues casts doubt on the conclusions made by Chen-Kiang et al.[16] It is, however, important to bear in mind that the Ad2 late transcriptional unit encodes some 18 or 19 individual mRNA species, and is processed in such a way that only one mature mRNA molecule can be fashioned from a transcript that includes sequences for all (see Reference 17 and Chapter 8). Thus, this may represent an unusual example in which methylation plays a specific role and is itself regulated, such that different mRNA sequences are methylated in different precursor molecules, as suggested by Chen-Kiang et al.[16]

Considerably more work is clearly needed, not only to elucidate the mechanisms of internal methylation but also to address the question of its function during subsequent mRNA maturation.

### B. Splicing of mRNA Precursors

The major rearrangement of cellular mRNA precursors takes place during a late step in their maturation, the removal of noncoding, intervening sequences by splicing reactions. The RNA species that emerge are apparently indistinguishable from cytoplasmic mRNA and must only be shuttled to the cytoplasm to be expressed.

Splicing appears to be the rule among transcripts of genes of higher eukaryotes, but is not universal (see, for example, References 23 and 24). This conclusion, deduced from the organization and structure of an increasing but nevertheless limited number of genes, is reinforced by the observation that when large cDNA, transcribed from mouse brain poly(A)-containing RNA, is hybridized to either genomic DNA or large nuclear RNA, most of the cDNA is reduced in size upon subsequent nuclease S1 digestion of the hybrids. Under the conditions employed, hybrids formed between the cDNA and mRNA or genomic DNA and the single-copy fraction prepared from it are not susceptible.[25] Exceptional colinear genes found among those of higher eukaryotes examined include those specifying histones,[26,27] members of the interferon α and β families,[28-34] the adenovirus protein IX,[35] and the herpes simplex type 1 virus thymidine kinase genes.[36,37] Products of transcription of the majority of histone genes in all eukaryotic organisms examined are also unusual in being devoid of 3′-terminal poly(A) and could therefore be considered to be subject to an unusual processing scheme. However, the other unspliced mRNA species mentioned do contain poly(A) so it can be concluded that splicing is not an obligatory step in the production of mature, poly(A)-containing mRNA. Indeed, it could not be, for although intervening sequences are present in the genomes of lower eukaryotes, they do not appear to be so common. In yeast, for example, genes for actin[37,39] and a subset of those encoding ribosomal proteins[40] contain intervening sequences, but many others do not. Despite these apparent differences in the frequencies of discontinuous genes among lower and higher eukaryotes, it is clear that eukaryotic genomes quite generally contain intervening sequences and therefore must also specify the enzymatic machinery to excise them.

The substrates of splicing reactions are normally polyadenylated nuclear RNA molecules; poly(A)-containing transcripts of specific genes that are not spliced are frequently and readily observed in nuclear RNA preparations whereas their spliced but nonpolyadenylated counterparts are not.[41-53] Thus, splicing appears to be last in the series of enzymatic reactions that fashion a mature mRNA species. Whether splicing is the rate-determining step in RNA processing is not unequivocally established, but the available evidence suggests that it is. Poly(A)-containing mRNA molecules take on the order of 15 to 20 min to reach the cytoplasm, whereas poly(A)-containing but unspliced RNA molecules can be observed in labeling times as short as 5 min (see, for examples, References 54 and 55). Such kinetic observations, coupled with the relative ease with which poly(A)-containing unspliced or partially spliced molecules can be discerned, argue that splicing is a time-consuming process.

*1. Properties of Intervening Sequences and Splice Junctions*

Our present understanding of splicing reactions and mechanisms derives to a great extent from knowledge of the structures of a relatively large number of genes that contain intervening sequences, and of their corresponding mRNA species. The number of intervening sequences that may be present in a gene ranges from the obvious minimum of one to surprisingly large numbers. Genes that contain a single intervening sequence (IVS) include a rat insulin gene,[56,57] actin genes of various organisms,[38,39,58,59] and silkmoth chorion genes.[60] At the other extreme, the chicken oviduct and ovomucoid genes each include eight intervening sequences,[61-68] the *Xenopus laevis* vitellogenin gene at least 33,[69,70] and the chicken α 2 type 1 collagen gene more than 50.[71-72] Not only do eukaryotic genes vary so strikingly in the number of intervening sequences that may be present, but also in their size and location. Intervening sequences are frequently located close to the 5′ ends of mRNA coding sequences, for example in α- and β-globin genes of several species,[73] a human β-tubulin gene,[74] and those encoding the precursors to the several peptide hormones.[75,76] Moreover, they often demarcate functional domains within polypeptides,[77,78] for example, in the globin,[79,80] chicken lysozyme,[81] and human immunoglobulin heavy-chain genes,[82-87] among others. Nevertheless, there exist no obvious preferred locations of intervening sequences with respect, for example, to the sites of other posttranscriptional modifications. Similarly, the sizes of intervening sequences are quite variable, ranging from tens of nucleotides in such cases as the small IVS of globin genes[73] to several kilobases in, for example, human β-tubulin[74] and insulin.[88] Furthermore, the intervening sequences within any one gene and its primary transcript show little homology either to one another or to those present in the corresponding gene of a different species: seven intervening sequences of the many present in the chicken α 2 type 1 collagen gene that have been compared in detail are variable in size and exhibit little sequence homology.[89] The structure of globin genes from several species, including the mouse, rabbit, goat, and man, have been characterized in quite exhaustive detail. Although the locations of the two intervening sequences have been conserved among members of the α- and β-globin gene families, little conservation of the nucleotide sequence of these intervening sequences is observed even among α- or β-like globin genes within one species.[73,90-93]

The facts cited in the previous paragraph strongly suggest that the greater part of an IVS is dispensable when it comes to recognition by splicing enzymes. This deduction has been demonstrated directly by construction of deletion mutants of SV40; deletion of intervening sequences of the early or late genes to within 10 to 12 nucleotides of the boundaries between coding and intervening sequences (hereafter designated splice junctions) does not abolish the production of accurately spliced mRNA species.[94-99] A similar result has been obtained with mutant mouse $\beta^{maj}$-globin genes from which all but 18 nucleotides of the larger IVS had been removed.[100] The frequency with which different splice junctions are recognized can, however, be influenced by deletion of internal portions of an IVS.[97,98] This latter result is not too surprising because the nature of an IVS, its size as well as its nucleotide sequence, would be expected to be an important determinant of the conformation of an mRNA-precursor molecule in the region neighboring splice junctions.

The nucleotide sequences that span splice junctions of many genes have now been determined. When these are aligned at the boundaries between coding and intervening sequences, all can be described by the generalized structure shown in Figure 1, the splice junction consensus sequences.[101-105] Actual splice junctions sequences that define the borders of an IVS differ in one to a few nucleotides from the consensus sequences shown in Figure 1, the most highly conserved feature being the dinucleotides GU and AG that demarcate

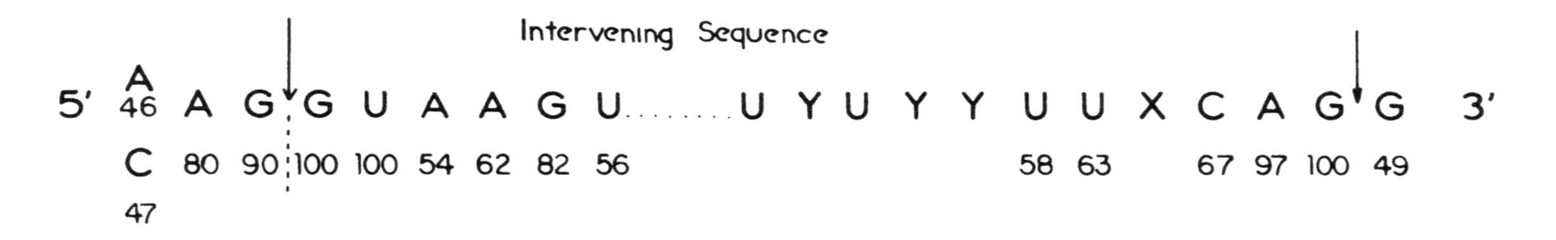


FIGURE 1. Splice junction consensus sequences in eukaryotic precursor mRNA. The splice junction consensus sequences deduced from comparison of the sequences of a large number of eukaryotic genes are drawn in the 5′ to 3′ direction with vertical arrows marking the boundaries between coding and intervening sequences. Y and X denote a pyrimidine or any nucleotides, respectively. The frequency with which each residue shown occurs is given below the sequence.

the 5′ and 3′ ends, respectively, of an IVS.[101] The great majority of genes that have been sequenced in the regions including splice junctions possess duplications that preclude precise location of splice junctions. Nevertheless, the description given in Figure 1 does appear to be more than a formality, for three intervening sequences of the chicken ovomucoid gene possess no such duplications, yet conform to the GU . . . AG boundary rule.[77] The duplications observed at most splice junction boundaries might be imagined to provide the splicing reaction with a small degree of freedom, in the sense that the same final coding sequence could be created by one of several potential ligation reactions (see Reference 24 for a discussion).

These splice junction sequences are highly conserved in the eukaryotic kingdom and have been found in genes of yeast,[38,39] insects,[59,78] and plants[106,107] as well as in the numerous avian and mammalian genes mentioned previously. Such conservation underlines the conclusion discussed previously: that it is the splice junction sequences themselves that are recognized by the splicing machinery. It would therefore be predicted that mutation of splice junction sequences would abolish, or at least reduce, mRNA production. This prediction has been confirmed by the discovery of naturally occurring mutants of mammalian globin genes that cannot be expressed and indeed differ from their functional counterparts by single base changes within the regions of the splice junction consensus sequences shown in Figure 1.[108-113] Similarly, site-directed alterations of one of three potential 5′ splice sites within the type C adenovirus E1A gene prevent synthesis of the corresponding mRNA species.[114-116] Thus, there exists compelling evidence to support the notion that the splicing machinery, or some component of it, recognizes relatively short nucleotide sequences that span the 5′ and 3′ junctions between coding and intervening sequences. Some implications of this mode of recognition will be discussed subsequently.

A second inference based on the conservation of splice junction sequences, that the splicing machinery, whatever its nature, must also be highly conserved has also been tested, at least partially: numerous examples of trans-species splicing, that is, correct splicing of a pre-mRNA molecule of one organism when its gene is introduced into cells of a second, have been reported. Correct expression has been achieved when mammalian genes are transferred to a different mammalian species, for example, mouse to monkey, or to amphibian oocytes, or when avian genes are placed within mammalian cells (see, for examples, References 117 to 122).

Such remarkable conservation when considered in the context of RNA processing raises a number of fundamental questions most obvious of which concerns the mechanisms that guarantee that the splicing of any pre-mRNA molecule containing more than one IVS is a productive process. The answers to this question are far from obvious. Nor do we yet know a great deal about either the mechanism whereby one IVS is excised or the constituents of the splicing machinery.

*2. Removal of an IVS: Recognition of Splice Junction Sequences*

All the properties of intervening and splice junction sequences discussed in the previous section point to recognition of the highly conserved splice junction sequences by the splicing apparatus of eukaryotic cells. Such consensus sequences have no inherent self-complementarity that might align the coding sequences on either side of an IVS that must be ligated (see Figure 1). They, do, however, possess the potential to form base pairs with sequences present in some small nuclear RNA species, found within the cell in the form of small, nuclear ribonucleoprotein particles, snRNP.[123] Such potential base-pairing interactions were first recognized for U1-RNA.[104-105] This RNA species, as illustrated in Figure 2A, includes a region near its 5′ end that could base-pair with the 5′ splice site consensus sequence and two regions of potential complimentarity to the 3′ splice junction consensus sequence, nucleotides 13 to 22 or nucleotides 131 to 138.[104,105,124] Ohshima and colleagues[124] have suggested that the second set of nucleotides listed is less likely to be sequestered in intramolecular base-pairing within U1 RNA than the former.

Several lines of indirect evidence lend support to the idea that U1 snRNP might participate in pre-mRNA splicing reactions as a guide molecule. The snRNP species themselves appear to be ubiquitous, abundant in all eukaryotic cells examined, and display very similar sequences;[104,105,123,125,126] snRNP containing U1 RNA is more abundant in actively growing cells[104,107] and can be isolated in association with large ribonucleoprotein forms that contain pre-mRNA, so-called hnRNP.[104,128-133] The ability of derivatives of psoralen to cross link U1 RNA to high-molecular-weight RNA in intact cells[134] strongly suggests that this association is not simply an artifact, for example, of nuclear lysis or RNP isolation procedures; intermediates created during the excision of one particular IVS from the chicken α2 type 1 collagen gene each include a new 3′ splice junction consensus sequence, created as a result of the preceding splicing step,[135] such that all intermediates made could also interact with U1 snRNA in the fashion depicted in Figure 2. Finally, antibodies that recognize snRNP particles that contain U1 RNA, such as anti-Sm and anti-RNP,[127] inhibit the splicing of adenoviral early pre-mRNA species in isolated nuclei.[136] These viral RNA species contain splice junction sequences that cannot be distinguished from cellular transcripts in their relationship to the consensus sequences. As antibodies that recognize other classes of snRNP or small cytoplasmic RNP[190] are not inhibitory, this observation is also consistent with the proposed role of U1 snRNP in splicing. The possibility that such U1 snRNP-splice junction sequence interactions as depicted in Figure 2 are essential to some process upon which splicing is in turn dependent seems to be eliminated by the inhibitory effect of anti-Sm and anti-RNP antibodies in a soluble splicing system.[258]

In light of the extreme conservation of splice junction sequences, any guide function of U1 snRNP must be universal, that is, U1 snRNP would recognize all splice junctions known. Thus, U1 snRNP could not distinguish among intervening sequences to aid the orderly and correct excision from the numerous pre-mRNA species that contain more than a single IVS. It is in this context that additional interactions postulated to occur between splice junction sequences and other small RNA molecules might be significant. The nuclear species U2 RNA, for example, could also form base pairs with sequences that border splice sites, as illustrated in Figure 2C. By contrast to the putative U1 RNA-consensus sequence interactions shown in Figure 2A and 2B, those in which U2 RNA might participate rely heavily on coding sequences immediately adjacent to splice sites.[124] Because U1 and U2 snRNA would be complementary to different sets of sequences near splice junctions, Ohshima and colleagues[124] have suggested that the two types of interaction would be cooperative, rather than exclusive. Moreover, the recognition by U2 RNA of coding sequences, which must be unique for each pre-mRNA species, could provide a mechanism to explain the correct pairing of the 5′ and 3′ ends of a particular IVS only with each other. Such a mode of recognition, that is, participation of coding sequences, could also ensure that the nucleotides to be ligated would be held in close proximity once the endonucleolytic cleavages that liberate the IVS had taken place.

This model implies that U2 RNA could base-pair in the manner shown in Figure 2C to a restricted class of intervening sequences in pre-mRNA species, those bordered by compatible coding sequences. Surprisingly, perhaps, the list of specific intervening sequences suggested by Ohshima and colleagues[124] to be compatible with U2 RNA is quite long.

Small RNA transcripts of the highly repeated Alu sequence have also been proposed to form base-paired structures, up to 11 nucleotides in length, with specific examples of splice junction sequences;[138,139] these proposed structures, too, rely heavily on pre-mRNA coding sequences. Although Alu transcripts can be found in association with nuclear and cytoplasmic, poly(A)-containing[140-142] RNA there exists no direct experimental evidence implicating these species, or indeed U2 RNA, in splicing. Thus, final evaluation of models that

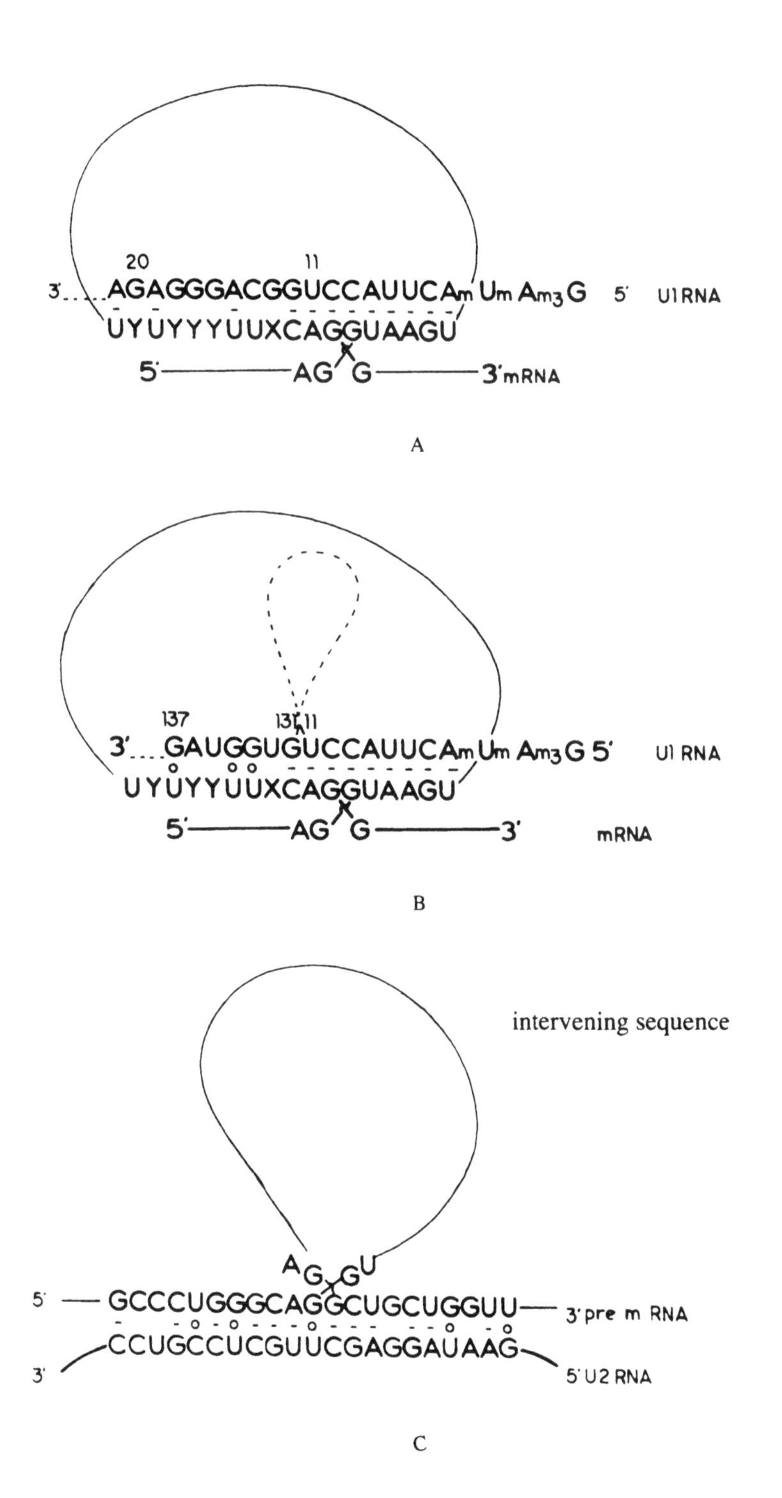


FIGURE 2. Potential base-pairing between splice junction consensus sequences and small, nuclear RNA species. Parts A and B depict two potential base-pairing interactions (see text) between U1 snRNA and splice junction consensus sequences within an mRNA precursor, whereas part C shows a potential interaction of the same sequences with U2 RNA. In all parts of the figure, Watson-Crick and non-Watson-Crick base-pairs are indicated by - and 0, respectively. The sources of these models are given in the text.

invoke small RNA molecules of different base-pairing potential as guides to the splicing reaction must await the development of reconstituted splicing systems, to test directly, for example, that snRNPs are necessary for faithful and accurate splicing or that mutations in the regions of U1 snRNA believed to be complementary to splice junction sequences alter the rate or fidelity of splicing reactions.

Excised intervening sequences have not been subject to much scrutiny but the available data suggest that removal of an IVS does not follow an invariant pathway. Thus, intervening sequence-specific probes have identified RNA species that are the size predicted for an excised, full-length, linear IVS of the adenoviral E2A gene[143] and the chicken ovalbumin gene.[144] The existence of such species implies both that the intervening sequences in question are excised in one splicing reaction that joins the coding sequences adjacent to that IVS and that the intervening sequences mentioned are not circularized during excision. By contrast, several examples of stepwise removal of an IVS have been reported. The excision of an IVS from the chicken α2 type 1 collagen gene that requires three splicing steps[135] has been mentioned previously. Similarly, removal of the larger of the two intervening sequences present in mouse and rabbit β-globin pre-mRNA transcripts appears to be mediated by at least two splicing reactions,[54,145] while the splicing reactions that fashion mature adenoviral late mRNA species generate numerous intermediates.[146]

It is too early to conclude that excision of transcribed intervening sequences is generally a multistep reaction, for the number of intervening sequences whose splicing has been examined in detail is very small. Nevertheless, the fact that stepwise excision has been observed raises a number of obvious questions, such as whether all splicing reactions have inherent polarity or directionality, and what is the nature of the mechanism(s) that ensure that stepwise splicing does not continue beyond the end of an IVS? The answer to the first question appears to be negative: opposite splicing polarities have been observed in different splicing reactions. The three-step reaction mentioned previously that removes an IVS from the chicken α2 type 1 collagen pre-mRNA, for example, displays 3′ → 5′ polarity and each splicing reaction recreates a sequence that resembles the 3′ splice junction consensus sequence.[135] By contrast, the excision of both the small and large intervening sequences of rabbit β-globin gene transcripts proceeds in the 5′ → 3′ direction[145] as does maturation of the adenovirus type 2 late, tripartite leader segment.[146] Whether new 5′ splice sites closely related to the consensus sequence are created during each splicing step in these cases is not yet established; the sequences necessary to such a mechanism are clearly present in rabbit β-globin transcripts.[145] If such stepwise excision does prove to be both common and generally mediated by recreation of 5′ or 3′ splice site sequences at each step, then one would expect that those internal sequences of an IVS recognized during this process would be conserved. Indeed, conservation of one such segment that could regenerate a 5′ splice site located 47 to 65 nucleotides upstream of the 3′ end of the large IVS of the rabbit β-globin and human β-, δ-, and γ-globin genes has been noted.[145]

The answer to the second question posed is even more opaque. Removal of each transcribed IVS must be precise in the great majority of instances, otherwise nuclear RNA preparations would be littered with aberrant products from which coding segments have been deleted. Little or no evidence for the existence of such products, created were stepwise splicing to continue beyond the boundary of an IVS, has appeared from examination of intermediates created during splicing of specific transcripts (see Section II.B.4). In principle, exclusion of sequences that resemble 5′ or 3′ splice sites from coding regions of a transcript would provide a simple mechanism to preclude this possibility. Not only are the odds against such an exclusion small, but it is also clear that cryptic splicing signals can be activated when those normally recognized in a given transcript are no longer present.[99,147]

The existence of viral mRNA species that share some coding regions and differ in internal splice sites, discussed in Chapter 8, clearly demonstrates that coding segments can include

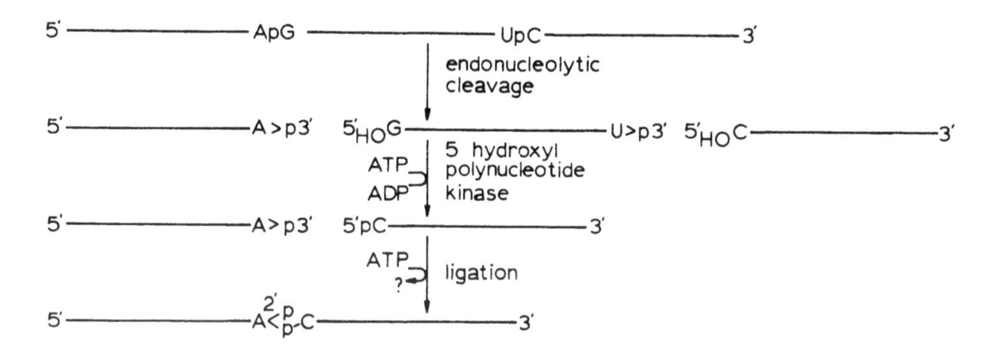


FIGURE 3. Action of wheat-germ enzymes that form a 2'-phosphomonoester, 3',5'-phosphodiester linkage. The RNA substrates of the wheat-germ endonuclease, 5'-hydroxyl polynucleotide kinase, and ligase are depicted as straight lines in which the bases designated at cleavage points were chosen at random and do not represent any known specificity of the endonuclease. This summary is based on that of Konarska et al.[154]

splicing signals, although it is possible that such cases are exceptional and in fact represent evolution from the stepwise splicing mechanisms (see Reference 24). In any case, it is very unlikely that the primary sequence of an unspliced pre-mRNA is the sole parameter that influences its splicing: the substrate for all processing reactions discussed in Section II is not RNA but RNP (see Section III.A). It seems most reasonable to suppose that the RNP proteins would play a crucial role in determination of the secondary and tertiary structures of an RNA molecule, and thus could influence profoundly the splicing process. Be that as it may, it is clear from even this brief discussion that we are far from understanding even the simplest aspect of splicing, the mechanism(s) whereby a single IVS is excised.

### *3. The Splicing Machinery*

Identification of the components of the enzymatic machinery that splices eukaryotic pre-mRNA has been hampered by the slow progress in the development of soluble splicing systems. Within the last year, extracts of mammalian cells have been reported to splice either transcripts synthesized by the extract in the presence of appropriate DNA templates[145a,148] or RNA molecules added exogenously.[149] As yet, such systems have provided little information about the requirements or properties of pre-mRNA splicing for the assays have been performed under conditions optimized for transcription. Nor has the efficiency of splicing in vitro been assessed quantitatively, although it appears to be quite good in both cases. The complex nature of these systems and the failure in other experiments of similar extracts to splice (see, for example, References 150 and 151) suggest that it will not be easy to dissect the splicing machinery. Nevertheless, in vitro splicing systems represent the only viable route to identification and characterization of the molecular entities that mediate pre-mRNA splicing and thus to the answers to such questions as whether the endonuclease and ligase that must participate in splicing are discrete physical entities or whether guide molecules are indeed elements essential to the recognition of splicing signals.

One enzyme, an unusual RNA ligase, that might participate in splicing has recently been identified and characterized using short RNA molecules as exogenous substrates. This ligase, initially described in wheat-germ extracts,[152] differs from conventional RNA ligase[153] in that it forms a 2'-phosphomonoester, 3',5'-phosphodiester bond,[152] as illustrated in Figure 3. This wheat-germ RNA ligase requires a 2',3' cyclic phosphate and a 5' phosphate at the 3' and 5' termini, respectively, of the molecules it will join.[154] Wheat-germ extracts possess a 5' hydroxyl polynucleotide kinase,[154] as do extracts of mammalian cells.[155-157] The necessary 3′ terminal structure might be created as a result of the splicing endonucleolytic cleavage, as appears to be the case during yeast pre-tRNA splicing.[158] The sequence of reactions postulated[154] is summarized in Figure 3; the most striking feature of this series of reactions is their unusual requirements and products, which could serve to distinguish both the termini of RNA molecules to be spliced and the phosphodiester bonds created as a result of splicing.

It is, however, not yet clear whether such activities play any role in pre-mRNA splicing. Detailed characterization of cleavage and ligation products of yeast pre-tRNA molecules in yeast and wheat-germ extracts,[159,160] for example, implicates the reaction series shown in Figure 3 in pre-tRNA splicing. On the other hand, the internal labeling of polyadenylated RNA molecules that are larger than 12S when mouse L-cell nuclei are incubated in the presence of $\gamma$-$^{32}$P-ATP by a reaction that is not the result of recycling of the label to $\alpha$-positions of ribonucleoside triphosphates,[157] raises the possibility that a similar mechanism (i.e., requiring a 5′ terminal monophosphate that is incorporated into the final bond — see Figure 3) pertains to other classes of pre-RNA molecules.

*4. Splicing of Transcripts that Contain Multiple Intervening Sequences: Processing Pathways*

Transcripts of many genes of higher eukaryotes contain not one but several intervening sequences, immediately raising the question of how their removal is carried out in orderly fashion, as it must be to ensure that the majority of spliced RNA molecules do not comprise misspliced coding segments. In the cases of those transcripts mentioned in Section II.B.1 that contain very large numbers of intervening sequences, the task of synthesizing a functional mRNA appears formidable, yet it is completed with high efficiency.

One solution to this problem might be to invoke the action of those snRNA species that could interact primarily with coding sequences: under certain conditions, such interactions could ensure that the 3′ end of one coding segment is ligated only to the 5′ end of the next in a pre-mRNA chain. However, this model does not seem to be tenable for it would require both a large number of small RNA species to accommodate all the combinations of coding sequences that border intervening sequences and probably also that a *different* snRNA species should interact with each set of splice junction sequences within a transcript that contains multiple intervening sequences. Although quite a few snRNA species have been described,[123,133] their number is not legion. Furthermore, Ohshima and colleagues[124] have suggested that U2 RNA possesses the potential to base-pair with both the smaller and larger intervening sequences of mouse and human $\beta$-globin transcripts.

A second readily conceived mechanism that would permit orderly splicing proposes that splicing displays polarity, proceeding from one or other end of the pre-mRNA molecule to the opposite end, or at least advances processively in both directions from an internal point. If either the splicing machinery or the pre-mRNA substrate were tethered, such directionality might be achieved quite readily. While an accumulating body of evidence suggests that newly synthesized RNA is attached to the nuclear matrix in eukaryotic cells (see Section III.B), all that is presently known of the pathways whereby transcripts of individual genes are spliced indicate that splicing is not processive, nor, indeed, does it necessarily follow an invariant pathway.

Initial characterization of pulse-labeled nuclear RNA species complementary to mouse $\beta^{maj}$-globin cDNA suggested that this statement is true even in the case of a relatively simple transcript that contains but two intervening sequences.[54] The lengths of the RNA segments protected from RNase A digestion when labeled intermediate globin RNA was hybridized to globin cDNA species of 1030 and 900 nucleotides led Kinniburgh and Ross[54] to propose the splicing pathway depicted in Figure 4A, in which the 1030-nucleotide intermediate would be created by removal of a large portion of the larger IVS. This intermediate appeared to be further spliced to the 900-nucleotide species by two alternative pathways. In one of these,

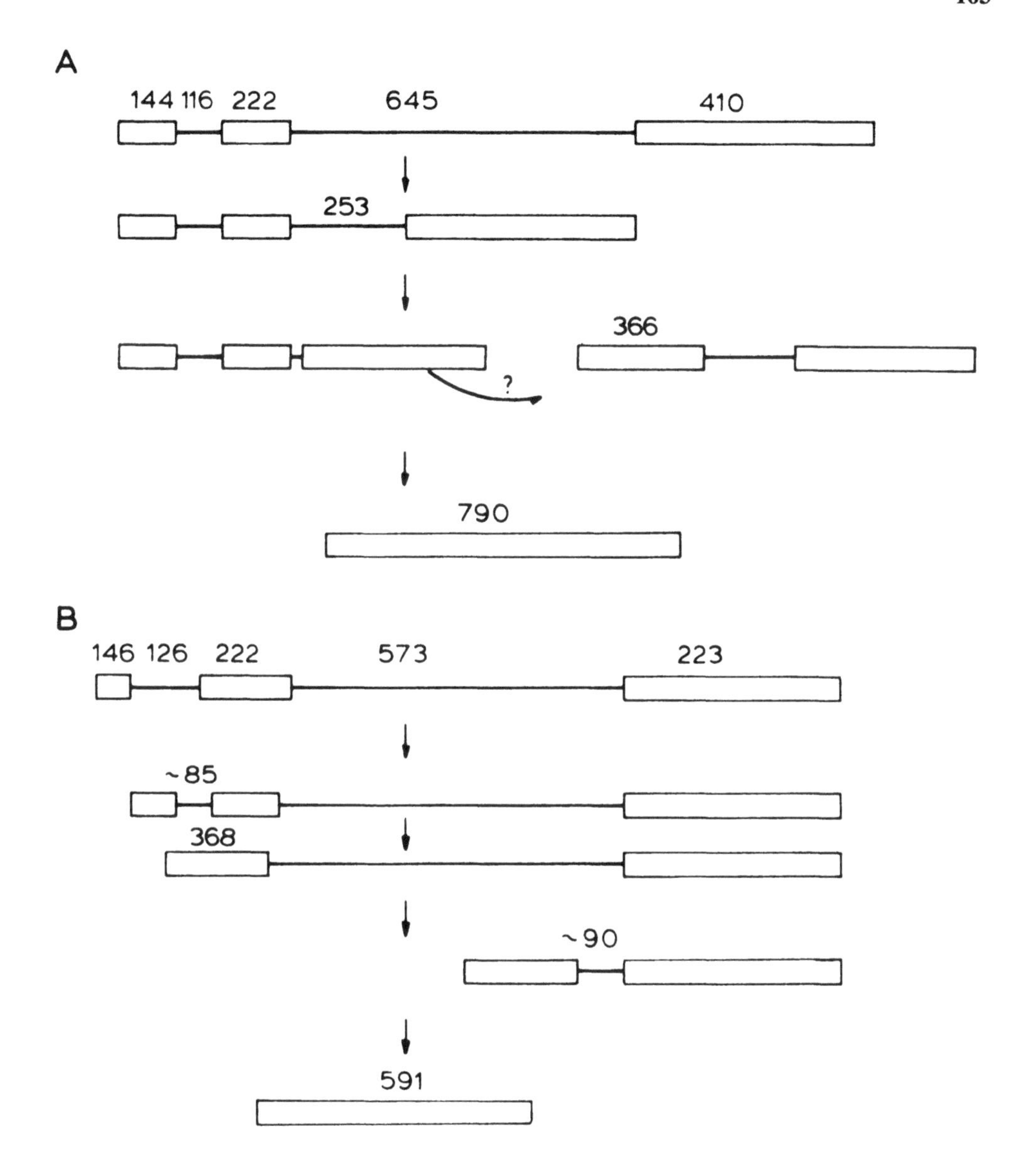


FIGURE 4. Proposed pathways of splicing of mouse and rabbit β-globin mRNA precursors. Parts A and B depict the pathways of splicing of mouse and rabbit β-globin, poly(A)-containing mRNA precursors, respectively. In both parts of the figure, coding and intervening sequences are shown as open boxes and horizontal lines, respectively, whose lengths, in nucleotides, are shown above each segment. The data upon which these pathways are based are discussed in the text.

a second portion of the larger IVS would be excised next, whereas in the other the smaller intervening sequence would be removed (see Figure 4A). Unfortunately, kinetic evidence to establish that these two species of about 900 nucleotides in length are indeed generated independently could not be obtained, for the β-globin transcripts are processed very rapidly in the system employed.

The mode of splicing of transcripts of a second β-globin gene, that of the rabbit, has also been examined in some detail.[145] Interestingly, the processing pathway deduced using the nuclease S1 assay to determine the structure of unlabeled, nuclear, β-globin RNA species (shown in Figure 4B) is quite different from that deduced for mouse cells. Most notably, the first step in the splicing of rabbit β-globin transcripts is the excision of the smaller IVS, that is, intermediates containing the smaller IVS but lacking all or part of the larger IVS are not observed in rabbit bone marrow cells.[145] Although Kinniburgh and Ross[54] characterized newly synthesized RNA whereas Grosveld and colleagues[145] examined the total β-globin RNA population, it is difficult to see how such procedural differences would account for the radically different processing pathways reported. Thus, it seems likely that β-globin transcripts follow different routes to fully spliced mRNA in the rabbit and the mouse. It should be noted, however, that this conclusion rests upon the assumption that the globin RNA species intermediate in size between unspliced precursor and mature mRNA are indeed normal processing intermediates, an assumption that remains to be proven formally.

A second striking feature of the pathway reported for rabbit β-globin RNA splicing is that it is actually processive, displaying 5′ → 3′ polarity (see Figure 4B). This, however, appears to be an unusual example. Processing of transcripts of the chicken oviduct ovomucoid and ovalbumin genes has also been investigated by pulse-chase experiments; by hybridization of fractionated nuclear RNA, transferred to an appropriate support, to cloned DNA fragments; or by heteroduplex analysis. The first approach can provide a direct demonstration that putative precursor or intermediate RNA species are indeed kinetic precursors to mRNA, whereas the latter two, with probes specific for various parts of the gene, can permit detailed elucidation of the structures of the RNA species of interest.[49,52] Both these genes include eight intervening sequences[61-69] so it is not surprising that total nuclear RNA or the poly(A)-containing fraction include a relatively large number of RNA species longer than the mature mRNA species; at least six and seven complementary to the ovalbumin and ovomucoid genes, respectively, can be readily detected.[152] Although the coding and intervening sequences represented in every one of these nuclear-specific RNA molecules have not yet been identified, the results of both kinetic and structural studies indicate that the intervening sequences of both transcripts are excised in a preferred order that is clearly not processive. Moreover, while there does not appear to exist a single, unique splicing route, preferred pathways of intervening sequence excision can be deduced.[152] Similar conclusions have been drawn for the processing of nuclear RNA species transcribed from the vitellogenin gene of *Xenopus laevis*; the structures of the RNA species observed suggest that excision of the large number of intervening sequences transcribed from this gene does not occur sequentially from either end of the poly(A)-containing pre-mRNA and that the mature mRNA can be fashioned by not one, but rather several preferred pathways of splicing.[53]

These observations exclude sequential or processive splicing as the usual mechanisms of removal of intervening sequences from a transcript that contains several. On the other hand, several cases of processive excision of individual intervening sequences have been documented, as discussed previously — observations that might imply that either the substrate or the splicing machinery can be tethered, at least transiently. We therefore possess no clear picture of the parameters that govern splicing of pre-mRNA species that are a complicated mosaic of coding and intervening sequences. In this context, it is probably important to recall that RNA molecules that must be spliced, or indeed processed in any way, are not naked, but in fact exist within the nucleus in the form of RNP. Presumably, each transcript adopts a unique conformation that is determined both by its RNA sequence and the associated proteins (see Section III.A). It is therefore not too difficult to imagine that in such a structure not all RNA segments are equally accessible to the splicing machinery and thus some splicing reactions might initially be favored. This state of affairs could readily explain the observations described in previous paragraphs of this section. This idea could also be extended to account for complete and orderly processing of an mRNA precursor by the postulate that removal of the most accessible intervening sequence, or sequences, induced sufficient conformational change to expose additional sets of splice sites to the splicing machinery. Finally, if the initial nucleoprotein structure both precluded accessibility of intervening sequences that are

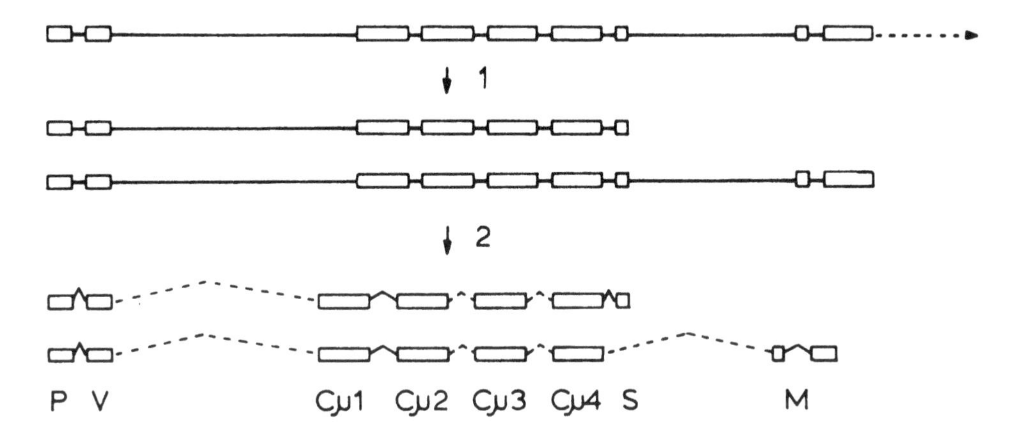


FIGURE 5. Schematic representation of the maturation of immunoglobulin heavy μ-chain mRNA species. The coding and intervening sequences of transcripts of immunoglobulin heavy μ-chain genes are shown by open boxes and horizontal lines, respectively. The symbols P, V, Cμ1, Cμ2, Cμ3, and Cμ4 indicate the promoter site, the variable region, and domains 1 to 4 of the constant region of a heavy μ-chain, respectively. S and M denote the 3′-terminal exons present on the mRNA species encoding the secreted and membrane-bound products, respectively.

adjacent within the primary sequence and ensured that exposed intervening sequences were located quite far apart in the primary sequence, then the likelihood of missplicing events would be quite small. While models of this kind have some attractive features, too little is yet known of the consequences of assembly of pre-mRNA species in ribonucleoprotein structures to assess their validity.

*5. Splicing and the Regulation of Gene Expression*

It is well established that expression of certain genes of several animal viruses can be regulated posttranscriptionally: in these cases, it is the use of a particular poly(A)-addition site or 5′ or 3′ splice site from, in each case, a limited set of potential sites that determines which of several possible mRNA species is actually fashioned from a given primary transcript (see Reference 161 and Chapter 8 of this volume). Furthermore, the frequency with which different potential splice sites in particular are 'chosen' can vary during an infectious cycle (see, for example, References 162 to 166), suggesting the existence of some form of active regulatory mechanism.

The best-documented example of posttranscription regulation of expression of cellular genes is provided by those encoding immunoglobulin heavy-chains. The IgM μ-chain, for example, is synthesized as a membrane-bound and secreted protein during different stages of differentiation and, as illustrated in Figure 5, the mRNA species encoding these two forms of the protein differ in both their poly(A) addition site and splicing patterns near the 3′ end of the mRNA chain.[167-169] The expression of other classes of heavy-chain genes is also mediated by alternative RNA processing pathways as well as by DNA sequence re-arrangements.[170-173]

Other cases of a single gene encoding more than one mRNA species include the mouse amylase and dihydrofolate reductase genes; at least two mRNA species, one characteristic of salivary gland and one found in the liver, complementary to cloned amylase DNA have been described,[174,175] whereas at least four poly(A)-containing dihydrofolate mRNA species of differing size have been identified.[176] The latter differ primarily in the length of the 3′ untranslated region they carry, and therefore in the sites in the initial transcript to which poly(A) was added,[176] whereas each of the former pair of species carries a unique set of 5′ terminal sequences.[175] Whether this latter difference reflects tissue-specific transcription and/

or processing is not yet established. However, differentially-spliced forms of rat prolactin mRNA have been detected by sequencing prolactin cDNA clones.[177] Similarly, there exists strong evidence that transcripts of the rat calcitonin gene follow one of two splicing pathways to fashion mRNA species encoding calcitonin itself or a second, calcitonin-like protein, termed pseudo-calcitonin.[178,179] Although such examples are rare at present, they do establish that the synthesis of cellular mRNA species can be governed at the posttranscriptional level by mechanisms that are akin to those displayed by several animal viruses.

### C. Processing of Transcripts that Contain No Intervening Sequences

In addition to the mosaic transcripts of intervening and coding sequences whose splicing has been discussed in Section II.B.4, transcripts that contain no intervening sequences must be synthesized and matured. When considering their processing, it is convenient to place such transcripts in two classes, those that become polyadenylated and those that do not.

The only well-defined members of the latter class are histone mRNA species of a wide variety of organisms; some typical, poly(A)-containing histone mRNA species have been found, for example, in yeast and *Xenopus* oocytes,[180,181] but these are exceptional. The available evidence, reviewed recently by Hentschel and Birnstiel,[27] indicates that transcription of histone genes initiates at a site that corresponds to the sequence encoding the 5′-terminus of mRNA and ends at discrete sites a short distance — 20 to 50 nucleotides, depending on the gene — beyond the last codon. The region in which termination occurs includes blocks of nucleotides that are both highly conserved among histone genes and reminiscent of prokaryotic transcription termination sites (see Reference 27 for details). It therefore seems most likely that the majority of histone mRNA species are not cleaved from large precursors, but rather defined by the sites at which transcription initiates and terminates. Indeed, large nuclear RNA species containing histone sequences do not appear to be the rule and those that have been observed in certain systems can be explained by a low level of read-through transcription. Thus, it would appear that the only processing reaction to which histone mRNA precursors are routinely subject is the formation of 5′-terminal caps.[14,15] Whether this very limited processing of nuclear histone RNA molecules is in any way related to their rapid exit from the nucleus is not known.

A variety of specific genes, such as the majority of *Drosophila melanogaster* heat-shock genes, various human interferon genes, *Dictyostelium* actin genes, and probably the majority of yeast genes, are known to be devoid of intervening sequences and expressed as poly(A)-containing mRNA. Little is yet known of the primary products of transcription of these genes or of the ways in which they are processed. On the other hand, sequences of several of these genes, including flanking sequences, have been completely or partially determined[28,32,182,183] and include typical sequences related to the canonical $TATA^{A}_{T}\,A^{A}_{T}$ sequence[78] on the order of 30 nucleotides upstream from sequences that correspond to mRNA 5′ termini. These observations argue strongly that the only processing necessary to mature 5′ termini is cap formation. Because the sites of initiation of transcription correspond to mRNA 5′ termini and coding sequences within these genes are co-linear, the only potentially dispensible sequences in their transcripts must lie beyond the sites to which poly(A) is added. The major questions to be answered therefore concern the extent and frequency of transcription beyond poly(A)-addition sites, questions that, of course, also apply to genes that possess intervening sequences.

There is, as described in Chapter 6 in this volume, clear evidence that transcription of genes expressed in mammalian cells does continue downstream from sites of poly(A) addition. This state of affairs, while most likely typical, has been demonstrated experimentally for only a few genes (see Chapter 6), all of which do in fact contain intervening sequences: the primary product(s) of transcription of no gene in the class under consideration has been identified. Consequently, it is not even clear that their transcripts are homogeneous, that is, that transcription terminates at discrete sites. The apparent underrepresentation in nuclear RNA populations of transcripts of sequences that lie immediately distal to sites of poly(A) addition in those cases examined, presumably a reflection of their instability, suggests that it may not prove easy to come by answers to such questions.

A mechanism of poly(A)-addition that requires cleavage at specific sites to liberate the 3′ termini to which poly(A) is actually attached requires the existence of at least one specific endoribonuclease involved in 3′-terminal processing. Purified hnRNP particles have been reported to contain an endoribonuclease that cuts double-stranded RNA regions (an RNase D)[184,185] and that can introduce a limited number of nicks into RNA packaged in hnRNP.[186,187] Interestingly, this enzyme displays some preference for double-stranded regions that are rich in pyrimidine residues.[184] However, the biological relevance of this activity remains to be established.

## III. ADDITIONAL POSTTRANSCRIPTIONAL EVENTS

The modifications that alter the sequence of an mRNA precursor, outlined in the previous section, are accompanied by additional alterations, such as RNP formation and transport of mRNA from the nucleus to the cytoplasm. While these are not processing in the sense in which the term is usually applied, such changes certainly participate in the formation of mature, functional mRNA species and will therefore be considered briefly.

### A. RNP Formation and RNA Processing

Examination of transcriptionally active chromatin in the electron microscope,[188-198] and biochemical experiments[199-203] have established firmly that nuclear RNA is present in the cell in ribonucleoprotein structures (RNP). Such RNP contain RNA sequences that have the characteristics of hnRNA[202,204-208] and are constructed upon nascent RNA that is still attached to chromatin (see, for example, References 194, 196, 198, 209, 210). By contrast to the general agreement about these properties, quite wide variations in both the number and apparent molecular weights of polypeptides associated with RNP have been reported. (See References 211 and 212 for reviews.) It seems likely that such differences reflect, at least in part, variations in the methods of RNP preparation employed in different experiments. Those in wide use release not only large RNP structures that contain RNA chains quite similar in length distribution to hnRNA, but also 30 to 55S particles, the so-called RNP core or 40S particles, the products of breakage of the RNA chain by endogenous nucleases. It is now clear that such core RNP particles comprise six major proteins, exhibiting apparent molecular weights of 32 to 44 kd and migrating in SDS-polyacrylamide gels as three pairs of closely spaced doublets.[211,213] Similar proteins were observed as major components of RNP preparations in previous experiments. The properties of these core polypeptides strongly argue that they are the major structural proteins of RNP, serving to package the RNA. The significance and function of the majority of other polypeptides that have been ascribed to hnRNP have not yet been assessed, although most appear to be confined to the nucleus. This group might well include the polypeptide constituents of the enzymes that must perform the processing reactions discussed previously.

As RNP is built during the act of transcription it must be that the normal substrates of the posttranscriptional processing reactions discussed in Section II are RNP molecules. The role, if any, of RNP structure and conformation in such processing reactions remains to be established. Nevertheless, the mere fact that these RNP molecules are processed throws the spotlight on the question of whether the association of RNA with proteins in RNP is sequence-specific to any degree. Quite a substantial body of evidence favoring some elements of sequence specificity in the RNA-protein association has now been amassed. In HeLa cell hnRNP, for example, RNA regions that participate in intramolecular base-pairing are considerably more sensitive to nuclease digestion than RNA sequences that remain single-stranded.[212] Thus, such regions must be more accessible, i.e., less tightly associated with proteins in RNP. Similarly, both the 5′ terminal caps[215,216] and 3′ terminal poly(A) tracts[208,217-219] are associated with specific polypeptides; such observations provide strong precedents for sequence-specific interactions between RNA and proteins in RNP.

Cloned DNA fragments have been employed to establish that both precursor and spliced RNA species complementary to specific genes are packaged in RNP,[220,221] but this kind of approach has not been applied to the question of sequence-specific organization. However, that such an organization is not uncommon can be deduced from the specific patterns of RNP fibril organization observed when transcriptionally-active chromatin is viewed in the electron microscope:[196,210,222-224] when, for example, *Drosophila* "sister chromatid" transcriptional units are compared, they can be seen to exhibit similar patterns of RNP fibril folding and particle distribution. Moreover, many transcripts appear to undergo cleavage at specific sites before their synthesis is complete.[196,210,223-225] The relationship of such cleavages to the processing reactions discussed in Section II remains to be elucidated.

In concert, these tantalizing findings argue strongly that the association between RNA molecules and polypeptides in RNP includes some element of sequence specificity. Further work is clearly necessary to establish the precise relationship of the sequences of specific transcripts whose modes of processing are known to be binding to RNP polypeptides or protein subunits. Because the majority of newly synthesized hnRNA sequences are protected against nuclease digestion,[214,220,221] it is very unlikely that coding sequences are specifically packaged whereas intervening sequences or those that lie downstream from poly(A)-addition sites are not. This conclusion does not, however, preclude the possibility that regions, of greater or lesser accessibility, may exist within RNP somewhat analogous to DNase I, hypersensitive and relatively resistant regions in chromatin. It is of considerable importance to determine whether a description of RNP organization couched in such terms bears any obvious relation to RNA processing signals.

## B. Translocation of Fully Processed RNA to the Cytoplasm

Once an mRNA species has been fashioned within the nucleus, it must be shuttled to the cytoplasm where it can be translated. This process is poorly understood, perhaps because its complex nature has rendered its study more difficult than other aspects of mRNA biogenesis. Nevertheless, this step is essential to mRNA function.

The very fact that cytoplasmic and nuclear RNA populations can be distinguished by a variety of properties indicates that transport is selective. The most obvious discrimination is, of course, between mature mRNA species and their precursors, the latter being restricted to the nucleus. While the mechanism(s) underlying selective transport of mRNA species is not yet fully appreciated, there exists some evidence to support the notion of specific transport signals inherent to an RNA sequence.

Variants of several mammalian cellular and viral genes that completely lack their normal intervening sequences have now been constructed and reintroduced into cells so that their expression might be investigated. When experiments such as these were originally performed with SV40 and mouse $\beta^{maj}$-globin genes, transcription of the mutated DNA segments could be detected readily, yet little stable RNA accumulated in the cytoplasm.[94,99,117,119,226,227] These observations, and the ability to 'rescue' such mutants and achieve synthesis of stable, cytoplasmic RNA by introduction of heterologous introns at sites thought to be inappropriate,[228,229] led to the proposal that splicing is in some way essential to transport of mRNA from the nucleus to the cytoplasm, at least of those mRNA species that should normally be spliced. It now seems that this interpretation is too simple: mutants of SV40 that induce the synthesis of large amounts of unspliced, late 19S mRNA in infected cells have been isolated.[230] Similarly, a recombinant SV40-rat preproinsulin gene that lacks any intervening sequence directs the synthesis of unspliced, stable, cytoplasmic RNA.[231] The cytoplasmic,

SV40 late 19S mRNA retains intervening sequences, whereas none are available in the SV40-preproinsulin recombinant gene. These two genes do have in common transcription from promoter sites upstream in the SV40 genome from those normally recognized during transcription of the SV40 late region,[230,231] but whether such unusual 5′-terminal sequences play any role in the transport of these RNA species remains to be determined. However, the notion of a simple transport signal can account for all the observations discussed in the previous paragraph only if it is postulated that the sequence-signalling transport, normally created during splicing, is present in the unusual 5′-terminal sequences of these unspliced mRNA species.

Comparisons of the RNA sequences comprising polyribosomal and nuclear RNA populations at different stages of sea urchin development have provided strong evidence for the selection of specific sets of sequences to serve as mRNA from among a more complex nuclear RNA population, that is, posttranscriptional regulation of gene expression.[232-234] The fate of transcripts of specific genes that are expressed as mRNA during only one developmental stage has not yet been examined, so it is not known whether mRNA selection is mediated by selective processing or whether transport of mature mRNA from the nucleus to the cytoplasm might be selective. One striking example of regulation that appears to operate at this level has emerged from investigations of the mechanisms whereby human adenoviruses inhibit cellular gene expression during the late phase of a productive infection: in such circumstance, the newly synthesized RNA sequences that enter the cytoplasm are exclusively viral in origin.[235,236] As transcripts of nonadenoviral genes are capped, polyadenylated, and spliced apparently normally,[236] it must be concluded that the virus infection does not disrupt normal processing reactions. These findings, in conjunction with the observation of a specific effect of adenovirus infection upon transport of 28S rRNA from the nucleus to the cytoplasm,[237] point to an alteration of the normal transport mechanisms following adenovirus infection. In this context, it is of no little interest that the changes in hnRNA and mRNA metabolism, observed when growth of anchorage-dependent fibroblasts is inhibited by transfer of the cells to suspension culture, appear to be reminiscent of the changes observed in adenovirus-infected cells: synthesis of hnRNA continues, but the appearance in the cytoplasm of newly made mRNA is drastically reduced.[236,237]

The results obtained with adenovirus-infected cells imply that complete processing of an RNA molecule, while perhaps necessary, is not sufficient for transport to the cytoplasm. A similar conclusion can be drawn from the experiments with intervening sequence-less genes discussed previously. None of these studies, however, addresses the question of the role of RNP structure and conformation in transport.

The actual mechanics of mRNA movement are also obscure, although it is generally assumed that nuclear pore complexes form the gateways through which RNP containing mature mRNA must pass (see References 240 and 241 for reviews). When considering this question, it is important to recall that mRNA and pre-mRNA molecules, in the form of their respective RNP complexes, are not generally free to diffuse about the cell at random. Thus, polyribosomal mRNA appears to be tethered to the cytoskeleton via the mRNA itself.[242-246] Similarly, nascent, nuclear RNA sequences have been recognized to be attached to the nuclear matrix for some time, a result that has been confirmed recently, in several cases by examination of specific transcripts.[247-250] It is certainly easier to envision directed movement of RNP structures attached to a skeletal framework, perhaps by a process resembling 'treadmilling',[253] than of 'free-floating' entities. Moreover, typical matrix preparations include nuclear pore complexes physically linked to the nuclear skeleton.[254-256] Experimental knowledge of the mechanics of mRNA transport, as opposed to attractive speculations, is, however, scanty, despite the quite considerable efforts devoted to attempts to establish in vitro transport systems (the reader is referred to References 257 and 258 for reviews of this confusing field).

The problems discussed in Section III, which can be considered as a group concerning the relationship of structural elements (RNP forms, the nuclear matrix and pores) to the process of functional mRNA biogenesis, are among the most fascinating facing molecular biology today. They are probably also some of the most difficult to approach experimentally, although it is to be hoped that continued analysis of the fate of transcripts of specific cellular (or viral) genes will provide as many answers as have been forthcoming in recent years.

## REFERENCES


1. **Perry, R. P.,** Processing of RNA, *Ann. Rev. Biochem.*, 45, 605, 1976.
2. **Lewin, B.,** *Gene Expression*, Vol. 2, John Wiley & Sons, New York, 1980, chap. 25.
3. **Bajszar, G., Samarina, O. P., and Georgiev, G.,** On the nature of 5′-termini in nuclear pre-mRNA of Erlich carcinoma cells, *Cell*, 9, 323, 1976.
4. **Perry, R. P., Kelley, D. E., Friderici, K. H., and Rottman, F. M.,** Methylated constituents of heterogeneous nuclear RNA: presence in blocked 5′-terminal structures, *Cell*, 6, 13, 1975.
5. **Schibler, U. and Perry, R.,** Characterization of 5′-termini of hnRNA in mouse L cells: implication for processing and cap formation, *Cell*, 9, 121, 1976.
6. **Salditt-Georgieff, M., Jelinek, W., Darnell, J. E., Furichi, Y., Morgan, M., and Shatkin, A. J.,** Methyl labelling of HeLa cell hnRNA: a comparison with mRNA, *Cell*, 7, 227, 1976.
7. **Wei, C. M., Gerschowitz, A., and Moss, B.,** 5′-Terminal and internal methylated nucleotide sequences in HeLa cell mRNA, *Biochemistry*, 15, 397, 1976.
8. **Furichi, Y., Morgan, M., Shatkin, A. J., Jelinek, W., Salditt-Georgieff, M., and Darnell, J. E.,** Methylated, blocked 5′-termini in HeLa cell mRNA, *Proc. Natl. Acad. Sci. U.S.A.*, 72, 1904, 1976.
9. **Wei, C. W., Gershowitz, A., and Moss, B.,** Methylated nucleotides block 5′-terminus of HeLa cell mRNA, *Cell*, 4, 379, 1975.
10. **Lavi, U., Fernandez-Munoz, R., and Darnell, J. E.,** Content of N6 methyladenylic acid in hnRNA and mRNA of HeLa cells, *Nucleic Acids Res.*, 4, 63, 1977.
11. **Desrosiers, R., Friderici, K., and Rottmann, F.,** Identification of methylated nucleosides in messenger RNA from Novikoff hepatoma cells, *Proc. Natl. Acad. Sci. U.S.A.*, 71, 3971, 1974.
12. **Wei, C. M. and Moss, B.,** Nucleotide sequences at the $N^6$ methyladenosine sites of HeLa cell mRNA, *Biochemistry*, 16, 1672, 1977.
13. **Schibler, U., Kelley, D. E., and Perry, R.,** Comparison of methylated sequences in mRNA and hnRNA from mouse L cells, *J. Mol. Biol.*, 115, 695, 1977.
14. **Moss, B., Gerschowitz, A., Weber, L. A., and Baglioni, C.,** Histone mRNAs contain blocked and methylated 5′-termini but lack methylated nucleosides at internal positions, *Cell*, 10, 113, 1977.
15. **Stein, J. L., Stein, G. S., and McGuire, P. M.,** Histone messenger RNA from HeLa cells: evidence for modified 5′-termini, *Biochemistry*, 16, 2207, 1977.
16. **Chen-Kiang, S., Nevins, J. R., and Darnell, J. E.,** N-6-methyladenosine in adenovirus type 2 nuclear RNA is conserved in the formation of messenger RNA, *J. Mol. Biol.*, 135, 733, 1979.
17. **Flint, S. J. and Broker, T.,** Lytic infection by adenovirus, in *Molecular Biology of Tumor Viruses, Part II*, Tooze, J., Ed., Cold Spring Harbor Laboratory, Cold Spring Harbor, New York, 1980, 443.
18. **Sommer, S., Salditt-Georgieff, M., Bachenheimer, S., Darnell, J. E., Furichi, Y., Morgan, M., and Shatkin, A. J.,** The methylation of adenovirus-specific nuclear and cytoplasmic RNA, *Nucleic Acids Res.*, 3, 749, 1976.
19. **Lockard, R. E. and RajBhandary, U. L.,** Nucleotide sequences of the 5′-termini of rabbit α and β globin mRNA, *Cell*, 9, 747, 1977.
20. **Cheng, T-C. and Kazazian, H. H.,** The 5′-terminal structures of murine globin mRNA, *J. Biol. Chem.*, 252, 1758, 1977.
21. **Heckle, W. L., Frenton, R. G., Wood, T. G., Merkel, C. G., and Lingrel, J. B.,** Methylated nucleosides in globin mRNA from mouse nucleated erythroid cells, *J. Biol. Chem.*, 252, 1764, 1977.
22. **Yang, N. S., Manning, R. F., and Gage, L. P.,** The blocked and methylated 5′-terminal sequence of a specific cellular messenger: the mRNA for silk fibroin of *Bombyx mori*, *Cell*, 7, 339, 1976.
23. **Breathnach, R. and Chambon, P.,** Organization and expression of eukaryotic split genes coding for proteins, *Ann. Rev. Biochem.*, 50, 349, 1981.
24. **Flint, S. J.,** RNA splicing *in vitro*, in *Enzymes of Nucleic Acid Synthesis and Modification*, Jacob, S. T., Ed., CRC Press, Boca Raton, Fla., 1983, 75.

25. **Maxwell, I. H., Maxwell, F., and Hahn, W. E.,** General occurrence and transcription of intervening sequences in mouse genes expressed via polyadenylated mRNA, *Nucleic Acids Res.*, 8, 5874, 1980.
26. **Kedes, L. H.,** Histone messengers and histone genes, *Cell*, 8, 321, 1976.
27. **Hentschel, C. H. and Birnstiel, M. L.,** The organization and expression of histone gene families, *Cell*, 25, 301, 1981.
28. **Derynck, R., Content, J., DeClerq, E., Volckaert, G., Tavernier, J., Devos, R., and Fiers, W.,** Isolation and structure of a human fibroblast interferon gene, *Nature (London)*, 285, 542, 1980.
29. **Goeddel, D. V., Leung, D. W., Dull., T. T., Gross, M., Lawn, R. M., McCandliss, R., Seeburgh, P. H., Ullrich, A., Yelverton, E., and Gray, W.,** The structure of eight distinct cloned human leukocyte interferon cDNAs, *Nature (London)*, 290, 20, 1980.
30. **Streuli, M., Nagata, S., and Weissmann, S.,** At least three human type interferons: structure of α 2, *Science*, 209, 1343, 1980.
31. **Lawn, R. M., Adelman, J., Francke, A. E., Houck, C. M., Gross, M., Najarian, R., and Goeddel, D. V.,** Human fibroblast interferon gene lacks introns, *Nucleic Acids Res.*, 9, 1045, 1981.
32. **Lavin, R. M., Adelman, J., Dull, T. J., Gross, M., Oredell, D., and Ullrich, A.,** DNA sequences of two closely linked human leukocyte interferon genes, *Science*, 212, 1159, 1981.
33. **Houghton, M., Jackson, I. J., Porter, A. G., Doel, S. J., Catlin, G. H., Barber, C., and Carey, N. H.,** The absence of introns within a human fibroblast interferon gene, *Nucleic Acids Res.*, 9, 247, 1981.
34. **Tavernier, J., Denynck, R., and Fiers, W.,** Evidence for a unique human fibroblast interferon (IFN-β1) chromosomal gene devoid of intervening sequences, *Nucleic Acids Res.*, 9, 461, 1981.
35. **Aleström, P., Akusjärvi, G., Perricaudet, M., Mathews, M. B., Klessig, D. F., and Pettersson, U.,** The gene for polypeptide IX of adenovirus type 2 and its unspliced messenger RNA, *Cell*, 19, 671, 1980.
36. **McKnight, S. L.,** The nucleotide sequence and transcript map of the herpes simplex virus thymidine kinase gene, *Nucleic Acids Res.*, 8, 5949, 1980.
37. **Wagner, M. J., Sharp, J. A., and Summers, W. C.,** Nucleotide sequence of the thymidine kinase gene of HSV type 1, *Proc. Natl. Acad. Sci. U.S.A.*, 78, 1441, 1981.
38. **Gallwitz, D. and Sures, I.,** Structure of a yeast split gene: complete nucleotide sequence of the actin gene in Saccharomyces cerevisiae, *Proc. Natl. Acad. Sci. U.S.A.*, 77, 2546, 1980.
39. **Ng, R. and Abelson, J.,** Isolation and sequence of the gene for actin in *Saccharomyces cerevisiae, Proc. Natl. Acad. Sci. U.S.A.*, 77, 3912, 1980.
40. **Rosbash, M., Harris, P. K. W., Woolford, J., and Taem, J. L.,** The effect of temperature-sensitive RNA mutants on transcription products from cloned ribosomal protein genes of yeast, *Cell*, 24, 679, 1981.
41. **Ziff, E. and Evans, R.,** Coincidence of the promoter and capped 5′-terminus of mRNA from the adenovirus 2 major late transcriptional unit, *Cell*, 23, 1463, 1978.
42. **Curtis, P. J., Mantei, N., and Weissmann, E.,** Characterization and kinetics of synthesis of 15S β-globin RNA, a putative precursor of β-globin mRNA, *Cold Spring Harbor Symp. Quant. Biol.*, 42, 971, 1977.
43. **Nevins, J. R. and Darnell, J. E.,** Steps in the processing of Ad2 mRNA: poly(A) and nuclear RNA sequences are conserved and poly(A) addition precedes splicing, *Cell*, 15, 1477, 1978.
44. **Schibler, U., Marcu, M. B., and Perry, R. B.,** The synthesis and processing of mRNAs specifying heavy and light chain immunoglobulins in MPC-11, *Cell*, 15, 1495, 1978.
45. **Gilmore-Herbert, M. and Wall, R.,** Nuclear RNA precursors in the processing pathway of MOPC 21k light chain messenger RNA, *J. Mol. Biol.*, 135, 879, 1979.
46. **May, E., Kress, M., and May, P.,** Characterization of two SV40 early mRNAs: evidence for a nuclear 'prespliced' RNA species, *Nucleic Acids Res.*, 5, 3083, 1978.
47. **Strair, R. K., Skoultchi, A. L., and Shafritz, D. A.,** A characterization of globin mRNA sequences in the nucleus of duck inmature red blood cells, *Cell*, 12, 133, 1977.
48. **Ross, J. and Knecht, D. A.,** Precursors of α and β globin messenger RNAs, *J. Mol. Biol.*, 119, 1, 1978.
49. **Nordstrom, J. L., Roop, D. R., Tsai, M-J., and O'Malley, B. W.,** Identification of potential ovomucoid mRNA precursors in chick oviduct nuclei, *Nature (London)*, 278, 328, 1979.
50. **Kinniburgh, A. J., Mertz, J. E., and Ross, J.,** The precursor to mouse β-globin mRNA contains two intervening sequences, *Cell*, 14, 681, 1978.
51. **Tilghman, S. M., Curtis, P. J., Tiemeier, D. C., Leder, P., and Weissman, C.,** The intervening sequence of a mouse β-globin gene is transcribed within the 15S β-globin mRNA precursor, *Proc. Natl. Acad. Sci. U.S.A.*, 75, 1309, 1978.
52. **Tsai, M.-J., Ting, A. C., Nordstrom, J. L., Zimmer, W., and O'Malley, B. W.,** Processing of high molecular weight ovalbumin and ovomucoid precursor RNAs to messenger RNA, *Cell*, 22, 219, 1980.
53. **Ryfell, G. U., Wyler, T., Muellner, D. B., and Weber, R.,** Identification, organization and processing intermediates of the putative precursors of Xenopus vitellogenin messenger RNA, *Cell*, 19, 53, 1980.
54. **Kinniburgh, A. J. and Ross, J.,** Processing of the mouse β-globin mRNA precursor: at least two cleavage ligation reactions are necessary to excise the larger intervening sequence, *Cell*, 17, 915, 1979.
55. **Nevins, J. R.,** Processing of late adenovirus nuclear RNA. Kinetics of formation of intermediates and demonstration that all events are nuclear, *J. Mol. Biol.*, 130, 493, 1979.

56. **Cordell, B., Bell, G., Tischer, E., deNoto, F. M., Ullrich, A., Pictet, R., Rutter, W. J., and Goodman, H. M.,** Isolation and characterization of a cloned rat insulin gene, *Cell,* 18, 533, 1979.
57. **Lomedico, P., Rosenthal, N., Efstatiadis, A., Gilbert, W., Kolodner, R., and Tizard, R.,** The structure and evolution of two non-allelic rat preproinsulin genes, *Cell,* 18, 545, 1979.
58. **Durica, D. S., Schloss, J. A., and Crain, R.,** Organization of actin gene sequences in the sea urchin: molecular cloning of an intron-containing DNA sequence coding for cytoplasmic actin, *Proc. Natl. Acad. Sci. U.S.A.,* 77, 5683, 1980.
59. **Fryberg, E. A., Kindle, K. L., Davidson, N., and Sodja, J. A.,** The actin genes of Drosophila: a dispersed multigene family, *Cell,* 19, 365, 1980.
60. **Jones, C. W. and Kafatos, F.,** Structure, organization and evolution of developmentally regulated chorion genes in silkmoth, *Cell,* 22, 855, 1980.
61. **Dugaiczyk, A., Woo, S. L. C., Lai, E. C., Mace, M. L., McReynold, L., and O'Malley, B. W.,** The natural ovalbumin gene contains seven intervening sequences, *Nature (London),* 274, 328, 1978.
62. **Dugaiczyk, A., Woo, S. L. C., Colbert, D. A., Lai, E. C., Mace, M. L., and O'Malley, B. W.,** The ovalbumin gene: cloning and molecular organization of the entire natural gene, *Proc. Natl. Acad. Sci. U.S.A.,* 76, 2253, 1976.
63. **Lai, E. C., Woo, S. L. C., Dugaiczyk, A., Catterall, J. F., and O'Malley, B. W.,** The ovalbumin gene: structural sequences in native chicken DNA are not contiguous, *Proc. Natl. Acad. Sci. U.S.A.,* 75, 2205, 1978.
64. **Garapin, A. C., Lepennec, J. P., Roskam, W., Perrin, F., Carri, B., Krust, A., Breathnach, R., Chambon, P., and Kourilsky, P.,** Isolation by molecular cloning of a fragment of the split ovalbumin gene, *Nature (London),* 273, 349, 1978.
65. **Mandel, J. L., Breathnach, R., Gerlinger, P., LeHeur, M., Gannon, F., and Chambon, P.,** Organization of coding and intervening sequences in chicken ovalbumin split gene, *Cell,* 14, 641, 1978.
66. **Woo, S. L. C., Dugaiczyk, A., Tsai, M.-J., Lai, E. C., Catterall, J. R., and O'Malley, B. W.,** The ovalbumin gene: cloning of the natural gene, *Proc. Natl. Acad. Sci. U.S.A.,* 75, 3688, 1978.
67. **Gannon, F., Hare, K. O., Perrin, F., LePennec, J. P., Benoist, C., Cochet, M., Breathnach, R., Royal, A., Garapin, A., Carri, B., and Chambon, P.,** Organization and sequence at the 5′ end of a cloned complete ovalbumin gene, *Nature (London),* 278, 428, 1979.
68. **Lai, E. C., Stein, J. P., Catterall, J. F., Woo, S. L. C., Mace, M. L., Means, A. R., and O'Malley, B. W.,** Molecular structure and flanking nucleotide sequences of the natural chicken ovomucoid gene, *Cell,* 18, 829, 1979.
69. **Wahli, W., Dawid, I. B., Wyler, T., Weber, R., and Ryffel, U.,** Comparative analysis of the structural organization of two closely-related vitellogenin genes, *Cell,* 20, 107, 1980.
70. **Wahli, W., Dawid, I. B., Ryffel, G. U., and Weber, R.,** Vitellogenesis and the vitellogenin gene family, *Science,* 212, 298, 1981.
71. **Ohkubo, H., Vogeli, G., Mudry, M., Avvedimento, V. E., Sullivan, M., Pastan, I., and de-Crombrugghe, B.,** Isolation and characterization of overlapping genomic clones containing the chicken α2 (type 1) collagen gene, *Proc. Natl. Acad. Sci. U.S.A.,* 77, 7059, 1980.
72. **Wozney, J., Hanahan, D., Morimoto, R., Boedter, H., and Doty, P.,** Fine structural analysis of the chicken proα2 collagen gene, *Proc. Natl. Acad. Sci. U.S.A.,* 78, 712, 1981.
73. **Efstratiadis, A., Posakony, J. W., Maniatis, T., Lawn, R. M., O'Connell, C., Spritz, R. A., deRiel, J., Forget, B. G., Weissman, S. M., Slightom, J. L., Blechl, A. E., Smithies, O., Baralle, F. E., Shoulders, C. C., and Proudfoot, N. J.,** The structure and evolution of the human β-globin gene family, *Cell,* 21, 653, 1980.
74. **Cowan, N. J., Wilde, C. D., Chow, L. T., and Wefold, F. C.,** Structural variation among human β-tubulin genes, *Proc. Natl. Acad. Sci. U.S.A.,* 78, 4877, 1981.
75. **Fiddes, J. C., Seeberg, P. H., deNoto, F. M., Hallewell, R. A., Baxter, J. D., and Goodman, H. M.,** Structure of genes for human growth hormone and chorionic somatomammotropin, *Proc. Natl. Acad. Sci. U.S.A.,* 76, 4294, 1979.
76. **Nakanishi, S., Teranishi, Y., Noda, M., Notake, M., Watanabe, Y., Kakidami, H., Jingami, H., and Numa, S.,** The protein coding sequence of the bovine ACTH-βLPH precursor gene is split near the signal peptide, *Nature (London),* 287, 752, 1980.
77. **Stein, J. P., Catterall, J. F., Kristo, P., Means, A. R., and O'Malley, B. M.,** Ovomucoid intervening sequences specify functional domains and generate protein polymorphism, *Cell,* 21, 681, 1980.
78. **Benyajati, C., Place, A. R., Powers, D. A., and Sofer, W.,** Alcohol dehydrogenase gene of Drosophila melanogaster: relationship of intervening sequences to functional domains in the protein, *Proc. Natl. Acad. Sci. U.S.A.,* 78, 2717, 1981.
79. **Blake, C. C. F.,** Exons and the structure, function and evolution of haemoglobin, *Nature (London),* 291, 616, 1981.

80. **Gō, M.**, Correlation of DNA exonic regions with protein structural units in haemoglobin, *Nature (London)*, 291, 90, 1981.
81. **Jung, A., Sippel, A. E., Grez, M., and Schutz, G.**, Exons encode functional and structural units of chicken lysozyme, *Proc. Natl. Acad. Sci. U.S.A.*, 77, 5759, 1980.
82. **Sakano, H., Rogers, J. H., Huppi, K., Brack, C., Traunecker, A., Maki, R., Wall, R., and Tonegawa, S.**, Domains and the hinge region of an immunoglobulin heavy chain are encoded in separate DNA fragments, *Nature (London)*, 277, 627, 1979.
83. **Kataoka, T., Yamawaki-Kataoka, Y., Yamagishi, H., and Honjo, T.**, Cloning immunoglobulin γ2b chain gene of mouse: characterization and partial sequence determination, *Proc. Natl. Acad. Sci. U.S.A.*, 76, 4240, 1979.
84. **Gough, N. M., Kemp, D. J., Tyler, B. M., Adams, J. M., and Cory, S.**, Intervening sequences divide the gene for constant region of mouse immunoglobulin μ chains into segments, each encoding a domain, *Proc. Natl. Acad. Sci. U.S.A.*, 77, 554, 1980.
85. **Honjo, T., Obata, M., Yamawaki-Kataoka, Y., Kataoka, T., Kawakami, T., Takahashi, N., and Mano, Y.**, Cloning and complete nucleotide sequence of mouse immunoglobulin γ1 chain gene, *Cell*, 18, 559, 1959.
86. **Early, P. W., Davis, M. M., Kabak, D. B., Davidson, N., and Hood, L.**, Immunoglobulin heavy chain gene organization in mice: analysis of a myeloma genomic clone containing variable and Cμ constant regions, *Proc. Natl. Acad. Sci. U.S.A.*, 76, 857, 1979.
87. **Calame, K., Rogers, J., Early, P., Davis, M., Livant, D., Wall, R., and Hood, L.**, Mouse C heavy chain immunoglobulin gene segment contains three intervening sequences separating domains, *Nature (London)*, 284, 452, 1980.
88. **Perler, F., Efstradiadis, A., Lomedico, P., Gilbert, W., Kolodner, R., and Dodgson, J.**, The evolution of genes: the chicken preproinsulin gene, *Cell*, 20, 555, 1980.
89. **Yamada, Y., Avvidimento, E., Mudryi, M., Ohkubo, H., Vogeli, G., Irani, M., Pastan, I., and deCrombrugghe, B.**, The collagen gene: evidence for its evolutionary assembly by amplification of a DNA segment containing an exon of 54 bp, *Cell*, 22, 887, 1980.
90. **Hardison, R. C., Butler, E. T., Lacy, E., Maniatis, T., Rosenthal, N., and Efstratiadis, A.**, The structure and transcription of four linked rabbit β-like globin genes, *Cell*, 18, 1285, 1979.
91. **Konkel, D. A., Maizel, J. V., and Leder, P.**, The evolution and sequence comparison of two recently diverged mouse chromosomal β-globin genes, *Cell*, 18, 856, 1979.
92. **Leder, P., Hansen, J. N., Konkel, D., Leder, A., Nishioka, Y., and Talkington, C.**, Mouse globin system: a functional and evolutionary analysis, *Science*, 209, 1336, 1980.
93. **Lawn, R. M., Efstratiadis, A., O'Connell, C., and Maniatis, T.**, The nucleotide sequence of a human β-gene, *Cell*, 21, 647, 1980.
94. **Volckaert, G., Feuntuen, J., Crawford, L. V., Berg, P., and Fiers, W.**, Nucleotide sequence deletions within the coding region for small t-antigen of SV40, *J. Virol.*, 30, 674, 1979.
95. **Thimmappaya, B. and Shenk, T.**, Nucleotide sequence analysis of viable deletion mutants lacking segments of the SV40 genome coding for small t-antigen, *J. Virol.*, 30, 668, 1979.
96. **Contreras, R., Cole, C., Berg, P., and Fiers, W.**, Nucleotide sequence analysis of two SV40 mutants with deletions in the late region of the genome, *J. Virol.*, 29, 789, 1979.
97. **Villareal, L. P., White, R. T., and Berg, P.**, Mutational alterations within the SV40 leader segment generate altered 16S and 19S mRNAs, *J. Virol.*, 29, 209, 1979.
98. **Khoury, G., Gruss, P., Dhar, R., and Lai, J. C.**, Processing and expression of early SV40 mRNA: a role for RNA conformation in splicing, *Cell*, 18, 85, 1979.
99. **Subramanian, K. N.**, Segments of SV40 DNA spanning most of the leader sequence of the major late mRNA are dispensible, *Proc. Natl. Acad. Sci. U.S.A.*, 76, 2556, 1979.
100. **Hamer, D. H. and Leder, P.**, Splicing and the formation of stable RNA, *Cell*, 18, 1299, 1979.
101. **Breathnach, R., Benoist, C., O'Hare, K., Gannon, F., and Chambon, P.**, Ovalbumin gene: evidence for a leader segment in mRNA and DNA sequences at exon-intron boundaries, *Proc. Natl. Acad. Sci. U.S.A.*, 75, 4853, 1978.
102. **Seif, I., Khoury, G., and Dhar, R.**, BKV splice sequences based on analysis of preferred donor and acceptor sites, *Nucleic Acids Res.*, 6, 3387, 1979.
103. **Seif, I., Khoury, G., and Dhar, R.**, The genome of human papovavirus BKV, *Cell*, 18, 963, 1979.
104. **Lerner, M. R., Boyle, J. A., Mount, S. M., Wolin, S. L., and Steitz, J. A.**, Are snRNPs involved in splicing? *Nature (London)*, 283, 220, 1980.
105. **Rogers, J. and Wall, R.**, A mechanism of RNA splicing, *Proc. Natl. Acad. Sci. U.S.A.*, 77, 1877, 1980.
106. **Jensen, E. O., Paladan, K., Hyldig-Nielsen, J. J., Jorgensen, P., and Marcker, K. A.**, The structure of a chromosomal leghemoglobin gene from soybean, *Nature (London)*, 291, 677, 1981.
107. **Sun, S. M., Slightom, J. L., and Hall, T. C.**, Intervening sequences in a plant gene: comparison of the partial sequence of cDNA and genomic DNA of French bean phaseolin, *Nature (London)*, 289, 37, 1981.

108. **Proudfoot, N. and Maniatis, T.**, The structure of a human α-globin pseudogene and its relationship to α-globin gene duplication, *Cell*, 21, 537, 1980.
109. **Lacy, E. and Maniatis, T.**, The nucleotide sequence of a rabbit β-globin pseudogene, *Cell*, 21, 545, 1980.
110. **Spritz, R. A., Jagadeeswaran, P., Choudray, P. V., Biro, P. A., Elder, J. T., deRiel, J. K., Manley, J. L., Gefter, M. L., Forget, B. G., and Weissman, S. M.**, Base substitution in an intervening sequence of a $\beta^+$ thalassemia human globin gene, *Proc. Natl. Acad. Sci. U.S.A.*, 78, 2455, 1981.
111. **Baird, M., Driscoll, C., Schreiner, H., Sciarratta, V., Sensone, G., Viazi, G., Ramirez, F., and Bank, H.**, A nucleotide change at a splice junction in the human β-globin gene is associated with β-thalassemia, *Proc. Natl. Acad. Sci. U.S.A.*, 78, 4218, 1981.
112. **Orkin, S. H., Goff, S. C., and Hechtman, L. L.**, Mutation in intervening sequence splice junction in man, *Proc. Natl. Acad. Sci. U.S.A.*, 78, 5041, 1981.
113. **Busslinger, M., Moschonas, N., and Flavell, R. A.**, $\beta^+$ Thalassemia: aberrant splicing results from a single point mutation in an intron, *Cell*, 27, 289, 1981.
114. **Carlock, L. R. and Jones, N.**, Transformation-defective mutant of adenovirus type 5 containing a single altered Ela mRNA species, *J. Virol.*, 40, 657, 1981.
115. **Solnick, D.**, An adenovirus mutant defective in splicing, *Nature (London)*, 291, 508, 1981.
116. **Montell, C., Fisher, E. F., Caruthers, M. H., and Berk, A. J.**, Resolving the functions of overlapping viral genes by site-specific mutagenesis at a mRNA splice site, *Nature (London)*, 295, 380, 1982.
117. **Hamer, D. H., Smith, K. D., Boyer, S. H., and Leder, P.**, SV40 recombinants carrying rabbit β-globin gene, *Cell*, 17, 725, 1979.
118. **Wold, B., Wigler, M., Lacy, E., Maniatis, T., Silverstein, S., and Axel, R.**, Introduction and expression of a rabbit β-globin gene in mouse fibroblasts, *Proc. Natl. Acad. Sci. U.S.A.*, 76, 5684, 1979.
119. **Hamer, D. H. and Leder, P.**, Expression of the chromosomal mouse $^{\beta maj}$-globin gene cloned in SV40, *Nature (London)*, 281, 35, 1979.
120. **Breathnach, R., Mantei, N., and Chambon, P.**, Correct splicing of a chicken ovalbumin gene transcript in mouse L cells, *Proc. Natl. Acad. Sci. U.S.A.*, 77, 740, 1980.
121. **Lai, E. C., Woo, S. L. C., Bordelon-Riser, M. E., Fraser, T. H., and O'Malley, B. W.**, Ovalbumin synthesized in mouse cells transformed with the natural chicken ovalbumin gene, *Proc. Natl. Acad. Sci. U.S.A.*, 77, 244, 1980.
122. **Rusconi, S. and Schaffner, W. S.**, Transformation of frog embryos with a rabbit β-globin gene, *Proc. Natl. Acad. Sci. U.S.A.*, 78, 5051, 1981.
123. **Reddy, R. and Busch, H.**, U snRNAs of nuclear snRNPs, *Cell Nucl.*, 8, 261, 1981.
124. **Ohshima, Y., Itoh, M., Okada, N., and Miyata, T.**, Novel models for RNA splicing that involve a small nuclear RNA, *Proc. Natl. Acad. Sci. U.S.A.*, 78, 4471, 1981.
125. **Branalant, C., Krol, A., Ebel, J.-P., Lazar, E., Gallinaro, H., Jacob, M., Sri-Wadada, J., and Jeanteur, P.**, Nucleotide sequences of nuclear U1a RNAs from chicken, rat and man, *Nucleic Acids Res.*, 8, 4143, 1980.
126. **Chung, S. Y., Cone, R., and Wooley, J.**, personal communication, 1981.
127. **Lerner, M. R. and Steitz, J. A.**, Antibodies to small nuclear RNAs complexed with proteins are produced by patients with systemic lupus erythematosus, *Proc. Natl. Acad. Sci. U.S.A.*, 76, 5495, 1979.
128. **Zieve, G. and Penman, S.**, Small RNA species of the HeLa cell: metabolism and subcellular location, *Cell*, 8, 19, 1976.
129. **Deimel, D., Louis, C., and Sekeris, C. E.**, The presence of small molecular weight RNAs in nuclear RNP carrying hnRNA, *FEBS Lett.*, 73, 80, 1977.
130. **Northemann, W., Scheurlen, M., Gross, V., and Heinrich, P. C.**, Circular dichroism of ribonucleoprotein complexes from rat liver nuclei, *Biochem. Biophys. Res. Commun.*, 76, 1130, 1977.
131. **Howard, E. F.**, Small, nuclear RNA molecules in nuclear ribonucleoprotein complexes from mouse erythroleukemia cells, *Biochemistry*, 17, 3228, 1978.
132. **Flytzanis, R., Alonso, A., Louis, C., Krieg, L., and Sekeris, C. E.**, Association of small nuclear RNA with HnRNA isolated from nuclear RNP complexes carrying HnRNA, *FEBS Lett.*, 96, 201, 1981.
133. **Zieve, G. W.**, Two groups of small stable RNAs, *Cell*, 25, 296, 1981.
134. **Calvert, J. P. and Pederson, T.**, Base-pairing interactions between small nuclear RNAs and nuclear RNA precursors as revealed by psoralen cross-linking *in vivo*, *Cell*, 26, 363, 1981.
135. **Avvedimento, V. E., Vogeli, G., Yamada, Y., Maizel, J. V., Pastan, I., and deCrombrugghe, B.**, Correlation between splicing sites within an intron and their sequence complementarity with U1 RNA, *Cell*, 21, 689, 1980.
136. **Yang, V. W., Lerner, M. R., Steitz, J. A., and Flint, S. J.**, A small, nuclear ribonucleoprotein is required for splicing of adenoviral early RNA sequences, *Proc. Natl. Acad. Sci. U.S.A.*, 78, 1371, 1981.
137. **Lerner, M. R., Boyle, J. A., Hardin, J. A., and Steitz, J. A.**, Two novel classes of small ribonucleoproteins detected by antibodies associated with lupus erythematosus, *Science*, 211, 400, 1981.

138. **Harada, F. and Kato, N.,** Nucleotide sequences of 4.5S RNAs associated poly(A)-containing RNAs of mouse and hamster cells, *Nucleic Acids Res.*, 8, 1273, 1980.
139. **Krayev, A. S., Kramerov, D. A., Skryabin, K. G., Ryskov, A. P., Bayer, A. A., and Georgiev, G. P.,** The nucleotide sequence of the ubiquitous, repetitive DNA sequence B1 complementary to the most abundant class of mouse fold-back RNA, *Nucleic Acids Res.*, 8, 120, 1980.
140. **Peters, G. G., Harada, F., Dahlberg, J. E., Panet, A., Haseltine, W. A., and Baltimore, D.,** Low molecular weight RNAs of Moloney murine leukemia virus: identification of the primer for RNA-directed DNA synthesis, *J. Virol.*, 21, 1031, 1977.
141. **Jelinek, W. and Leinwand, L.,** Low molecular weight RNAs hydrogen bonded to nuclear and cytoplasmic poly(A)-terminated RNA from cultured Chinese hamster ovary cells, *Cell*, 15, 205, 1978.
142. **Harada, F., Kato, N., and Hoshino, H.,** Series of 4.5S RNAs associated with poly(A)-containing RNAs of rodent cells, *Nucleic Acids Res.*, 7, 909, 1979.
143. **Blanchard, J. M., Weber, J., Jelinek, W., and Darnell, J. E.,** *In vitro* RNA-RNA splicing in adenovirus mRNA formation, *Proc. Natl. Acad. Sci. U.S.A.*, 75, 5344, 1978.
144. **Chambon, P., Benoist, C., Breathnach, R., Cochet, M., Gannon, F., Gerlinger, P., Krust, A., Lemeur, M., LePennec, J. P., Mandel, J. L., O'Hare, K., and Perrin, F.,** Structural organization and expression of ovalbumin and related chicken genes, in *From Gene to Protein: Information Transfer in Normal and Abnormal Cells*, Russell, T. R., Brew, K., Faber, H., and Schultz, J., Eds., Academic Press, New York, 1979, 55.
145. **Grosveld, G. C., Koster, A., and Flavell, R. A.,** A transcription map for the rabbit β-globin gene, *Cell*, 23, 573, 1981.
146. **Berget, S. M. and Sharp, P. A.,** Structure of the late adenovirus 2 heterogeneous nuclear RNA, *J. Mol. Biol.*, 129, 547, 1979.
147. **Kohoury, G., Alwine, J., Goldman, N., Gruss, P., and Jay, G.,** New chimeric splice junction in adenovirus 2 — SV40 hybrid mRNA, *J. Virol.*, 36, 143, 1980.
148. **Weingärtner, B. and Keller, W.,** Transcription and processing of adenoviral RNA by extracts from HeLa cells, *Proc. Natl. Acad. Sci. U.S.A.*, 78, 4092, 1981.
148a. **Kolo, R. and Weissman, S.,** Accurate *in vitro* splicing of human β-globin RNA, *Nucleic Acids Res.*, 10, 5429, 1982.
149. **Goldenberg, C. J. and Raskas, H. J.,** *In vitro* splicing of purified precursor RNAs specified by early region 2 of the adenovirus 2 genome, *Proc. Natl. Acad. Sci. U.S.A.*, 78, 5430, 1981.
150. **Manley, J. L., Fire, A., Campo, A., Sharp, P. A., and Gefter, M. L.,** DNA-dependent transcription of adenovirus genes in a soluble whole-cell extract, *Proc. Natl. Acad. Sci. U.S.A.*, 77, 3855, 1980.
151. **Cepko, C. L., Hansen, U., Handa, H., and Sharp, P. A.,** Sequential transcription-translation of SV40 by using mammalian cell extracts, *Mol. Cell. Biol.*, 1, 919, 1981.
152. **Konarska, M., Dilipowicz, W., Domdey, H., and Gross, H. J.,** Formation of a 2′-phosphomonoester, 3′,5′-phosphodiester linkage by a novel RNA ligase in wheat germ, *Nature (London)*, 293, 112, 1981.
153. **Silber, R., Malathi, V. G., and Hurwitz, J.,** Purification and properties of bacteriophage T4-induced RNA ligase, *Proc. Natl. Acad. Sci. U.S.A.*, 69, 3009, 1972.
154. **Konarska, M., Filipowicz, W., and Gross, H. J.,** RNA ligation via 2′-phosphomonoester, 3′,5′-phosphodiester linkage: requirement of 2′,3′-cyclic phosphate termini and involvement of a 5′-hydroxyl polynucleotide kinase, *Proc. Natl. Acad. Sci. U.S.A.*, 79, 1474, 1982.
155. **Winicov, I.,** RNA phosphorylation: a polynucleotide kinase function in mouse L cell nuclei, *Biochemistry*, 16, 4233, 1977.
156. **Shurman, S. and Hurwitz, J.,** 5′ Hydroxyl polynucleotide kinase from HeLa cell nuclei, *J. Biol. Chem.*, 254, 10396, 1979.
157. **Winicov, I. and Bulton, J. D.,** unpublished observations, 1982.
158. **Peebles, C. L., Gegenheimer, P., and Abelson, J. N.,** unpublished observations, 1982.
159. **Greer, C., Gegenheimer, P., Schwartz, R. C., Peebles, C. L., and Abelson, J. N.,** unpublished observations, 1982.
160. **Gegenheimer, P., Peebles, C. L., Greer, C., Schwartz, R. C., and Abelson, J.,** unpublished observations, 1982.
161. **Flint, S. J.,** Splicing and the regulation of viral gene expression, *Curr. Top. Microbiol. Immunol.*, 93, 47, 1981.
162. **Spector, D. J., McGrogan, M., and Raskas, H. J.,** Regulation of the appearance of cytoplasmic RNAs from region 1 of the adenovirus genome, *J. Mol. Biol.*, 126, 395, 1978.
163. **Chow, L. T., Broker, T. R., and Lewis, J. B.,** Complex splicing patterns of RNAs from the early regions of adenovirus 2, *J. Mol. Biol.*, 134, 265, 1979.
164. **Esche, H., Mathews, M. B., and Lewis, J. B.,** Proteins and messenger RNAs of the transforming region of wild-type and mutant adenoviruses, *J. Mol. Biol.*, 142, 399, 1980.

165. **Akusjärvi, G. and Persson, H.,** Controls of RNA splicing and termination in the major late adenovirus transcription unit, *Nature (London)*, 292, 420, 1981.
166. **Nevins, J. R. and Wilson, C.,** Regulation of adenovirus 2 gene expression and the levels of transcriptional termination and RNA processing, *Nature (London)*, 290, 113, 1981.
167. **Early, P., Rogers, J., Davis, M., Calame, K., Bond, M., Wall, R., and Hood, L.,** Two mRNAs can be produced from a single immunoglobulin $\mu$ gene by alternative RNA processing pathways, *Cell*, 20, 313, 1980.
168. **Roger, J., Early, P., Carter, C., Calame, K., Bond, M., Hood, L., and Wall, R.,** Two mRNAs with different 3′ ends encode membrane-bound and secreted forms of immunoglobulin $\mu$ chain, *Cell*, 20, 303, 1980.
169. **Singer, P. A., Singer, H. H., and Williamson, A. R.,** Different species of messenger RNA encode receptor and sensory IgM $\mu$ chains differing at their carboxy termini, *Nature (London)*, 285, 294, 1980.
170. **Liu, C. P., Tucker, P. W., Mushinski, J. F., and Blattner, F. R.,** Mapping of the heavy chain genes for mouse immunoglobulins M and D, *Science*, 209, 1348, 1980.
171. **Moore, K. W., Rogers, J., Hunkapiller, T., Early, P., Nottenburg, C., Weissman, I., Bazin, I., Wall, R., and Hood, L.,** Expression of IgD may use both DNA rearrangement and RNA splicing mechanisms, *Proc. Natl. Acad. Sci. U.S.A.*, 78, 1800, 1981.
172. **Maki, R., Roeder, W., Traunecker, A., Sidman, C., Waki, M., Raschke, W., and Tonegawa, S.,** The role of DNA rearrangement and alternative RNA processing in the expression of immunoglobulin delta genes, *Cell*, 24, 353, 1981.
173. **Tyler, B. M., Cowman, A. F., Adams, J. M., and Harris, A. W.,** Generation of long mRNA for membrane immunoglobulin $\gamma$2a chain by differential splicing, *Nature (London)*, 293, 406, 1981.
174. **Young, R. A., Hagenbuchle, O., and Schibler, U.,** A single mouse $\alpha$-amylase gene specifies two different, tissue-specific mRNAs, *Cell*, 23, 451, 1981.
175. **Hagenbuchle, O., Tosi, M., Schibler, W., Bovey, R., Wellauer, P. K., and Young, R. A.,** Mouse liver and salivary gland $\alpha$-amylase mRNAs differ only in 5′ non-translated sequences, *Nature (London)*, 289, 643, 1981.
176. **Setzer, D. R., McGrogan, M., Nunberg, J. H., and Schimke, R. T.,** Size heterogeneity in the 3′ end of dihydrofolate reductase messenger RNA in mouse cells, *Cell*, 22, 361, 1980.
177. **Cooke, E. and Baxter, J. D.,** Structural analysis of the prolactin gene suggests a separate origin for its 5′ end, *Nature (London)*, 297, 603, 1982.
178. **Amara, S. G., Jonas, V., Rosenfeld, M. G., Ong, E. S., and Evans, R. M.,** Alternative RNA processing in calcitonin gene expression generates mRNAs encoding different polypeptide products, *Nature (London)*, 298, 240, 1982.
179. **Fahrner, K., Yarger, L., and Hereford, L.,** Yeast histone mRNA is polyadenylated, *Nucleic Acids Res.*, 8, 5725, 1980.
180. **Zernik, M., Heintz, M., Boime, I., and Roeder, R. G.,** *Xenopus laevis* histone genes: variant H1 genes are present in different clusters, *Cell*, 22, 807, 1980.
181. **Holmgren, R., Corces, V., Morimoto, R., Blackman, R., and Meselson, M.,** Sequence homologies in the 5′ regions of four *Drosophila* heat-shock genes, *Proc. Natl. Acad. Sci. U.S.A.*, 78, 3775, 1981.
182. **McKeown, M. and Firtel, R.,** Differential expression and 5′ end mapping of actin genes in *Dicytostelim*, *Cell*, 24, 799, 1981.
183. **Molnar, J., Bajszar, G., Marczinovits, I., and Szabo, G.,** Cleavage of premRNA sequences by ribonucleases bound to nuclear RNP particles of rat liver, *Mol. Biol. Rep.*, 4, 157, 1978.
184. **Rech, J., Brunel, C., and Jeaateur, P.,** HnRNP from HeLa cells contain a ribonuclease active on double-stranded RNA, *Biochem. Biophys. Res. Commun.*, 88, 422, 1979.
185. **Paolitti, J., Rech, J., Brunel, C., and Jeanteur, P.,** Constrained configuration of double-stranded RNA in HeLa hnRNP and its relaxation by ribonuclease D, *Biochemistry*, 14, 5223, 1980.
186. **Rech, J., Cathala, G., and Jeanteur, P.,** Isolation and characterization of a ribonuclease activity specific for double-stranded RNA (RNAase D) from Krebs II ascites cells, *J. Biol. Chem.*, 255, 6700, 1980.
187. **Gall, J.,** Lampbrush chromosomes from oocyte nuclei of the newt, *J. Morphol.*, 94, 283, 1954.
188. **Callan, H. G. and Lloyd, L.,** Lampbrush chromosomes of crested newts, *Philos. Trans. R. Soc. B*, 243, 135, 1960.
189. **Stevens, B. J. and Swift, H.,** RNA transport from the chironomus salivary glands, *J. Cell Biol.*, 31, 55, 1966.
190. **Monneron, A. and Bernhard, W.,** Fine structural organization of the interphase nucleus of some mammalian cells, *J. Ultrastruct. Res.*, 27, 266, 1969.
191. **Miller, O. L. and Beatty, B. R.,** Visualization of nucleolar genes, *Science*, 164, 955, 1969.
192. **Sommerville, J.,** Ribonucleoprotein particles derived from lampbrush chromosomes of newt oocytes, *J. Mol. Biol.*, 78, 487, 1973.

193. **McKnight, S. L. and Miller, O. L.**, Ultrastructural patterns of RNA synthesis during early embryogenesis of *Drosophila melanogaster, Cell,* 8, 305, 1976.
194. **McKnight, S. L. and Miller, O. L.**, Post-replicative, non-ribosomal transcriptional units in *D. melanogaster* embryos, *Cell,* 17, 551, 1979.
195. **Laird, C. D. and Chooi, W. Y.**, Morphology of transcription units in *D. melanogaster, Chromosoma,* 58, 193, 1976.
196. **Laird, C. D., Wilkinson, L. E., Foe, F. C., and Chooi, W. Y.**, Analysis of chromatin-associated fiber arrays, *Chromosoma,* 58, 169, 1979.
197. **Foe, V. E., Wilkinson, L. E., and Laird, C. D.**, Comparative organization of active transcription units in *Oncopelteus fasciatus, Cell,* 9, 131, 1976.
198. **Samarina, O. P., Lukanidin, E. M., Molnar, J., and Georgiev, G. P.**, Structural organization of nuclear complexes containing DNA-like RNA, *J. Mol. Biol.,* 33, 251, 1969.
199. **Martin, T., Billings, P., Levey, A., Ozarslan, S., Quinlan, T., Swift, H., and Urbas, L.**, Some properties of RNA: protein complexes from the nucleus of eukaryotic cells, *Cold Spring Harbor Symp. Quant. Biol.,* 38, 921, 1973.
200. **Malcom, D. B. and Sommerville, J.**, The structure of chromosome-derived ribonucleoprotein in oocytes of *Triturus cristatus cernifex. Chromosoma,* 48, 137, 1974.
201. **Pederson, T.**, Proteins associated with heterogeneous nuclear RNA in eukaryotic cells, *J. Mol. Biol.,* 83, 163, 1974.
202. **Pederson, T.**, Gene activation in eukaryotes: are nuclear acidic proteins the cause or effect? *Proc. Natl. Acad. Sci. U.S.A.,* 71, 617, 1974.
203. **Georgiev, G. P. and Samarina, O. P.**, D-RNA containing ribonucleoprotein particles, *Adv. Cell Biol.,* 2, 47, 1971.
204. **Martin, T. E. and McCarthy, B. J.**, Synthesis and turnover of RNA in the 30S nuclear ribonucleoprotein complexes of mouse ascites cells, *Biochim. Biophys. Acta,* 277, 354, 1972.
205. **Firtel, R. A. and Pederson, T.**, Ribonucleoprotein particles containing heterogeneous nuclear RNA in the cellular slime mold *Dictyostelium discoideum, Proc. Natl. Acad. Sci. U.S.A.,* 72, 301, 1974.
206. **Kinniburgh, A. and Martin, T. E.**, Detection of mRNA sequences in nuclear 30S ribonucleoprotein subcomplexes, *Proc. Natl. Acad. Sci. U.S.A.,* 73, 2725, 1976.
207. **Sommerville, J. and Malcom, D. B.**, Transcription of genetic information in amphibian oocytes, *Chromosoma,* 55, 183, 1976.
208. **Miller, O. L. and Hamkalo, B. A.**, Visualization of RNA synthesis on chromosomes, *Int. Rev. Cytol.,* 33, 1, 1972.
209. **Scheer, U., Spring, H., and Trendelenburg, M. F.**, Organization of transcriptionally active chromatin in lampbrush chromosome loops, in *The Cell Nucleus,* Vol. 7, Busch, H., Ed., Academic Press, New York, 1979, 3.
210. **Beyer, A. L., Christensen, M. E., Walker, B. W., and LeSturgeon, W.**, Identification and characterization of the packaging proteins of core 40S hnRNP, *Cell,* 11, 127, 1977.
211. **Stevinin, J., Gattoni, R., Gallimore-Matringe, H., and Jacob, M.**, Nuclear RNP particles contain specific proteins and unspecific non-histone nuclear proteins, *Eur. J. Biochem.,* 84, 541, 1978.
212. **LeStourgeon, W. M., Lothstein, L., Walker, B. W., and Beyer, A. L.**, The composition and general topology of RNA and protein in monomer 40S ribonucleoprotein particles, in *The Cell Nucleus,* Vol. 9, Busch, H., Ed., Academic Press, New York, 1981, 49.
213. **Calvert, J. P. and Pederson, T.**, Nucleoprotein organization of inverted repeat DNA transcripts in heterogeneous nuclear RNA — ribonucleoprotein particles from HeLa cells, *J. Mol. Biol.,* 122, 361, 1978.
214. **Sonenberg, N., Morgan, M. A., Merrick, W. C., and Shatkin, A. J.**, A polypeptide in eukaryotic initiation factors that crosslinks specifically to the 5′-terminal cap in mRNA, *Proc. Natl. Acad. Sci. U.S.A.,* 75, 4843, 1979.
215. **Sonenberg, N.**, ATP/$Mg^{++}$-dependent cross-linking of cap binding proteins to the 5′ end of eukaryotic mRNA, *Nucleic Acids Res.,* 9, 1643, 1981.
216. **Kish, V. and Pederson, T.**, Ribonucleoprotein organization of polyadenylate sequences in HeLa cell heterogeneous nuclear RNA, *J. Mol. Biol.,* 95, 227, 1975.
217. **Schwartz, H. and Darnell, J. E.**, The association of protein with the polyadenylic acid of HeLa cell messenger RNA: evidence for a 'transport' role of a 75,000 molecular weight polypeptide, *J. Mol. Biol.,* 104, 833, 1976.
218. **Greenberg, J. R.**, Proteins cross-linked to messenger RNA by irradiating polyribosomes with ultraviolet light, *Nucleic Acids Res.,* 8, 5685, 1980.
219. **Pederson, T. and Davis, R. G.**, Messenger RNA processing and nuclear structure: isolation of nuclear ribonucleoprotein particles containing β-globin messenger RNA precursors, *J. Cell Biol.,* 87, 47, 1980.
220. **Munroe, S. H. and Pederson, T.**, Messenger RNA sequences in nuclear ribonucleoprotein particles are complexed with protein as shown by nuclease protection, *J. Mol. Biol.,* 147, 437, 1981.

221. **Beyer, A. L., Miller, O. L., and McKnight, S. L.,** Ribonucleoprotein structure in nascent hnRNA is non-random and sequence-dependent, *Cell,* 20, 75, 1980.
222. **Beyer, A. L., Bouton, A. H., Hodge, L. P., and Miller, O. L.,** Visualization of the major late r strand transcription unit of adenovirus serotype 2, *J. Mol. Biol.,* 147, 269, 1981.
223. **Beyer, A. L., Bouton, A. H., and Miller, O. L.,** Correlation of hnRNP structure and nascent transcript cleavage, *Cell,* 26, 155, 1981.
224. **Old, R., Callan, H. G., and Gross, K. W.,** Localization of histone gene transcripts in newt lampbrush chromosomes by in situ hybridization, *J. Cell Sci.,* 27, 57, 1977.
225. **Lai, C. J. and Khoury, G.,** Deletion mutants of SV40 defective in biosynthesis of late viral mRNA, *Proc. Natl. Acad. Sci. U.S.A.,* 76, 71, 1979.
226. **Buchman, A. R. and Berg, P.,** Studies of the structure and function of intervening sequences, *J. Supramol. Struct. Cell Biochem.,* Suppl. 5, 443, 1981.
227. **Mulligan, R., Howard, B. H., and Berg, P.,** Synthesis of β-globin in cultured monkey kidney cells following infection with an SV40-β-globin recombinant genome, *Nature (London),* 227, 108, 1979.
228. **Gruss, P. and Khoury, G.,** Rescue of a splicing defective mutant by insertion of a heterologous intron, *Nature (London),* 286, 634, 1980.
229. **Ghosh, P. K., Roy, P., Barkan, A., Mertz, J. E., Weissman, S. M., and Leibowitz, P.,** Unspliced, functional late 19S mRNAs containing intervening sequences are produced by a late leader mutant of SV40, *Proc. Natl. Acad. Sci. U.S.A.,* 78, 1861, 1981.
230. **Gruss, P., Efstratiadis, A., Karathanasis, S., Konig, M., and Khoury, G.,** Synthesis of stable, unspliced mRNA from an intronless SV40-rat preproinsulin gene recombinant, *Proc. Natl. Acad. Sci. U.S.A.,* 78, 6091, 1981.
231. **Keene, K. C. and Humphreys, T.,** Similarity of hnRNA sequences in blastula and pleteus stage sea urchin embryos, *Cell,* 12, 143, 1977.
232. **Wold, B. J., Klein, W. H., Hough-Evans, B. R., Britten, R. J., and Davidson, E.,** Sea urchin embryo mRNA sequences expressed in the nuclear RNAs of adult tissue, *Cell,* 14, 941, 1978.
233. **Shepert, G. W. and Nemur, M.,** Developmental shifts in frequency distribution of polysomal mRNA and their post-transcriptional regulation in the sea urchin embryo, *Proc. Natl. Acad. Sci. U.S.A.,* 77, 4653, 1980.
234. **Beltz, G. A. and Flint, S. J.,** Inhibition of HeLa cell protein synthesis during adenovirus infection: restriction of cellular messenger RNA sequences to the nucleus, *J. Mol. Biol.,* 131, 353, 1979.
235. **Beltz, G. A., Plunkett, J. J., and Flint, S. J.,** unpublished observations.
236. **Castiglia, C. L. and Flint, S. J.,** The effect of adenovirus infection upon ribosomal RNA synthesis and processing, *Mol. Cell. Biol.,* 3, 662, 1983.
237. **Benecke, B.-J., Ben-Ze'ev, A., and Penman, S.,** The control of mRNA production, translation and turnover in suspended and reattached anchorage-dependent fibroblasts, *Cell,* 14, 931, 1978.
238. **Ben-Ze'ev, A., Farmer, S. R., and Penman, S.,** Protein synthesis requires cell-surface contact while nuclear events respond to cell shape in anchorage-dependent fibroblasts, *Cell,* 21, 365, 1980.
239. **Wunderlich, F., Berezney, R., and Klunig, H.,** The nuclear envelope, in *Biological Membranes,* Chapman, D. and Wallach, D. F. H., Eds., Academic Press, New York,, 1976, 241.
240. **Wunderlich, F.,** Nucleocytoplasmic transport of ribosomal subparticles: interplay with the nuclear envelope, in *The Cell Nucleus,* Vol. 9, Busch, H., Ed., Academic Press, New York, 1981, 249.
241. **Wolosewick, J. J. and Porter, K. R.,** Microtrabecular lattice of the cytoplasmic ground substance: artifact or reality? *J. Cell Biol.,* 82, 114, 1979.
242. **Porter, K. R., Byers, H. R., and Ellsman, M. H.,** The cytoskeleton, *Neurosci. Res. Prog.,* 4, 703, 1979.
243. **Lenk, R., Ransom, L., Kaufman, Y., and Penman, S.,** A cytoskeletal structure with associated polyribosomes obtained from HeLa cells, *Cell,* 10, 67, 1977.
244. **Cervera, M., Dreyfuss, G., and Penman, S.,** Messenger RNA is translated when associated with the cytoskeletal framework in normal and VSV-infected HeLa cells, *Cell,* 23, 113, 1981.
245. **van Venrooij, W. J., Sillekens, P. T. G., van Eekelen, C. A. G., and Reinders, R. J.,** On the association of mRNA with the cytoskeleton in uninfected and adenovirus-infected human KB cells, *Exp. Cell Res.,* 135, 79, 1981.
246. **Herman, R., Weymouth, L., and Penman, S.,** Heterogeneous nuclear RNA-protein fibres in chromatin-depleted nuclei, *J. Cell Biol.,* 78, 663, 1978.
247. **Miller, T. E., Huang, C., and Pogo, A. O.,** Rat liver nuclear skeleton and ribonucleoprotein complexes containing hnRNA, *J. Cell Biol.,* 76, 675, 1978.
248. **van Eekelen, C. A. G. and van Venrooij, W. J.,** HnRNA and its attachment to a nuclear protein matrix, *J. Cell Biol.,* 88, 554, 1981.
249. **Jackson, D. A., McCready, S. J., and Cook, P. R.,** RNA is synthesized at the nuclear cage, *Nature (London),* 292, 352, 1981.

250. **Jackson, D. A., Caton, A. J., McCready, S. J., and Cook, P. A.,** Influenza virus RNA is synthesized at fixed sites in the nucleus, *Nature (London)*, 296, 366, 1982.
251. **Mariman, E. C. M., van Eekelen, A. G., Reinder, R. J., Berns, A. J. M., and van Venrooij, W. J.,** Adenovirus heterogeneous nuclear RNA is associated with the host cell nuclear matrix during splicing, *J. Mol. Biol.*, 154, 103, 1982.
252. **Margolis, R. L. and Wilson, L.,** Microtubule treadmills — possible molecular machinery, *Nature (London)*, 293, 705, 1981.
253. **Riley, D. E. and Keller, J. M.,** The polypeptide composition and ultrastructure of nuclear ghosts isolated from mammalian cells, *Biochim. Biophys. Acta*, 444, 899, 1976.
254. **Hodge, L. D., Mancini, P., Davis, F. M., and Heywood, P.,** Nuclear matrix of HeLa $S_3$ cells, *J. Cell Biol.*, 72, 194, 1977.
255. **Aaronson, R. P. and Blobel, G.,** Isolation of nuclear pore complexes in association with a lamina, *Proc. Natl. Acad. Sci. U.S.A.*, 72, 1007, 1975.
256. **Roy, R. K., Sarkar, S., Guha, C., and Munro, H. N.,** RNP particles involved in release of *in vitro* synthesized poly(A)-containing RNA in isolated nuclei, in *The Cell Nucleus*, Vol. 9, Busch, H., Ed., Academic Press, New York, 1981, 289.
257. **Ortegui, C. and Patterson, R. J.,** RNA metabolism in isolated nuclei: processing and transport of immunoglobulin light chain sequences, *Nucleic Acids Res.*, 9, 4767, 1981.
258. **Sharp, P. A.,** personal communication.

Chapter 8

# ANIMAL VIRUS RNA PROCESSING

**Thomas R. Broker**

## TABLE OF CONTENTS

## I. INTRODUCTION

The stages of eukaryotic messenger RNA metabolism include promotion, elongation, capping and methylation of the 5′ end, modification of internal bases, polyadenylation of the 3′ end, splicing, transport to the cytoplasm, association with ribosomes and translation, loss of poly(A), and degradation. Processing of primary transcripts into messenger RNAs should be regarded as a collective strategy for extracting information from the genome and regulating its expression. The distinctive properties of eukaryotes and their mRNAs, as compared to prokaryotes, include:

1. Covalent splicing of most eukaryotic mRNAs, with preservation of 5′- and 3′-terminal sequences and some internal segments of the primary transcript and deletion of intervening sequences
2. Transportation of RNAs from their sites of synthesis and processing in the nucleus to their site of translation on the cytoplasmic ribosomes, in contrast to the coupled transcription-translation of prokaryotic mRNAs
3. The longer lifetimes of eukaryotic messages conferred by capping 5′ ends and polyadenylating 3′ ends
4. The selective translation of only the 5′ proximal open reading frame of most eukaryotic mRNAs, including those which are physically polycistronic, in contrast to the functional polycistrony of prokaryotic mRNAs
5. The dramatically higher genomic complexity of eukaryotes
6. The differentiated gene expression of eukaryotes with respect to developmental stage and cell type.

RNA splicing is a form of genetic recombination. Its fundamental distinctions are the involvement of single-stranded RNA rather than double-stranded DNA and the somatic rather than germline reassociation of polynucleotides. Splicing reassembles bits of genetic information dispersed into a fractured array during evolution or selected from pools of peptide-coding regions, to make more extensive use of the genome than otherwise would be possible.

## II. SPLICING PATTERNS IN THE ANIMAL VIRUSES

The animal viruses are of great biological, medical, and economic importance and they provide superb model systems for the patterns of eukaryotic transcription and RNA processing. The majority of these events were discovered and characterized in viral transcripts. In the years preceding recombinant DNA technology, the only genes that were easily isolated in homogeneous DNA segments were viral chromosomes. They afforded a unique opportunity for hybridization-selection of specific messages from bulk cellular RNA. The transcription of viruses continues to present a degree of complexity and regulatory interplay that is not readily available in cloned cellular genes. The viral DNAs are also ideal substrates for site-directed mutagenesis since they can be reintroduced to cells at high efficiency for in vivo analysis of transcription and regulation.

Processing of eukaryotic viral RNAs is a broad topic. To present a comprehensive analysis, this review emphasizes the human adenoviruses. Additional descriptions are presented on SV40, polyoma, and papilloma viruses of the papovavirus family, on minute virus of the mouse (MVM) and adeno-associated virus (AAV) of the parvovirus family, and on Rous sarcoma virus of the RNA retrovirus group. Most of their messenger RNAs are spliced. Herpes simplex virus and the single-stranded RNA virus influenza exhibit only limited RNA splicing. At present, there is no evidence for splicing in messages of vaccinia virus (double-stranded DNA chromosome), vesicular stomatitis virus (negative-strand RNA chromosome),

Sindbis virus (a togavirus with a positive strand RNA chromosome), polio and encephalomyocarditis (EMC) viruses (picornaviruses), also with a positive-strand RNA chromosome, nor in those of reovirus (double-stranded RNA chromosome). Since splicing is primarily if not exclusively a nuclear event, it is pertinent that vaccinia, VSV, EMC, and polio, for instance, are replicated and transcribed in the cytoplasm.

The fundamental role of RNA splicing in the expression of these different viruses is best visualized by aligning the structures of the messenger RNAs with diagrams of open translation frames determined from DNA sequences. The detailed conclusions about the effects of splicing on the translational properties of the messages and on the regulation of expression are based on these maps.

### A. Adenoviruses

The chromosome of human adenovirus is a linear, duplex DNA of 36,200 nucleotide pairs (36.2 kb), a length sufficient for intriguing organizational complexities but still manageable for isolation, analysis, and genetic manipulation. Its expression is programed into early, intermediate, and late phases during the lytic infection of human cells in culture. Six "early" regions of the chromosome from both DNA strands are transcribed from five promoters during the first hours of infection (Figure 1).[4,26,27] The early proteins they encode permit DNA replication, which continues throughout intermediate times and into the late phase of infection. Additional genes are activated at intermediate times; some of these gene products may be needed to allow full expression of the late genes. The precursor for most of the late mRNAs is synthesized from the rightward-transcribed (r-) strand starting from a single promoter.[28] Each is cleaved at one of seven 3′ end determinants,[1,3,29,30] then is repeatedly spliced until an identical 5′ leader[30,32] is assembled and joined to one of a dozen possible coding regions (Figure 1). Analysis of the late transcription unit of adenoviruses resulted in the original description of the phenomenon of RNA splicing.[2,33,34] Most of the late mRNAs are translated into capsid proteins needed for virion assembly. The cell dies two to three days after infection.

As with other eukaryotic genes encoding proteins, the adenoviruses are transcribed by the host RNA polymerase II.[35,36] They are capped and methylated with $^{7me}G^{5'}ppp^{me}N$ at the 5′ end[37-41] and are polyadenylated at the 3′ end.[42] The primary transcripts from each of the viral promoters are processed into families of related mRNAs that share common 5′ and 3′ ends but differ by alternative splicing, except for peptide IX mRNA, which is not spliced.[2,43] Detailed aspects of adenoviral gene expression and RNA processing will be presented in later sections and have also been the subject of numerous reviews.[44-48]

### B. SV40, Polyoma, and BK Viruses

Studies of RNA splicing in SV40 and polyoma were among the earliest and most definitive because their small circular double-stranded DNAs were the first eukaryotic viral chromosomes completely sequenced; their lytic cycles and interactions with host cells have been the most intensively studied of all systems. Their circular chromosomes make site-directed mutagenesis relatively straightforward, and this has been a useful tool in testing ideas about RNA synthesis and splicing. The molecular biology of these papovaviruses has been reviewed in *DNA Tumor Viruses*.[49] Figure 2 summarizes the structures of the polyoma virus early and late transcripts, as superimposed on the control signals and open translation frames evident from the DNA sequences.

SV40, polyoma, and BKV have quite similar organizations of genes[50,55-58] and transcription patterns.[43-45,51-53,59-67] The early and the late promoters flank the origin of DNA replication and transcription proceeds in opposite directions. A "TATA" sequence (Honess-Goldberg element) is necessary for early transcription,[68,69] with RNA initiation about 25 nucleotides beyond the element, but the late transcription unit has no "TATA" sequence. "Enhancers"

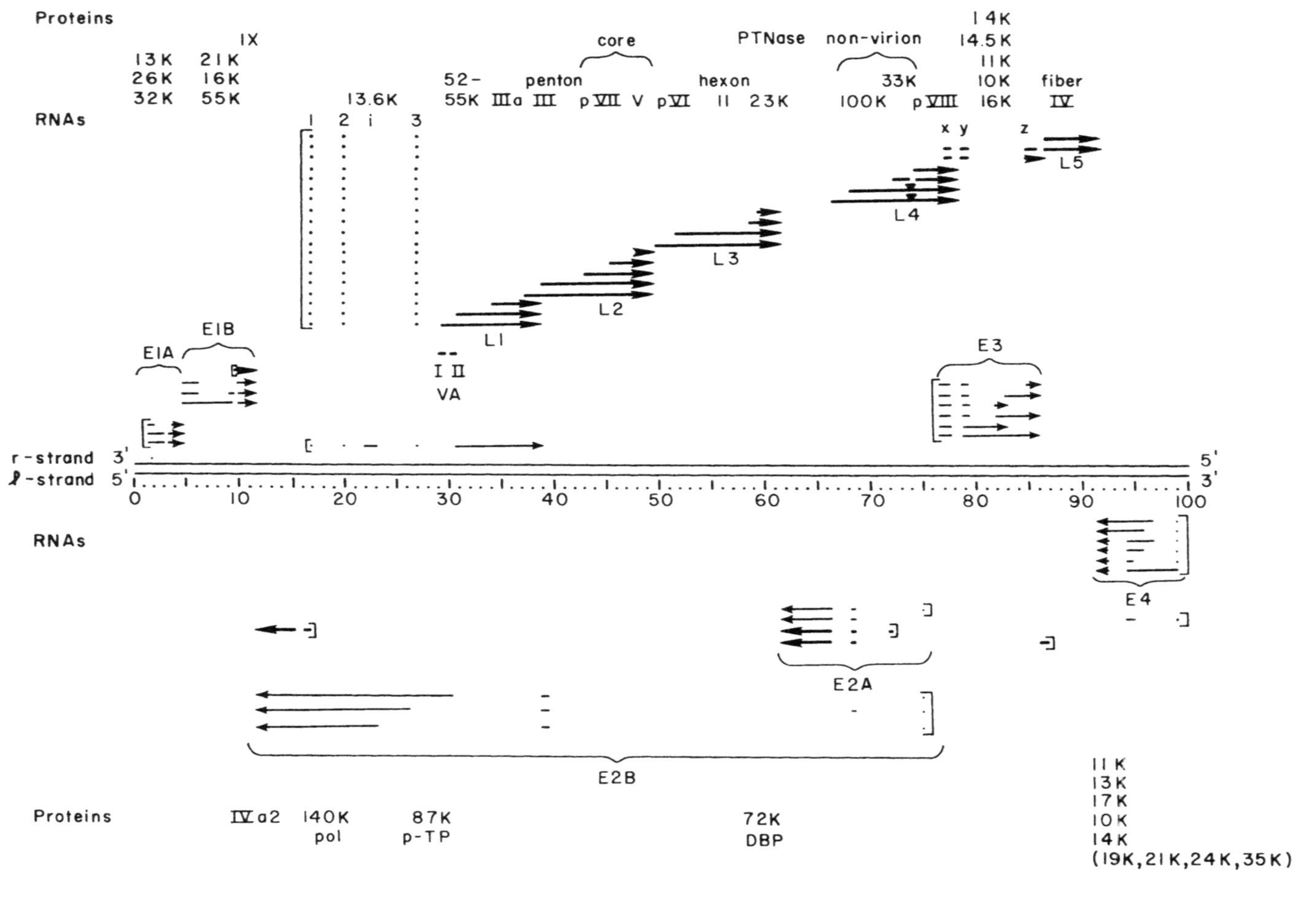
Proteins
IX
13K 26K 32K
21K 16K 55K
13.6K
52-55K
IIIa
penton III
core
pVII
V
pVI
hexon II
23K
PTNase
non-virion
100K
33K
pVIII
14K 14.5K 11K 10K 16K
fiber IV
RNAs
1 2 i 3
x y
z
L1
L2
L3
L4
L5
E1A
E1B
I II
VA
E3
r-strand 3' 5'
l-strand 5' 3'
0 10 20 30 40 50 60 70 80 90 100
RNAs
E4
E2A
E2B
Proteins
IVa2
140K pol
87K p-TP
72K DBP
11K 13K 17K 10K 14K (19K,21K,24K,35K)

FIGURE 1

are also required for early transcription; these are *cis*-acting elements that stimulate RNA synthesis independently of their orientation and position relative to transcription units.[69-73]

Polyoma viruses generate three mRNAs from the early gene block (Figure 2). All have the same 5′ and 3′ ends, and the encoded proteins share the same amino terminal sequences. Use of alternative splice donors and acceptors in the primary transcript permits all three translation frames beyond the splice to be addressed for encoding the small-, middle-, and large-T antigens.[52,54] SV40 and BKV encode only small- and large-T antigens. The early and the late RNAs have substantial heterogeneity in their 5′ cap sites.[53,74-77] The polyoma late transcripts have a most unusual feature: once initiated, RNA polymerase II continues repeatedly around the circular chromosome up to four or more times, creating a primary transcript of tandem duplications.[78-79] These can be spliced into mature, functional RNAs that retain a single coding region but have multiple copies of a poly leader (one kept from each tandem reiteration in the primary transcript.[51,80]

Three capsid proteins are encoded by the polyoma late RNAs.[81] VP2 and VP3 share the same carboxy terminus, but the leader in the VP3 mRNA is spliced well into the open reading frame, making VP3 shorter than VP2. The coding region for the carboxy termini of VP2 and VP3 overlaps that for the amino terminus of VP1. Thus, as with the early products, three proteins are derived from three overlapping messages.

The utility of SV40 for mutagenesis studies can be illustrated with several interesting sets of experiments. Khoury et al.[82] and Kamen et al.[83] isolated and characterized mutations near the SV40 early RNA splice sites, and Piatek et al.[84] did similar experiments at the late splice sites. Each group found that, in some cases, the deletions could extend to within a dozen nucleotides of a junction without blocking splicing. Nonetheless, the relative distributions of the primary transcript into alternative products could be changed and, surprisingly, so could the 5′ ends at which transcription is initiated. Gruss et al.[85] constructed a deletion mutant in which the intervening sequence of VP1 was precisely excised. The mutant failed to produce stable message. But when a heterologous sequence including donor and acceptor splice sites of the mouse beta-globin gene were recombined into the late transcription unit, splicing occurred and stable RNA was produced.[86,87] These results suggested that the act of splicing is essential for maturation and transport to the cytoplasm of RNAs that are normally spliced. On the other hand, Ghosh et al.[88] created a deletion in the late RNA leader-coding region that blocked splicing of the leader to the main body in the VP2 RNA, yet the RNA was successfully transported and translated. Furthermore, splicing of the leader to the VP1

FIGURE 1. The human adenovirus-2 cytoplasmic RNA transcripts characterized by electron microscopy of RNA:DNA heteroduplexes.[1-5] The chromosome of 36,000 base pairs is divided into 100 map units. Arrows indicate the direction of transcription along the r-strand or ℓ-strand of DNA. The 5′ ends of the cytoplasmic RNAs are capped and indicate the locations of transcriptional promoters (shown by vertical brackets).[6-9] The conserved segments constituting early RNAs are depicted by thin arrows and those in late RNAs by thick arrows. The promoter for the late RNAs at coordinate 16.5 is also active at early times but cytoplasmic RNA is primarily that for the 52—55 kd protein from family L1. Gaps in arrows represent intervening sequences removed from the cytoplasmic RNAs by splicing. Early regions 1A, 1B, 2A, 2B, 3 and 4 are labeled. At intermediate and late times, region 2 is expressed from several additional promoters, and region 3 RNA can be made under the direction of the major r-strand late promoter. All products of the late r-strand transcript have the same tripartite leader, the segments of which are labeled 1, 2, and 3. The i leader may also be present, particularly at early and intermediate times. The late messages form families (L1 → L5) of 3′ coterminal transcripts. Some of the RNAs for fiber can also contain any combination of ancillary leader segments x, y, and z. The ▼ marks optional splices. The viral-associated (VA) RNAs were mapped by nucleotide sequencing the RNAs and the genomic DNA.[19] The correlations of mRNAs with encoded proteins, as indicated directly above (r-strand) or below (ℓ-strand) the coding region, were based on cell-free translations of RNA selected by hybridization to DNA restriction fragments. In the cases of overlapping genes such as in E1A, the encoded proteins and their RNAs are in corresponding stacks. IIIa is the peripentonal hexon-associated protein, and IX, pVI, and pVIII are precursors of hexon-associated proteins; pol is the DNA polymerase, p-TP the precursor to the DNA terminal protein, and DBP the DNA binding protein.[5,11-24] Proteins are designated by K (1000 daltons molecular weight) or by Roman numerals (virion components).[25]

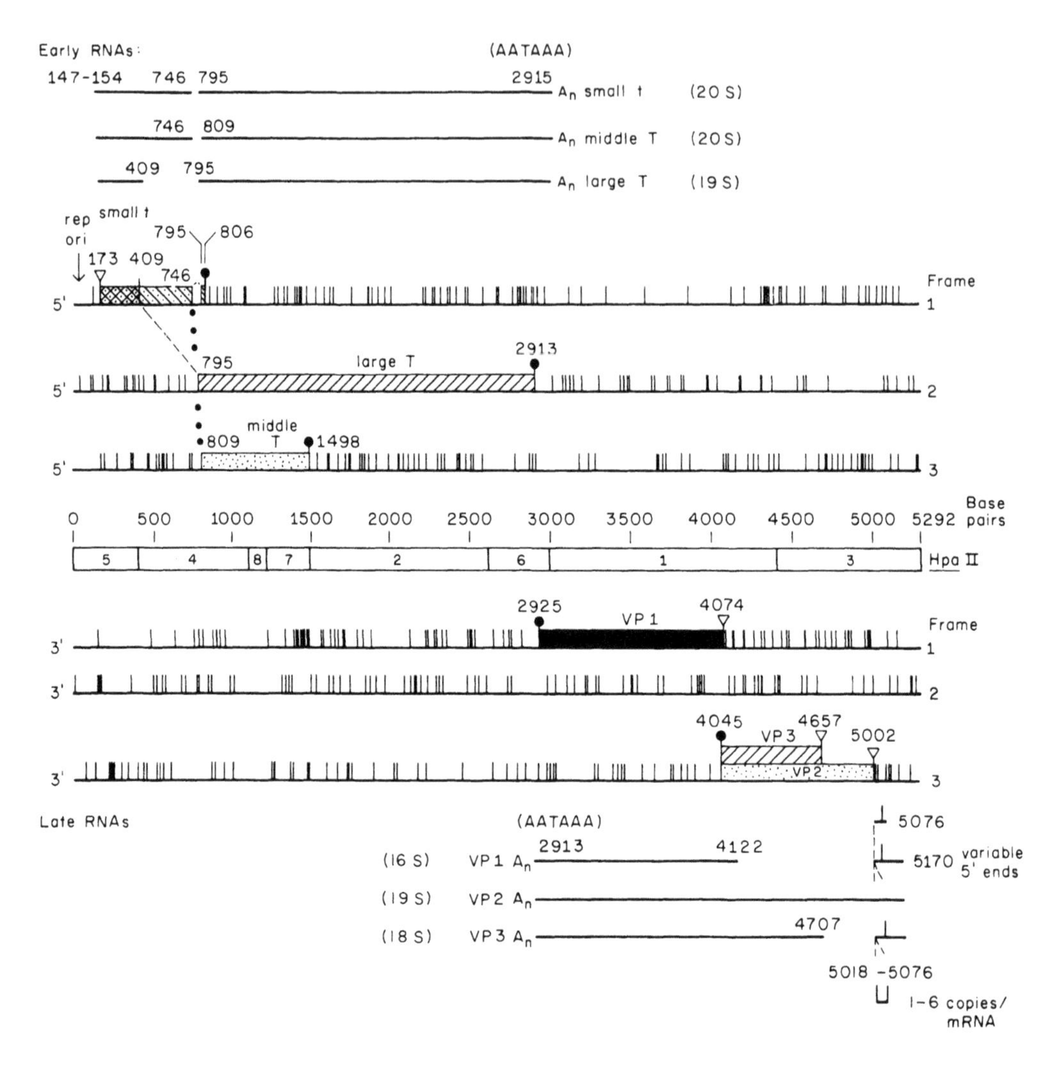


FIGURE 2. Polyoma virus. The complete DNA sequence of the A-2 strain was reported and analyzed by Soeda et al.[50] The mRNA 5′ cap sites and splice sites were determined by two-dimensional S1-nuclease mapping.[51-53] The early proteins were identified by Smart and Ito.[54] The *Hpa*II restriction sites are indicated. Vertical lines in each potential reading phase of both DNA strands indicate termination codons (redrawn from S.M. Dilworth in Appendix B of *DNA Tumor Viruses*[49]). The first nucleotides of transcription and translation signals and RNA splice boundaries are numbered. The position of the replication origin is shown; ▽ signifies utilized AUG protein initiation codons and ⚲ the utilized termination codons. Rectangles indicate the open reading frames used for the different proteins. Those for the small ( ▧ ), middle ( ▒ ) and large ( ▨ ) T-antigens involve the same initiation codon and overlap until splicing (dashed or dotted lines) switches translation to different reading frames. At late times and in the presence of large T-antigen, early region RNAs exhibit additional 5′ ends derived from the short interval between nucleotides 5260 and 20. The 5′ ends of late RNAs vary over a 100-nucleotide region, and the RNAs can have multiple copies of the sequence 5018 to 5076. This tandem repetition arises by splicing very long primary RNAs produced by continuous transcription repeatedly around the circular chromosome. The VP3 coding region is contained within VP2.

coding region still occurred, revealing that sequence alterations can have profound effects on the choice and efficiency of splice combinations.

## C. Papilloma Viruses

The *pa*pilloma (wart) viruses are classified with *po*lyoma and simian *va*cuolating virus (SV40) in the *papova* family on the basis of their similar capsid structures and circular,

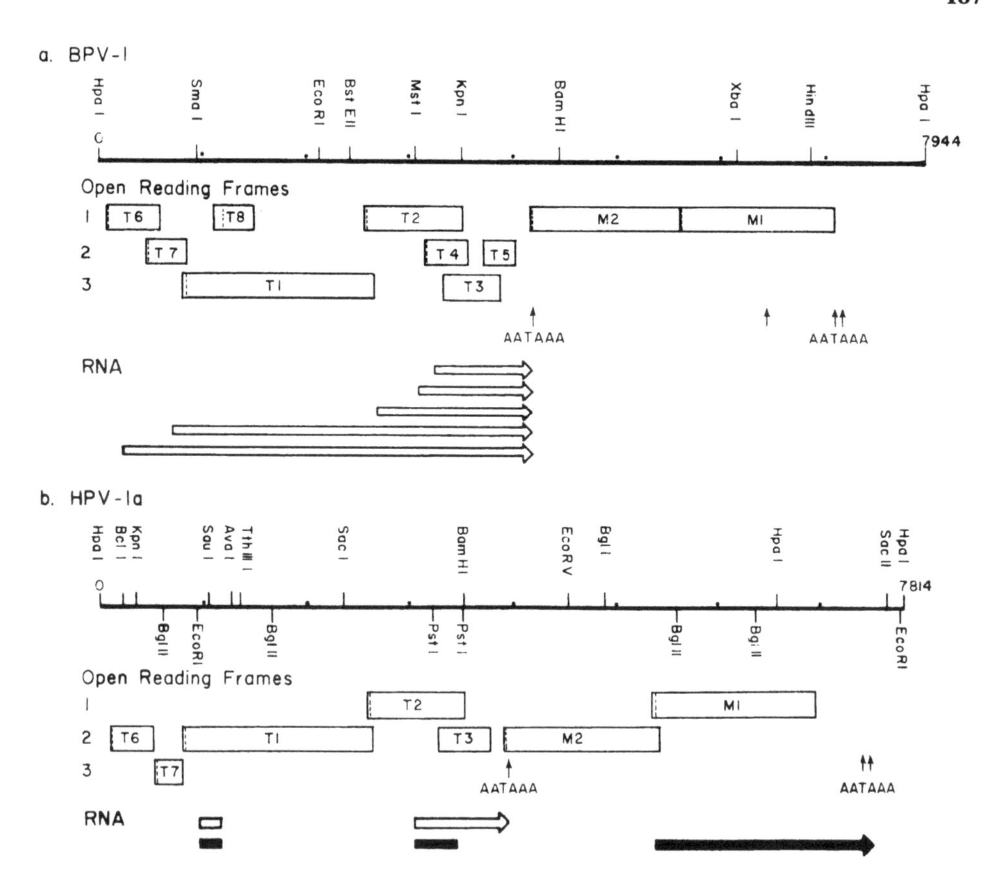


FIGURE 3. Bovine and human papilloma viruses. Open reading frames in the three translation phases of BPV-1, sequenced and analyzed by Chen et al.,[90] and of HPV-1, sequenced and analyzed (with revisions) by Danos et al.,[89,250] are shown by open boxes. The circular DNAs are represented as linearized at comparable *Hpa*I restriction sites. The RNAs isolated from BPV-transformed mouse cells[91] are indicated with open arrows. L1 is the gene for the major capsid protein.[248] The human papilloma virus RNAs were isolated from monkey COS cells acutely transfected with HPV-SV40 expression/shuttle vectors and mapped by electron microscopical heteroduplex analysis.[92] They were initiated at an SV40 late promoter present adjacent to the *Eco*RI cloning site at nucleotide 5240, but immediately entered into HPV-specific sequences and underwent polyadenylation and splicing, using HPV-1 specific sequences.

double-stranded DNA chromosomes. At the genetic level, they are totally distinct. An inability to propagate the papilloma viruses in any known cell culture system had retarded studies until DNA cloning enabled preparation of large amounts of homogeneous material. Human papilloma virus (HPV) type 1[89,250] and bovine papilloma virus (BPV) type 1[90] have been sequenced recently. The genomic organizations of the viruses are remarkably analogous. All appreciable open reading frames are along only one of the two DNA strands (Figure 3). Viral RNAs from BPV-transformed mouse cells form a family of 3′-coterminal transcripts.[91] Natural HPV transcripts are extremely rare and hard to analyze. To circumvent this problem, Chow and Broker[92] joined an SV40 replication origin to various positions in cloned HPV-1 DNA. Certain recombinants produced RNAs after transfection into SV40-transformed monkey COS cells.[93] Most had 5′ ends that were derived from SV40 sequences and spliced to HPV-1 sequences. The similarity of the HPV-specific portions of the RNAs, irrespective of whether the surrogate early or late promoter of SV40 was used, lends credence to the view that the HPV RNA splice sites, main bodies, and 3′ ends are bona fide. This expectation is supported by the correspondence of the conserved HPV-1 RNA sequences with the open reading frames and poly(A) signals in the HPV-1 DNA sequence and by the similarity to

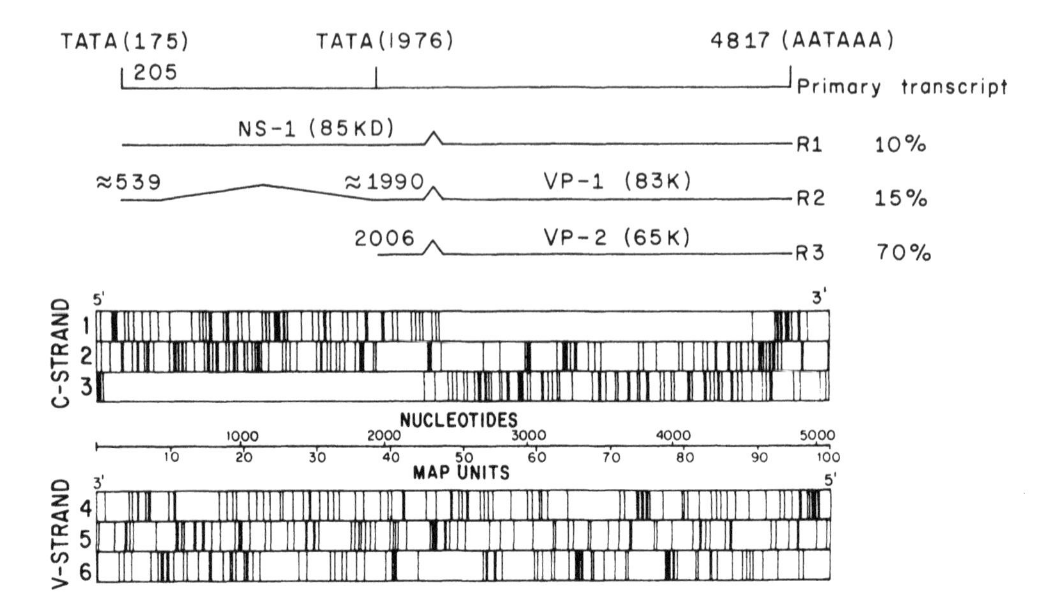


FIGURE 4. Minute virus of the mouse. The open reading frames were deduced from the DNA sequence[94] and the several spliced RNAs were mapped by using $S_1$-nuclease.[245] Protein assignments were based on cell-free translations of selected mRNAs and comparison of their immunological properties and tryptic peptides with proteins made in vivo. NS-1 is a nonstructural protein rather highly conserved among parvoviruses, while VP1 and VP2 are closely related capsid proteins.

some of the BPV-1 RNAs isolated from transformed mouse cells. The location(s) of the papilloma virus promoter(s) and the structures of mRNAs leading to expression of several of the other open reading frames remain to be determined.

## D. Parvoviruses

The parvoviruses have a single-stranded DNA chromosome; that of the minute virus of mouse (MVM) is 5084 bases long.[94] Inverted duplications at each end of the DNA cause the single strand to form terminal hairpins through which replication is primed. MVM can carry out autonomous lytic infection, whereas adenovirus-associated virus (AAV) depends on co-infection with an adenovirus. Analysis of the MVM sequence reveals open translation frames only on the C (complementary) strand, and none of significance on the V (viral strand) (Figure 4). Several spliced RNA species in infected cells have been mapped using $S_1$-nuclease.[245] A TATA promoter sequence near the 5′ end of the chromosome and an AATAAA signal for 3′ end determination are found in the DNA. Protein analysis indicates capsid proteins VP1 and VP2 share extensive homology.[95] In addition, it is likely that VP-1 relies on the same initiation codon as NS-1, but that RNA splicing eliminates much of the NS-1 information. The second splice may result in both a frameshift and a deletion of termination codons, allowing extension of translation into the second long open reading frame.

AAV generates an analogous, but larger, set of RNA species.[96,97] Most have a single splice between coordinates 40 and 48 and a common 3′ poly(A) end at coordinate 96. AAV may have as many as four promoters at coordinates 6, 13, 20, and 39, based on the staggered 5′ ends observed. Alternatively, several of these main bodies could share a short 5′ leader and there might be only two promoters, as with MVM. Three viral proteins VP1 (86 kd), VP2 (72 kd), and VP3 (62 kd) have extensive peptide homology and are encoded primarily between coordinates 48 and 96. Presumably the size differences reflect the extent of coding before the splice in the overlapping messages.

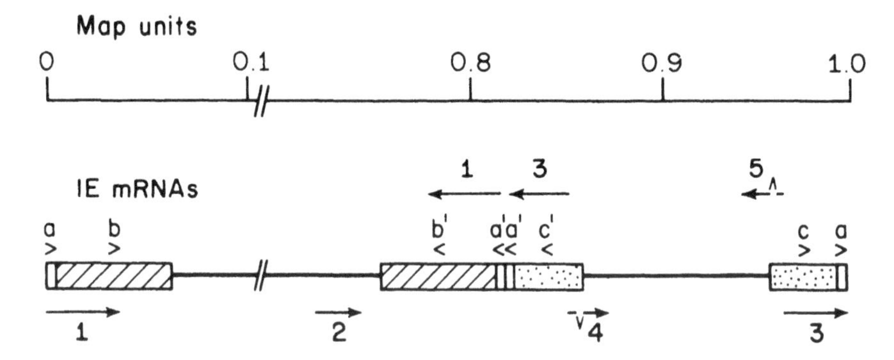


FIGURE 5. Herpes simplex immediate early mRNA-4 and mRNA-5. The promoters for mRNA-4 and mRNA-5 transcription are located in the internal and terminal inverted repetitions $IR_S$ and $TR_S$ flanking the short unique region ($U_S$). The primary transcripts extend into $U_S$, copying two different genes. Each RNA undergoes deletion of identical intervening sequences representing part of the inverted repetition.[99,100] Thus, the donor and acceptor splice sites are within the repetition. The 5′ leader segment, plus a few nucleotides between the splice acceptor and the junctions with unique sequences, are identical in mRNAs-4 and -5. Minor differences in $IR_s$ and $TR_s$ exist between different Herpes strains, which affect the length of the splices.

## E. Herpes Simplex Virus

The herpes simplex virus type 1 has a double-stranded DNA chromosome of about 150,000 base pairs. The chromosome is partitioned into long 105-kb ($U_L$) and short 15-kb ($U_S$) unique sequences, each flanked by inverted duplications that promote the inversions of the unique segments relative to one another. The transcription patterns of HSV-1 have been reviewed recently.[98] The RNAs appear through a sequential program of immediate early (alpha), early (beta), and late (gamma) times, distinguished by various requirements for host or viral protein synthesis and DNA replication. The RNAs are synthesized and processed in the nucleus in rather conventional ways and using the standard sequence signals, but few are spliced. Some form 3′ co-terminal families arising from different promoters. In other cases, failure to terminate can yield 5′ co-terminal RNAs with different 3′ ends.

Transcripts from homologous promoters in the inverted repetitions flanking the short unique sequence $U_S$ extend into the unique sequences from opposite ends and have different 3′ ends, according to their direction of entry (Figure 5). Sequences entirely within the $R_s$ region are spliced from each primary transcript.[99,100] The net result is that a common leader segment is joined to two different main body sequences and the respective protein initiation codons are located just downstream of the splices in the unique regions. At low frequency, at least one and perhaps several families of late RNAs can have some members with short segments spliced out near their 5′ ends.[98,101] But, the substantial majority of HSV-1 transcripts are collinear with the DNA.

## F. Vaccinia Virus

The pox virus vaccinia DNA consists of 180,000 bp and has the unusual property that the two DNA strands are terminally linked to form a continuous loop.[102] The 10,000 bp at the ends are terminal repetitions and contain within them two short sequences tandemly duplicated numerous times. Vaccinia DNA replication, as well as RNA transcription and processing, takes place in the cytoplasm of infected cells,[103] providing an interesting contrast to many of the DNA tumor viruses. The virus encapsidates many of the enzymes necessary for its transcription and RNA processing[104,105] — notably an RNA polymerase,[106] capping and methylating enzymes,[107] and polyadenylating enzymes.[108,108a] A vaccinia virus core-associated endoribonuclease has been described which can trim high-molecular-weight RNA,[109] as has a vaccinia RNA kinase that can regenerate 5′ di- and triphosphate-terminated RNAs,[110] the substrate for capping enzymes. In this regard, studies of vaccinia have been particularly valuable in providing sources of these enzymes for in vitro studies.

The retention of the β-phosphate group of the initiating ribonucleoside triphosphate in the cap structure of several mRNAs examined and the correspondence of the cap sites with the initiation sites for mRNA synthesis suggest that their 5′ ends are not determined by endonucleolytic cleavage of precursors.[110a] The AAUAAA signal seen near the 3′ end of most polyadenylated RNAs[111] does not appear in at least some vaccinia RNAs.[112] Nonetheless, the vaccinia RNAs are polyadenylated. Attempts to search for splicing in vaccinia early or late RNAs by nuclease mapping and electron microscopical heteroduplex analysis have so far given negative results.[113,114] Perhaps splicing activities are confined to the nucleus and viruses that are transcribed and replicated in the cytoplasm evolved without the need for RNA splicing.

### G. Influenza Virus

Human influenza virus type A, an orthomyxovirus, has a genome consisting of eight single-stranded RNA segments with a combined length of 14,000 nucleotides. Each RNA segment is anti-sense (the negative-strand with respect to messenger RNA). Segments 1 through 6 each contain a single gene encompassing nearly the entire chromosome and produce collinear (unspliced) mRNAs. Segments 7 and 8 each encode one unspliced and one spliced mRNA and two corresponding gene products.

In a novel variation on promotion of transcription, the flu virus messages are primed by 5′ capped RNA oligonucleotides derived from various host cell mRNAs.[115,116] The reaction has been recreated in vitro using, for instance, globin and reovirus RNAs as primers.[117] 3′-Polyadenylation occurs at a poly(A) tract encoded near the ends of RNA segments 7 and 8 and apparently involves neither an AAUAAA signal nor a cleavage reaction.

RNA segment 7 encodes two membrane proteins, $M_1$ and $M_2$, that are specified by different open reading frames[118,119] (Figure 6a). The $M_1$ message is an unspliced RNA collinear with essentially the entire template RNA segment. $M_2$ mRNA is spliced. The $M_2$ and $M_1$ proteins start at the same initiation codon and share nine amino-terminal amino acids before the splice.[121,122] Following the splice, the two coding regions overlap in different frames. $M_2$ coding continues almost to the poly(A) site encoded near the end of the segment. The $M_1$ and $M_2$ proteins are quite highly conserved among three serotypes sequenced, particularly in the region of overlapping coding.[118,122] A third RNA shares the same 5′ and 3′ ends, but an alternative splice donor site precedes the initiation codon. This RNA theoretically can encode a peptide of nine amino acids, starting from an AUG following the splice and terminating at the same site as $M_1$.

RNA segment 8, the smallest flu chromosome, encodes two largely unrelated nonstructural proteins, $NS_1$ and $NS_2$, which are translated from different frames in overlapping messages, analogous to the expression of segment 7[123,124] (Figure 6b). Comparison of the overlap region beyond the splice from two different influenza virus strains shows that the greater selective pressure was on conservation of the $NS_2$ protein.[125]

### H. Vesicular Stomatitis Virus

The chromosome of VSV, a rhabdovirus, is a single-stranded RNA of negative polarity (reverse sense to the messenger RNAs) consisting of about 10,000 bases. Transcription is by a virion-associated RNA polymerase.[126] Two models for VSV transcription and DNA processing have been proposed, as reviewed by Ball and Wertz.[127] Transcription is progressive from the 3′ end of the genome.[128,129] Five capped, polyadenylated monocistronic mRNAs are produced, with each RNA less abundant than the upstream species.[130] The majority of transcripts seem to arise by *de novo* initiation at the beginning of each gene and termination at the end of each gene, a "start-stop" mechanism.[131-133] Several different short

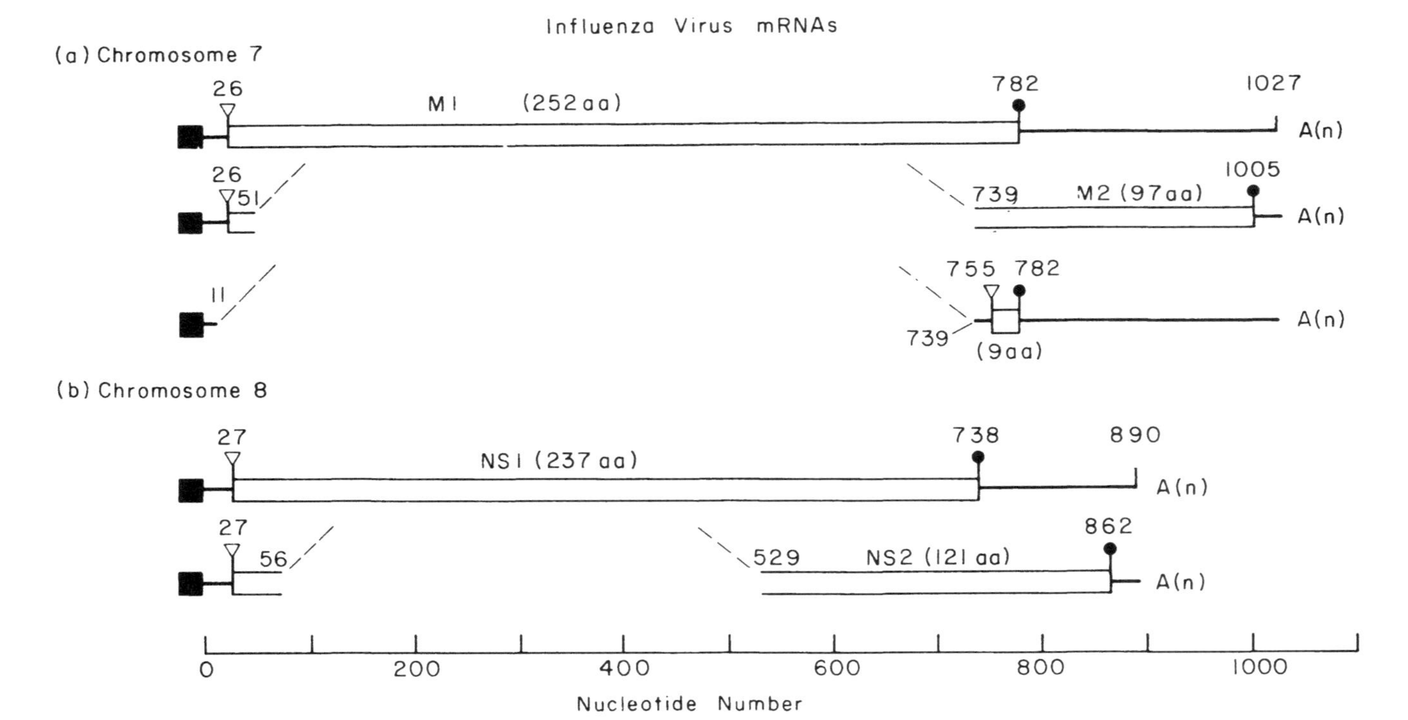


FIGURE 6. Influenza virus RNA segments 7 and 8. Transcripts of flu RNA segments 7 and 8 can be spliced to form a family of overlapping messages translated in part from identical and in part from different reading frames. Solid boxes at the 5′ ends of the RNAs represent heterogeneous primer sequences derived from cellular mRNAs. Thin lines are untranslated portions of each message, while open boxes are the coding regions; ▽ indicates the AUG initiation codon and ● the termination codon for each message. (a) Segment 7: (Adapted from Reference 120.) M1 and M2 are viral matrix proteins. The splice junctions adhere to the usual sequence requirements. The interpretation is based on DNA sequence analysis showing two overlapping open reading frames,[118,121,122] on characterization of the M1 and M2 proteins by hybridization-arrested translation of RNA and tryptic peptide analyses[119] and by sequence studies of the mRNAs.[120] (b) Segment 8:[123] (Adapted from Reference 123.) NS1 and NS2 are nonstructural proteins induced by the virus.[124] The RNAs are 3′-coterminal, differing by a splice in the NS2 message and having overlapping reading frames.[123] The structures of the chromosones and mRNAs were confirmed by polynucleotide sequencing.[125]

"leader" RNAs with polyphosphorylated 5′ ends have been isolated which are complementary to the 5′ end of various genes.[134,135] These may be primers — trimmed from the final message — whose elongation is possibly triggered by transcription through the upstream genes. On the other hand, some polycistronic transcripts distinguishable from replication intermediates by the unusual presence of intercistronic poly(A) tracts suggest the possibility of message production via endonucleolytic processing. In each intercistronic region, there is a sequence 3′ UAGCUUUUUUUNAUUGUC 5′[136,137] where the RNA polymerase apparently pauses and cycles, or "chatters", perhaps by repeated slippage, to insert the poly(A) tracts 200 to 500 nucleotides long.[138] The polymerase then reengages and progresses through the next gene and builds in another poly(A) segment. Inhibition of methylation with *S*-adenosyl methionine increases the extent of poly(A) synthesis.[139] Presumably the precursor is cleaved at the 3′ end of each poly(A) tract and the new monophosphoryl-5′ ends are capped, although this has yet to be demonstrated. Thus, in contrast to RNA splicing, which removes from the primary transcript nucleotides encoded in the chromosome, VSV can occasionally polymerize nonencoded nucleotides into a fraction of the primary transcripts.

### I. Rous Sarcoma Virus

There are a large number of similarly organized RNA tumor viruses — the retrovirus group — which have a positive (sense-strand) single-stranded RNA chromosome in the virion. Upon entry into a permissive host cell, the RNA is converted by a viral-encoded polymerase (reverse transcriptase) to a double-stranded DNA copy which can integrate into the host DNA. The provirus has long terminal repetitions (LTRs) which contain transcriptional regulatory signals. These complex events are reviewed in *RNA Tumor Viruses*.[140] Some of the retroviruses such as the Rous sarcoma virus (RSV) specify an oncogene (v-onc) responsible for host cell transformation and oncogenicity in vivo. These viral genes seem to have been adapted from normal cellular genes (c-onc). In several cases recently examined, c-*onc* contains intervening sequences, but they are absent in the viral analog.[141,246,251-253] This implies that the progenitor virus originally integrated next to c-*onc* and that a long transcript originating at the viral promoter extended through c-*onc* and underwent splicing and ultimately became the transducing virus chromosome. Support for this model comes from recent reports of retrovirus vectors converting cloned genes such as the mouse alpha-globin[142] and the human alpha-chorionic gonadotropin,[143] each with intervening sequences, to copies in the transducing retroviral RNAs identical to their spliced RNA products.

Transcription of RSV from the integrated state and processing to mRNAs are illustrated in Figure 7. Each copy of the $U_3$ portion of the LTR has a TATTTAA-promoter sequence and an AATAAA cleavage-polyadenylation signal. The functional promoter is in the left (5′) LTR and the functional cleavage signal is provided by the 3′ LTR. Ten or more proteins are encoded by just three co-terminal mRNAs, one of which is the unspliced full-length transcript, and two of which are derived from it by splicing. Each of the primary translation products (except $pp60^{src}$), is cleaved to yield two or more mature proteins; $pp60^{src}$ apparently arises by initiation at an internal AUG codon, even though its RNA includes the same AUG (from the 5′ leader) that functions for initiation of the gag, gag-pol, and env poly-proteins. The most remarkable product is the $Pr180^{gag\text{-}pol}$ fusion protein, which is made despite being encoded in two different reading frames. So far, there is no indication of a mini-splice between the gag and pol coding regions to effect the frameshift. The best present interpretation invokes a $-1$ frameshift during translation of the RNA, where one nucleotide is used as part of two overlapping triplets. Thus, several distinct strategies for producing proteins are utilized by the retroviruses.

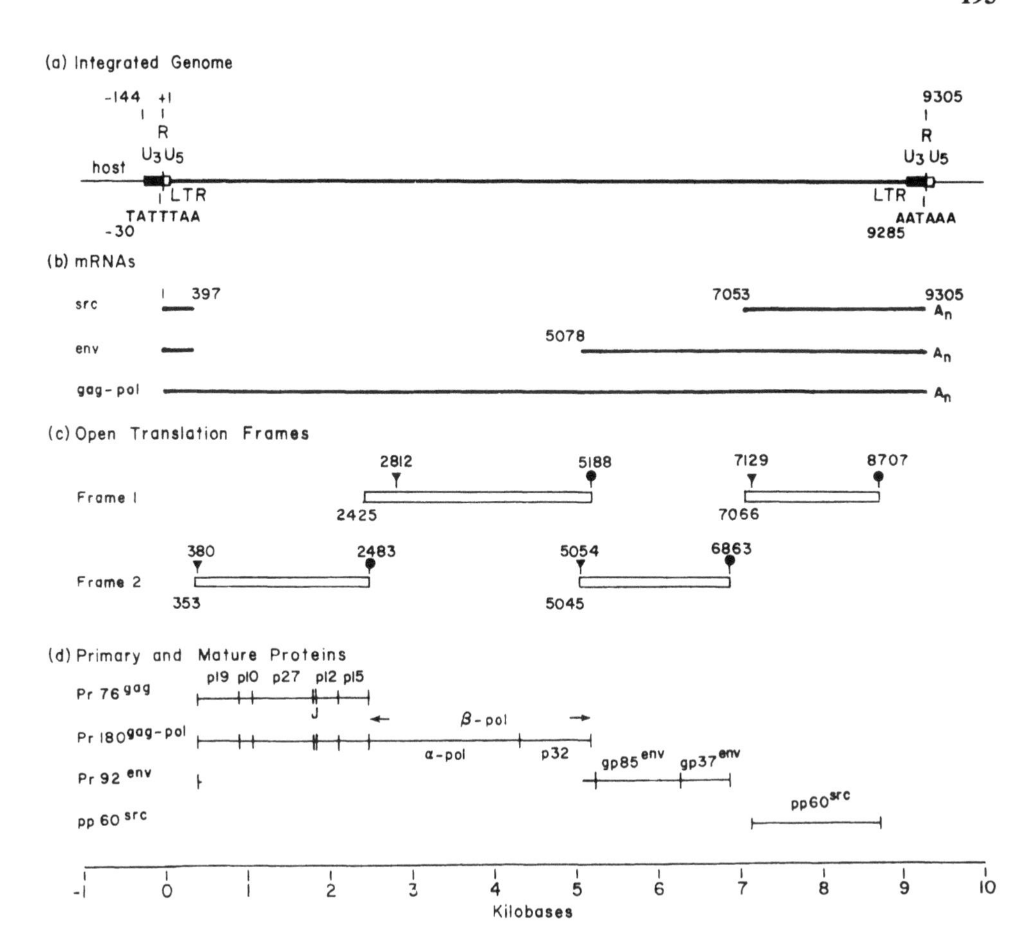


FIGURE 7. Rous sarcoma virus. The Pr-C strain of RSV was sequenced by Schwartz, et al.[249] The sequence itself and additional contributions to the sequence are described in Appendix E1 of Reference 140. (a) The mRNAs are transcribed from the integrated DNA copy of RSV (or the viral RNA chromosome may itself serve as a gag-pol message). R is a direct repetition present at both ends of mature viral RNA and is also part of the long terminal repetition (LTR) flanking the integrated cDNA copy. U3 is adjacent to R at the 3′ end and U5 adjacent to R at the 5′ end of viral RNA and both are present as part of the integrated LTR. U3 encodes both the promoter (TATTTAA) for transcription and the poly(A) signal AATAAA. (b) Of the viral messenger RNAs, one is collinear with the viral chromosome and two are spliced from a common donor to alternative acceptors. (c) Boxes indicate the extent of open translation frames flanked by in-phase termination codons: ▼ marks the first AUG in each open frame, and ● marks the closing termination codons. Translation into the gag-pol fusion protein requires a frame shift to connect the open frames, which could be achieved either by a short RNA splice in a fraction of the RNA or a frameshift stutter during translation. The low abundance of the gag-pol protein relative to the gag protein would be consistent with either possibility. The precursor peptide to the envelope (env) proteins is initiated by the same AUG used for the gag and gag-pol proteins. This AUG is also present in the leader segment of the mRNA for the sarcoma (src) protein, but src initiates at an internal AUG (at nucleotide 7129). (d) All primary peptides but pp60src are processed into two or more functional proteins. The group-specific antigenic determinants (gag products) are structural components of the viral core. The beta-subunit of the polymerase (reverse transcriptase) arises as a cleavage product of the pr180$^{gag\text{-}pol}$ peptide. In turn, some beta-pol is cleaved to alpha-pol (to yield a functional alpha-$_2$ beta-$_2$ tetrameric enzyme) and p32, an endonuclease. Pr92$^{env}$ precursor is processed into gp85$^{env}$ and gp37$^{env}$, two virion envelope proteins. The amino terminal of the precursor is the same as for pr76$^{gag}$, but is processed away. The src mRNA contains the same 5′ leader, but if translation initiated at the AUG at nucleotide 380, the nascent peptide would terminate at nucleotide 7063, just past the downstream splice site (7053). Instead, pp60$^{src}$ originates via internal initiation at the second AUG from the 5′ end, at nucleotide 7129. The protein is not further cleaved.

# III. MECHANISMS OF RNA SPLICING

Although there is considerable understanding of the general patterns of viral RNA processing, and splicing of nascent transcripts has been achieved in isolated nuclei[144-146] and in vitro using crude extracts,[147,148] relatively little is known about the molecular details of the process: the enzymes, the cofactors, the higher-order structures of the RNA substrate, and possible contributions of the nuclear matrix and membrane.

In lieu of an enzymatic approach, and to pave the way for it, studies of mechanism have focused on the structures of the primary transcripts which serve as substrates for splicing, on apparent splicing intermediates that have some but not all the deletions, on the product mRNAs, and on infection conditions and viral mutants which lead to the accumulation of intermediates by disturbing the splicing process.

## A. Substrate and Polyadenylation

The substrate for splicing is a primary transcript collinear with the genomic DNA. In general, the RNAs are polyadenylated, capped, and methylated prior to splicing, as discussed below. Splicing takes place during or very shortly after RNA synthesis. Beyer et al.,[149] using electron microscopic observations of adenovirus transcription complexes, found some nascent transcripts still attached to the DNA by RNA polymerase which were shortened, presumably by splicing (rather than by breakage), but most RNAs had no evidence of deletion. Among adenovirus late RNAs isolated from infected cell nuclei, few primary transcripts were found in which the first and second leader segments were not yet spliced together.[150,151] In early region 2A RNA, the splice between the 5′ proximal leader and the internal leader occurs immediately after the polyadenylation step.[152] These observations indicate the first splices occur very shortly after release from the template.

The 5′ end of a spliced mRNA is always derived from a promoter-proximal region, indicating the 5′ end of the primary transcript is conserved. It exerts a major influence on the selection of donor splice sites. For instance, in the early form of adenovirus early region 2A mRNA, the promoter-proximal leader at genome coordinate 75 is spliced to the internal leader from coordinate 68 (Figure 1).[4,26,27] The potential splice donor site at coordinate 71.8 appears to function only at intermediate to late times, when the RNA is initiated from the late promoter at coordinate 72.[4] It is deleted as part of the intervening sequence at early times. Similarly, the minor late promoter for early region 2A at coordinate 87 yields a primary transcript that passes through the early and late promoters and their nearby donor splice sites.[4] These sites are eliminated from the mature mRNA when the leader from coordinate 86.2 is spliced directly to the common internal leader at coordinate 68. The most straightforward interpretation of these observations invoke mutually exclusive secondary structures or splicing activities that recognize capped 5′ ends and potentiate nearby splice sites.

Bina et al.[153] proposed a role for poly(A) in aligning splice sites. But Zeevi et al.[154,155] found that cordycepin, an inhibitor of poly(A) synthesis, blocked neither splicing nor transport to the cytoplasm. Nonetheless, short poly(A) tracts sufficient for these events might have been present, since cordycepin is a chain-terminating competitor for ATP pools. In any case, the creation and polyadenylation of the 3′ ends of mRNAs play critical roles in the processing of primary transcripts into mature messengers. Some of these have been considered in Chapter 6 of this volume.

The 3′ ends of RNA pol II transcripts of many viruses are determined by endonucleolytic cleavage of primary transcripts at sites some 10 to 25 nucleotides to the 3′ side of an AAUAAA signal.[111,157] Primary transcripts often extend well beyond the signal, as do those of the SV40 late genes,[158] adenovirus early region 2 and 4,[159] and the adenovirus major late gene block,[160] which includes at least seven such signals.[161] From a kinetic study, Nevins and Darnell[162] reported that cleavage at a site occurs in the nascent RNA shortly after RNA polymerase II passed a potential polyadenylation site, but that the enzyme progresses to near the right end of the DNA, irrespective of cleavage behind it. Conversely, the electron microscopic study of transcription complexes by Beyer et al.[149] indicated many late Ad-2 transcripts are completed to the end of the chromosome before cleavage.

For primary transcripts that include several possible poly(A) sites, the general perception has been that cleavage would occur at one site per transcript and this event permanently determined the 3′ end of the RNA and, accordingly, the main body coding region that would be preserved during RNA splicing. However, it is possible that cleavage and polyadenylation are in dynamic competition with splicing and are able to take place more than once per primary transcript if splicing does not first remove other upstream potential cleavage sites remaining in the nascent transcript. The problem posed by multiple poly(A) sites is the achievement of an equitable distribution of a primary transcript into processed mRNAs from each of the families. Possible mechanisms are that: (1) RNA endonuclease is present in limited amount and RNAs are only likely to be cut once; (2) some structural feature such as nascent 3′ polyadenylate inhibits further endonuclease activity on that RNA; (3) the endonuclease is an RNA polymerase cofactor that is present in a complex at the time of initiation and travels in association with the polymerase, dropping off to function once per primary transcript on a random basis or at a preordained site; and (4) viral-encoded factors modulate the efficiency of activation or suppression of cleavage at a particular poly(A) signal.

The author favors possibility (1) and proposes that the 3′ endoribonuclease activity is sufficient to cut at all possible sites at early times after viral infection, but that it becomes limiting at later times as the protein turns over. (Host RNA and protein synthesis stop late after adenovirus infection.) Thereafter, cutting would become random and incomplete. Various distal coding regions will remain associated with the 5′ proximal sequences for longer than at early times, during which period splicing is more likely to join them to the leaders. If, for any reason, splicing is delayed or the amount of late RNA synthesis is reduced, cleavage at residual AAUAAA sites in the nascent transcript would regain a relative advantage and the mRNAs produced would be those from families closer to the 5′ end.

At early times it is rare for a mature cytoplasmic RNA to extend beyond the first available AAUAAA site. Even among transcripts originating from the adenovirus major late promoter (16.6) at early times[37,163-165] or in the presence of drugs that block DNA replication and hold transcription in an early mode,[4,166] almost all mature cytoplasmic RNA represent late family 1, even though primary transcripts in the nucleus extend through the late family 2 and 3 poly(A) sites.[167] Polyadenylation of adenovirus early region 3 RNAs is particularly instructive, for there are a series of five or more open reading frames that must be addressed.[168,169] Two alternative poly(A) sites are utilized with similar efficiency.[1,4] The first potential cut site at coordinate 83 has a variant sequence AUUAAA,[161,170] which may make it less competitive and allows many of the primary transcripts to be spared cleavage in favor of the AAUAAA site at coordinate 86.

In summary, RNA cleavage and polyadenylation may interplay with splicing to determine the distribution of alternative products of composite transcription units. Thereafter, as described in Chapter 6, the poly(A) stands watch over the product mRNAs by guarding against premature degradation.

## B. Methylation

The 5′ ends of eukaryotic mRNAs are capped with G(5′)ppp(5′)N and methylated on the bases and sugars.[37] Those of reovirus,[171] vaccinia virus,[172] and vesicular stomatitis virus,[173] all of which replicate in the cytoplasm, and those of SV40[174] and adenovirus[34,38-41] are similarly methylated.

Internal methylation is not found in viruses that are transcribed in the cytoplasm and do not exhibit splicing.[171,172,175,176] On the other hand, SV40[174] and adenoviral[39] mRNAs contain internal $N^6$-methyl adenosine on the average once every 400 bases. Most $m^6A$ seems to be in the sequence GA*C or AA*C in the rRNAs and mRNAs that are transcribed in the nucleus.[177-179] Internal methylation occurs in primary transcripts and is largely conserved in spliced cytoplasmic mRNA.[180,181] Inhibition of methylation with drugs blocks the processing of nuclear RNA.[182,183] Consequently, internal methylation has been considered a possible marker in the RNA at the boundaries of conserved sequences. However, examination of a large number of adenoviral splice junctions defined by cDNA analysis reveals no simple or reproducible placement of GAC and AAC sequences with respect to the junctions. Methylation and other nucleotide modification merit additional examination, but the analysis is severely hampered by the lack of methods for studying modifications at specific sites in small amounts of long RNA molecules present in heterogeneous mixtures.

## C. RNA Secondary Structures

Potential secondary structures of viral DNAs and RNAs have been calculated in the neighborhoods of transcriptional promoters,[28,184] splice donors and acceptors,[185] and polyadenylation sites.[22] But they have not been determined over long ranges (such as the entire length of a primary transcript), or even by pairwise analysis of alternative splice donor and acceptor sequences to search for possible interactions. In any case, such analyses do not take into account the potential contributions of bound proteins or ribonucleoprotein complexes.

Zain et al.[185] calculated the optimal secondary structures of the leader-coding regions of adenovirus-2 late r-strand transcripts. In each case, quite stable secondary interactions seem possible that leave the splice donor and acceptor sites exposed in open loops. In these configurations, they remain free for endonucleolytic cleavage or interaction with remote splice partners or complementary adapter molecules such as the small nuclear RNA-protein complexes[145,186,187] as described in more detail in Chapter 7 of this volume.

RNA secondary structure also appears to be important in 3′ poly(A) site determination. A feature of the convergent early and late transcripts of adenoviruses, SV40, polyoma, and BKV is the presence of 3′ cleavage-poly(A) signals on opposite strands in close proximity (AAUAAA..$N_x$..UUUAUU) (Figures 1 and 2). In these cases, double-stranded template DNA might establish cruciforms, or nascent RNA could form stem-loop structures with the 3′ cleavage sites available to nicking and polyadenylating enzymes.

## D. Splice Sites

Splice sites in a variety of viral mRNAs have been determined by comparing the sequences of cDNA copies of the RNAs with the parent genomic DNA sequences. At all splice junctions, the first two nucleotides of the deleted intervening sequence are /GU and the last two are AG/. A large data base indicates that AG/GU$^{AA}_{GG}$ GU is a frequent but not exclusive sequence of donors, and $U_n$PuCAG/ is typical of many splice acceptors.[186,187,189,190] Examination of the alternative partners of common splice donors or common acceptors in adenoviruses and papovaviruses reveals no obvious similarities in their primary sequences, a situation which makes generalizations difficult. Even analogous sites in different adenoviral serotypes show variations within a few nucleotides of splice junctions.

Probably the most remarkable aspect of RNA splicing is the extraordinary accuracy of these site-specific recombination events at sites seemingly short and rather variable. Splicing is faithful to the nucleotide, appropriately creating or preserving open reading frames.

Unfortunately, neither donor nor acceptor sequence is sufficiently unique to serve as a good guide to the location of splice sites in genomic DNA unless accurate nuclease or EM mapping of RNA coordinates or direct cDNA sequencing are available. Educated guesses based on this information can be improved by consideration of actual protein sizes compared to open reading frames with prospective AUG initiation codons, of the maintenance of open frames when splicing is internal to a gene, and of evolutionary conservation when comparing related genomes.

Using site-directed mutagenesis in the vicinity of splice sites, Montell et al.[191] altered the /GU at the donor end of an Ad-2 early region 1A intervening sequence (IVS) and destroyed its ability to participate in splicing. Solnick[192] altered the fifth and sixth nucleotides into the same IVS and also blocked splicing. Deletions (effectively, substitutions) have been created in the conserved sequences to within a few nucleotides of SV40 splice sites without eliminating the splicing, although the efficiency and relative selection of alternative splice partners were altered.[82,84,88,193] It appears that the 6 to 11 nucleotides at the ends of the IVS that form the splice donor and acceptor consensus sequences cannot be changed significantly without seriously affecting splicing. Thus the boundaries of the deleted IVS seem more important than those of the retained sequences in spliced mRNAs.

### E. Polarity and Splicing Intermediates

Without exception, the segments ultimately conserved in a spliced RNA are in the same relative order as in the primary transcript. The basis for this is not entirely inherent in the polarity of RNA, for folded RNA structures can, in principle, be formed in which downstream and upstream segments are shuffled. Splicing is not necessarily processive along the primary transcript from $5' \rightarrow 3'$ or $3' \rightarrow 5'$, although there is some tendency to observe completion of splices near the 5′ end in partially spliced nuclear RNAs.[150-152]

In the adenovirus transcripts, a plurality of structures involving the same main body can be observed. It is not clear to what extent the various species represent functional products, processing intermediates, or dead-end aberrations. In any case, these can have downstream splices without certain upstream splices seen in closely related RNA species. If they are indeed intermediates, they also suggest that the splices can occur in multiple steps or in alternative orders. For instance, the i leader segment (between the second and third leaders of the major r-strand transcription unit) and/or long third leader segments may remain in the RNA despite correct coupling of the third leader to main body acceptors.[150,166,194] In other cases, the ancillary x, y, and z leaders may remain between the 5′ tripartite leaders and the fiber gene,[3,32,195] or additional ancillary leaders can precede many of the other main body acceptors,[247] again revealing that both downstream and upstream splices had taken place without some of the internal splices.

Additional putative splicing intermediates are observed when human Ad-2 or Ad-5 abortively infect monkey cells.[196] Cytoplasmic RNAs are 20 times underrepresented for fiber transcripts.[197,198] About half the fiber region RNAs observed contain segments of what should have been deleted as IVS, usually from late region 4 and early region 3 (Figure 1), and these must block translational access to the fiber-coding sequences. Adenovirus mutants have been isolated[199] that map in the DNA binding protein (DBP) gene[199-201] and that confer the ability to grow in monkey cells because normal amounts and structures of fiber mRNAs are achieved.[196] One interpretation is that transcription and/or splicing of fiber RNA is particularly slow in monkey cells and that alterations in the DBP improve the efficiency, perhaps by making DBP more compatible with cellular factors with which it might interact. If so, the DBP can be regarded as a probable component of the RNA transcription and splicing apparatus.

## IV. BIOLOGICAL CONSEQUENCES OF RNA SPLICING

The animal viruses present a succinct but surprisingly comprehensive microcosm of the attributes of splicing — possibly because their limited genetic capacity has forced them to optimize its utilization. Careful evaluation of viral DNA and RNA sequences, encoded proteins, and the changing RNA transcription and processing patterns throughout the infection cycle has revealed a remarkable range of genetic consequences of RNA splicing that is testimony to its importance in amplifying the addressable coding information of eukaryotic DNA.

### A. Access to an Array of Genes

One of the most noticeable features of the eukaryotic viruses that exhibit RNA splicing are the linear arrays of related genes under the coordinated control of a single promoter. In the most extreme example, the adenovirus major late promoter initiates synthesis of a 29,000-nucleotide-long primary transcript that is spliced into messages encoding 13 or more proteins, most of them involved in capsid morphogenesis (Figure 1). Early regions 2A and 2B transcribed from the same promoter[5] encode several functions necessary for viral DNA replication: the single-stranded DNA binding protein,[202] a protein primer that becomes covalently linked to the nascent DNA strand,[5,203-205] and a DNA polymerase.[206-208] Similarly the same leader segment of SV40, polyoma, and BKV late RNAs is joined to at least three regions encoding capsid proteins. These are the functional equivalents of bacterial operons. But unlike prokaryotic ribosomes, which can translate a series of individual proteins from a polycistronic messenger RNA, eukaryotic ribosomes are, with the exceptions noted below, able to utilize only the first AUG found by polarized scanning from the 5′ end of the mRNA.[209] Accordingly, a major function of RNA splicing is to translocate each interior coding region in a polycistronic precursor to a site of 5′ proximity, while maintaining the 5′ cap group and the 3′ poly(A).

Several viruses have devised other strategies to address individual translation frames in a linear array of genes. Some manage to initiate translation at internal AUG codons. Polio starts translation at the ninth AUG from the 5′ end,[210,211] which is the first with a favorable sequence environment (see below).[209] There is an AUG very near the 5′ end of the leader sequence for SV40 late messages which is used to encode the "agno-protein", but the predominant stable products of the 19 S and 16 S RNAs are VP1, VP2, and VP3, each initiated at the second AUG present in their respective messages.[212,213] Adenovirus early region 1B RNAs can give rise to completely different proteins, depending on whether the first or second AUG is recognized.[214] Infectious pancreatic necrosis virus of fish, a double-stranded RNA virus, produces a single RNA transcript from which three proteins seem to be initiated independently.[215] The RNA retroviruses proteolytically excise individual products from a poly-protein made as the primary translation product of the *gag, gag-pol* and *env* genes (Figure 7). And many viruses such as Herpes and vaccinia have separate promoters for most or all of their genes, apparently preferring to regulate promotion rather than alternative 3′-termination and splicing.

### B. Modulation of Translational Efficiency by RNA Splicing

#### *1. 5′ Ends*

Upon sequencing and identifying the first leader segment of late adenovirus-2 RNAs, Ziff and Evans[28] noted that the 5′ end of the leader had the potential to form base pairs with the 3′ end of 18S ribosomal RNA in a fashion similar to that proposed for prokaryotic mRNAs.[216] A similar conclusion was reached when the 5′ end of the Ad-5 mRNA for protein IX was analyzed.[217] Subsequently, the cap sites of most of the Ad-2 transcripts were determined.[9] Generally, there is a significant opportunity for establishing base pairs with the 3′ end of the 18S rRNA, irrespective of whether the 5′ end is a nontranslated leader or is immediately adjacent to the coding regions in the primary transcript.

It is not clear why one mRNA is more efficiently translated than another. Wolgemuth et al.[218] measured the translation efficiencies in vitro of different Ad2 mRNAs that shared the same tripartite leader sequences. The various messages had different levels of activity. Conversely, SV40 messages with different 5′ leaders but the same coding regions had very similar translation efficiencies, suggesting that the major determinants of efficiency lie in or near the main coding regions of the genes.

*2. Initiation Codons*

Kozak[209] has surveyed large numbers of eukaryotic initiation codons and found that certain nucleotides optimize translational efficiency:

1. A > G ⩾ pyrimidines at the $-3$ position preceding $^{+1}$AUG.
2. A, G ⩾ C ⩾ U at the $+4$ position immediately following the $^{+1}$AUG triplet.

The 52 to 55 kd genes of late family 1 of adenovirus-2[10,165] and adenovirus-7[219] and the fiber gene of late family 5 of Ad-2[185] have a C at position $-3$ in the primary transcript. In each case, the late tripartite leader *(AAG*/GUA) is spliced to the coding region immediately preceding the initiation codon (CAG/*AUG)*, thereby replacing the genomic sequences upstream from the initiation codon and introducing an A from the third leader segment to the $-3$ position. Since its last three nucleotides are purines, the donor sequence of the third leader can only create or maintain a strong translation signal upon splicing. Some fiber messages contain ancillary leaders — usually the y segment and, less frequently, z, yz, or xyz segments — situated between the tripartite leader and the fiber coding region.[3] The y[32] or z[195] leaders result in the substitution of a G for the C at $-3$ in the primary transcript.

Examination of the messages from the early region 1B transformation genes encoding the 21-kd proteins of Ad-2, -5 and -7, and the 55-kd proteins of -2, -5, -7, and -12 reveals that all have a pyrimidine at $-3$ preceding the AUG. None of these are replaced by splicing. E1B is unusual in that the 55-kd proteins arise by internal initiation (at the second AUG triplet) in the E1B messages.[214] The weak environment around the AUG initiation codon for the 21-kd proteins from Ad-2, -5, and -7 probably contributes to making this possible by allowing some percentage of ribosomes to slip on by to the second AUG used for the 55-kd proteins.

Splicing of a leader containing an AUG to a main body with an open reading frame does not necessarily mask the proper initiation codon. The first leader present in all adenovirus-2 early region 3 transcripts and the x leader (which is a portion of the early region 3 first leader), in some fiber RNAs at late times, contain an AUG followed by a significant open translation frame, but it is preceded by a U at position $-3$.[170] apparently weakening it and enabling most translation to initiate at "the proper" internal AUGs, all of which have the more favorable environment.

Splicing can create the opportunity for translation by coupling an AUG to an open reading frame without one. The most clear-cut examples are the IVa2 proteins of adenovirus types -2, -5, -7, and -12,[220-223] in which the 5′ leader of the RNA encodes only the initiating methionine and three other amino acids before splicing to the main body (Figure 8). Interestingly, those four amino acids are effectively plucked out of a sequence comprising the carboxy terminus of the overlapping 140 kd DNA polymerase.[206-208] A similar situation with an AUG in a leader segment is observed with the envelope protein of Rous sarcoma virus (Figure 7).

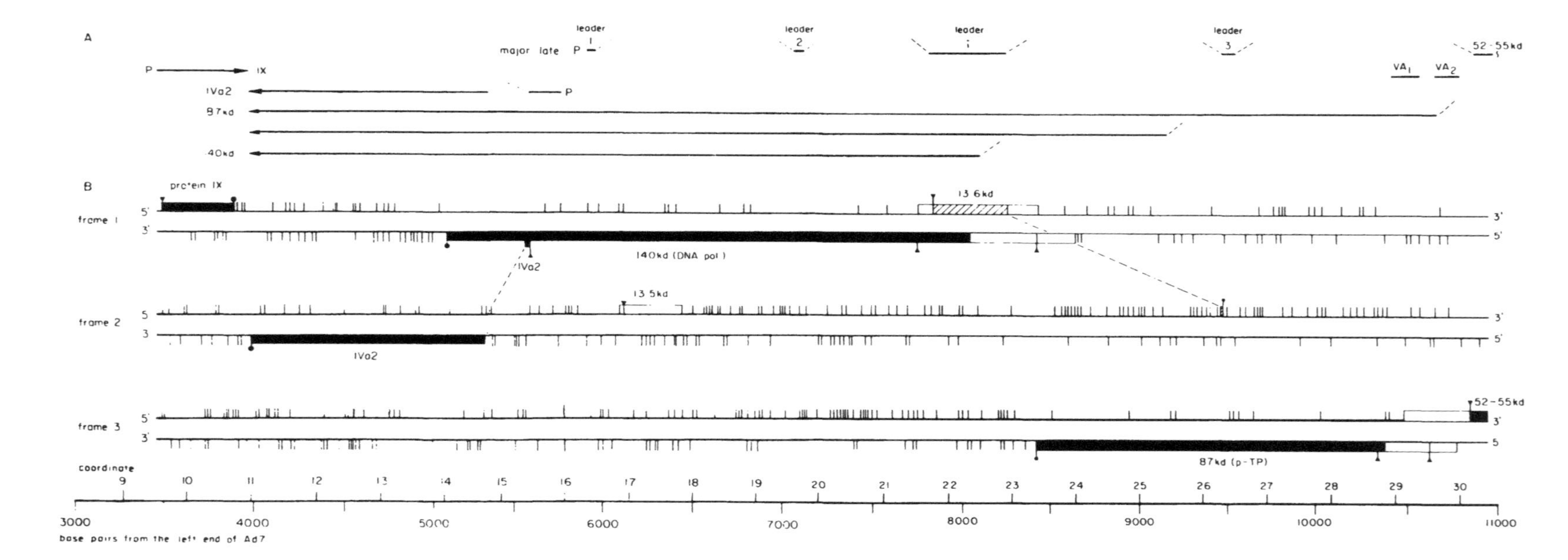


FIGURE 8. Adenovirus-7 DNA sequence analysis between coordinates 9.5 and 31.[219,221] (A) The distance from the left end of Ad-7 is given in base pairs and in chromosome coordinate. Thin lines designate early mRNAs from region E2B of Ad-2;[5] thick lines show the positions of the promoters and leader segments for the late mRNAs for protein IX, IVa2, the i leader, the viral-associated (VA) RNA genes, and the 5′ end of the main body for late region 1 (52–55 kd).[194] (B) Vertical lines show the positions of stop codons in each potential translation frame. The first ATG in each major reading frame is shown with a ▼ and the first stop codon with a ●. Frames that are thought to encode proteins are shown as open or filled rectangles; those that are highly conserved between Ad-7 and Ad-2 and Ad-5[220,222,223] are shown as filled rectangles. The hatched rectangle shows the position of the presumed 13.6-kd polypeptide encoded by the i and third leaders, equivalent to that identified in Ad-2 by Virtanen et al.[24]

### C. Assembly of Translation Frames

RNA splicing has several additional ways of contributing to the assembly of open translation frames of messenger RNAs. Termination codons that would otherwise end polypeptide synthesis can be spliced out of the mature messenger as, for instance, in adenovirus early region 1A,[224-226] the IVa2 protein (Figure 8), the 33 kd protein,[227] and probably in the proteins encoded near the 3′ end of early region 4[4,164,222,228] as well as in the large T-antigen messages of SV40 and polyoma (Figure 2) and in the VP-1 and VP-2 messages of minute virus of the mouse (Figure 4). In many of these examples, the RNA deletion takes out 3n ± 1 nucleotides and effects a frameshift, allowing separate open translation frames to be joined.

The counterpart to joining initiation codons to open frames is splicing UGA, UAG, or UAA termination codons into upstream open translation frames to truncate peptide synthesis. In the polyoma virus small t-antigen, only four amino acids are encoded beyond the splice acceptor. The adenovirus i leader segment can be retained in a 1-2-i-3 leader assembly spliced to the 5′ end of various r-strand messages, particularly at early and intermediate times after infection (Figure 1). The i leader is now known to be a gene in itself, based on cell-free translations of RNA[14] and on DNA and RNA sequences.[24,219,222,223] Its 13.6-kd product would have been substantially longer without splicing. But, after splicing, it terminates a few amino acids into the third leader. In some rare RNAs derived from the late r-strand promoter, the i leader is spliced to "long" third leaders which include various amounts of RNA upstream from the normal splice acceptor at coordinate 26.5.[166,194] There are a couple of possibilities for fusion proteins — partly encoded by i, and partly by the different long third leaders. It is intriguing to speculate that the 13.6-kd protein plays an integral role in guiding the early → late transition in transcription.

The spectacular assembly of a message from large numbers of conserved coding sequences, as occurs with many nuclear genes such as ovalbumen[189] and collagen, is not seen in viruses. This, in part, is a reflection on their limited packaging capacity. Among the viral mRNAs, there is no known example where protein coding is contributed from more than two or three RNA segments, although they may have several additional, nontranslated leader segments. A number of AA*UAA*A sequences which serve as signals for 3′ end cleavage and polyadenylation also double as the utilized termination codon *UAA* in the gene they follow (e.g., in the adenovirus IVa2[220-223] and fiber[228] genes).

### D. Regulation of Gene Expression

When a primary transcript is spliced into a family of overlapping mRNAs encoding related proteins, the proteins can have important differences as well as functional domains in common. The E1A proteins are regulatory factors.[229,230] Genetic dissection of the adenovirus E1A coding regions suggests there are different properties for the peptide segments encoded by the sequences upstream and downstream of the splice and for those segments distinguished by the alternative donor splice sites.[224-226] The segment encoded by the 13S but not by the 12S RNA is much more hydrophobic than the common amino- and carboxy-terminal portions, and this region confers on its product the specific ability to regulate positively the E2 promoter and possibly other early promoters, an ability not shared by the 12S RNA.[191] Interestingly, at late times during infection, the predominant E1A RNA[4,231] lacks the middle portions unique to the 13S RNA and also those shared by the 13S and 12S RNAs upstream of the splice and encodes primarily the carboxy-terminal portion of the protein. Thus, differential RNA splicing is a candidate agent in the reversal of regulatory properties when early promoters yield to the late promoters.

The primary product of the major r-strand transcription unit at early and intermediate times after infection is the late family message (coordinates 30 to 39) for the 52- to 55-kd protein, the longer L1 message.[4,163,165] At later times, the 5′ leaders are spliced main bodies with coordinates 34 to 39 (IIIa protein mRNA) and coordinates 37 to 49 (penton mRNA). Thus there are two changes at late times: the synthesis of messages beyond late family 1 and the alteration in preferred splice sites within late family 1. This may be due to the induction of specific regulatory factors that select (or repress) certain sites in preference to others.

Another attribute of RNA splicing in regard to gene regulation is in permitting the same gene to be linked to alternative promoters that are differentially regulated. The delayed early promoter for the Ad-2 DNA replication proteins at coordinate 75 (Figure 1) depends on the product of the 13 S message of early region 1A.[191] A rather small amount of RNA and protein is made, sufficient to allow some DNA replication. Later, when replication is in full swing and more DBP, terminal protein, and polymerase are needed, a strong promoter at coordinate 72 (Figure 1) becomes turned on,[4,166] apparently due to changes in the relative amounts of the E1A regulatory proteins. Although the cap site and the 5′ end sequences of the RNA are totally different, this alternative leader is spliced to the same internal leader segment utilized at early times and, together, they are joined to the same coding segments. Hence, the DNA binding protein and other replication factors persist and dramatically increase at late times.[232,233]

## V. EFFECT OF HOST CELL DIFFERENTIATION ON VIRAL RNA SPLICING

A fascinating aspect of viral RNA processing is the close tie between viral development and the host cell type and degree of differentiation. Obviously every virus has its host range with respect to susceptible species and target tissues. But in certain cell culture systems, a virus can establish a latent infection of a stem cell and only gain the ability to be expressed when the cell undergoes differentiation towards its terminal state.

SV40 and polyoma do not express detectable levels of early cytoplasmic RNAs and proteins in undifferentiated murine teratocarcinoma stem cells,[234-236] even though the primary transcripts are synthesized and polyadenylated. There is a defect in RNA splicing.[237] Upon cell differentiation after retinoic acid treatment, the viruses emerge from their cryptic state to express the early products.[238] Vasseur et al.[239] characterized mutants of polyoma that have gained the ability to grow in undifferentiated embryonal carcinoma cells. They all affect a region recently shown to contain ''enhancer'' activity which stimulates early viral transcription.[69,73] It is possible that these mutants overproduce early RNAs in sufficient amount to overcome a deficiency splicing and to establish a productive infection.

Ectopic murine C-type retroviruses[240,241] and minute virus of the mouse,[242] which rely on splicing, also fail to grow in undifferentiated embryonal carcinoma cells and gain the ability to propagate in their differentiated derivatives. Conversely, Sindbis virus, encephalomyocarditis virus, and vaccinia virus, none of which exhibit RNA splicing, grow in the undifferentiated cells.[241]

In human epithelium, a single layer of basal epithelial cells gives rise to a progression of increasingly differentiated keratinocytes that are pushed upward in the epidermis until their nuclei resorb and they become metabolically inert. Many viruses infect this tissue. Herpes virus — with very few spliced RNAs — propagates well in basal cells.[243] Human adenoviruses on the other hand, can infect the basal layer but do not express detectible early antigens. Only after basal cell division do adenoviral antigens appear in the (rather immature) daughter keratinocytes.[243] The exact stage at which viral expression is blocked awaits characterization.

Papilloma viruses represent a fastidious extreme. They are harbored as free episomes in basal epithelium, but do not express detectible proteins.[244] Even in young keratinocytes — the extent of differentiation presently attainable in culture — the viruses are ostensibly inert. In epidermis *in situ*, papilloma viral proteins are made in the more fully differentiated keratinocytes. In these several examples, there appears to be a strong correlation between permissivity for viral expression and the degree of host cell differentiation. The intriguing possibility is that control of the viral transcription mimics regulation on certain host cell messages that depend on differentiation.

## VI. CONCLUSIONS

Viral RNA synthesis in infected eukaryotic cells is a complex and fascinating process. In the course of its elucidation, the remarkable diversity and economy of viral genomes and the many stages of RNA metabolism shared with most eukaryotic systems have emerged in considerable detail. The classical Beadle and Tatum dictum of ''one gene — one enzyme'' does not generally apply to most transcription units of eukaryotes and their viruses. These may be regarded as ''one gene—one family of proteins'' and also as ''multiple gene segments—one protein''. The implications of this fundamental difference between prokaryotes and eukaryotes are enormous. Changes in messenger RNA abundance and activity hint at an exquisite network of regulatory controls of gene expression imposed by both the host and the virus. This interplay will command a center of research attention in forthcoming years.

## ACKNOWLEDGMENTS

I thank Louise Chow for many valuable discussions and considerable assistance in the preparation of this manuscript. I also appreciate the communication of unpublished results from a number of colleagues. Research in our laboratory is supported by a National Cancer Institute Program Project Grant and an American Cancer Society Research Grant.

## REFERENCES

1. **Chow, L. T., Roberts, J. M., Lewis, J. B., and Broker, T. R.,** A map of cytoplasmic RNA transcripts from lytic adenovirus type 2, determined by electron microscopy of RNA:DNA hybrids, *Cell,* 11, 819, 1977.
2. **Chow, L. T., Gelinas, R. E., Broker, T. R., and Roberts, R. J.,** An amazing sequence arrangement at the 5′ ends of adenovirus-2 messenger RNA, *Cell,* 12, 1, 1977.
3. **Chow, L. T. and Broker, T. R.,** The spliced structures of adenovirus-2 fiber message and the other late mRNAs, *Cell,* 15, 497, 1978.
4. **Chow, L. T., Broker, T. R., and Lewis, J. B.,** Complex splicing patterns of RNAs from the early regions of adenovirus-2, *J. Mol. Biol.,* 134, 265, 1979.
5. **Stillman, B. W., Lewis, J. B., Chow, L. T., Matthews, M. B., and Smart, J. E.,** Identification of the gene and mRNA for the adenovirus terminal protein precursor, *Cell,* 23, 497, 1981.
6. **Evans, R. M., Fraser, N., Ziff, E., Weber, J., Wilson, M., and Darnell, J. E.,** The initiation sites for RNA transcription in Ad2 DNA, *Cell,* 12, 733, 1977.
7. **Berk, A. J. and Sharp, P. A.,** Ultraviolet mapping of the adenovirus 2 early promoters, *Cell,* 12, 45, 1977.
8. **Wilson, M. C., Fraser, N. W., and Darnell, J. E.,** Mapping of RNA initiation sites by high doses of UV irradiation: Evidence for three independent promoters within the left 11% of the Ad2 genome, *Virology,* 94 175, 1979.
9. **Baker, C. C. and Ziff, E. B.,** Promoters and heterogeneous 5′-termini of the messenger RNAs of adenovirus serotype 2, *J. Mol. Biol.,* 149, 189, 1981.
10. **Akusjärvi, G., Mathews, M. B., Anderson, P., Vennström, B., and Petterson, U.,** Structure of genes for virus-associated RNA I and RNA II of adenovirus type 2, *Proc. Natl. Acad. Sci. U.S.A.,* 77, 2424, 1980.

11. **Lewis, J. B., Atkins, J. F., Anderson, C. W., Baum, P. R., and Gesteland, R. F.,** Mapping of late adnovirus genes by cell-free translation of RNA selected by hybridization to specific DNA fragments, *Proc. Natl. Acad. Sci. U.S.A.*, 72, 1344, 1975.
12. **Lewis, J. B., Anderson, C. W., and Atkins, J. F.,** Further mapping of late adenovirus genes by cell-free translation of RNA selected by hybridization to specific DNA fragments, *Cell*, 12, 37, 1977.
13. **Harter, M. L. and Lewis, J. B.,** Adenovirus type 2 early proteins synthesized *in vitro* and *in vivo:* identification in infected cells of the 38,000- to 50,000-molecular-weight protein encoded by the left end of the adenovirus type 2 genome, *J. Virol.*, 26, 736, 1978.
14. **Lewis, J. B. and Mathews, M. B.,** Control of adenovirus early gene expression: a class of immediate early products, *Cell*, 21, 303, 1980.
15. **Ross, S. R., Flint, S. J., and Levine, A. J.,** Identification of the adenovirus early proteins and their genomic map positions, *Virology*, 100, 419, 1980.
16. **Esche, H., Mathews, M. B., and Lewis, J. B.,** Proteins and messenger RNAs of the transforming region of wildtype and mutant adenoviruses, *J. Mol. Biol.*, 142, 399, 1980.
17. **Pettersson, U. and Mathews, M. B.,** The gene and messenger RNA for adenovirus polypeptide IX, *Cell*, 12, 741, 1977.
18. **Green, M., Wold, W. S. M., Brackmann, K. H., and Cartas, M. A.,** Identification of families of overlapping polypeptides coded by early "transforming" gene region 1 of human adenovirus type 2, *Virology*, 97, 275, 1979.
19. **Spector, D. J., Crossland, L. D., Halbert, D. N., and Raskas, H. J.,** A 28k polypeptide is the translation product of a 9 S RNA encoded by region 1A of adenovirus 2, *Virology*, 102, 218, 1980.
20. **van der Eb, A. J., van Ormondt, H., Schrier, P. I., Lupker, J. H., Jochemsen, H., van den Elsen, P. J., DeLeys, R. J., Maat, J., van Beveren, C. P., Dijkema, R., and deWaard, A.,** Structure and function of the transforming genes of human adenoviruses and SV40, *Cold Spring Harbor Symp. Quant. Biol.*, 44, 383, 1979.
21. **Miller, J. S., Ricciardi, R. P., Roberts, B. E., Paterson, B. M., and Mathews, M. B.,** Arrangement of messenger RNAs and protein coding sequences in the major late transcription unit of adenovirus-2, *J. Mol. Biol.*, 142, 455, 1980.
22. **Kruijer, W., van Schaik, F. M. A., and Sussenbach, J. S.,** Nucleotide sequence analysis of a region of adenovirus 5 DNA encoding a hitherto unidentified gene, *Nucleic Acids Res.*, 8, 6033, 1980.
23. **Akusjärvi, G., Zabielski, J., Perricaudet, M., and Pettersson, U.,** The sequence of the 3′ noncoding region of the hexon mRNA discloses a novel adenovirus gene, *Nucleic Acids Res.*, 9, 1, 1981.
24. **Virtanen, A., Aleström, P., Persson, H., Katze, M. G., and Pettersson, U.,** An adenovirus agnogene, *Nucleic Acids Res.*, 10, 2539, 1982.
25. **Ishibashi, M. and Maizel, J. V., Jr.,** The polypeptides of adenovirus. V. Young virions, structural intermediates between top components and aged virions, *Virology*, 57, 409, 1974.
26. **Kitchingman, G. R., Lai, S.-P., and Westphal, H.,** Loop structures in hybrids of early RNA and the separated strands of adenovirus DNA, *Proc. Natl. Acad. Sci. U.S.A.*, 74, 4392, 1977.
27. **Berk, A. J., and Sharp, P. A.,** Structure of the adenovirus 2 early mRNAs, *Cell*, 14, 695, 1978.
28. **Ziff, E. B. and Evans, R. M.,** Coincidence of the promoter and capped 5′ terminus of RNA from the adenovirus 2 major late transcription unit, *Cell*, 15, 1463, 1978.
29. **McGrogan, M. and Raskas, H. J.,** Two regions of the adenovirus-2 genome specify families of late polysomal RNAs containing common sequences, *Proc. Natl. Acad. Sci. U.S.A.*, 75, 625, 1978.
30. **Nevins, J. R. and Darnell, J. E.,** Groups of adenovirus type 2 mRNAs derived from a large primary transcript: probable nuclear origin and possible common 3′ ends, *J. Virol.*, 25; 811, 1978.
31. **Akusjärvi, G. and Petterson, U.,** Sequence analysis of adenovirus DNA. IV. The genomic sequences encoding the common tripartite leader of late adenovirus messenger RNA, *J. Mol. Biol.*, 134, 143, 1979.
32. **Zain, S., Sambrook, J., Roberts, R. J., Keller, W., Fried, M., and Dunn, A. R.,** Nucleotide sequence analysis of the leader segments in a cloned copy of the adenovirus 2 fiber mRNA, *Cell*, 16, 851, 1979.
33. **Berget, S. M., Moore, C. C., and Sharp, P. A.,** Spliced segments at the 5′ terminus of adenovirus-2 late mRNA, *Proc. Natl. Acad. Sci. U.S.A.*, 74, 3171, 1977.
34. **Klessig, D. F.,** Two adenovirus mRNAs have a common 5′ terminal leader sequence encoded at least 10 kb upstream from their main coding regions, *Cell*, 12, 9, 1977.
35. **Price, R. and Penman, S.,** Transcription of the adenovirus genome by an alpha-amanitine-sensitive ribonucleic acid polymerase in HeLa cells, *J. Virol.*, 9, 621, 1972.
36. **Wallace, R. D. and Kates, J.,** State of adenovirus 2 deoxyribonucleic acid in the nucleus and its mode of transcription: Studies with isolated viral deoxyribonucleic acid-protein complexes and isolated nuclei, *J. Virol.*, 9, 627, 1972.
37. **Perry, R. P. and Kelley, D. E.,** Kinetics of formation of 5′-terminal caps in mRNA, *Cell*, 8, 433, 1976.
38. **Moss, B. and Koczot, F.,** Sequence of methylated nucleotides at the 5′-terminus of adenovirus-specific RNA, *J. Virol.*, 17, 385, 1976.

39. **Sommer, S., Salditt-Georgieff, M., Bachenheimer, S., Darnell, J. E., Furuichi, Y., Morgan, M., and Shatkin, A. J.,** The methylation of adenovirus-specific nuclear and cytoplasmic RNA, *Nucleic Acids Res.*, 3, 749, 1976.
40. **Wold, W. S. M., Green, M., and Munns, T. W.,** Methylation of late adenovirus 2 nuclear and messenger RNA, *Biochem. Biophys. Res. Commun.*, 68, 643, 1976.
41. **Gelinas, R. E. and Roberts, R. J.,** One predominant 5′-undecanucleotide in adenovirus 2 late messenger RNAs, *Cell*, 11, 533, 1977.
42. **Philipson, L., Wall, R., Glickman, G., and Darnell, J. E.,** Addition of polyadenylated sequences to virus-specific RNA during adenovirus replication, *Proc. Natl. Acad. Sci., U.S.A.*, 68, 2806, 1971.
43. **Aleström, P., Akusjärvi, G., Perricaudet, M., Mathews, M. B., Klessig, D. F., and Pettersson, U.,** The gene for polypeptide IX of adenovirus type 2 and its unspliced messenger RNA, *Cell*, 19, 671, 1980.
44. **Chow, L. T. and Broker, T. R.,** The elucidation of RNA splicing in the adenoviral system, in *Gene Structure and Expression*, Dean, D. H., Johnson, L. F., Kimball, P. C., and Perlman, P. S., Eds., Ohio State University Press, Columbus, 1980, 175.
45. **Ziff, E. B.,** Transcription and RNA processing by the DNA tumour viruses, *Nature (London)*, 287, 491, 1980.
46. **Nevins, J. R. and Chen-Kiang, S.,** Processing of adenovirus nuclear RNA to mRNA, in *Advances in Virus Research*, Vol. 26, Academic Press, New York, 1981, 1.
47. **Darnell, J. E., Jr.,** Variety in the level of gene control in eukaryotic cells, *Nature (London)*, 297, 365, 1982.
48. **Flint, S. J.,** Expression of adenoviral genetic information in productively infected cells, *Biochim. Biophys. Acta*, 651, 175, 1982.
49. **Tooze, J., Ed.,** *DNA Tumor Viruses, Molecular Biology of Tumor Viruses*, 2nd ed., Part 2, (rev.), Cold Spring Harbor Laboratory, Cold Spring Harbor, New York, 1981, 1073pp.
50. **Soeda, E., Arrand, J. R., Smolar, N., Walsh, and Griffin, B. E.,** Coding potential and regulatory signals of the polyoma virus genome, *Nature (London)*, 283, 445, 1980.
51. **Treisman, R.,** Characterisation of polyoma late mRNA leader sequences by molecular cloning and DNA sequence analysis, *Nucleic Acids Res.*, 8, 4867, 1980.
52. **Treisman, R., Cowie, A., Favaloro, J., Jat, P., and Kamen, R.,** The structures of the spliced mRNAs encoding polyoma virus early region proteins, *J. Mol. Appl. Genet.*, 1, 83, 1981.
53. **Cowie, A., Tyndall, C., and Kamen, R.,** Sequences at the capped 5′ -ends of polyoma virus late region mRNAs: an example of extreme terminal heterogeneity, *Nucleic Acids Res.*, 9, 6305, 1981.
54. **Smart, J. E. and Ito, Y.,** Three species of polyoma virus tumor antigens share common peptides probably near the amino termini of the proteins, *Cell*, 15, 1427, 1978.
55. **Reddy, V. V., Thimmappaya, B., Dhar, R., Subramanian, K. N., Zain, B. S., Pan, J., Ghosh, P. K., Celma, M. L., and Weissman, S. M.,** The genome of simian virus 40, *Science*, 200, 494, 1978.
56. **Fiers, W., Contreras, R., Haegeman, G., Rogiers, R., van de Voorde, A., van Herreweghe, J., Volckaert, G., and Ysebaert, M.,** Complete nucleotide sequence of SV40 DNA, *Nature (London)*, 273, 113, 1978.
57. **Yang, R. C. A. and Wu, R.,** BK virus DNA: Complete nucleotide sequence of a human tumor virus, *Science*, 206, 456, 1979.
58. **Seif, I., Khoury, G., and Dhar, R.** The genome of human papovavirus BKV, *Cell*, 18, 963, 1979.
59. **Aloni, Y., R. Dhar, O. Laub, M. Horowitz, and G. Khoury,** Novel mechanism for RNA maturation: The leader sequences of simian virus 40 mRNA are not transcribed adjacent to the coding sequences, *Proc. Natl. Acad. Sci. U.S.A.*, 74, 3686, 1977.
60. **Hsu, M.-T. and Ford, J.,** Sequence arrangement of the 5′ ends of simian virus 40 16 S and 19 S mRNAs, *Proc. Natl. Acad. Sci. U.S.A.* 74, 4982, 1977.
61. **Lavi, S. and Groner, Y.,** 5′-Terminal sequences and coding region of late simian virus 40 mRNAs are derived from noncontiguous segments of the viral genome, *Proc. Natl. Acad. Sci.*, 74, 5323, 1977.
62. **Bratosin, S., Horowitz, M., Laub, O., and Aloni, Y.,** Electron microscropic evidence for splicing of SV40 late mRNAs, *Cell*, 13, 783, 1978.
63. **Berk, A. J. and Sharp, P. A.,** Spliced early mRNAs of simian virus 40, *Proc. Natl. Acad. Sci. U.S.A.*, 75, 1274, 1978.
64. **Haegeman, G. and Fiers, W.,** Evidence for "splicing" of SV40 16S mRNA, *Nature (London)*, 273, 70, 1978.
65. **Lai, C.-J., Dhar, R., and Khoury, G.,** Mapping the spliced and unspliced late lytic SV40 RNAs, *Cell*, 14, 971, 1978.
66. **Reddy, V. B., Ghosh, P. K., Lebowitz, P., Piatak, M., and Weissman, S. M.,** Simian virus 40 early mRNAs. I. Genome localization of 3′ and 5′ termini and two major splices in mRNA from transformed and lytically infected cells, *J. Virol.*, 30, 279, 1979.
67. **Manaker, R. A., Khoury, G., and Lai, C.-J.,** The spliced structure of BK virus mRNAs in lytically infected and transformed cells, *Virology*, 97, 112, 1979.

68. **Ghosh, P. K., Lebowitz, P., Frisque, R., and Gluzman, Y.,** Identification of a promoter component involved in positioning the 5′ termini of simian virus 40 early mRNAs, *Proc. Natl. Acad. Sci. U.S.A.*, 78, 100, 1981.
69. **Benoist, C. and Chambon, P.,** In vivo sequence requirements of the SV40 early promoter region, *Nature (London)*, 290, 304, 1981.
70. **Gruss, P., Dhar, R., and Khoury, G.,** Simian virus 40 tandem repeated sequences as an element of the early promoter, *Proc. Natl. Acad. Sci. U.S.A.*, 78, 943, 1981.
71. **Banerji, J., Rusconi, S., and Schaffner, W.,** Expression of a beta-globin gene is enhanced by remote SV40 DNA sequences, *Cell*, 27, 299, 1981.
72. **de Villiers, J. and Schaffner, W.,** A small segment of polyoma virus DNA enhances the expression of a cloned beta-globin gene over a distance of 1400 base pairs, *Nucleic Acids Res.*, 9, 6251, 1981.
73. **Tyndall, C., La Mantia, G., Thacker, C. M., Favaloro, J., and Kamen, R.,** A region of the polyoma virus genome between the replication origin and late protein coding sequences is required in *cis* for both early gene expression and viral DNA replication, *Nucleic Acids Res.*, 9, 6231, 1981.
74. **Ghosh, P. K., Reddy, V. V., Swinscoe, J., Lebowitz, P., and Weissman, S. M.,** Heterogeneity and 5′-terminial structures of the late RNAs of simian virus 40, *J. Mol. Biol.*, 126, 813, 1978.
75. **Haegeman, G. and Fiers, W.,** Characterization of the 5′-terminal capped structures of late simian virus 40-specific mRNA, *J. Virol.*, 25, 824, 1978.
76. **Cowie, A., Jat, P., and Kamen, R.,** Determination of sequences at the capped 5′ ends of polyoma virus early region transcripts synthesized in vivo and in vitro demonstrates an unusual microheterogeneity, *J. Mol. Biol.*, 159, 225, 1982.
77. **Canaani, D., Kahana, C., Mukamel, A., and Groner, Y.,** Sequence heterogeneity at the 5′-termini of late simian virus 40 19 S and 16 S mRNAs, *Proc. Natl. Acad. Sci.*, 76, 3078, 1979.
78. **Rosenthal, L. J., Salomon, C., and Weil, R.,** Isolation and characterization of poly(A)-containing intranuclear polyoma-specific "giant" RNAs, *Nucleic Acids Res.*, 3, 1167, 1976.
79. **Acheson, N. H.,** Polyoma giant RNAs contain tandem repeats of the nucleotide sequence of the entire viral genome, *Proc. Natl. Acad. Sci.*, 75, 4754, 1978.
80. **Legon, S., Flavell, A. J., Cowie, A., and Kamen, R.,** Amplification in the leader sequence of late polyoma virus mRNAs, *Cell*, 16, 373, 1979.
81. **Siddell, S. G. and Smith, A. E.,** Polyoma virus has three late mRNAs: One for each virion protein, *J. Virol.*, 27, 427, 1978.
82. **Khoury, G., Gruss, P., Dhar, R., and Lai, C.-J.,** Processing and expression of early SV40 mRNA: a role for RNA conformation in splicing, *Cell*, 18, 85, 1979.
83. **Kamen, R., Jat, P., Treisman, R., Favaloro, J., and Folk, W. R.,** 5′ Termini of polyoma virus early region transcripts synthesized *in vivo* by wild-type virus and viable deletion mutants, *J. Mol. Biol.*, 159, 189, 1982.
84. **Piatak, M., Subramanian, K. N., Roy, P., and Weissman, S. M.,** Late messenger RNA production by viable simian virus 40 mutants with deletions in the leader region, *J. Mol. Biol.*, 153, 589, 1981.
85. **Gruss, P., Lai, C.-J., Dhar, R., and Khoury, G.,** Splicing as a requirement for biogenesis of functional 16S mRNA of simian virus 40, *Proc. Natl. Acad. Sci. U.S.A.*, 76, 4317, 1979.
86. **Hamer, D. H. and Leder, P.,** Splicing and the formation of stable RNA, *Cell*, 18, 1299, 1979.
87. **Gruss, P. and Khoury, G.** Rescue of a defective splicing mutant by insertion of an heterologous intron, *Nature (London)*, 286, 634, 1980.
88. **Ghosh, P. K., Roy, P., Barken, A., Mertz, J. E., Weissman, S. M., and Lebowitz, P.,** Unspliced functional late 19S mRNAs containing intervening sequences are produced by a late leader mutant of simian virus 40, *Biochemistry*, 78, 1386, 1981.
89. **Danos, O., Katinka, M., and Yaniv, M.,** Human papillomavirus 1a complete DNA sequence: a novel type of genome organization among papovaviridae, *EMBO J.*, 1, 231, 1982.
90. **Chen, E. Y., Howley, P. M., Levinson, A.D., and Seeburg, P. H.,** The primary structure and genetic organization of the bovine papillomavirus (BPV) type I genome, *Nature (London)*, 299, 529, 1982.
91. **Heilman, C. A., Engel, L., Lowy, D. R., and Howley, P. M.,** Virus-specific transcription in bovine papillomavirus-transformed mouse cells, *Virology*, 119, 22, 1982.
92. **Chow, L. T. and Broker, T. R.,** Human papilloma virus type 1 RNA transcription and processing in COS-1 cells, in *Gene Transfer and Cancer*, Eds., Pearson, M. L. and Steinberg, N. L., Raven Press, New York, in press.
93. **Gluzman, Y.,** SV40-transformed simian cells support the replication of early SV40 mutants, *Cell*, 23, 175, 1981.
94. **Astell, C. R., Thomson, M., Chow, M. B., and Ward, D. C.,** Structure and replication of minute virus of mice DNA, *Cold Spring Harbor Symp. Quant. Biol.*, 47, 751, 1982.

95. **Tattersall, P., Shatkin, A. J., and Ward, D. C.,** Sequence homology between the structural polypeptides of minute virus of mouse, *J. Mol. Biol.*, 111, 375, 1977.
96. **Laughlin, C. A., Westphal, H., and Carter, B. J.,** Spliced adenovirus-associated virus RNA, *Proc. Natl. Acad. Sci. U.S.A.*, 76, 5567, 1979.
97. **Jay, F. T., Laughlin, C. A., and Carter, B. J.,** Eukaryotic translational control: Adeno-associated virus protein synthesis is affected by a mutation in the adenovirus DNA-binding protein, *Proc. Natl. Acad. Sci. U.S.A.*, 78, 2927, 1981.
98. **Wagner, E. K.,** Transcription patterns in HSV infections, *Adv. Viral Oncol.*, 3, 239, 1983.
99. **Watson, R. J., Sullivan, M., and Vande Woude, G. F.,** Structures of two spliced herpes simplex virus type 1 immediate-early mRNAs which map at the junctions of the unique and reiterated regions of the virus DNA S component, *J. Virol.*, 37, 431, 1981.
100. **Rixon, F. J. and Clements, J. B.,** Detailed structural analysis of two spliced HSV-1 immediate-early mRNAs, *Nucleic Acids Res.*, 10, 2241, 1982.
101. **Frink, R. J., Anderson, K. P.,; and Wagner, E. K.,** Herpes simplex virus type 1 *Hin*dIII fragment L encodes spliced and complementary mRNA species, *J. Virol.*, 39, 559, 1981.
102. **Baroudy, B. M., Venkatesan, S., and Moss, B.,** Incompletely base-paired flip-flop terminal loops link the two DNA strands of the vaccinia virus genome into one uninterrupted polynucleotide chain, *Cell*, 28, 315, 1982.
103. **Oda, K. and Joklik, W. K.,** Hybridization and sedimentation studies on "early" and "late" vaccinia messenger RNA, *J. Mol. Biol.*, 27, 395, 1967.
104. **Munyon, W., Paoletti, E., and Grace, J. T., Jr.,** RNA polymerase activity in purified infectious vaccinia virus, *Proc. Natl. Acad. Sci. U.S.A.*, 58, 2280, 1967.
105. **Kates, J. R. and McAuslan, B. R.,** Poxvirus DNA-dependent RNA polymerase, *Proc. Natl. Acad. Sci. U.S.A.*, 58, 134, 1967.
106. **Nevins, J. R. and Joklik, W.,** Poly(A) sequences of vaccinia virus messenger RNA: nature, model of addition and function during translation *in vitro* and *in vivo*, *Virology*, 63, 1, 1975.
107. **Martin, S. A., Paoletti, E., and Moss, B.,** Purification of mRNA guanylyltransferase and mRNA (guanine-7-) methyltransferase from vaccinia virions, *J. Biol. Chem.*, 250, 9322, 1975.
108. **Kates, J. R. and Beeson, J.,** Ribonucleic acid synthesis in vaccinia virus. II. Synthesis of polyriboadenylic acid, *J. Mol. Biol.*, 50, 19, 1970.
108a. **Moss, B. Rosenblum, E. N., and Gershowitz, A.,** Characterization of a polyriboadenylate polymerase from Vaccinia virions, *J. Biol. Chem.*, 250, 4722, 1975.
109. **Paoletti, E. and Lipinskas, B. R.,** Soluble endoribonuclease activity from vaccinia virus: specific cleavage of virion-associated high-molecular-weight RNA, *J. Virol.*, 26, 822, 1978.
110. **Spencer, E., Loring, D., Hurwitz, J., and Monroy, G.,** Enzymatic conversion of 5′-di- and triphosphate-terminated RNA, *Proc. Natl. Acad. Sci. U.S.A.*, 75, 4793, 1978.
110a. **Venkatesan, S., and Moss, B.,** *In vitro* transcription of the inverted terminal repetition of the Vaccinia virus genome: Correspondence of initiation and cap sites, *J. Virol.*, 37, 738, 1981.
111. **Proudfoot, J. J. and Brownlee, G. G.,** 3′ Noncoding region sequence in eukaryotic mRNA, *Nature (London)*, 263, 211, 1976.
112. **Venkatesan, S., Baroudy, B. M., and Moss, B.,** Distinctive nucleotide sequences adjacent to multiple initiation and termination sites of an early vaccinia virus gene, *Cell*, 125, 805, 1981.
113. **Cooper, J. A., Wittek, R., and Moss, B.,** Hybridization-selection and cell-free translation of mRNA's encoded within the inverted terminal repetition of the vaccinia virus genome, *J. Virol.*, 37, 284, 1981.
114. **Wittek, R. and Moss, B.,** Colinearity of RNAs with the vaccinia virus genome: Anomalies with two complementary early and late RNAs result from a small deletion or rearrangement within the inverted terminal repetition, *J. Virol.*, 42, 447, 1982.
115. **Krug, R. M., Broni, B. A., and Bouloy, M.,** Are the 5′ ends of influenza viral mRNAs synthesized *in vivo* donated by host mRNAs? *Cell*, 18, 329, 1979.
116. **Plotch, S. J., Bouloy, M., and Krug, R. M.,** Transfer of 5′-terminal cap of globin mRNA to influenza viral complementary RNA during transcription, *in vitro*, *Proc. Natl. Acad. Sci. U.S.A.*, 76, 1618, 1979.
117. **Bouloy, M., Morgan, M. A., Shatkin, A. J., and Krug, R. M.,** Cap and internal nucleotides of reovirus mRNA primers are incorporated into influenza viral complementary RNA during transcription in vitro, *J. Virol.*, 32, 895, 1979.
118. **Allen, H., McCauley, J., Waterfield, M., and Gething, M.-J.,** Influenza virus RNA segment 7 has the coding capacity for two polypeptides, *Virology*, 107, 548, 1980.
119. **Lamb, R. A. and Choppin, P. W.,** Identification of a second protein ($M_2$) encoded by RNA segment 7 of influenza virus, *Virology*, 112, 729, 1981.
120. **Lamb, R. A., Lai, C.-J., and Choppin, P. W.,** Sequences of mRNAs derived from genome RNA segment 7 of influenza virus: Collinear and interrupted mRNAs code for overlapping proteins, *Proc. Natl. Acad. Sci. U.S.A.*, 78, 4170, 1981.

121. **Winter, G. and Fields, S.,** Cloning of influenza DNA into M13: the sequence of the RNA segment encoding the A/PR/8/34 matrix protein, *Nucleic Acids Res.*, 8, 1965, 1980.
122. **Lamb, R. A. and Lai, C.-J.,** Conservation of the influenza virus membrane protein ($M_1$) amino acid sequence and an open reading frame of RNA segment 7 encoding a second protein ($M_2$) in H1N1 and H3N2 strains, *Virology*, 112, 746, 1981.
123. **Lamb, R. A., Choppin, P. W., Chanock, R. M., and Lai, C.-J.,** Mapping of the two overlapping genes for polypeptides $NS_1$ and $NS_2$ on RNA segment 8 of influenza virus genome, *Proc. Natl. Acad. Sci. U.S.A.*, 77, 1857, 1980.
124. **Inglis, S. C., Barrett, T., Brown, C. M., and Almond, J. W.,** The smallest genome RNA segment of influenza virus contains two genes that may overlap, *Proc. Natl. Acad. Sci. U.S.A.*, 76, 3790, 1979.
125. **Lamb, R. A., and Lai, C.-J.,** Sequence of interrupted and uninterrupted mRNAs and cloned DNA coding for the two overlapping nonstructural proteins of influenza virus, *Cell*, 21, 475, 1980.
126. **Baltimore, D., Huang, A. S., and Stampfer, M.,** Ribonucleic acid synthesis of vesicular stomatitis virus, II. An RNA polymerase in the virion, *Proc. Natl. Acad. Sci. U.S.A.*, 66, 572, 1970.
127. **Ball, L. A. and Wertz, G. W.,** VSV RNA synthesis: How can you be positive? *Cell*, 26, 143, 1981.
128. **Ball, L. A. and White, C. N.,** Order of transcription of genes of vesicular stomatitis virus, *Proc. Natl. Acad. Sci. U.S.A.*, 73, 442, 1976.
129. **Abraham, G. and Banerjee, A. K.,** Sequential transcription of the genes of vesicular stomatitis virus, *Proc. Natl. Acad. Sci. U.S.A.*, 73, 1504, 1976.
130. **Villarreal, L. P., Breindle, M., and Holland, J. J.,** Determination of molar ratios of vesicular stomatitis virus induced RNA species in BHK cells, *Biochemistry.*, 15, 1663, 1976.
131. **Testa, D., Chanda, P. K., and Banerjee, A. K.,** Unique mode of transcription *in vitro* by vesicular stomatitis virus, *Cell*, 21, 267, 1980.
132. **Iverson, L. E. and Rose, J. K.,** Localized attenuation and discontinuous synthesis during vesicular stomatitis virus transcription, *Cell*, 23, 477, 1981.
133. **Lazzarini, R. A., Chien, I., Yang, F. M., and Keene, J. D.,** The metabolic fate of independently initiated VSV mRNA transcripts, *J. Gen. Virol.*, 58, 429, 1982.
134. **Chanda, P. K. and Banerjee, A. K.,** Identification of promoter-proximal oligonucleotides and a unique dinucleotide, pppGpC, from *in vitro* transcription products of vesicular stomatitis virus, *J. Virol.*, 39, 93, 1981.
135. **Pinney, D. F. and Emerson, S. U.,** Identification and characterization of a group of discrete initiated oligonucleotides transcribed *in vitro* from the 3′ terminus of the *N*-gene of vesicular stomatitis virus, *J. Virol.*, 42, 889, 1982.
136. **McGeoch, D. J.,** Structure of the gene N: gene NS intercistonic junction in the genome, *Cell*, 17, 673, 1979.
137. **Rose, J. K.,** Complete intergenic and flanking gene sequences from the genome of vesicular stomatitis virus, *Cell*, 19, 415, 1980.
138. **Herman, R. C., Adler, S., Lazzarini, R. A., Colonno, R. J., Banerjee, A. K., and Westphal,** Intervening polyadenylated sequences in RNA transcripts of vesicular stomatitis virus, *Cell*, 15, 587, 1978.
139. **Herman, R. C., Schubert, M., Keene, J. D., and Lazzarini, R. A.,** Polycistronic vesicular stomatitis virus RNA transcripts, *Proc. Natl. Acad. Sci. U.S.A.*, 77, 4662, 1980.
140. **Weiss, R., Teich, N., Varmus, H., and Coffin, J., Eds.,** *RNA Tumor Viruses, Molecular Biology of Tumor Viruses*, 2nd ed., Part 1), Cold Spring Harbor Laboratory, Cold Spring Harbor, New York, 1982, 1396 pp.
141. **Klempnauer, K.-H., Gonda, T. J., and Bishop, J. M.,** Nucleotide sequence of the retrovirus leukemia gene V-*myb* and its cellular progenitor C-*myb*: the architecture of a transduced oncogene, *Cell*, 31, 453, 1982.
142. **Shimotohno, K. and Temin, H. M.,** Loss of intervening sequences in genomic mouse alpha-globin DNA inserted in an infectious retrovirus vector, *Nature (London)*, 299, 265, 1982.
143. **Sorge, J. and Hughes, S. H.,** The splicing of intervening sequences introduced into an infectious retroviral vector, *J. Mol. Appl. Genet.*, 1, 547, 1982.
144. **Blanchard, J.-M., Weber, J., Jelinek, W., and Darnell, J. E.,** *In vitro* RNA-RNA splicing in adenovirus 2 mRNA formation, *Proc. Natl. Acad. Sci. U.S.A.*, 75, 5344, 1978.
145. **Yang, V. W., Lerner, M- R., Steitz, J. A., and Flint, S J.,** A small nuclear ribonucleoprotein is required for splicing of adenoviral early RNA sequences, *Proc. Natl. Acad. Sci. U.S.A.*, 78, 1371, 1981.
146. **Manley, J. L., Sharp, P. A., and Gefter, M. L.,** RNA synthesis in isolated nuclei, Processing of adenovirus serotype 2 late messenger RNA precursors, *J. Mol. Biol.*, 159, 581, 1982.
147. **Weingärtner, B. and Keller, W.,** Transcription and processing of adenoviral RNA by extracts from HeLa cells, *Proc. Natl. Acad. Sci. U.S.A.*, 78, 4092, 1981.

148. **Goldenberg, C. J., Addenberg, C. T., and Hauser, S. D.,** Accurate and efficient *in vitro* splicing of purified precursor RNAs specified by early region 2 of the adenovirus 2 genome, *Nucleic Acids Res.*, 11, 1337, 1983.
149. **Beyer, A. L., Bouton, A. H., Hodge, L. D., and Miller, O. L., Jr.,** Visualization of the major late R strand transcription unit of adenovirus serotype 2, *J. Mol. Biol.*, 147, 269, 1981.
150. **Berget, S. M. and Sharp, P. A.,** Structure of late adenovirus 2 heterogeneous nuclear RNA, *J. Mol. Biol.*, 129, 547, 1979.
151. **Keohavong, P., Gattoni, R., LeMoullec, J. M., Jacob, M., and Stevenin, J.,** The orderly splicing of the first three leaders of the adenovirus-2 major late transcript, *Nucleic Acids Res.*, 10, 1215, 1982.
152. **Weber, J., Blanchard, J.-M., Ginsberg, H., and Darnell, J. E.,** Jr., Order of polyadenylic acid addition and splicing events in early adenovirus mRNA formation, *J. Virol.*, 33, 286, 1980.
153. **Bina, M., Feldman, R. J., and Deeley, R. G.,** Could poly(A) align the splicing sites of messenger RNA precursors? *Proc. Natl. Acad. Sci. U.S.A.*, 77, 1278, 1980.
154. **Zeevi, M., Nevins, J. R., and Darnell, J. E., Jr.** Nuclear RNA is spliced in the absence of poly(A) addition, *Cell*, 26, 39, 1981.
155. **Zeevi, M. Nevins, J. R., and Darnell, J. E., Jr.** Newly formed mRNA lacking polyadenylic acid enters the cytoplasm and the polyribosomes but has a shorter half-life in the absence of polyadenylic acid, *Mol. Cell. Biol.*, 2, 517, 1982.
156. **Nevins, J. R.,** Poly(A) in eukaryotic mRNA, in *RNA Processing*, Apirion, D., Ed., CRC Press, Boca Raton, Fla., 1983, chap. 6.
157. **Fitzgerald, M. and Shenk, T.,** The sequence 5′-AAUAAA-3′ forms part of the recognition site for polyadenylation of late SV40 mRNAs, *Cell*, 24, 251, 1981.
158. **Ford, J. P. and Hsu, M.-T.,** Transcription pattern of *in vivo*-labeled late simian virus 40 RNA: Equimolar transcription beyond the mRNA 3′ terminus, *J. Virol.*, 28, 795, 1978.
159. **Nevins, J. R., Blanchard, J.-M., and Darnell, J. E., Jr.,** Transcription units of adenovirus type 2: Termination of transcription beyond the poly(A) addition site in early regions 2 and 4, *J. Mol. Biol.*, 144, 377, 1980.
160. **Fraser, N. W., Nevins, J. R., Ziff, E., and Darnell, J. E., Jr.,** The major late adenovirus type-2 transcription unit: termination is downstream from the last poly(A) site, *J. Mol. Biol.*, 129, 643, 1979.
161. **Fraser, N. W., Baker, C. C., Moore, M. A., and Ziff, E. B.,** Poly(A) sites of adenovirus serotype 2 transcription units, *J. Mol. Biol.*, 155, 207, 1982.
162. **Nevins, J. R. and Darnell, J. E., Jr.,** Steps in the processing of Ad2 mRNA: poly(A)$^+$ nuclear sequences are conserved and poly(A) addition precedes splicing, *Cell*, 15, 1477, 1978.
163. **Shaw, A. R. and Ziff, E. B.,** Transcripts from the adenovirus-2 major late promoter yield a single family of 3′ coterminal mRNAs during early infection and five families at late times, *Cell*, 22, 905, 1980.
164. **Kitchingman, G. R. and Westphal, H.,** The structure of adenovirus 2 early nuclear and cytoplasmic RNAs, *J. Mol. Biol.*, 137, 23, 1980.
165. **Akusjärvi, G. and Persson, H.,** Controls of RNA splicing and termination in the major late adenovirus transcription unit, *Nature (London)*, 292, 420, 1981.
166. **Chow, L. T., Lewis, J. B., and Broker, T. R.,** RNA transcription and splicing at early and intermediate times after adenovirus-2 infection, *Cold Spring Harbor Symp. Quant. Biol.*, 44, 401, 1979.
167. **Nevins, J. R. and Wilson, M. C.,** Regulation of adenovirus gene expression at the level of transcriptional termination and RNA processing, *Nature (London)*, 290, 113, 1981.
168. **Hérissé, J., Courtois, G., and Galibert, F.,** Nucleotide sequence of the *Eco*RI D fragment of adenovirus 2 genome, *Nucleic Acids Res.*, 8, 2173, 1980.
169. **Hérissé, J. and Galibert, F.,** Nucleotide sequence of the *Eco*RI E fragment of adenovirus-2 genome, *Nucleic Acids Res.*, 9, 1229, 1981.
170. **Ahmed, C. M. I., Chanda, R., Stow, N., and Zain, S.,** The sequence of 3′ termini of mRNAs from early region III of Adenovirus 2, *Gene*, 19, 297, 1982.
171. **Furuichi, Y., Morgan, M., Muthukrishnan, S., and Shatkin, A. J.,** Reovirus messenger RNA contains a methylated, blocked 5′-terminal structure: $m^7G(5')ppp(5')G^mpCp$-, *Proc. Natl. Acad. Sci. U.S.A.*, 72, 362, 1975.
172. **Wei, C. W. and Moss, B.,** Methylated nucleotides block 5′-terminus of vaccinia virus messenger RNA, *Proc. Natl. Acad. Sci. U.S.A.*, 72, 318, 1975.
173. **Abraham, G., Rhodes, D. P., and Banerjee, A. K.,** The 5′ terminal structure of the methylated mRNA synthesized in vitro by vesicular stomatitis virus, *Cell*, 5, 51, 1975.
174. **Lavi, S. and Shatkin, A. J.,** Methylated simian virus 40-specific RNA from nuclei and cytoplasm of infected BSC-1 cells, *Proc. Natl. Acad. Sci. U.S.A.*, 72, 2012, 1975.
175. **Furuichi, Y. and Miura, K. I.,** A blocked structure at the 5′ terminus of mRNA from cytoplasmic polyhedrosis virus, *Nature (London)*, 253, 374, 1975.

176. **Urichibara, T., Furuichi, Y., Nishimura, C., and Kiura, K.,** A modified structure at the 5′-terminus of mRNA of vaccinia virus, *FEBS Lett.*, 49, 385, 1975.
177. **Maden, B. E. H. and Salim, M.,** The methylated nucleotide sequences in HeLa cell ribosomal RNA and its presursors, *J. Mol. Biol.*, 88, 133, 1974.
178. **Wei, C. W. and Moss, B.,** Nucleotide sequences at the $N^6$-methyladenosine sites of HeLa cell messenger ribonucleic acid, *Biochemistry*, 16, 1672, 1977.
179. **Dimock, K. and Stoltzfus, D. M.,** Sequence specificity of internal methylation in B77 avian sarcoma virus RNA subunits, *Biochemistry*, 16, 471, 1977.
180. **Schibler, U., Kelley, D. E., and Perry, R. P.,** Comparison of methylated sequences in messenger RNA and heterogeneous nuclear RNA from mouse L cells, *J. Mol. Biol.*, 115, 695, 1977.
181. **Chen-Kiang, S., Nevins, J. R., and Darnell, J. E., Jr.,** N-6-methyl-adenosine in adenovirus type 2 nuclear RNA is conserved in the formation of messenger RNA, *J. Mol. Biol.*, 135, 733, 1979.
182. **Caboche, M. and Bachellerie, J. P.,** RNA methylation and control of eukaryotic RNA biosynthesis. Effects of cycloleucine, a specific inhibitor of methylation, on ribosomal RNA maturation, *Eur. J. Biochem.*, 74, 19, 1977.
183. **Caboche, M. and La Bonnardiere, C.,** Vesicular stomatitis virus mRNA methylation in vivo: effect of cycloleucine, an inhibitor of S-adenosylmethione biosynthesis, on viral transcription and translation, *Virology*, 93, 547, 1979.
184. **Engler, J. A., Chow, L. T., and Broker, T. R.,** Sequences of human adenovirus Ad3 and Ad7 DNAs encoding the promoter and first leader segment of late RNAs, *Gene*, 13, 133, 1980.
185. **Zain, S., Gingeras, T. R., Bullock, P., Wong, G., and Gelinas, R. E.,** Determination and analysis of adenovirus-2 DNA sequences which may include signals for late messenger RNA processing, *J. Mol. Biol.*, 135, 413, 1979.
186. **Rogers, J. and Wall, R.,** A mechanism for RNA splicing, *Proc. Natl. Acad. Sci. U.S.A.*, 77, 1877, 1980.
187. **Lerner, M. R., Boyle, J. A., Mount, S. M., Wolin, S. L., and Steitz, J. A.,** Are snRNPs involved in splicing? *Nature (London)*, 283, 220, 1980.
188. **Flint, S. J.,** In *RNA Processing*, Apirion, D., Ed., CRC Press, Boca Raton, Fla., 1983, chap. 7.
189. **Breathnach, R., Benoist, C., O'Hare, K., Gannon, F., and Chambon, P.,** Ovalbumin gene: Evidence for a leader sequence in mRNA and DNA sequences at the exon-intron boundaries, *Proc. Natl. Acad. Sci.*, 75, 4853, 1978.
190. **Seif, I., Khoury, G., and Dhar, R.,** BKV splice sequences based on analysis of preferred donor and acceptor sites, *Nucleic Acids Res.*, 6, 3387, 1979.
191. **Montell, C., Fisher, E. F., Caruthers, M. H., and Berk, A. J.,** Resolving the functions of overlapping viral genes by site-specific mutagenesis at a mRNA splice site, *Nature (London)*, 295, 380, 1982.
192. **Solnick, D.** An adenovirus mutant defective in splicing RNA from early region 1A, *Nature (London)*, 291, 508, 1981.
193. **Ghosh, P. K., Piatak, M., Mertz, J. E., Weissman, S. M., and Lebowitz, P.,** Altered utilization of splice sites and 5′ termini in late RNAs produced by leader region mutants of simian virus 40, *J. Virol.*, 44, 610, 1982.
194. **Kilpatrick, B. A., Gelinas, R. E., Broker, T. R., and Chow, L. T.,** Comparison of late mRNA splicing among class B and class C adenoviruses, *J. Virol.*, 30, 899, 1979.
195. **Uhlen, M., Svensson, C., Josephson, S., Aleström, P., Chattapadhyaya, J. B., Pettersson, U., and Philipson, L.,** Leader arrangement in the adenivorus fiber mRNA, *EMBO J.*, 1, 249, 1982.
196. **Klessig, D. F. and Chow, L. T.,** Deficient accumulation and incomplete splicing of several late viral RNAs in monkey cells infected by human adenovirus type 2, *J. Mol. Biol.*, 139, 221, 1980.
197. **Klessig, D. F. and Anderson, C. W.,** Block to multiplication of adenovirus serotype 2 in monkey cells, *J. Virol.*, 16, 1650, 1975.
198. **Farber, M. and Baum, S. G.,** Transcription of adenovirus RNA in permissive and nonpermissive infections, *J. Virol.*, 27, 136, 1978.
199. **Klessig, D. F. and Grodzicker, T.,** Mutations that allow human Ad2 and Ad5 to express late genes in monkey cells map in the gene encoding the 72K DNA binding protein, *Cell*, 17, 957, 1979.
200. **Kruijer, W., van Schaik, F. M. A., and Sussenbach, J. S.,** Structure and organization of the gene coding for the DNA binding protein of adenovirus type 5, *Nucleic Acids Res.*, 9, 4439, 1981.
201. **Kruijer, W., Van Schaik, F. M. A., and Sussenbach, J. S.,** Nucleotide sequence of the gene encoding adenovirus type 2 DNA binding protein, *Nucleic Acids Res.*, 10, 4493, 1982.
202. **van der Vliet, P. C., Levine, A. J., Ensinger, M. J., and Ginsberg, H. S.,** Thermolabile DNA binding proteins from cells infected with a temperature-sensitive mutant of adenovirus defective in viral DNA synthesis, *J. Virol.*, 15, 348, 1975.
203. **Rekosh, D. M., Russell, W. C., Bellett, A. J. D., and Robinson, A. J.,** Identification of a protein covalently linked to the ends of adenovirus DNA, *Cell*, 11, 283, 1977.

204. **Challberg, M. D., Desiderio, S. V., and Kelly, T. J., Jr.**, Adenovirus DNA replication *in vitro:* characterization of a protein covalently linked to nascent DNA strands, *Proc. Natl. Acad. Sci. U.S.A.*, 77, 5105, 1980.
205. **Smart, J. E. and Stillman, B. W.**, Adenovirus terminal protein precursor: Partial amino acid sequence and the site of covalent linkage to virus DNA, *J. Biol. Chem.*, 257, 13499, 1982.
206. **Enomoto, T., Lichy, J. H., Ikeda, J. E., and Hurwitz, J.**, Adenovirus DNA replication *in vitro:* Purification of the terminal protein in a functional form, *Proc. Natl. Acad. Sci. U.S.A.*, 78, 6779, 1981.
207. **Lichy, J. H., Nagata, K., Friefeld, B. R., Enomoto, T., Field, J., Guggenheimer, R. A., Ikeda, J. E., Horwitz, M. S., and Hurwitz, J.**, Isolation of proteins involved in the replication of adenoviral DNA *in vitro, Cold Spring Harbor Symp. Quant. Biol.*, 47, 731, 1983.
208. **Stillman, B. W. and Tamanoi, F.**, Adenovirus DNA replication: DNA sequences and enzymes required for initiation *in vitro, Cold Spring Harbor Symp. Quant. Biol.*, 47, 741, 1983.
209. **Kozak, M.**, Possible role of flanking nucleotides in recognition of the AUG initiator codon by eukaryotic ribosomes, *Nucleic Acids Res.*, 9, 5233, 1981.
210. **Kitamura, N., Semler, B. L., Rothberg, P. G., Larson, G. R., Adler, C. J., Dorner, A. J., Emini, E. A., Hanacak, R., Lee, J. J., van der Werf, S., Anderson, C. W., and Wimmer, E.**, Primary structure, gene organization and polypeptide expression of poliovirus RNA, *Nature (London)*, 291, 547, 1981.
211. **Racaniello, V. R. and Baltimore, D.**, Molecular cloning of poliovirus DNA and determination of the complete nucleotide sequence of the viral genome, *Proc. Natl. Acad. Sci. U.S.A.*, 78, 4887, 1981.
212. **Dhar, R., Subramanian, K. N., Pan, J., and Weissman, S. M.**, Structure of a large segment of the genome of simian virus 40 that does not encode known proteins, *Proc. Natl. Acad. Sci. U.S.A.*, 74, 827, 1977.
213. **Jay, G., Nomura, S., Anderson, C. W., and Khoury, G.**, Identification of the SV40 agnogene product: a DNA binding protein, *Nature (London)*, 291, 346, 1981.
214. **Bos, J. L., Polder, L. J., Bernards, R., Schrier, P. I., van den Elsen, P. J., van der Eb, A. J., and van Ormondt, H.**, The 2.2Kb E1B mRNA of human Ad12 and Ad5 codes for two tumor antigens starting at different AUG triplets, *Cell*, 27, 121, 1981.
215. **Mertens, P. P. C. and Dobos, P.**, Messenger RNA of infectious pancreatic necrosis virus is polycistronic, *Nature (London)*, 297, 243, 1982.
216. **Shine, J. and Delgarno, L.**, The 3′-terminal sequence of *Escherichia coli* 16s ribosomal RNA: complementarity to nonsense triplets and ribosome binding sites, *Proc. Natl. Acad. Sci. U.S.A.*, 71, 1342, 1974.
217. **Maat, J., van Beveren, C. P., and van Ormondt, H.**, The nucleotide sequence of adenovirus type 5 early region E-1: the region between map positions 8.0 (*Hin*dIII site) and 11.8 (*Sma*I site), *Gene*, 10, 27, 1980.
218. **Wolgemuth, D. J., Uy, H.-Y., and Hsu, M.-T.**, Studies on the relationship between 5′ leader sequences and initiation of translation of adenovirus 2 and simian virus 40 late mRNAs, *Virology*, 101, 363, 1980.
219. **Engler, J. A., Hoppe, M. S., and van Bree, M. P.**, The nucleotide sequence of the genes encoded in early region 2b of human adenovirus type 7, *Gene*, 21, 145, 1983.
220. **van Beveren, C. P., Maat, J., Dekker, B. M. M., and van Ormondt, H.**, The nucleotide sequence of the gene for protein $IVa_2$ and of the 5′ leader segment of the major late mRNAs of adenovirus type 5, *Gene*, 16, 179, 1981.
221. **Engler, J. A. and van Bree, M.**, The nucleotide sequence of the gene encoding protein IVa2 in human adenovirus type 7, *Gene*, 19, 71, 1982.
222. **Gingeras, T. R., Sciaky, D., Gelinas, R. E., Jiang, B.-D., Yen, C. E., Kelly, M. M., Bullock, P. A., Parsons, B. L., O'Neill, K. E., and Roberts, R. J.**, Nucleotide sequences from the adenovirus-2 genome, *J. Biol. Chem.*, 257, 13475, 1982.
223. **Aleström, P., Akusjärvi, G., Pettersson, M., and Pettersson, U.**, DNA sequence analysis of the region encoding the terminal protein and the hypothetical N-gene product of adenovirus type 2, *J. Biol. Chem.*, 257, 13492, 1982.
224. **van Ormondt, H., Maat, J., and Dijkema, R.**, Comparison of nucleotide sequences of the early E1a regions for subgroups A, B, and C human adenoviruses, *Gene*, 12, 63, 1980.
225. **Perricaudet, M., Akusjärvi, G., Virtanen, A., and Pettersson, U.**, Structure of two spliced mRNAs from the transforming region of human subgroup C adenoviruses, *Nature (London)*, 281, 694, 1979.
226. **Kimura, T., Sawada, Y., Shinawawa, M., Shimizu, Y., Shiroki, K., Shimojo, H., Sugisaki, H., Takanami, M., Uemizu, Y., and Fujinaga, K.**, Nucleotide sequence of the transforming early region E1b of adenovirus type 12 DNA: structure and gene organization, and comparison with those of adenovirus type 5 DNA, *Nucleic Acids Res.*, 9, 6571, 1981.
227. **Oosterom-Dragon, E. A. and Anderson, C. W.**, Polypeptide structure and encoding location of the adenovirus serotype 2 late, non-structural 33K protein, *J. Virol.*, 45, 251, 1983.

228. **Hérissé, J. Rigolet, M., Dupont de Dinechin, S., and Galibert, F.,** Nucleotide sequence of adenovirus 2 DNA fragment encoding for the carboxylic region of the fiber protein and the entire E4 region, *Nucleic Acids Res.*, 9, 4023, 1981.
229. **Jones, N. and Shenk, T.,**An adenovirus type 5 early gene function regulates expression of other early viral genes, *Proc. Natl. Acad. Sci. U.S.A.*, 76, 3665, 1979.
230. **Berk, A. J., Lee, F., Harrison, T., Williams, J., and Sharp, P. A.,** Pre-early adenovirus 5 gene product regulates synthesis of early viral messenger RNAs, *Cell*, 17, 935, 1979.
231. **Spector, D. J., McGrogan, M., and Raskas, H. J.,** Regulation of the appearance of cytoplasmic RNAs from region 1 of the adenovirus 2 genome, *J. Mol. Biol.*, 126, 395, 1978.
232. **Goldenberg, C. J., Rosenthal, R., Bhaduri, S., and Raskas, H.,** Coordinate regulation of two cytoplasmic RNA species transcribed from early region 2 of the adenovirus 2 genome, *J. Virol.*, 38, 932, 1981.
233. **Binger, M.-H., Flint, S. J., and Rekosh, D. M.,** Expression of the gene encoding the adenovirus DNA terminal protein precursor in productively infected and transformed cells, *J. Virol.*, 42, 488, 1982.
234. **Swartzendruber, D.E. and Lehman, J. M.,** Neoplastic differentiation: interaction of simian virus 40 and polyoma virus with murine teratocarcinoma cells in vitro, *J. Cell. Physiol.*, 85, 179, 1975.
235. **Topp, W., Hall, J. D., Rifkin, D., Levine, A. J., and Pollack, R.,** The characterization of SV40-transformed cell lines derived from mouse teratocarcinoma: growth properties and differentiated characteristics, *J. Cell. Physiol.*, 93, 269, 1977.
236. **Boccara, M. and Kelly, F.,** Expression of polyoma virus in heterocaryons between embryonal carcinoma cells and differentiated cells, *Virology*, 90, 147, 1978.
237. **Segal, S., Levine, A. J., and Khoury, G.,** Evidence for non-spliced SV40 RNA in undifferentiated murine teratocarcinoma stem cells, *Nature (London)*, 280, 335, 1979.
238. **Segal, S. and Khoury, G.,** Differentiation as a requirement for simian virus 40 gene expression in F-9 embryonal carcinoma cells, *Proc. Natl. Acad. Sci. U.S.A.*, 76, 5611, 1979.
239. **Vasseur, M., Katinka, M., Herbomel, P., Yaniv, M., and Blangy, D.,** Physical and biological features of polyoma virus mutants able to infect embryonal carcinoma cell lines, *J. Virol.*, 43, 800, 1982.
240. **Peries, J., Alves-Cardoso, E., Canivet, M., Debons-Guillemin, M., and Lasneret, J.,** Lack of multiplication of ectopic murine C-type viruses in mouse teratocarcinoma primitive cells, *J. Natl. Cancer Inst.*, 59, 463, 1977.
241. **Teich, N. M., Weiss, R., Matin, G. R., and Lowy, D. R.,** Virus infection of murine teratocarcinoma stem cell lines, *Cell*, 12, 973, 1977.
242. **Miller, R., Ward, D., and Ruddle, F.,** Embryonal carcinoma cells (and their somatic cell hybrids) are resistant to infection by the murine parvovirus MVM, which does infect other teratocarcinoma derived cell lines, *J. Cell. Physiol.*, 91, 393, 1977.
243. **LaPorta, R. F. and Taichman, L. B.,** Adenovirus type-2 infection of human keratinocytes: Viral expression dependent upon the state of cellular maturation, *Virology*, 110, 137, 1981.
244. **LaPorta, R. F. and Taichman, L. B.,** Human papilloma viral DNA replicates as a stable episome in cultured epidermal cells, *Proc. Natl. Acad. Sci. U.S.A.*, 79, 3393, 1982.
245. **Pintell, D., Dadachanji, D., Astell, C. R., and Ward, D. C.,** The genome of minute virus of mice, an autonomous parvovirus, encodes two overlapping transcription units, *Nucleic Acids Res.*, 11, 1109, 1983.
246. **Perbal, B., Cline, J. M., Hillyard, R. L., and Baluda, M. A.,** Organization of chicken DNA sequences homologous to the transforming gene of avian myeloblastosis virus. II. Isolation and characterization of λ *proto-amv* DNA recombinant clones from a library of leukemic chicken DNA, *J. Virol.*, 45, 925. 1983.
247. **Chow, L. T. and Broker, T. R.,** unpublished observation, 1980. (See also Reference 44.)
248. **Engel, L., Heilman, C., and Howley, P.,** personal communication.
249. **Schwartz, D. E., Tizard, R., and Gilbert, W.,** Nucleotide sequence of Rous sarcoma virus, *Cell*, 32, 853, 1983.
250. **Danos, O., Engel, L. W., Chen, E. Y., Yaniv, M., and Howley, P. M.,** Comparative analysis of the human type 1a and bovine type 1 papilloma viruses, *J. Virol.*, 46, 557, 1983.
251. **Takeya, T. and Hanafusa, H.,** Structure and sequence of the cellular gene homologous to the RSV *src* gene and the mechanism for generating the transforming virus, *Cell*, 32, 881. 1983.
252. **Van Beveren, C., van Straaten, F., Curran, T., Müller, R. and Verma, I. M.,** Analysis of FBJ-MuSV provirus and *c-fos* (mouse) gene reveals that viral and cellular *fos* gene products have different carboxy termini, *Cell*, 32, 1241, 1983.
253. **Watson, D. K., Reddy, E. P., Duesberg, P. H. and Papas, T. S.,** Nucleotide sequence analysis of the chicken *c-myc* gene reveals homologous and unique coding regions by comparison with the transforming gene of avian myelocytomatosis virus MC29, Δ *gag-myc*, *Proc. Natl. Acad. Sci. U.S.A.*, 80, 2146. 1983.

Chapter 9

# RIBOSOMAL RNA PROCESSING IN EUKARYOTES

**Robert J. Crouch**

## TABLE OF CONTENTS

# I. INTRODUCTION

Transcription of ribosomal genes in both eukaryotic and prokaryotic organisms results in the formation of an RNA molecule which, through a series of processing steps, is converted to mature ribosomal RNAs. The general outline for these events was sketched during the 1960s and early 1970s and has been reviewed by Perry.[1] A more recent review on the structural organization of ribosomal genes has been written by Long and Dawid.[2] Processing of ribosomal RNA to form active ribosomes can be considered as a multiple series of events. However, this review will discuss processing in terms of the nucleolytic scissions leading to 5.8S, 18S, and 28S rRNA production.*

Elucidation of the biochemistry of rRNA processing in eukaryotes has been hampered by a variety of factors. The large size of the transcript (10 to 12 kbases in higher eukaryotes), a lack of understanding of the number and sites of cleavage, and the nature of the substrate (i.e., a possible requirement for a ribonucleoprotein for processing) have all contributed to the technical difficulty in studies of details of eukaryotic rRNA processing.

The use of cloned fragments of ribosomal RNA genes is beginning to reveal details of processing of ribosomal RNA transcripts. The information from such analyses will be described in this review along with other results concerning enzymes involved in rRNA processing as well as some aberrant processing systems. However, no attempt has been made to include all of the many papers which demonstrate changes in rRNA processing as a result of some type of cellular perturbation.

# II. TRANSCRIPTIONAL UNIT

Figure 1 shows a diagrammatic representation of the ribosomal transcript of a eukaryote. Transcription initiates with the formation of a *L*eader *E*xternal *T*ranscribed *S*pacer (LETS), followed by 18S RNA, an *I*nternal *T*ranscribed *S*pacer (ITS-1), then 5.8S rRNA, a second ITS (ITS-2), 28S RNA and concludes with a *T*erminal *E*xternal *T*ranscribed *S*pacer (TETS). Variations in the generalized transcript as represented in Figure 1 are of two types: (1) the size of each transcribed spacer region (LETS and ITS) varies considerably from one species to another, and (2) interruptions by intervening sequences in the 28S rRNA gene (as exemplified in *Tetrahymena*[3] and *Drosophila*[4]).

A triphosphorylated 5′ terminus can be used to define the point of transcriptional initiation as has been done for full-length transcripts of yeast,[5] *Xenopus laevis*,[6] and mouse.[7] The 3′ terminus is more difficult to define since no unique chemical structure differentiates between transcription termination and a processing event. Recent sequence analysis of RNA from yeast,[8] mouse,[9,10] and *T. thermophila*[11] have shown that some precursor RNA molecules have nucleotides at their 3′ termini which are absent in mature 28S rRNAs. These TETS sequences are 7 for yeast,[8] 14 for *T. thermophila*,[11] and approximately 30 for mouse.[9,10] Coincidence of 3′ termini for 28S rRNA and the initial transcript has been reported for *X. laevis*[12] and *D. melanogaster*.[13] It is possible that differences in rates of processing at the 3′ terminius account for the discrepancy between the transcriptional termination being different from, or identical to, the 3′ terminus of 28S rRNA.

# III. NUMBER AND ORDER OF PROCESSING SITES

The standard transcriptional unit shown in Figure 1 requires six cleavages to produce the three mature ribosomal RNAs (positions denoted by integer numbers). In some systems,

* Mature rRNAs are generally referred to as 5.8S, 18S, and 28S even though these sizes are not always those for a particular organism.

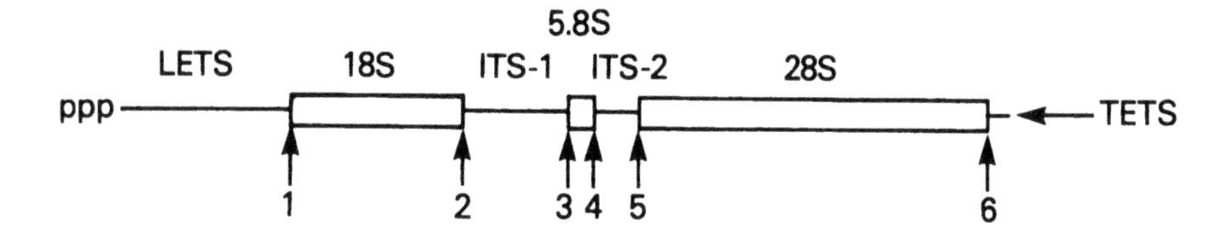


FIGURE 1. Eukaryotic ribosomal RNA transcriptional unit. rRNA transcriptional unit for a "typical" eukaryotic organism. LETS, ITS-1, ITS-2, and TETS are defined in text. Sites 1 through 6 are designations of the ends of mature rRNA species.

additional sites of processing have been described which would increase the total number of intermediates. The order in which each site is cleaved varies from one organism to another and may be different within a given cell,[1] although the actual use of all pathways to produce mature rRNAs has not been demonstrated.

To date, the best-defined pathways of eukaryotic rRNA processing come from studies of cloned DNA fragments of yeast,[14] *Drosophila,*[15] and mouse[16] rRNA. Analysis in the yeast system is more extensive since the entire sequence of the transcriptional unit is known as well as the specific location of the termini of mature and intermediate rRNAs.[14,17] Both mouse and *D. melanogaster* rRNA pathways have been studied using hybridization of DNA to RNA fractionated by size on agarose gels.

Veldman et al.[14] have described the yeast rRNA processing pathway depicted in Figure 2A. The assigned order is based on detection of intermediates which are most abundant and does not rule out alternate routes. In addition to the minimal six processing sites shown in Figure 1, two cleavages occur, one at $A_2$ and one at $C_2$.

The pathways for *D. melanogaster,* shown in Figure 2B, are based on the processing intermediates observed by Long and Dawid.[15] In addition to sites at the ends of the mature rRNA molecules, one processing site (2.5) is detected. In fact, it appears that this site is, in some molecules, the first to be cleaved. Additional nucleolytic cleavage of *D. melanogaster* rRNA produces a modified 5.8S (two RNAs, 5.8S + 2S) and two fragments of 28S rRNA.

The mouse rRNA processing pathways as determined by Bowman et al.[16] require only cleavages at the ends of mature rRNA species (Figure 2C). One of these pathways is very similar to that described for HeLa cells (see Perry[1]). In addition to the sites described by Bowman et al.,[15] two other cleavages appear to take place, one which removes the 3′ terminal ≈30 nucleotides found in the precursor[9,10] and a second which has been described to occur during in vitro transcription of rDNA.[18,19] A corresponding in vivo site has been found subsequently. The latter site is about 650 nucleotides from the initiation of transcription and is indicated in Figure 2C as 0.5.

## IV. SEQUENCES AT rRNA CLEAVAGE SITES

DNA sequencing of rRNA processing sites has been performed for DNA from a number of organisms and the results are presented in Figure 3. Examination of these reveals that mature rRNA sequences are highly conserved, whereas little conservation is seen for regions immediately adjacent to those of mature rRNA sequences. Several explanations of the manner by which a processing enzyme might recognize these sequences can be put forward:

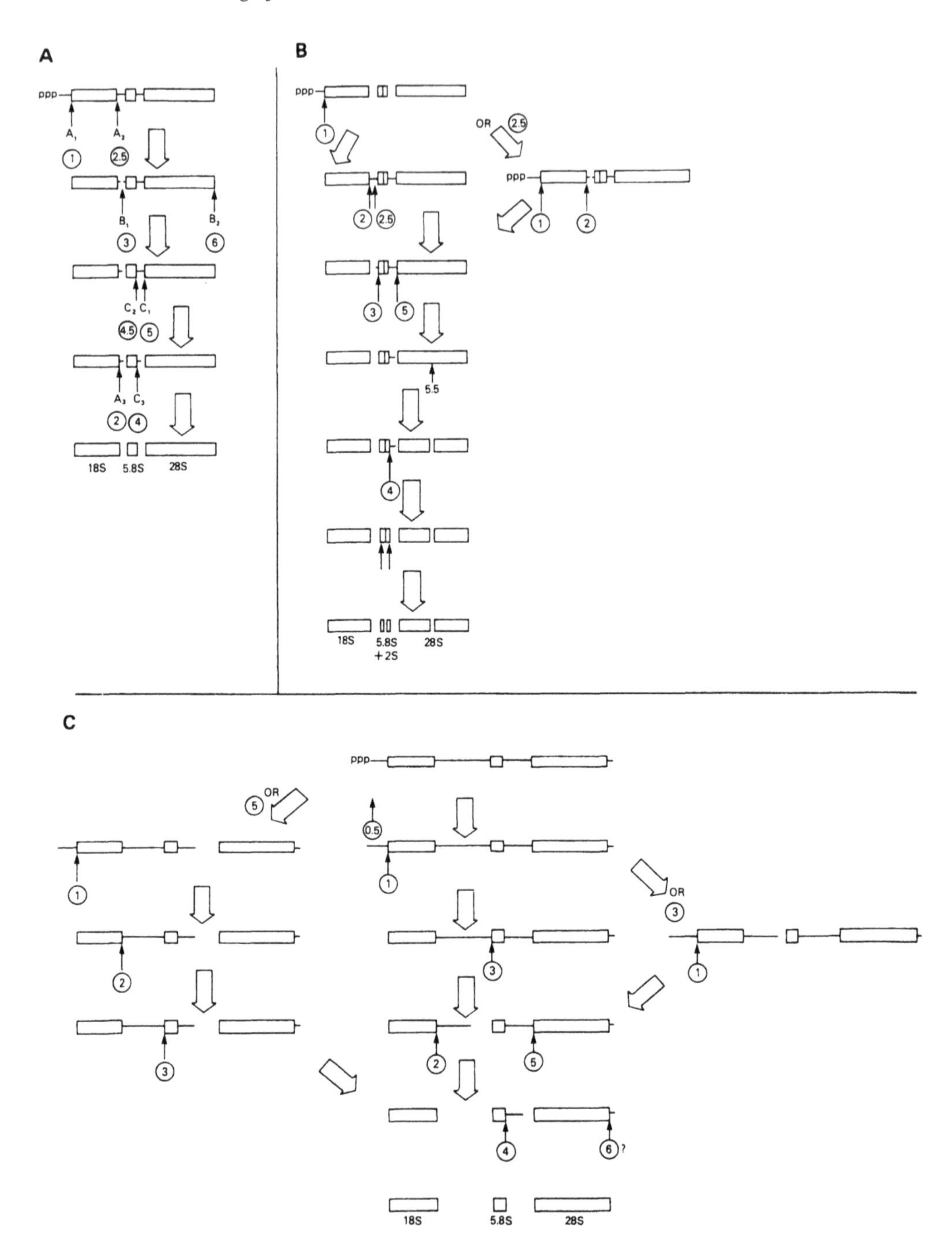


FIGURE 2. rRNA processing pathways. rRNA processing pathways for yeast (A), *Drosophila melanogaster* (B), and mouse (C). Numbers representing sites of cleavage-integers are the same as in Figure 1. Fractional numbers are assigned for sites between ends of mature rRNA. For yeast, the designation of sites are shown as described by Veldman et al.[8]

1. The relevant sequences are not always those at the ends of mature rRNAs. For example, the sequences underlined adjacent to the 5′ and 3′ termini of yeast 5.8S rRNA have been observed in "precursor" molecules. It may be important to know if such extensions are found in other "precursor 5.8S rRNAs" before a specific sequence recog-

nition site can be determined. Muramatsu and colleagues[20] have shown that 5′ termini of nuclear and cytoplasmic 28S rRNA have different terminal nucleotides. A difference is also found for the 3′ terminus.[21] The final generation of mature rRNAs could result from a limited digestion from these pre RNAs.

2. Processing enzymes could recognize sequences in the highly conserved mature rRNAs and cleavage outside these regions would generate the precursors and/or the mature rRNAs. This would be analogous to the recognition by RNA polymerase III[22] of a transcriptional initiation site within the 5S RNA gene. Conservation in evolution of some sequences of mature rRNA may reflect a requirement for ribosomal activity or for processing, or both. A dual role of these sequences in rRNA processing and ribosomal activity would be difficult to determine without an in vitro processing system.
3. Multiple enzymes are involved in processing — with specificities for one site or another. In this line of explanation it is conceivable that in evolution the processing enzymes also change to recognize new sites.
4. Sequence or primary structure alone does not produce a processing site but some combination of sequence at the actual site of cleavage and the secondary structure of a large segment, in which the cleavage site is located, comes into play.

Weldman et al.[14] have suggested a common sequence-secondary structure model for yeast rRNA processing (see Figure 4). An interesting feature of this model is that 5′ processing sites have one consensus sequence, whereas 3′ processing sites have a second. Part of this model is based on their observation of temporally related cleavages at sites A1 and A2, or B1 and B2, or C1 and C2. However, the remaining processing sites seem to result from a single cleavage site (A3 and C3). Although these consensus sequences can be found in certain instances in other rRNAs (for example the 3′ end of 18S rRNA of *Bombyx mori* contains the same sequence as the 3′ end of yeast 17S rRNA), they are not universal. 5′ Terminal yeast consensus sequences are not maintained in other eukaryotic rRNAs (compare the 5′ terminal regions adjacent to and including 28S rRNAs in Figure 3). Processing pathways differ for yeast and other organisms and the "consensus" sequences need not be the same in all organisms.

## V. SPACER SIZE VARIATION

The size of mature rRNAs among different organisms varies little in comparison to variations of two of the spacer regions (LETS and ITS).[23] Transcripts of human ribosomal RNA genes have a LETS of about 3 kb and an ITS of 2 kb,[24] while chicken transcripts show a shorter LETS (~1 kb) and a longer ITS (~3 kb).[25] Almost all of the difference in size between the LETS region of human and chicken can be attributed to fold-back structures that have been observed by electron microscopy. The same is true for the ITS regions. We have determined the sequence for one of these fold-back structures in chick pre-rRNA, the region between 5.8S and 28S (ITS-2).[26] The sequence consists of ~650 nucleotides which are extremely rich in $G^+C$ (85%). A structure can be drawn with the aid of a computer algorithm which resembles that observed by electron microscopy. This structure serves to maintain close spatial relationship between 5.8S rRNA and 28S rRNA in spite of the separation by 650 nucleotides of primary sequence. The size of ITS-2 varies widely from organism to organism. Any size may be tolerable to the cell so long as the spatial relationships are maintained. The suggestion that 5.8S rRNA corresponds to the 5′ terminal region of prokaryotic 23S rRNA[27-29] fits well with this idea. Once the evolutionary event occurred

```
SITE 1  (5' 18S rRNA)

        LETS                              18S rRNA

Y.      UGUUGCUUCUUCUUUUAAGAUAGUUAU|CUGGUUGAUCCUGCCAGUAG
X.l.    CGCGCCGGGCCCGGGAAAGGUGGCUAC|CUGGUUGAUCCUGCCAGUAG


SITE 2  (3' 18S rRNA)

        18S rRNA                        ITS 1

Y.         GGUGAACUCGCGGAAGGAUCAUUA|AGAAAUUUAAUAAUUUUGAAAAU
B.m.    GUAGGUAACCUGCGGAAGGAUCAUUA|ACGGGUGAUGGGAAGAAA
D.m.    UAGGUGAACCUGCGGAAGGAUCAUUA|UUGUAUAAUAUCCUUACCGUUAAU


SITE 3  (5' 5.8S rRNA)

          ITS 1                           5.8S rRNA

Y.      AUUUUCGUAACUGGAAAUUUUAAAAUAUUAA|AAACUUUCAACAACGGAUCUC
N.c.    AUGCUCUCUGAGUAAACUUUUAAAUAAGUCA|AAACUUUCAACAACGGAUCUC
X.l.   GAAAACCGACCGACGCGUCGGGGAGAGCUCG|CGACUCUUAGCGGUGGAUCAC
Ch.   CCGCGCGCGCGGGGCGGUGCCGAAAGUCAGA|CAACUCUUAGCGGUGGAUCAC
D.m.     CAUAAUUGACAUUAUAUAAAAAUGAAUUAUA|AAACUCUAAGCGGUGGAUCAC
S.c.     UUAACAAUAAAUUUGUAAAAAAAAUUUAAUU|CAACCCUAAGCGGGGGAUCAC
H.L.                                   U|CGACUCUUAGCGGUGGAUCAC
```

FIGURE 3. Sequences of ribosomal RNA transcripts at processing sites. Sites are listed 1 through 6 according to Figures 1 and 2. Mature rRNA sequences are enclosed. Sites 2.5 and 4.5 are from yeast. All sequences are written as RNA even though the sequence is not known to be a part of a transcript. 3′ Terminus of mouse 28S RNA is not precisely known. TETS sequences are marked with I to indicate 3′ terminus of 45S RNA. Abbreviations are Y. = yeast, N.C. = *Neurospora crassa,* X.1. = *Xenopus laevis,* D.m. = *Drosophila melanogaster,* B.m. = *Bombyx mori,* S.c. = *Sciara coprophila,* H. L. = HeLa cell, Ch. = chicken, T.t. = *tetrahymena thermophila.* M. =

| Site | Organism | Ref. | Site | Organism | Ref. |
|---|---|---|---|---|---|
| 1 | Y. | 8 | 4 | Y. | 8 |
| | X.l. | 75 | | N.c. | 74 |
| | | | | X.l. | 77 |
| 2 | Y. | 8 | | Ch. | 26 |
| | B.m. | 73 | | D.m. | 76 |
| | D.m. | 78 | | S.c. | 71 |
| 3 | Y. | 8 | 5 | Y. | 8 |
| | N.c. | 74 | | X.l. | 77 |
| | X.l. | 77 | | Ch. | 26 |
| | Ch. | 26 | | | |
| | D.m. | 76 | 6 | Y. | 8 |
| | S.c. | 71 | | T.t. | 11 |
| | H.L. | 72 | | M. | 9, 10 |
| | | | | D.m. | 13 |
| | | | | X.l. | 12 |
| | | | 2.5 | Y. | 8 |
| | | | 4.5 | Y. | 8 |

SITE 4 (3' 5.8S rRNA)

```
            5.8S rRNA                         ITS 2

Y.    GCAUGCCUGUUUGAGCGUCAUUUCCUUCUCAAACAUUCUGUUUGGUAGUGAGU
N.c.  GCAUGCCUGUUCGAGCGUCAUUUCAACCAUCAAGCUCUGCUUGCGUUGGG
X.l.  CCACGCCUGUCUGAGGGUCGCUCCGACGUCCAUCGCCCCCGCCGGGUCCCGUC
Ch.   CUACGCCUGCCUGAGCGUCGCUUGACGGUCAAUCGCCGAUGGCCGCCGUCCGC
D.m.  CUACAUAUGGUUGAGGGUUGUAAGACUAUGCUAAUUAAGUGCUUAUAAAUUUU
S.c.  CCACACAUGGUUGAGGGUCGUUAGAUUACUUAAUAAAAAUUGCAUUUUUAUUA
```

SITE 5 (5' 28S rRNA)

```
        ITS 2                            28S rRNA

Y.    UCUAGGCGAACAAUGUUCUUAAAGUUUGACCUCAAAUCAGGUAGGAGUACCCGCU
X.l.  CGCGCCCCCCCCCCCCCCACGACUCAGACCUCAGAUCAGACGCGGCGACCCGCU
Ch.   CCGGCGUUCGGCCGGCUUUCCGAUCGCGACCUCAGGUCAGACGUGGCGACCCGCG
```

SITE 6 (3' 28S rRNA)

```
        28S rRNA                          TETS

Y.    UAAGCCUUUGUUGUCUGAUUUGUUUUUUAUUUCUUUCUAAGUGGGUACUGGCAGGA
T.t.  UCAGCCCGUCUCCUUAGAUUUAUCUCAUCUCCCUUUAUUUUUUACUUCUGCUGGGG
X.l.  UCAUCCCUUGAGAAAAGCUUUUGUCGGAAGGAGCAGGCCGGAAGGGCGCCCCCGCC
M.    UCAGCCCUCGACACAAGGGUUUGUCUCUGCGGGCUUUCCCGUCGCACGCCCGCUCG
D.m.  UUAUGGUUUGCUUGAUGAUUCGAUAUAAAAUAAAUCGUUGCCAAACAGCUCGUCAU
```

SITE 2.5 (YEAST)

```
AUAGGACAAUUAAAACCGUUUCAAUACAACACACUGUGGAGUUUUC
```

SITE 4.5 (YEAST)

```
CUGCGUGCUUGAGGUAUAAUGCAAGUACGGUCGUUUUAGGUUUUACCAA
```

FIGURE 3 (continued)

which generated ITS-2, the only constraint placed on its size was the maintenance of proximity.

## VI. ALTERNATIVE STRUCTURES OF PRIMARY TRANSCRIPTS

In most eukaryotic systems described to date, the primary transcript is completed prior to processing. Exceptions to this rule are the cleavage of mouse site 0.5 (Figure 2C)[18,19] and *Dictyostelium*[30] in which processing seems to occur in the course of transcription. As

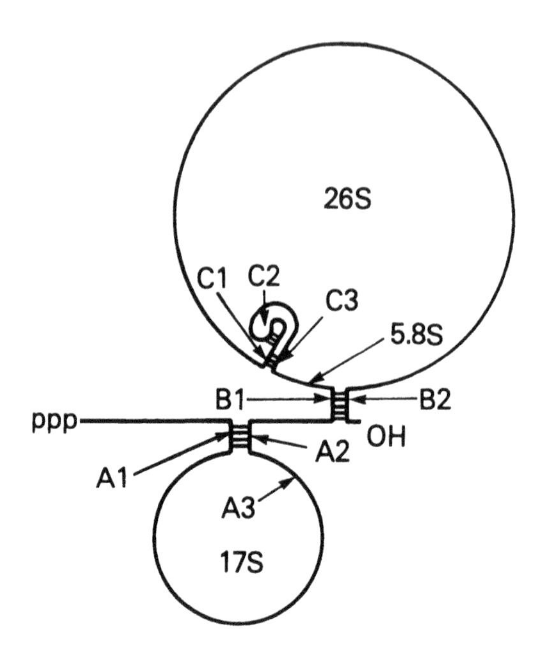


FIGURE 4. Yeast pre-rRNA. A model for interaction of yeast pre-rRNA to form processing sites (Veldman et al.[8]). "Consensus" sequences are

| 5′ Terminal sites | | 3′ Terminal sites | |
|---|---|---|---|
| Site 1 | UUUUAAGAUAGUUA | Site 2 | UCAUUA |
| Site 3 | UUUUAAAAUA-UUA | Site 4 | UCAUUU |
| Site 5 | UCUUAAAGUU-UGA | Site 6 | UGAUUU |
| | | Site 2.5 | UCAAUA |
| | | Site 4.5 | UCGUUU |

transcription proceeds it is reasonable to assume that the secondary structure of the RNA molecule undergoes a number of changes.[31] The number and rate of such changes would be influenced by a variety of factors including the primary sequence and protein nucleic acid interactions. In any case, interaction between two sequences is possible only if both sequences are present — a situation which is not always true during the course of transcription.

A case in point concerns 5.8S rRNAs. Evidence from a number of laboratories shows that 5.8S rRNA is hydrogen-bonded to 28S rRNA when these cytoplasmic RNAs are isolated under nondenaturing conditions.[32-34] However, in the pre-rRNA molecules which contain 5.8S rRNA, but not 28S rRNA, no similar interaction is possible. A "reasonable" secondary structure can be made in which the 5′ and 3′ termini of 5.8S rRNA and their flanking regions are hydrogen-bonded (see Figure 5).

Of particular interest in the possible existence of alternative forms of 5.8S rRNA sequences in pre-rRNA is the observation that the small nuclear RNA, U3, is hydrogen-bonded to 32S rRNA.[35,36] A search for sequences which could be sites of interaction between U3 RNA and 32S rRNA led to the structure shown in Figure 5. It is suggested from the two sites of interaction of U3 RNA with 32S rRNA that U3 RNPs might be involved in isomerizations of the RNA.[37] If multiple secondary structures can exist, cleavage of RNA in any one form could differ from that of alternate forms and, thereby, generate multiple processing pathways.

## VII. ENZYMES

The search for eukaryotic rRNA processing enzymes has taken two paths. One approach

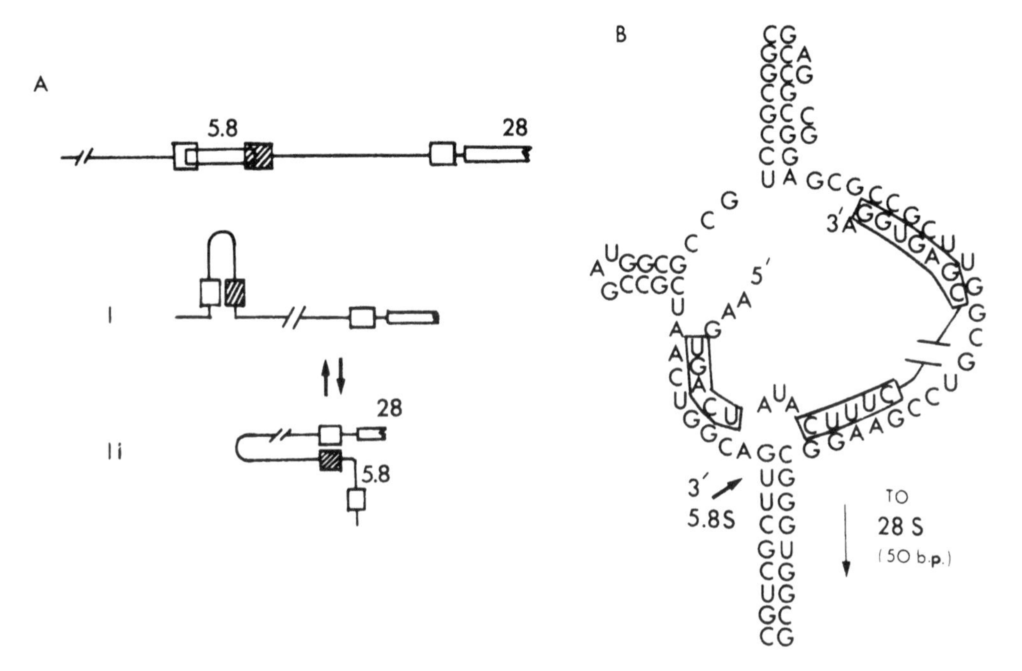


FIGURE 5. Alternate interactions of chick 5.8S rRNA. (A) Model of interaction of chick 5.8S rRNA. Form I shows a hairpin structure in which 5′ and 3′ of 5.8S rRNA plus flanking regions form a hairpin. Form II shows the 3′ region of 5.8S rRNA interacting with a region near the 5′ terminus of 28S rRNA. Open boxes represent sequences complementary to hatched boxes. (B), A potential hydrogen-bonding of U3RNA to Form II. U3 sequences are enclosed in boxes.

has been to study ribonucleases which are present in nucleoli, the site of rRNA processing. This approach was used by Mirault and Scherrer,[38] Winicov and Perry,[39] and Prestayko et al.[40] Each of these groups found an enzymatic activity (or activities) which cleave pre-rRNA to RNA which resembles the size of intermediates observed in in vivo processing. With the demonstration of the participation of RNase III in *E. coli* rRNA processing[41] (see the review in Chapter 2 of this volume), several laboratories described the isolation from eukaryotes of enzymes that degrade double-stranded RNA[42-49] (the property by which RNase III was initially isolated).[50] Several of these enzymes have been examined for processing activity on 45S rRNA, either as free RNA or in nucleolar extracts. Most of these activities yield RNA similar in size to known in vivo intermediates.

For example, we have described an enzyme which degrades double-stranded RNA isolated originally from total cell extracts of chick embryos.[42] This activity (or one very similar) was found in the nucleolus.[43] Both the nucleolar activity (either from chick embryo cell extracts, chick embryo nucleoli, or mouse Ehrlich ascites cell nucleoli) showed similar inhibition by $Mg^{++}$ and $(NH_4)_2SO_4$ for degradation of double-stranded RNA and "processing".[43] A similar activity from HeLa cells is reportedly an exonuclease,[47] and would be unable to produce the appropriate endonucleolytic cleavage to process pre-rRNA. Krebs ascites cells contain a double-stranded RNA specific nuclease which is found associated with heterogeneous nuclear RNP as well as nucleoli[45] and has been suggested to be involved in processing of RNA which is not ribosomal.[44,45]

A totally different approach has been used by Denoya et al.[51] Extraction of chick embryo with Sarkosyl and cold phenol/chloroform yields several proteins bound to RNA. Among these proteins is a ribonuclease which, by the criterium of cleaving 45S rRNA to the size of rRNA intermediates, processes rRNA. This activity is found both in the cytoplasm, associated with 50S ribosomal subunits, and in the nucleolus. In the presence of EDTA, or if the enzyme-associated RNA has been degraded, the activity becomes nonspecific and degrades 18S and 28S rRNAs.

It should again be mentioned that none of these putative processing enzymes has been shown to produce cleavages at sites identical to in vivo sites.

Discussion of enzymes involved in processing must take into consideration the nature of the substrate. Two extreme views can be taken. One assumes that only the enzymes involved in cleavage and the RNA are sufficient for the processing of rRNA. This is certainly the case for RNase III processing of *E. coli* ribosomal RNA[41,52] (see also review in Chapter 2 of this volume). The other extreme is that in ribosomal RNA processing of mature rRNA sequences are protected by proteins, leaving the remainder of the 45S rRNA exposed to degradation by whatever nucleases are around. Somewhere between these extremes would be the requirement for an RNP to produce a cleavage site for a specific ribonuclease. This latter situation is also found in *E. coli* ribosomal RNA processing[53-55] (also see Chapter 2). An in vitro system which correctly processes rRNA should define the substrate.

## VIII. SPLICING OF *TETRAHYMENA* RIBOSOMAL RNA PRECURSOR

Certain lower eukaryotes have been shown to have interruptions in their "active" 28S rRNA genes. Although this type of processing is not characteristic of rRNA processing in general, considerable progress has been made toward understanding the mechanism by which the splicing reaction is accomplished. Grabowski et al.[56] have presented evidence that the intervening sequence(IVS) is removed from pre-rRNA as a linear RNA(IVS) which is later circularized. Components in the reaction mixture required for the cleavage and ligation are simple. Pre-rRNA, $(NH_4)_2SO_4$, a divalent cation, and a guanosine compound are the only factors needed.[57] The rather remarkable fact that no enzyme is required for this splicing event suggests that such nonenzymatic processing must be considered as a viable mechanism in other types of processing.

## IX. ALTERATIONS IN rRNA PROCESSING

This review has focused on the nucleolytic events in rRNA processing in eukaryotes. Cellular changes that perturb the events involved in conversion of pre-rRNA to mature rRNAs in vivo need not directly affect the enzymological cleavage of the appropriate phosphodiester bond. For example, protein binding to pre-rRNA occurs in an ordered process,[58] probably reflecting the synthesis of rRNA to which each ribosomal protein is to bind. Any alterations which change binding of proteins, including proteins that modify RNA structure, can influence the fate of rRNA precursors.

Methylation of ribosomal RNA is a well-known event, with most of the methyl groups being found in mature rRNA sequences.[1] Failure to methylate pre-rRNA leads to improper processing. An interesting example of altered processing correlated with abnormal methylation is seen when CHO cells are heat shocked.[59] Upon shifting to 43°C, pre-rRNA synthesis leads to the formation of 45S and 39S RNAs, neither of these RNAs is methylated. Return of the cells to 37° leads to resumption of synthesis and processing rRNA. However, the 45S and 39S synthesized at 43° continue to be cleaved as they would have been at 43° (i.e., they are not converted to cytoplasmic 18 and 28S rRNAs). A mutant of BHK cells[60] which is temperature-sensitive for growth has an abnormal rRNA processing pattern. At the restrictive temperature, synthesis and processing of ribosomal RNA are normal up to the conversion of 32S rRNA to 28S and 5.8S rRNAs. Mutants with similar disturbances in rRNA processing have been described for yeast[61] and *N. crassa*.[62] Picolinic acid or 5-methylnicotinamide, an analog of NAD, elicits in rat kidney cells an rRNA processing pattern similar to that of the mutants.[63] In such cells, the fate of the 32S rRNA is unknown. It neither accumulates nor is converted to cytoplasmic 28S rRNA.

Another mutant of yeast is abnormally slow in converting 20S rRNA to mature 18S

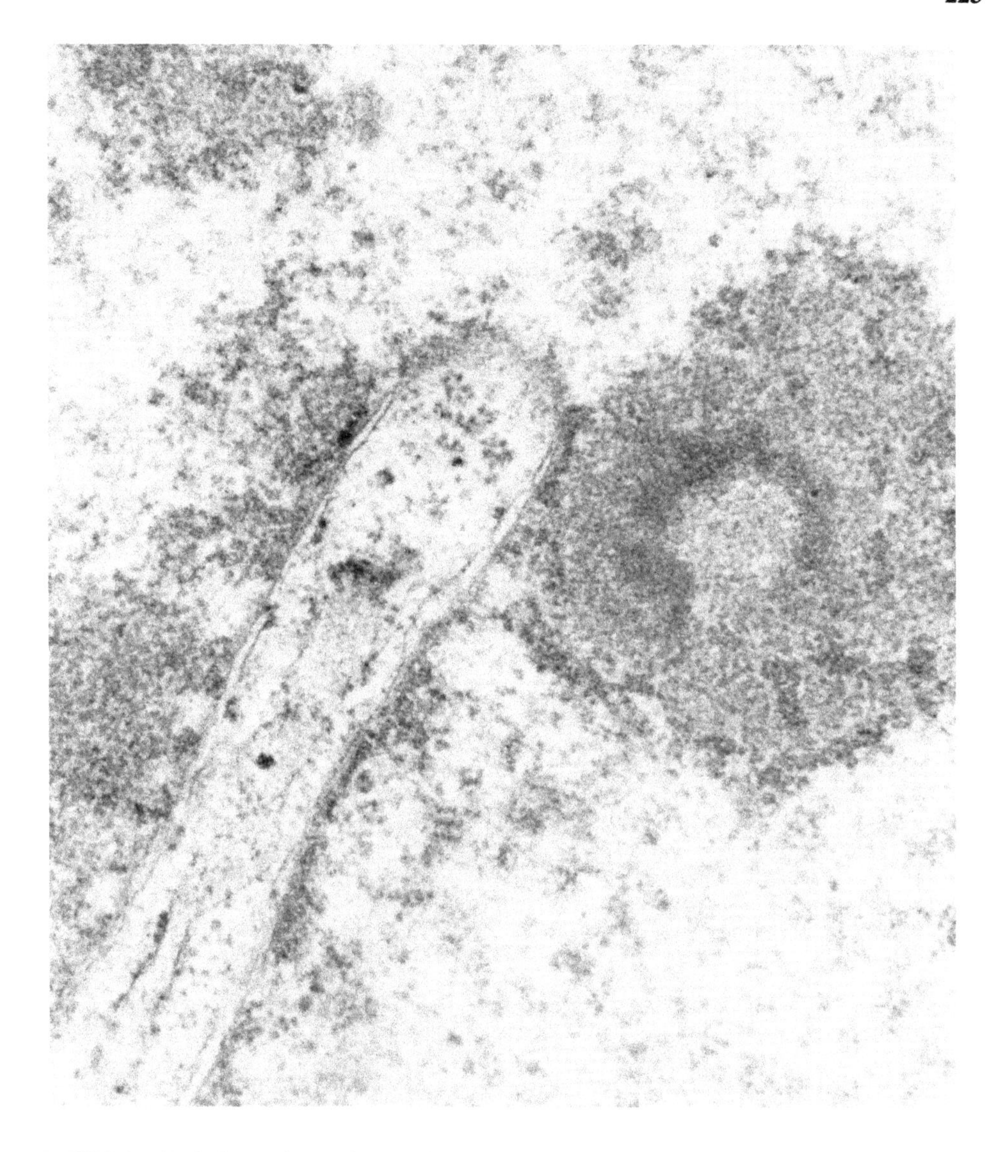

FIGURE 6. Nucleolar canals. An electron micrograph of a nucleolar canal. (Courtesy of Dr. C. A. Bourgeois, Laboratoire Pathologic Cellulaire, University of Paris.) Connection of the nucleolus with nuclear membrane can be seen to give nucleolar-cytoplasmic contact. See Reference 64 for more details of these structures.

rRNA[64] (conversion of 20S rRNA to 18S rRNA in yeast is a cytoplasmic event).[65] This mutant is not temperature sensitive but is slow in growth. Veldman et al.[8] have suggested that 5′ and 3′ consensus sequences reflect recognition sites for processing enzymes. Since the 3′ consensus sequences are similar for 18S and 20S, a common enzyme would necessarily be present in the nucleus and cytoplasm.

Several nucleoside analogs have been shown to change rRNA processing. Toyocamycin,[66,67] 5-fluorouridine,[67] and ethidium bromide[68] all inhibit rRNA processing by altering the RNA structure. This interferes with the interaction of either nucleases[69] or other rRNP proteins[67] and pre-rRNA.

This limited discussion of the many agents that alter rRNA processing is presented to point to a number of potential changes in substrate or enzymes involved in producing mature eukaryotic rRNAs which might prove to be important in defining in vitro rRNA processing events.

Transport of ribosomal RNAs to the cytoplasm signals, in most instances, the end of rRNA processing. Production of 18S rRNA is followed almost immediately by export to the cytoplasm. Failure to convert 32S rRNA to 5.8S and 28S RNA often leads to the destruction of the 32S RNA — suggesting that export to the cytoplasm saves 28S rRNA from a similar fate. Rapid export of ribosomal proteins from the nucleolus may occur through nuclear canals (Figure 6).[70] In considering rRNA processing in eukaryotes, it may also be important to consider the influence of such structures on rRNA processing.

## ACKNOWLEDGMENTS

I thank Dr. Marie-Luise Dirksen for critically reading the manuscript and Terri Broderick for preparation of the manuscript. Thanks also to Dr. I. Grummt for her unpublished sequences and to Drs. C. A. Bourgeois and M. Bouteille for the electron micrograph for Figure 6.

## REFERENCES

1. **Perry, R. P.,** Processing of RNA, *Ann. Rev. Biochem.*, 45, 605, 1976.
2. **Long, E. O. and Dawid, I. B.,** Repeated genes in eukaryotes, *Ann. Rev. Biochem.*, 49, 727, 1980.
3. **Wild, M. A. and Gall, J. G.,** An intervening sequence in the gene coding for 25S ribosomal RNA of Tetrahymena pigmentosa, *Cell*, 16, 565, 1979.
4. **Glover, D. M. and Hogness, D. S.,** A novel arrangement of the 18S and 28S sequences in a repeating unit of Drosophila melanogaster rDNA, *Cell*, 10, 167, 1977.
5. **Klootwijk, J., de Jonge, P., and Planta, R. J.,** The primary transcript of the ribosomal repeating unit in yeast, *Nucleic Acids Res.*, 6, 27, 1979.
6. **Reeder, R., Sollner-Webb, B., and Wahn, B.,H.,** Sites of transcription initiation in vivo on X. laevis rDNA, *Proc. Natl. Acad. Sci. U.S.A.*, 74, 5402, 1977.
7. **Bach, R., Grummt, I., and Allet, B.,** The nucleotide sequence of the initiation region of the ribosomal transcription unit from mouse, *Nucleic Acids Res.*, 9, 1559, 1981.
8. **Veldman, G. M., Klootwijk, J., de Jonge, P., Leer, R. J., and Planta, R. J.,** The transcription termination site of the ribosomal RNA operon in yeast, *Nucleic Acids Res.*, 8, 5179, 1980.
9. **Grummt, I.,** personal communication.
10. **Kominami, R., Mishima, Y., Urano, Y., Sakai, M., and Muramatsu, M.,** Cloning and determination of the transcription termination site of ribosomal RNA gene of the mouse, *Nucleic Acids Res.*, 10, 1963, 1982.
11. **Din, N., Engberg, J., and Gall, J. G.,** The nucleotide sequence at the transcription termination site of the ribosomal RNA gene in Tetrahymena thermophila, *Nucleic Acids Res.*, 10, 1503, 1982.
12. **Sollner-Webb, B. and Reeder, R. H.,** The nucleotide sequence of the initiation and termination sites for ribosomal RNA transcription in X. leavis, *Cell*, 18, 485, 1979.
13. **Mandal, R. K. and Dawid, I. B.,** The nucleotide sequence at the transcription termination site of ribosomal RNA in Drosophila melanogaster, *Nucleic Acids Res.*, 9, 1801, 1981.
14. **Veldman, G. M., Klootwijk, J., van Heerikhuizen, H., and Planta, R. J.,** The nucleotide sequence of the intergenic region between the 5.8S and 26S rRNA genes of the yeast ribosomal RNA operon. Possible implications for the interaction between 5.8S and 26S rRNA and the processing of the primary transcript, *Nucleic Acids Res.*, 9, 4847, 1981.
15. **Long, E. O. and Dawid, I. B.,** Alternative pathways in the processing of ribosomal RNA precursor in *Drosophila melanogaster*, *J. Mol. Biol.*, 138, 873, 1980.
16. **Bowman, L. H., Rabin, B., and Schlessinger, D.,** Multiple ribosomal RNA cleavage pathways in mammalian cells, *Nucleic Acids Res.*, 9, 4951, 1981.
17. **Bayer, A. A., Georgiev, C. I., Hadjiolov, A. A., Skryabin, K. G., and Zakharyev, V. M.,** The structure of the yeast ribosomal RNA genes. III. Precise mapping of the 18S and 25S rRNA genes and the structure of the adjacent regions, *Nucleic Acids Res.*, 9, 789, 1981.
18. **Grummt, I.,** Mapping of a mouse ribosomal DNA promoter by in vitro transcription, *Nucleic Acids Res.*, 9, 6093, 1981.
19. **Miller, K. G. and Sollner-Webb, B.,** Transcription of mouse rRNA genes by RNA polymerase I: in vitro and in vivo initiation and processing sites, *Cell*, 27, 165, 1981.
20. **Kominami, R., Hamada, H., Fujii-Kuriyama, Y., and Muramatsu, M.,** 5′-Terminal processing of 28S RNA, *Biochemistry*, 17, 3965, 1978.

21. **Hamada, H., Kominami, R., and Muramatsu, M.,** 3′-Terminal processing of ribosomal RNA precursors in mammalian cells, *Nucleic Acids Res.*, 8, 889, 1980.
22. **Sakonju, S., Bogenhagen, D. F., and Brown, D. D.,** A control region in the center of the 5S RNA gene directs specific initiation of transcription. I. The 5′ border of the region, *Cell*, 19, 13, 1980.
23. **Loening, U. E., Jones, K. W., and Birnstiel, M. L.,** Properties of the ribosomal RNA precursor in *Xenopus laevis;* comparison to the precursor in mammals and in plants, *J. Mol. Biol.*, 45, 343, 1969.
24. **Wellauer, P. K. and Dawid, I. B.,** Structure and processing of ribosomal RNA: a comparative electron microscopic study in three animals, *Brookhaven Symp. Biol.*, 26, 214, 1975.
25. **Schibler, U., Wyler, T., and Hagenbüchle, C.,** Changes in size and secondary structure of the ribosomal transcription unit during vertebrate evolution, *J. Mol. Biol.*, 94, 503, 1975.
26. **Earl, P., Feldmann, R. J., and Crouch, R. J.,** unpublished observations.
27. **Jacq, B.,** Sequence homologies between eukaryotic 5.8S rRNA and the 5′ end of prokaryotic 23S rRNA: evidences for a common evolutionary origin, *Nucleic Acids Res.*, 9, 2913, 1981.
28. **Cox, R. A. and Kelly, J. N.,** Mature 23S rRNA of prokaryotes appears homologous with the precursor of 25-28 rRNA of eukaryotes: comments on the evolution of 23-28 rRNA, *FEBS Lett.*, 130, 217, 1981.
29. **Nazar, R. N.,** A 5.8 S rRNA-like sequence in prokaryotic 23 S rRNA, *FEBS Lett.*, 119, 212, 1980.
30. **Grainger, R. M. and Maizels, N.,** Dictyostelium ribosomal RNA is processed during transcription, *Cell*, 20, 619, 1980.
31. **Kramer, F. R. and Mills, D. R.,** Secondary structure formation during RNA synthesis, *Nucleic Acids Res.*, 9, 5109, 1981.
32. **Pace, N. R., Walker, T. A., and Schroeder, E.,** Structure of the 5.8S RNA component of the 5.8S-28S ribosomal RNA junction complex, *Biochemistry*, 16, 5321, 1977.
33. **Nazar, R. N. and Sitz, T. C.,** Role of the 5′ terminal sequence in the RNA binding site of yeast 5.8S rRNA, FEBS Lett., 115, 71, 1980.
34. **Kelly, J. M. and Cox, R. A.,** The nucleotide sequence at the 3′ end of *Neurospora crassa* 25S-rRNA and the location of a 5.8S-rRNA binding site, *Nucleic Acids Res.*, 9, 1111, 1981.
35. **Prestayko, A. W., Tonato, M., and Busch, H.,** Low molecular weight RNA associated with 28S nucleolar RNA., *J. Mol. Biol.*, 47, 505, 1970.
36. **Zieve, G. and Penman, S.,** Small RNA species of the HeLa cell: metabolism and subcellular localization, *Cell*, 8, 19, 1976.
37. **Crouch, R. J., Feldmann, R., and Earl, P.,** unpublished observations.
38. **Mirault, M. E. and Scherrer, K.,** In vitro processing of HeLa preribosomes by a nucleolar endoribonuclease, *FEBS Lett.*, 20, 233, 1972.
39. **Winicov, I. and Perry, R. P.,** Characterization of a nucleolar endonuclease possibly involved in ribosomal ribonucleic acid maturation, *Biochemistry*, 13, 2908, 1974.
40. **Prestayko, A. W., Lewis, B. C., and Busch, H.,** Purification and properties of a nucleolar endoribonuclease from Novikoff hepatoma, *Biochim. Biophys. Acta*, 319, 323, 1973.
41. **Dunn, J. J. and Studier, F. W.,** T7 early RNAs and Escherichia coli ribosomal RNAs are cut from large precursor RNAs in vivo by ribonuclease III, *Proc. Natl. Acad. Sci., U.S.A.*, 70, 3296, 1973.
42. **Hall, S. H. and Crouch, R. J.,** Isolation and characterization of two enzymatic activities from chick embryos which degrade double-stranded RNA, *J. Biol. Chem.*, 252, 4092, 1977.
43. **Grummt, I., Hall, S. H., and Crouch, R. J.,** Localization of an endonuclease specific for double-stranded RNA within the nucleolus and its implication in processing ribosomal transcripts, *Eur. J. Biochem.*, 94, 437, 1979.
44. **Rech, J., Cathala, G., and Jeanteur, P.,** Partial purification of a double-stranded RNA specific ribonuclease (RNase D) from Krebs II ascites cells, *Nucleic Acids Res.*, 3, 2055, 1976.
45. **Rech, J., Cathala, G., and Jeanteur, P.,** Isolation and characterization of a ribonuclease activity specific for double-stranded RNA (RNase D) from Krebs II ascites cells, *J. Biol. Chem.*, 255, 6700, 1980.
46. **Ohtsuki, K., Groner, Y., and Hurwitz, J.,** Isolation and purification of double-stranded specific ribonuclease from calf thymus, *J. Biol. Chem.*, 252, 483, 1977.
47. **Saha, B. K. and Schlessinger, D.,** Separation and characterization of two activities from HeLa cell nuclei that degrade double-stranded RNA, *J. Biol. Chem.*, 253, 4537, 1978.
48. **Büsen, W. and Hausen, F.,** Distinct ribonuclease H activities in calf thymus, *Eur. J. Biochem.*, 52, 179, 1975.
49. **Shanmugan, G.,** Partial purification and characterization of ribonuclease III-like enzyme activity from cultured mouse cells, *Biochemistry*, 17, 5052, 1978.
50. **Robertson, H. D., Webster, R. E., and Zinder, N. D.,** Purification and properties of ribonuclease III from *Escherichia coli*, *J. Biol. Chem.*, 243, 82, 1968.
51. **Denoya, C., Costa Giomi, P., Schodeller, E. A., Vasquez, C., and LaTorre, J. L.,** Processing of naked 45S ribosomal RNA precursor in vitro by an RNA-associated endoribonuclease, *Eur. J. Biochem.*, 115, 375, 1981.

52. **Nikolaev, N., Silengo, L., and Schlessinger, D.,** A role for ribonuclease III in processing ribosomal ribonucleic acid and messenger ribonucleic acid precursors in *Escherichia coli, J. Biol. Chem.,* 248, 7967, 1973.
53. **Meyhack, B., Meyhack, I., and Apirion, D.,** Processing of precursor particles containing 17S rRNA in a cell-free system, *FEBS Lett.,* 49, 215, 1974.
54. **Hayes, F. and Vaseur, M.,** Processing of 17S *Escherichia coli* precursor RNA in the 27S pre-ribosomal particle, *Eur. J. Biochem.,* 61, 433, 1976.
55. **Dahlberg, A. E., Dahlberg, J. E., Lund, E., Tokimatsu, H., Rabson, A. B., Calvert, P. C., Reynolds, F., and Zahalak, M.,** Processing of the 5′ end of *Escherchia coli* 16S ribosomal RNA, *Proc. Natl. Acad. Sci. U.S.A.,* 75, 3598, 1978.
56. **Grabowski, P. J., Zaug, A. J., and Cech, T. R.,** The intervening sequence of the ribosomal RNA precursor is converted to a circular RNA in isolated nuclei of *Tetrahymena, Cell,* 23, 467, 1981.
57. **Cech, T. R., Zaug, A. J., and Grabowski, P. J.,** *In vitro* splicing of ribosomal RNA precursor of *Tetrahymena.* Involvement of a guanosine nucleotide in the excision of the intervening sequence, *Cell,* 27, 487, 1981.
58. **Lastick, S. M.,** The assembly of ribosomes in HeLa cell nucleoli, *Eur. J. Biochem.,* 113, 175, 1980.
59. **Bouche, G., Raynal, F., Amalric, F., and Zalta, J. P.,** Unusual processing of nucleolar RNA synthesized during a heat shock in CHO cells, *Mol. Biol. Rep.,* 7, 253, 1981.
60. **Toniolo, D., Meiss, H. K., and Basilico, C.,** A temperature sensitive mutation affecting 28S ribosomal RNA production in mammalian cells, *Proc. Natl. Acad. Sci. U.S.A.,* 70, 1273, 1973.
61. **Andrew, C., Hopper, A., and Hall, B. D.,** A yeast mutant defective in the processing of 27S r-RNA precursor, *Molec. Gen. Genet,* 144, 29, 1976.
62. **Loo, M. W., Schriker, N. S., and Russell, P. J.,** Heat-sensitive mutant strain of *Neurospora crassa,* 4M(t), conditionally defective in 25S ribosomal ribonucleic acid production, *Molec. Cell. Biol.,* 1, 199, 1981.
63. **Constantini, M. G. and Johnson, G. S.,** Disproportionate accumulation of 18S and 28S ribosomal RNA in cultured normal rat kidney cells treated with picolinic acid or 5-methylnicotinamide, *Exp. Cell Res.,* 132, 443, 1981.
64. **Carter, C. J. and Cannon, M.,** Maturation of ribosomal precursor RNA in *Saccharomyces cerevisiae.* A mutant with a defect in both the transport and terminal processing of the 20S species, *J. Mol. Biol.,* 143, 179, 1980.
65. **Udem, S. A. and Warner, J. R.,** The cytoplasmic maturation of a ribosomal precursor ribonucleic acid in yeast, *J. Biol. Chem.,* 248, 1412, 1973.
66. **Tavitian, A., Uretsky, S. C., and Acs, G.,** Selective inhibition of ribosomal RNA synthesis in mammalian cells, *Biochim. Biophys. Acta,* 157, 33, 1963.
67. **Hadjiolova, K. V., Naydenova, Z. G., and Hadjiolav, A. A.,** Inhibition of ribosomal RNA maturation in Friend erythroleukemia cells by 5-fluorouridine and toyocamycin, *Biochem. Pharmacol,* 30, 1861, 1981.
68. **Snyder, A. L., Kann, H. E., Jr., and Kotin, K. W.,** Inhibition of the processing of ribosomal precursor RNA by intercalating agent, *J. Mol. Biol.,* 58, 555, 1971.
69. **Gatoh, S., Nikolaev, N., Battaner, B., Birge, C. H., and Schlessinger, D.,** *Escherichia coli* RNaseIII cleaves HeLa cell nuclear RNA, *Biochem. Biophys. Res. Commun.,* 59, 972, 1974.
70. **Bourgeois, C. A., Hemon, D., and Bouteille, M.,** Structural relationship between the nucleolus and nuclear envelope, *J. Ultrastruct. Res.,* 68, 328, 1979.
71. **Jordan, B. R., Latil-Damotte, M., and Jourdan, R.,** Coding and spacer sequences in the 5.8S-2S region of *Sciara coprophila* ribosomal DNA, *Nucleic Acids Res.,* 8, 3565, 1980.
72. **Khan, N. S. N. and Maden, B. E. H.,** Nucleotide sequence relationships between vertebrate 5.8S ribosomal RNAs, *Nucleic Acids Res.,* 4, 2495, 1977.
73. **Samols, D. R., Hagenbuchle, O., and Gage, L. P.,** Homology of the 3′ terminal sequences of the 18S rRNA of *Bombyx mori* and the 16S rRNA of *Escherichia coli, Nucleic Acids Res.,* 7, 1109, 1979.
74. **Selker, E. and Yanofsky, C.,** Nucleotide sequence and conserved features of the 5.8S rRNA coding region of. *Neurospora crassa Nucleic Acids Res.,* 6, 2561, 1979.
75. **Salim, M. and Maden, B. E. H.,** Nucleotide sequence encoding the 5′ end of *Xenopus laevis* 18SrRNA, *Nucleic Acids Res.,* 8, 2871, 1980.
76. **Pavlakis, G. N., Jordan, B. R., Wurst, R. M., and Vournakis, J. N.,** Sequence and secondary structure of *Drosophila melanogaster* 5.8S and 2S rRNAs and the processing site between them, *Nucleic Acids Res.,* 8, 2213, 1979.
77. **Hall, L. M. C. and Maden, B. E. H.,** Nucleotide sequence through the 18S—28S intergene region of a vertebrate ribosomal transcriptional unit, *Nucleic Acids Res.,* 8, 5993, 1980.
78. **Jordan, B. R., Latil-Damotte, M., and Jourdan, R.,** Sequence of the 3′-terminal portion of *Drosophila melanogaster* 18S rRNA and of the adjoining spacer: comparison with the corresponding prokaryotic and eukaryotic sequences, *FEBS Lett.,* 117, 227, 1980.

Chapter 10

# RNA SYNTHESIS AND PROCESSING IN MITOCHONDRIA

**Giuseppe Attardi**

## TABLE OF CONTENTS

## I. INTRODUCTION

All eukaryotic cells possess an extrachromosomal genetic system, sequestered in their mitochondria, which, though of limited informational content, plays a crucial role in the biogenesis of these organelles.[1,2] Mitochondrial DNA (mtDNA) has been known for some time to code for the RNA components of the organelle-specific protein synthesizing apparatus (rRNAs and tRNAs) and for a small set of essential proteins; these include some subunits of enzyme complexes of the inner mitochondrial membrane involved in respiration and oxidative phosphorylation — in particular the three largest subunits of cytochrome *c* oxidase (CO), cytochrome *b* and one or two subunits, depending upon the organism, of the oligomycin-sensitive ATPase — and, at least in yeast, a protein associated with the small ribosomal subunit. Furthermore, the recent discovery of unidentified reading frames (URFs) in the mtDNAs from several organisms has revealed a greater than previously suspected informational content of mtDNA, pointing to possible new functions of this DNA.[3-10]

The first evidence of the essential role of mtDNA in the assembly of a functional inner mitochondrial membrane was provided, though not immediately recognized as such, by the discovery of the "petite" mutation in yeast[11] and the "poky" mutation in *Neurospora crassa.*[12] It is not surprising, therefore, that the study of the genetic role of the mitochondrial genome took an early start in yeast, facilitated by the availability of mutants and by the application of refined genetic techniques. By contrast, in other organisms, in particular in mammalian cells, the analysis of the genetic content of mtDNA was seriously hindered by the lack of genetic approaches easily applicable to these systems. In the last few years, however, the development of powerful DNA, RNA, and protein technologies has opened the way for a detailed structural analysis of mtDNA in these systems as well. It is a remarkable fact, and indeed a tribute to the power of the molecular approaches, that the most thoroughly investigated so far among the mitochondrial genomes are those of yeast and man, i.e., the eukaryotic systems most and least amenable, respectively, to genetic analysis.

The most surprising outcome of the investigations that have been carried out over the past 15 years on the mitochondrial genomes of various sources is that, while the essential genetic functions of mtDNA appear to have been preserved in great part in a vast spectrum of eukaryotic cells at different evolutionary levels, the structure and gene organization of the mitochondrial genome and the mechanisms of its expression have evolved in strikingly diverse ways in different organisms. In this review, the author will present an up-to-date account of what is known concerning the transcription and RNA processing patterns in mitochondria of different eukaryotic cells (the literature up to August 1982 is covered in this article).

## II. HUMAN MITOCHONDRIAL DNA

### A. Gene Organization

The determination of the complete sequence of the human mtDNA[3] has revealed an extremely compact gene organization. With the exception of the D-loop region, the entire length of mtDNA (~16.6 kb) is saturated by genes (Figure 1). In particular, the heavy (H)-strand, which is the main coding strand, contains the genes for the two rRNA species, 14 tRNA species, and 12 significant (⩾120 nucleotide [nt]) reading frames, while the light (L)-strand contains the genes for 8 tRNA species and one significant reading frame.

In the H-strand, the rRNA and tRNA genes and the significant reading frames are in most cases butt-jointed to each other or separated by a few nucleotides, and there is a nearly complete absence of noncoding stretches.[3,13-18] Thus, the initiator codon of each reading frame, which can be either AUG, AUA, or AUU, follows immediately, or with an interval of a few nucleotides, the gene adjacent to it on its 5′ side; furthermore, most reading frames lack a termination codon and exhibit either a T or a TA following the last sense codon and immediately preceding the adjacent gene on the 3′ side. As will be discussed below, completion of the termination (ochre) codon occurs at the time of RNA processing by polyadenylation of the mRNAs.[3,17] A striking feature of the gene organization in the H-strand is the interspersion of the tRNA genes, which separate with nearly absolute regularity the rRNA and protein coding genes.[13]

A correlation of the DNA sequences or corresponding RNA sequences with protein sequence data or with known yeast gene sequences has allowed the identification of five reading frames of human mtDNA as coding for subunits I, II, and III of cytochrome *c* oxidase (CO), cytochrome *b*, and a peptide homologous to yeast ATP ase subunit 6.[3,18] The other eight reading frames (URFs) have not yet been identified as to the polypeptides they code for. The observations that these URFs are conserved in bovine,[19] mouse,[20] and rat[21,22] mtDNA, and that they correspond to what appear to be bona fide mRNAs strongly suggest that they represent genes for proteins synthesized in the mitochondria.

### B. Complete Symmetrical Transcription

The most striking feature of HeLa cell mtDNA transcription, and the first to be discovered, was its complete symmetry.[23-25] Early experiments, in which the hybridization saturation level of the H-strand by mitochondrial RNA uniformly labeled with (5-$^3$H)uridine was measured, and in which the density in a $Cs_2SO_4$ gradient and the appearance in the electron microscope of RNA-DNA hybrids formed at saturation were analyzed, clearly showed that this strand is completely or almost completely transcribed.[23] In the same experiments, a very low level of hybridization (~2%) was observed with the L-strand. However, subsequent experiments, involving hybridization in solution of (5-$^3$H)uridine pulse-labeled mitochondrial RNA with separated mtDNA strands indicated that the L-strand is also extensively transcribed.[24] An explanation for the apparent discrepancy of results in the two types of experiments was provided by the observation that the ratio of pulse-labeled RNA hybridizable

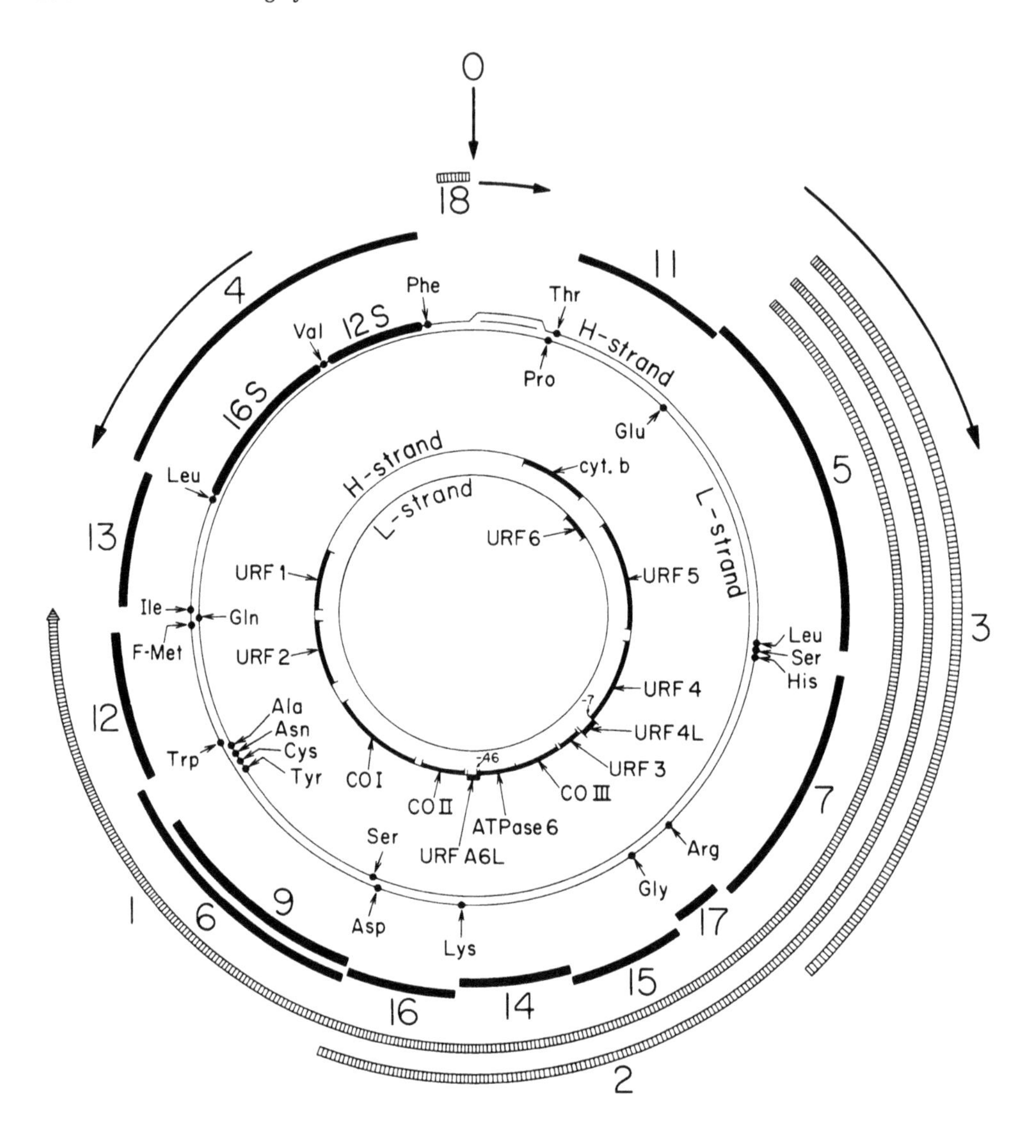


FIGURE 1. Genetic and transcription maps of the HeLa cell mitochondrial genome.[3,13] The two outer circles show the positions of the two rRNA genes as derived from mapping and RNA sequencing experiments,[14,15] and those of the tRNA genes as derived from the mtDNA sequence.[3,14] Mapping positions of the polyadenylated H-strand transcripts are indicated by the black bars, those of the L-strand transcripts by the hatched bars.[13] Left and right arrows indicate the direction of H- and L-strand transcription, respectively; the vertical arrow (marked O) and the rightward arrow at the top indicate the location of the origin and the direction of H-strand synthesis. The two inner circles show the positions of the mtDNA reading frames.[3] URF: unidentified reading frame. (Modified from Gelfand, R. and Attardi, G., *Molec. Cell Biol.*, 1, 497, 1981. With permission.)

with H-strands to that complementary to L-strands was lower the shorter the length of the pulse, pointing to a much greater instability of the L-transcripts as compared to the H-transcripts.[24] Thus, for very short pulses (2 min), the ratio of radioactivity incorporated in the L- and H-transcripts, taking into account the difference in their dA contents[23] and assuming the existence of a common nucleotide precursor pool, has indicated that the overall rate of transcription of the L-strand is two to three times higher than that of the H-strand.[26] By contrast, in the steady state, the great majority of mitochondrial RNA in HeLa cells is represented by H-strand transcripts, explaining the failure to detect significant hybridization of total mitochondrial RNA with the L-strand in the early experiments.[23]

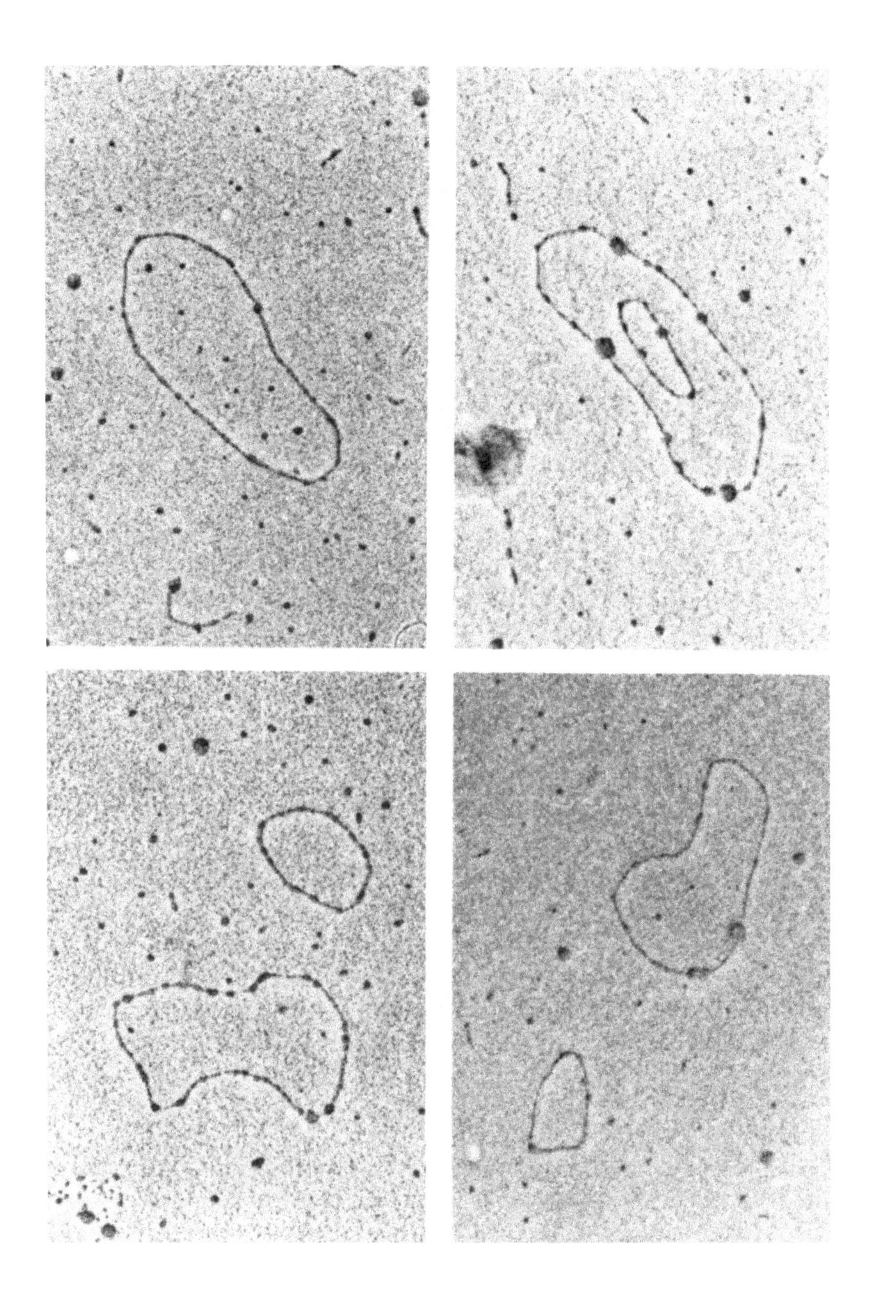

FIGURE 2. Electron micrographs of hybrids between L-strands of HeLa cell mtDNA and saturating amounts of partially purified L-strand transcripts, spread by the basic protein film technique. ØX174 RF DNA (5375 bp) is included for size reference. (With permission from Murphy, W. I., Attardi, B., Tu, C., and Attardi, G., *J. Mol. Biol.*, 99, 809, 1975. Copyright: Academic Press Inc. (London) Ltd.)

As to the extent of L-strand transcription, saturation hybridization experiments utilizing a partially purified preparation of L-transcripts clearly showed, both by measurements of density in a $Cs_2SO_4$ gradient and by electron microscopic analysis of the hybrids formed at saturation (Figure 2), that the L-strand is also completely or almost completely transcribed.[25]

The symmetrical transcription of HeLa cell mtDNA is in striking contrast with the highly asymmetric distribution of informational content between the two strands. Its significance is still elusive, although one can make some plausible speculations about it (see below).

### C. Discrete Transcription Products

Chromatography of mitochondrial RNA through oligo(dT)-cellulose separates an oligo(dt)-cellulose bound and a nonbound fraction.[27] The oligo(dT)-bound components carry at their 3′ end a poly(A) tail of ~55 residues, which is added post-transcriptionally.[28,29] The oligo(dT)-nonbound components either have a short stretch of one to nine A residues at their 3′ end (bulk of rRNA species[15]), or completely lack 3′-terminal A residues (tRNAs and a minor portion of the rRNAs[15]). A high resolution of mitochondrial RNA species has been achieved by electrophoresis through agarose slab gels in the presence of the strongly denaturing agent methylmercuric hydroxide ($CH_3HgOH$).[27] Figure 3 shows the autoradiogram, after electrophoresis through an agarose-$CH_3HgOH$ slab gel, of the poly(A)-containing RNA from HeLa cells. In A, the RNA was extracted from the mitochondrial fraction of cells labeled for 3 hr with [$^{32}$P]orthophosphate in the presence of 20 μg of camptothecin per mℓ (to inhibit nuclear RNA synthesis); in B, the RNA was extracted from mitochondria of cells labeled for 2.5 hr with [$^{32}$P]orthophosphate in the absence of inhibitors, after treatment of the mitochondrial fraction with micrococcal nuclease to destroy any contaminating extra-mitochondrial nucleic acids. All the discrete RNA species detected in the autoradiograms hybridize with mtDNA. The identity of the two patterns indicates that there is no detectable nuclear DNA-coded RNA component in human mitochondria. As shown in Figure 3 and Table 1, 18 discrete poly(A)-containing RNAs, covering a range of molecular sizes between ~215 and ~10,400 nt, have been identified in HeLa cell mitochondrial RNA.[27] Of these, the three largest species (RNAs 1, 2, and 3) and the smallest one (RNA 18) are L-strand coded, while the others are encoded in the H-strand (the origin of RNA 8 is unknown).

Of the H-strand polyadenylated transcripts, species 5, 7, 9, and 11 to 17, for their perfect correspondence to significant reading frames[3,16,17] (Figure 1), their presence in polysomes,[27] and their relative metabolic stability[30] (Table 2), are presumably the mRNAs of mitochondrially synthesized polypeptides. A correlation of the mapping positions of the RNAs with the reading frames of mtDNA and an alignment of their 5′ end proximal sequence with the $NH_2$-terminal sequence of known polypeptides[3,18] have allowed the identification of five of these polyadenylated RNAs as the mRNAs for cytochrome *c* oxidase subunits I, II, and III (RNAs 9, 16, and 15, respectively), cytochrome *b* (RNA 11), and a polypeptide homologous to ATPase subunit 6 (RNA 14) (Figure 1). The other putative mRNAs (RNAs 5, 7, 12, 13, and 17) correspond to URFs.

Human mitochondrial mRNAs have distinctive structural properties. In contrast to other eukaryotic mRNAs, they lack a "cap" structure at their 5′ end.[31] Furthermore, they do not exhibit another typical attribute of eukaryotic mRNAs, namely, the presence of a 5′-noncoding stretch, but start at or very near to the initiator codon(AUG or AUA or AUU[16]). This finding poses interesting questions concerning the mechanism whereby mitochondrial ribosomes attach themselves to these mRNAs. Also at their 3′ end, mitochondrial mRNAs have unique structural features. Namely, most mRNAs lack a 3′-noncoding stretch (the only exceptions being RNAs 5, 9, and 16); furthermore, these mRNAs have an incomplete stop codon, an ochre codon being generated by the poly(A) addition step[17] (see below).

Polyadenylated RNA 6, on account of its mapping position (Figure 1) and its relatively short half-life (Table 2), is probably a precursor of RNA 9 (CO I mRNA). RNA 10 represents a small fraction, polyadenylated, of the 16S rRNA.[27] The nature of RNA 4 will be discussed below.

Among the L-strand coded polyadenylated RNAs, species 1, 2, and 3, which are very short lived (with a half-life estimated to be less than 10 min, Table 2), may be intermediates

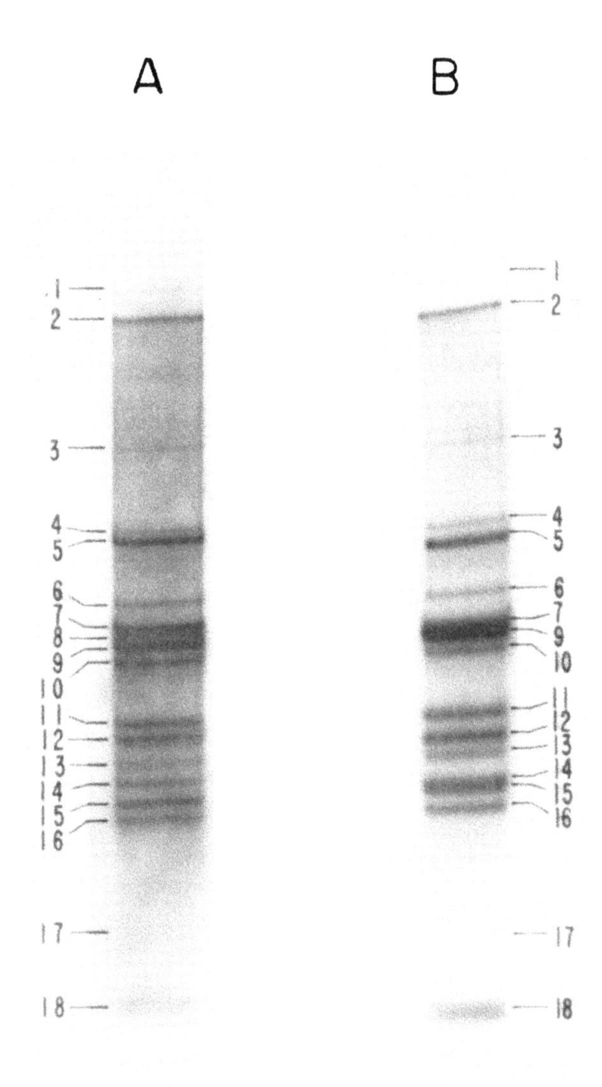


FIGURE 3. Autoradiograms, after electrophoresis through 1.4% agarose-$CH_3HgOH$ slab gels, of the oligo(dT)-bound RNA extracted from the mitochondrial fraction (untreated) of HeLa cells labeled for 3 hr with [$^{32}$P]orthophosphate in the presence of 20 μg/mℓ camptothecin (A), or from the micrococcal nuclease-treated mitochondrial fraction of cells labeled for 2.5 hr with [$^{32}$P]orthophosphate in the absence of inhibitors of nuclear RNA synthesis (B). See text for details. (With permission from Gelfand, R. and Attardi, G., *Molec. Cell. Biol.*, 1, 497, 1981. Copyright: Academic Press Inc. (London) Ltd.)

in the formation of L-strand coded tRNAs. The observation that their common 5′-end proximal 500-nucleotide stretch corresponds precisely to an L-strand unidentified reading frame (URF6)[3] strongly suggests that these RNA species or some as yet unidentified derivative of them functions as mRNA for this reading frame.

**Table 1**
**TRANSCRIPTS OF HeLa CELL MITOCHONDRIAL DNA**

| RNA species | Molecular length[a] (number of nucleotides) | Functional assignment | Template strand |
|---|---|---|---|
| Oligo(dT)-unbound | | | |
| 4S RNA | 59—75 | Includes 22 tRNAs | 14 tRNAs: H; 8 tRNAs L |
| 12S RNA | 954 | Structural components of ribosomes | H |
| 16S RNA | 1559 | | H |
| Oligo(dT)-bound | | | |
| 1 | ~10400 | URF6 mRNAs? | L |
| 2 | ~7070 | | L |
| 3 | ~4155 | | L |
| 4 | ~2700 | | H |
| 5 | 2410 | URF5 mRNA | H |
| 6 | 1938 | Precursor of RNA 9 | H |
| 7 | 1668 | (URF4L + URF4) mRNA | H |
| 9 | 1617 | CO I mRNA | H |
| 11 | 1141 | Cyt.*b* mRNA | H |
| 12 | 1042 | URF2 mRNA | H |
| 13 | 958 | URF1 mRNA | H |
| 14 | 842 | (URF A6L + ATPase 6) mRNA | H |
| 15 | 784 | CO III mRNA | H |
| 16 | 709 | CO II mRNA | H |
| 17 | 346 | URF3 mRNA | H |
| 18 (7S RNA) | ~215 | | L |

[a] Of nonpoly(A) or nonoligo(A) portion, determined from the length of the coding DNA sequences, except for RNAs 2, 3, and 18 (estimated from S1 protection data) and RNAs 1 and 4 (estimated from electrophoretic mobility).

RNA 18 (also called 7S RNA on the basis of its sedimentation constant) is a small polyadenylated RNA which maps in the region immediately upstream of the origin of H-strand synthesis.[32] Besides the rRNAs and tRNAs, it is, on a molar basis, the most abundant mtDNA transcript; it is the only polyadenylated RNA encoded in the L-strand which accumulates; it has a stability comparable to that of well-characterized mitochondrial mRNAs; and it is found in part in polysomes. These features are consistent with the possibility of its being the mRNA for a mitochondrially synthesized small polypeptide. There is indeed a small reading frame for a polypeptide of 23 to 24 amino acids near its 3′ terminus, but the lack of conservation of this sequence in bovine,[19] rat,[33] and mouse mtDNA[20] raises questions about its significance. The most intriguing feature of this RNA is the presence of an 11-nucleotide stretch that shows a perfect base-sequence complementarity to a stretch near the 3′ terminus of the 12S rRNA, in a way that is strikingly reminiscent in its details of the interaction between the Shine-Dalgarno sequence and bacterial mRNAs.[34] This 11-nucleotide stretch suggests that 7S RNA has in vivo an interaction with 12S rRNA either free or incorporated into the small ribosomal subunit. The location of the 7S RNA near the origin of replication and near an initiation site for L-strand transcription (see below) may be significant for its function, possibly in connection with the initiation of mtDNA replication or with the complete transcription of the L-strand.

The oligo(dT)-nonbound RNA components include what are quantitatively the two major classes of mitochondrial RNA, i.e., the two high-molecular-weight rRNA species 16S and 12S RNA, and 4S RNA. Mitochondrial rRNAs of human, and animal cells in general, are

**Table 2**
**STEADY-STATE AMOUNT, RATE OF SYNTHESIS, AND HALF-LIFE (IN MIN) OF INDIVIDUAL MITOCHONDRIAL RNA SPECIES**

| RNA species | Steady-state amount (no. molecules/cell) | Rate of synthesis (no. molecules/min/cell) | Half-life (in min): Kinetics of incorporation of label | Half-life (in min): Decay of label after cordycepin block |
|---|---|---|---|---|
| Oligo(dT)-unbound | | | | |
| 16S rRNA | n.d. | n.d. | 215 | 282 |
| 12S rRNA | 34,000 | 265 | 208 | n.d. |
| Oligo(dt)-bound | | | | |
| 2 | n.d. | n.d. | n.d. | 7 |
| 4 | 44 | 0.8 | n.d. | 39 |
| 5 | 165 | 3.4 | n.d. | 87 |
| 6 | 125 | 4.4 | n.d. | 16 |
| 7 | 960 | 15.0 | 47 | 112 |
| 9 CO I mRNA | 950 | 10.5 | 67 | 116 |
| 10 (16S rRNA) | 560 | 2.1 | 185 | n.d. |
| 11 Cyt.*b* mRNA | 570 | 7.0 | 53 | 56 |
| 12 | 720 | 6.1 | 73 | 51 |
| 13 | 650 | 18.0 | n.d. | n.d. |
| 14 ATPase 6 mRNA | 770 | 6.8 | 59 | 141 (14 and 15 combined) |
| 15 CO III mRNA | 980 | 9.4 | 71 | |
| 16 CO II mRNA | 1,190 | 10.0 | 77 | 191 |
| 17 | 225 | 4.5 | n.d. | n.d. |
| 18 (7S RNA) | 1,900 | 7.3 | 67 | n.d. |

From Attardi, G., Cantatore, P., Chomyn, A., Crews, S., Gelfand, R., Merkel, C., Montoya, J., and Ojala, D., in *Mitochondrial Genes*, Slonimski, P., Borst, P., and Attardi, G., Eds., Cold Spring Harbor Laboratory, Cold Spring Harbor, New York, 1982, 51. With permission.

the smallest known rRNA species, if one excludes the rRNAs of *Trypanosomatid* mitochondria (see below). Human and hamster mitochondrial rRNAs are methylated,[35-37] and, in hamster cells, the level of this methylation — only in ribose residues in the large rRNA (on the average, about 3 methyl groups per molecule[36]), in bases in the small rRNA (about 7 methyl groups per molecule[37]) — is considerably lower than that of cytoplasmic rRNAs (77 and 45 methyl groups per molecule in the large and small species, respectively). No pseudouridine has been detected in the hamster mitochondrial rRNA species, whereas 41 and 29 pseudouridines per molecule have been found in the large and small, respectively, cytoplasmic rRNA species from the same organism.[36] The 3′ ends of the 12S rRNA from human and hamster mitochondria are, in their majority, oligoadenylated, exhibiting mostly stretches of one to five A residues.[15] Similarly, most of the 16S rRNA molecules of human and hamster origin exhibit between one and nine A residues at their 3′ ends; furthermore, these ends are characteristically ragged[15] (Figure 5) (see below).

No 5S rRNA equivalent has been found in the human, and in general, animal mitochondrial ribosomes. However, a 23-nucleotide sequence is present at the 3′ end of the human 16S rRNA molecule that exhibits a 68% sequence homology and a similarity in structure to a portion of the *Bacillus subtilis* 5S rRNA;[38] these features and the location of the corresponding coding sequence relative to the mitochondrial rRNA genes, which is analogous to the position of the 5S rRNA gene in bacteria, have led to the suggestion that this segment represents a truncated 5S rRNA gene which has become a part of the 16S rRNA gene.

The 4S RNA components represent mostly, if not exclusively, mitochondria-specific tRNA species. On the basis of the mtDNA sequence, the 4S RNA is expected to include 22 species of tRNA specific for all amino acids, with two acceptors for serine and two for leucine.[3] The presence of at least 17 species has been directly demonstrated in HeLa cell mitochondria by hybridization experiments utilizing AA-tRNA complexes labeled in the amino acid moiety.[39] The tRNA sequences reveal only a partial agreement with the invariant tRNA pattern, with some striking deviations.[3] They are generally smaller than their cytoplasmic counterparts, varying in size in human mitochondria between 59 and 75 nucleotides. The 3′-terminal -CCA is not encoded in mtDNA and must be added post-transcriptionally.[3] In human and hamster mitochondrial tRNAs, the level of methylation — all in bases (on the average, ~2 methyl groups per molecule) — is about 30% of that of the cytoplasmic tRNAs and with significant differences in methylation pattern.[40]

### D. General Organization of Mitochondrial DNA Transcripts

A detailed picture of the organization of the mtDNA transcripts in HeLa cells has been provided by an analysis by the S1 protection mapping technique (Figure 1). This organization reflects in a faithful way the arrangement in the mtDNA sequence of the rRNA genes and reading frames and their interspersion with tRNA genes.[13] All the identified H- and L-strand transcripts are collinear with the DNA, pointing to a fundamental difference in gene organization in mammalian mtDNA as compared to the mtDNA from lower eukaryotes, i.e., the lack of intervening sequences in the genes. On the H-strand, the sequences coding for the discrete transcripts (including the expected tRNA gene transcripts) saturate the entire length of the strand, with the exception of a portion, corresponding to about 7% of the genome, around the origin of replication, from coordinate 98/100 to coordinate 5/100 relative to the origin taken as 100(0)/100. These observations confirm the earlier evidence concerning the complete or almost complete transcription of the H-strand in HeLa cells.[23] Excluding RNAs 4 (see below) and 6 (probable precursor of CO I mRNA), there is no apparent overlapping in the H-strand of the sequences coding for the rRNAs, tRNAs, and poly(A)-containing RNAs.

There is an almost perfect correspondence between individual reading frames of the H-strand and the H-strand-coded putative mRNAs.[3] In nearly all cases, each mRNA contains only one reading frame; the two exceptions are RNA 14, which contains a 5′-proximal reading frame of 207 nucleotides (URFA6L), overlapping out of phase by 46 nucleotides the reading frame for the ATPase subunit 6 polypeptide,[3] and RNA 7, which contains a 5′-proximal unidentified reading frame of 297 nucleotides (URF4L), overlapping out of phase by 7 nucleotides the unidentified reading frame URF4.[3] The existence of two pairs of reading frames overlapping out-of-phase, each pair being represented by a single mRNA, is preserved in the bovine,[19] mouse,[20] and rat[21] mitochondrial genomes. This situation raises intriguing questions as to the mechanism whereby the ribosome, which presumably enters the mRNA at its 5′ end, can read the second reading frame after translating the upstream cistron. A distinct possibility is that a frameshift occurring during the translation causes a premature termination at one of several nonsense codons just preceding out-of-phase ATPase 6 reading frame or URF4, followed by initiation at the start codon of the following reading frame. Such a mechanism has been suggested to operate in lysis gene expression of RNA phage MS2.[41]

The juxtaposition with the tRNA genes of the H-strand sequences coding for the rRNAs and poly(A)-containing RNAs, which has been indicated by the S1 mapping experiments, has been confirmed and refined by sequencing analysis of the 5′-end and 3′-end proximal segments of the putative mRNAs and of the rRNAs.[16,17] This analysis has shown that whenever, as is usually the case, a tRNA gene is situated on the 5′-side or 3′-side of the

FIGURE 4. Schematic representation of the 3′ ends of the transcribed moieties of HeLa cell mitochondrial 16S and 12S rRNAs. The arrows specify residues corresponding to 3′ termini of rRNA transcripts, with their length correlating roughly with the relative abundance of the various classes. Dots correspond to 16S rRNA transcripts occurring in trace amounts (<4%) and/or whose exact termini are uncertain. See text for details. (With permission from Dubin, D. T., Montoya, J., Timko, K. D., and Attardi, G., *J. Mol. Biol.*, 157, 1, 1982. Copyright: Academic Press Inc. (London) Ltd.)

mRNA- or rRNA-coding sequence, this coding sequence starts immediately after or, respectively, terminates gene immediately before the tRNA gene (Figures 4 and 5). The 3′ ends of the 16S rRNA transcripts have been found, however, to be ragged, with the 3′-termini of the two major types (which constitute about 90% of the total) corresponding to the nucleotide immediately adjacent to the $tRNA^{Leu}$ gene or to the preceding one, and the 3′ ends of the minor types forming with those of the major types a cluster of nucleotides, mostly contiguous, in the mtDNA sequence[15] (Figure 4). All species of the transcribed moiety have been shown to be oligoadenylated.

The 3′-end sequencing analysis of the putative mitochondrial mRNAs from HeLa cells has revealed a striking feature, which points to the crucial role of RNA processing and polyadenylation in the formation of the termination codons of the reading frames. In particular, all mRNAs, with the exception of RNA 5 (URF5 mRNA), RNA 9 (CO I mRNA), and RNA 16 (CO II mRNA), lack a stop codon at the end of the coding sequence and terminate with a U or UA, which follows the last sense codon. An ochre codon is created at the time of addition of a poly(A) tail to the 3′ end of the mRNA[17] (Figure 5). The three exceptions to this rule, RNAs 5, 9, and 16, do possess a complete termination codon (UAA, AGA, and UAG, respectively), at some distance from the 3′-terminus.

In the L-strand, the discrete transcripts identified so far correspond to only 65% of its length; in particular, they extend continuously from coordinate 12/100 to coordinate 75/100, and include only short transcripts (7S RNA and a few tRNAs) in the remainder of the map.[13] However, the earlier evidence of complete transcription of the L-strand[25] suggests that there must be transcripts corresponding to the as yet unaccounted for 35% of the map, which are in too low a concentration to be detected and/or are heterogeneous in length and therefore not recognizable as discrete species.

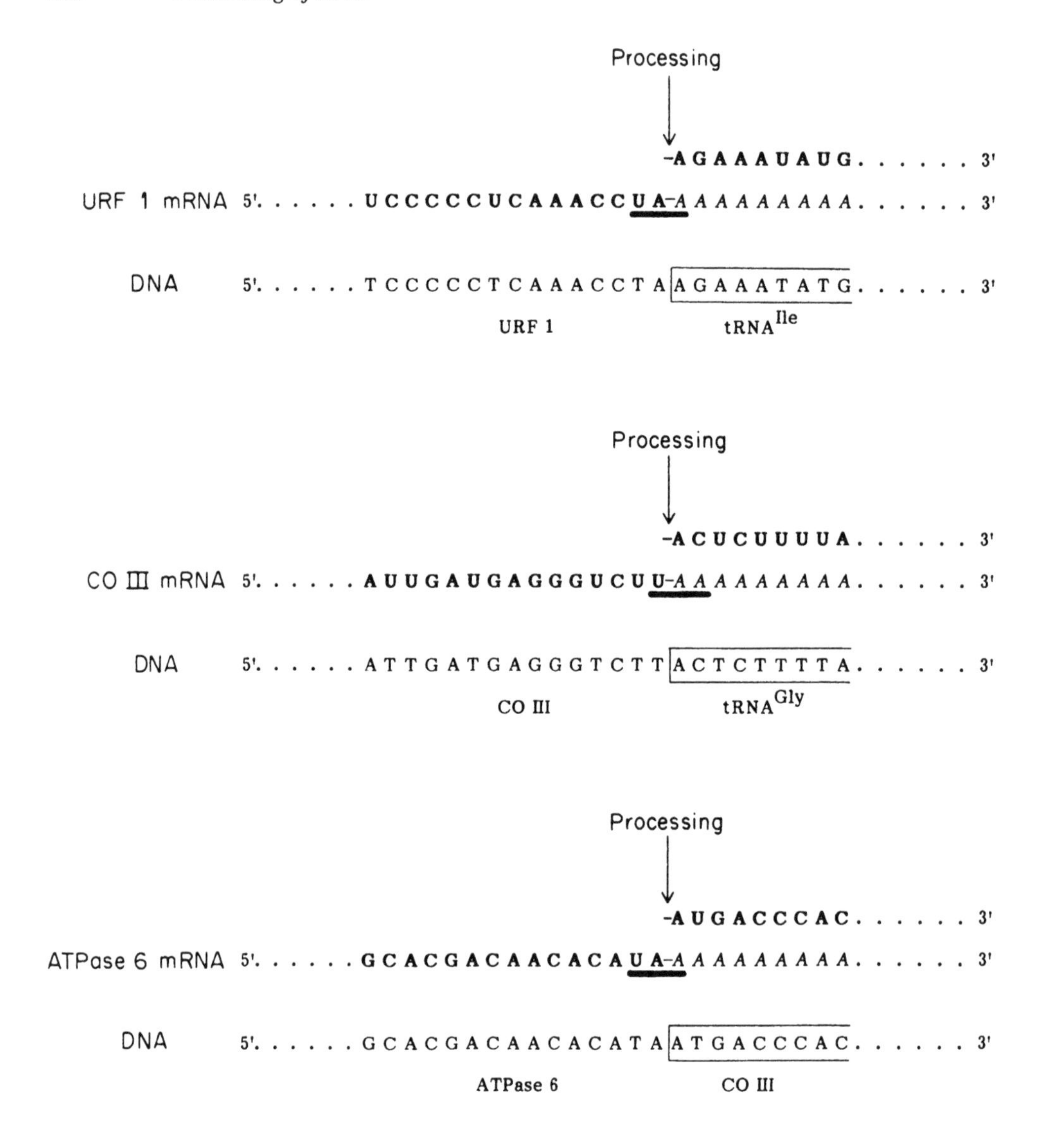


FIGURE 5. The 3′-end proximal segment of three mtDNA reading frames (URF1, CO III, and ATPase 6) and a 5′-end proximal segment of the adjacent tRNA gene or reading frame are shown, aligned with the corresponding transcripts, cleaved and polyadenylated at the 3′ end of the mRNAs. The ochre codons created by the polyadenylation are underlined. See text for details.

## E. The "tRNA Punctuation" Model of RNA Processing[17]

The mapping and sequencing data discussed above have indicated that the H-strand sequences coding for the rRNAs, poly(A)-containing RNAs, and tRNAs are immediately contiguous to each other, extending continuously through almost the entire length of the H-strand. This arrangement is consistent with a model of transcription of the H-strand in the form of a single molecule that is processed by precise endonucleolytic cleavages before and after each tRNA to yield, in most cases, nearly mature products,[17] with the polyadenylation (or oligoadenylation) of the mRNAs and rRNAs and the -CCA addition to the tRNAs starting concomitantly or subsequently to the cleavage step. As will be discussed below, recent mapping studies of nascent RNA molecules isolated from transcription complexes of HeLa

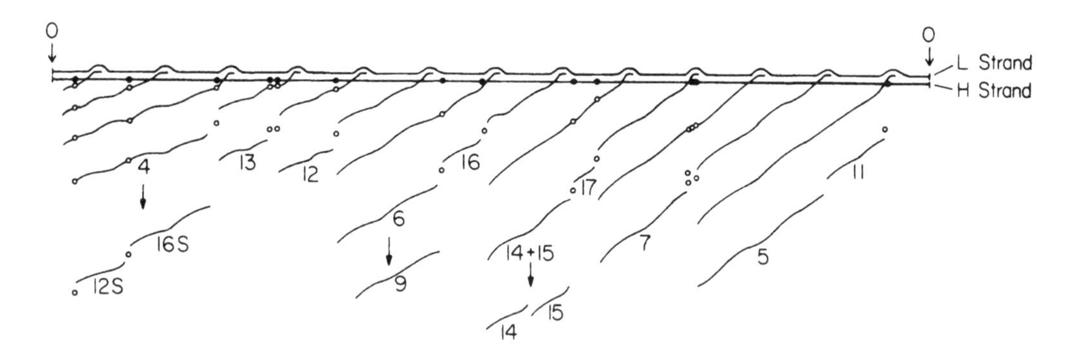


FIGURE 6. Proposed model for mtDNA H-strand transcription and processing of nascent mitochondrial RNA chains in the transcription complexes: ● tRNA gene; ○ mature tRNA. (Reprinted by permission from Ojala, D., Montoya, J., and Attardi, G., *Nature (London)*, 290, 470, 1981. Copyright © 1981 Macmillan Journals Limited.)

cell mtDNA and of in vitro "capped" RNA molecules have indicated the existence of two initiation sites for H-strand transcription near the origin of replication.[42] One or both of these sites may be the starting point of the single transcripts postulated here. Furthermore, the failure to detect any giant-size H-strand transcripts, either associated with transcription complexes or as discrete species, suggests that the processing of the H-strand transcripts may occur while they still reside in the mtDNA transcription complexes (Figure 6).

In the processing of the primary transcripts, the secondary structure of the tRNA sequences may represent the main recognition signal, providing the punctuation in the reading of mtDNA information.[17] It is possible, in fact, that these sequences acquire a cloverleaf configuration while they are still a part of the nascent transcripts, and that a processing enzyme(s) analogous to *Escherichia coli* RNase P[43,44] recognizes this structure or a portion of it, producing precise endonucleolytic cleavages at the 5′- and 3′-termini of the mature tRNAs.

The observation that all the RNA species other than tRNA which derive from the primary processing of the nascent H-strand transcripts are polyadenylated (putative mRNAs) or oligoadenylated (12S and 16S rRNAs) strongly suggests that adenylation may be linked in some way to, and possibly required for, the processing step that produces the 3′-termini of these RNA products, independent of the functional role of the products. The previously mentioned role of adenylation in completing the ochre codon of most of the mRNAs, and the immediate juxtaposition of the sequences coding for these mRNAs to a tRNA gene or to the initiator codon of another reading frame (as in the case of RNA 14 [Figure 1 and Figure 5]) demand that the endonucleolytic cleavages at the 3′ ends of these mRNAs be absolutely precise. In this connection, the heterogeneity of the 3′ end of the 16S rRNA may be particularly significant (see below). The observation that most mRNAs start directly with the initiator codon, and that their coding sequences are immediately flanked on their 5′-side by a tRNA gene or by the incomplete stop codon of another reading frame (as is the case of RNA 15 [Figure 1 and Figure 5]), likewise demands a high precision of cleavage of these mRNAs at their 5′ ends. A similar precision has been observed for the endonucleolytic cut produced by RNase P in the tRNA precursors in bacteria.[44]

There are a few processing sites in the H-strand transcripts where no tRNA sequences have been found, i.e., at the site where RNA 9 is cleaved out of RNA 6 and at the border between RNA 14 and RNA 15 and between RNA 5 and RNA 11 (Figure 1). It is conceivable that the processing enzyme(s) can recognize a secondary structure which shares some critical features with the cloverleaf structure of the tRNA (as the RNase P in bacteria, which can recognize the precursor of the 4.5S RNA[44]). A stem-loop structure which occurs at the border between the ATPase 6 and CO III reading frames in mouse mtDNA has been suggested as one which could replace a tRNA as a processing signal.[20] Furthermore, it should be mentioned that a sequence complementary to an L-strand-coded tRNA, which should therefore be susceptible to folding in a cloverleaf-like structure, occurs in the RNA 6 segment just preceding the cleavage site for RNA 9 and in the RNA 5 segment adjoining the 5′ end of RNA 11 (Figure 1).

The 5′-termini of the three large L-strand transcripts, apparently coincident as indicated by the S1 protection experiments, fall very near to the $tRNA^{Glu}$ gene on the L-strand (Figure 1). Similarly, the S1 data indicate that the 3′ end of RNA 2 falls very near to the $tRNA^{Ser}$ gene on the L-strand, and the 3′ end of RNA 3 corresponds closely to the $tRNA^{Arg}$ gene on the H-strand. Furthermore, the 3′ end of RNA 1, as judged from its size estimated from electrophoretic mobility, falls near the $tRNA^{Gln}$ gene on the L-strand. It thus seems possible that tRNA sequences also play a role as processing signals for the L-strand transcripts. In the case of the 3′ end of RNA 3, the putative processing enzyme(s) may recognize the anti-$tRNA^{Arg}$ sequence.

## F. Differential Expression of the rRNA and Protein-Coding Genes on the H-Strand

The model of transcription of the H-strand in the form of a single polycistronic molecule described above would place rRNA genes, tRNA genes, and protein coding genes in a single transcription unit. This situation is without precedence in nature, and raises the question of how a differential expression of the various genes is achieved in this system. In particular, how is the large excess of rRNA over the individual mRNAs (30- to >100-fold) that exists in the mitochondrial system[30] (Table 2), as in other systems, produced? An analysis of the metabolic properties of HeLa cell mitochondrial RNAs has revealed that the basis for the difference in amount of rRNAs relative to mRNAs is not primarily a difference in metabolic stability. In fact, as shown in the last two columns in Table 2, the half-life of the rRNAs, as exemplified by the 12S rRNA species, is only two to five times longer than that of the mRNAs.[30] Therefore, a difference in the rate of synthesis must be the main factor responsible for the large difference in amount between the rRNA and mRNA species. As a matter of fact, a direct determination of the rate of synthesis of the HeLa cell mitochondrial RNA species has fully confirmed this prediction.[30] As shown in Table 2 (second column), the rate constants of synthesis of the rRNA species, using as a representative figure the value obtained for 12S rRNA, are 15 to 60 times higher than those of most mRNAs.

The rates of synthesis discussed here represent the rates of appearance of the mature RNA species. However, they should reflect faithfully the rate of transcription of the corresponding stretches of DNA. In fact, assuming that the rate of transcription of the first 2653 nucleotides of the putative single transcription unit (including the 12s and 16s rRNA genes and the $tRNA^{Phe}$ and $tRNA^{Val}$ genes) is the same as the rate of synthesis found experimentally for 12S rRNA, and that the rate of transcription of the remainder of the transcription unit is approximately equal to the weighted average found for the rates of synthesis of the mRNA species, and further considering that the rate of transcription of the L-strand is at least two to three times higher than that of the H-strand,[26] one can estimate a total rate of transcription of mtDNA of at least $2.5 \times 10^6$ to $3.3 \times 10^6$ nucleotides per minute per cell.[45] This rate is fairly close to that previously estimated on the basis of the initial rate of [5-$^3$H]uridine incorporation into the mitochondrial fraction ($\sim 4 \times 10^6$ nucleotides per minute per cell).[45] This comparison strongly suggests that a substantial part, if not all, of the H-strand transcripts in HeLa cells are processed into mature RNA molecules. The relative rates of synthesis of rRNA and mRNA, therefore, should reflect fairly closely the rates of transcription of the corresponding DNA segments. Furthermore, considering that there are from one to several thousands of molecules of mtDNA per cell, one arrives at the striking conclusion that in

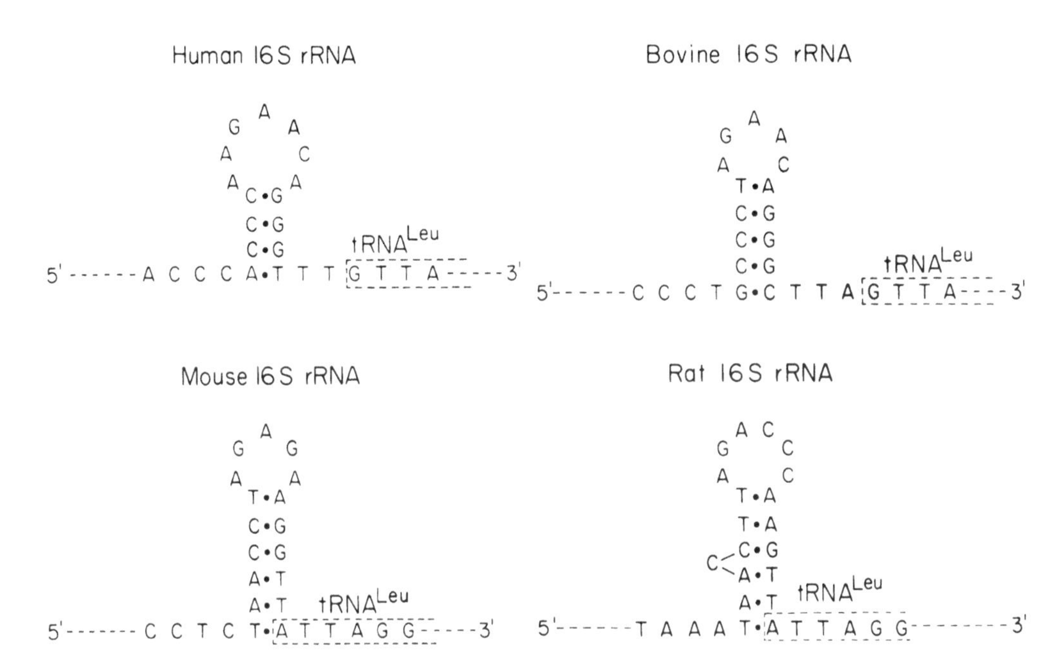


FIGURE 7. Potential stem-and-loop configurations of the 3′-terminal segment of the 16S rRNA gene. (From Attardi, G., Cantatore, P., Chomyn, A., Crews, S., Gelfand, R., Merkel, C., Montoya, J., and Ojala, D., in *Mitochondrial Genes*, Slonimski, P., Borst, P., and Attardi, G., Eds., Cold Spring Harbor Laboratory, Cold Spring Harbor, New York, 1982, 51. With permission.)

each mtDNA molecule the major portion of the H-strand which codes for the mRNAs and most tRNAs is transcribed infrequently, and possibly as rarely, on average, as twice to ten times per cell generation. The rDNA region is correspondingly transcribed, on average, 50 to 250 times per cell generation.[45]

**G. The "Alternative Pathway" Model of Regulation of rRNA and mRNA Synthesis**

In order to account for the great difference in rate of transcription between the first 2653 nucleotides and the remainder of the mtDNA molecule, still preserving the mode of transcription of the H-strand in the form of a single polycistronic molecule, it has been proposed that the postulated single polycistronic transcripts normally have a very high probability of premature termination at or near the 3′ end of the initial stretch of 2653 nucleotides.[17,46] The 3′-terminal region of the human 16S rRNA gene indeed shows a structure[15] resembling, in a rudimentary form, the hairpin-oligo (U) signal postulated for bacterial termination-attenuation[47] (Figure 7). A similar structure is exhibited by the 3′-terminal segment of the 16S rRNA gene of mouse, rat, and bovine mtDNA (Figure 7).[15] The imprecision mentioned above in the process leading to the formation of the 3′ end of 16S rRNA, which contrasts strikingly with the absolute precision of the endonucleolytic cleavages between mRNAs and downstream tRNA sequences, would also be consistent with the idea that the ragged ends of the large mitochondrial rRNA species arise by termination rather than by processing.

Two types of observations have been crucial in elucidating the process involved in the regulation of rRNA and mRNA synthesis. The first type of observation concerns the nature of polyadenylated RNA 4. Because of its mapping position, which corresponds precisely to that of the 12S and 16S rRNA species[13] (Figure 1), and because its half-life, as determined after cordycepin block of RNA synthesis (~40 min), is considerably shorter than that of the rRNA species (3 to 4 hr, Table 2), RNA 4 was originally interpreted to be a precursor

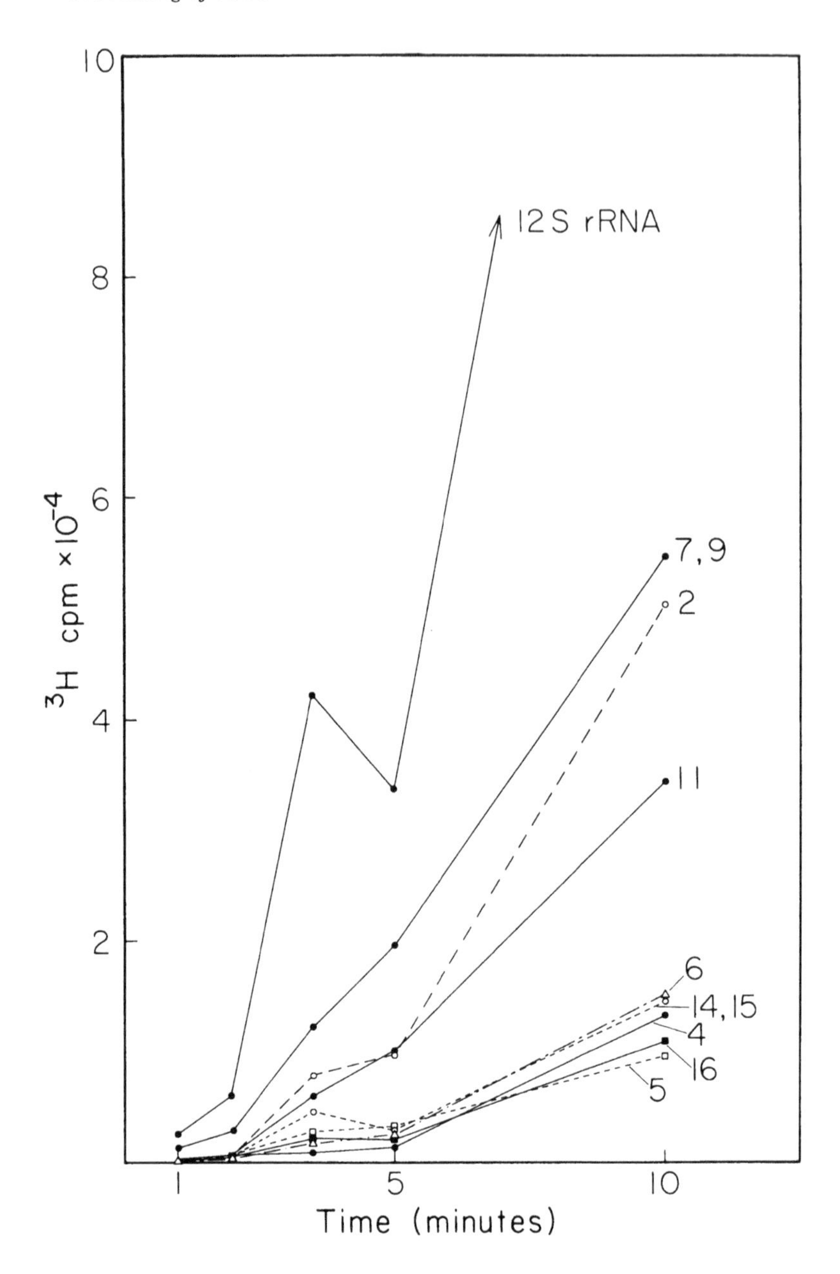


FIGURE 8. Flow of [5′-$^3$H] uridine into different mitochondrial RNA species during exposure of exponentially growing cells to different pulses of the precursor.

of the two rRNA species.[13] Further work, however, has shown that this functional identification was not correct. Figure 8 illustrates the flow of [5-$^3$H]uridine into different mitochondrial RNA species during exposure of exponentially growing HeLa cells to different pulses of the precursor. It is clear that the flow of $^3$H radioactivity through RNA 4 is not faster than the flow through all the other RNA species whose coding sequences lie downstream relative to the rRNA genes, and it is, furthermore, much slower than the labeling of 12S rRNA. These experiments have led to the unambiguous conclusion that RNA 4 is not the kinetic precursor of the great majority of the 12S and 16S rRNAs. They have suggested, on the contrary, that RNA 4 results from an alternative pathway of transcription of the rDNA region, which does not lead to the formation of the bulk of the rRNA, but represents, rather, a bypass joining the promoter near the origin to the main route running through the quasi-totality of the H-strand. At some stage, a decision is made as to whether transcription should go through the pathway leading to the obligatory processing of the transcripts to 12S rRNA and $tRNA^{Val}$ and to their termination at the 3′ end of the 16S rRNA gene, or proceed through the less-frequented pathway leading to the downstream regions of the transcription unit.

Recent experiments have provided evidence strongly suggesting that the decision discussed above is made at the step of initiation of transcription. In fact, mapping experiments using mitochondrial RNA molecules "capped" in vitro with [$\alpha$-$^{32}$P]GTP and guanylyltransferase, or using nascent RNA chains isolated from mtDNA transcription complexes, have identified two initiation sites for H-strand transcription.[42] Of these, one is situated very near the 5′ terminus of the 12S rRNA gene, and the other 90 to 110 nucleotides (nt) upstream of this site, i.e., 20 to 40 nt upstream of the $tRNA^{Phe}$ gene (Figure 9). It is a plausible idea that these two initiation sites for H-strand transcription are correlated with two different transcription events, one leading to the synthesis of a polycistronic molecule corresponding to almost the entire H-strand, and the other limited to the rDNA region and responsible for the synthesis of the bulk of rRNA (Figure 9). According to this "alternative pathway" model, a transcript has its fate determined by its having been started at one of the two sites. A scanning of the DNA sequence has not revealed any significant repeated sequence near the two initiation sites, suggesting the possible involvement of two structurally related or unrelated polymerases. The polymerase itself and/or the extra nucleotide stretch at the 5′ end, either directly or through sequential changes induced in the secondary structure of the rDNA transcript, would be involved in determining the termination or continuation of transcription at the 3′ end of the 16S rRNA gene. No data are available as yet to decide whether the promoter controlling the rRNA synthesis is the one close to the 5′ end of the 12S rRNA gene (Figure 9B), or the one upstream from it (Figure 9A). The two possibilities lead, however, to two different predictions concerning the rate of synthesis of the $tRNA^{Phe}$, which are directly testable. The model discussed above implies that RNA 4 results from the processing of the polycistronic transcript destined to run through the nearly total length of the H-strand. It seems likely that RNA 10 results from processing of RNA 4. On the contrary, there is no evidence as to whether any $tRNA^{Val}$ and 12S rRNA result from processing of RNA 4.

### H. L-Strand Transcription

The mapping study mentioned above has also revealed the occurrence of an initiation site for L-strand transcription very near to the 5′ end of 7S RNA.[42] It is not known what relationship exists between this initiation site and the synthesis of 7S RNA and/or the complete transcription of the L-strand or the synthesis of the RNA primer for H-strand synthesis.[48] It seems likely that the large L-strand-coded polyadenylated transcripts 1, 2, and 3 derive from processing of molecules initiated near the origin. If the initiation site detected here is involved in the formation of these primary transcripts, and if 7S RNA also derives from processing of these transcripts, such processing must be a relatively rare event, judged from the relatively low rate of formation of 7S RNA (Table 2) as compared to the high rate of transcription of the L-strand.

The reason for complete L-strand transcription is not easily explained by the need for expression of the few scattered genes encoded in this strand. Particularly intriguing is its high rate, which is at least two to three times higher than that of the H-strand, suggesting a special role for this transcription. One interesting possibility is that the L-strand transcripts, terminating preferentially at the processing sites for the H-strand polycistronic transcripts growing in the opposite direction, may facilitate this processing. Another possibility is that

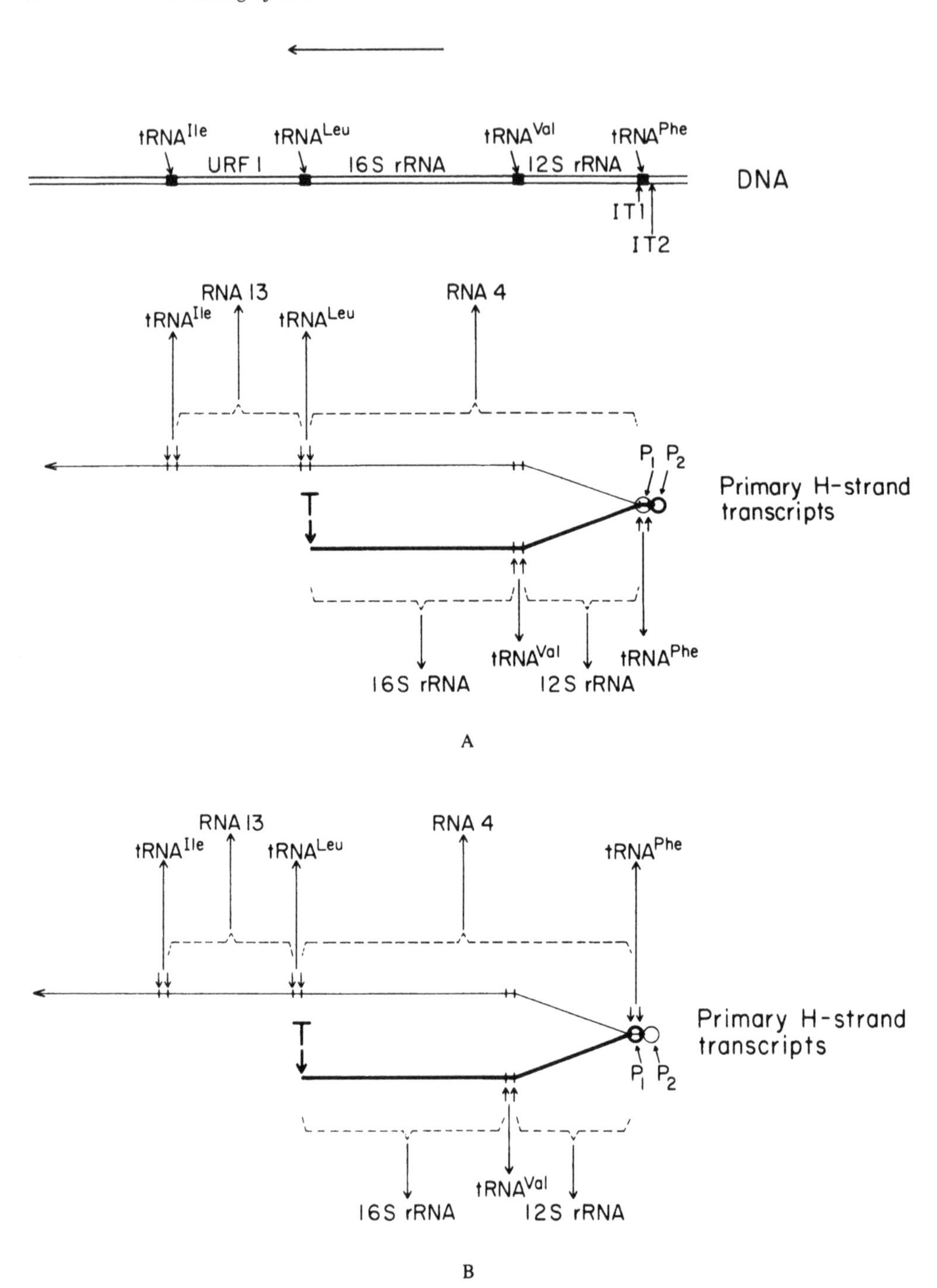


FIGURE 9. "Alternative pathway" model of regulation of mitochondrial rRNA and mRNA synthesis. IT: initiation of transcription; P: promoter; T: termination.

the complete L-strand transcription has the role of "activating" the DNA for the transcription of the H-strand. This function, which possibly is responsible for the loosening up of the normal compact configuration of mtDNA, may be similar to that which starts emerging for RNA synthesis in activating nuclear chromosomal domains.

# III. MITOCHONDRIAL DNA OF OTHER ANIMAL CELLS

Sequence analysis of mtDNA from several mammalian species (bovine,[19] mouse,[20] and rat[21,22,33,49,50]) has shown that, in spite of sequence differences, the detailed gene organization is substantially identical in these mtDNAs. Thus, it seems reasonable to extrapolate from these results and to predict that mtDNA from all mammals will be found to have essentially the same gene organization, in spite of the very rapid sequence divergence that has occurred in the mtDNA from different species. As an example of a minor difference, from the mouse mtDNA sequence it has been inferred that, in addition to AUG, AUA, and AUU, also AUC can function as initiator codon, in this mammal.[20] The substantial identity in gene organization suggests that the DNA transcription and RNA processing patterns observed in human mitochondria should apply to all mammals. The available evidence concerning the transcription maps of mouse[51] and rat[52] mtDNA is in agreement with this expectation. The only exception so far, i.e., the failure to detect in these organisms large L-strand transcripts or the equivalent of 7S RNA, is very probably due to technical factors.

The tRNA[53] and rRNA[54] gene organization revealed by electron microscopy of RNA-DNA hybrids in *Xenopus laevis* mtDNA is very similar, if not identical, to that of mammalian mtDNA; furthermore, the transcription map of *X. laevis* H-strand mtDNA[55] has been found to mirror closely in number and position of transcripts that of man and mouse. It seems, therefore, likely that in all vertebrates the mtDNA transcription patterns will be similar in its general features to that observed in mammals.

Most of the information on mtDNA transcription in invertebrates derives from studies on the mitochondrial genome of *Drosophila melanogaster*. The mitochondrial genome of species of the genus *Drosophila* is similar in size and structure to that of other animal cells, except for the presence of an AT-rich region; this region varies in size in the mtDNAs of different species from 1.0 kb in *D. ananassae* to 5.1 kb in *D. melanogaster*, accounting for the difference in size of the mtDNA molecules which contain them (from 15.9 kb to 19.5 kb, respectively).[56] In all *Drosophila* species analyzed (six), the AT-rich region contains the origin of replication,[56] and, in *D. melanogaster*, it has been shown to map close to the position of the two rRNA genes.[57]

A total of 11 oligo(dT)-cellulose-bound RNA species and one nonbound species have been identified in mitochondrial RNA preparations from *D. melanogaster*.[58,59] An identical set of species with the same binding properties to oligo(dT)-cellulose has been observed in *Aedes albopictus*.[60] In both organisms, the oligo(dT)-cellulose nonbound mitochondrial RNA species is the small rRNA component, while one of the oligo(dT)-cellulose bound RNA species is the large rRNA component. The almost quantitative binding to oligo(dT)-cellulose of the large mitochondrial RNA species from *Drosophila* and mosquito does not seem to be due to its high AT content (>82% in *D. melanogaster*); in fact, a poly(A)-tail of ~36 nt has been detected at the 3′ end of the large rRNA species from *A. albopictus*.[61] By contrast, the small rRNA species from the same source has been found to contain one or two 3′-terminal As.[61] In the absence of DNA sequence data, it is not possible to say whether the small rRNA species is oligoadenylated posttranscriptionally, like that from mammalian sources.

Electron microscopy of RNA-DNA hybrids and RNA transfer hybridization experiments have shown that the two rRNA genes map next to each other on the mtDNA of *D. melanogaster* with a gap of about 160 bp between them, and that the small rRNA gene may overlap the AT-rich region by as much as 200 bp.[57] Furthermore, RNA-DNA hybridization experiments, after digestion of appropriate restriction fragments with 3′ end- or 5′-end-specific exonucleases, have indicated that transcription, here, as in mammalian mtDNA (Figure 1), proceeds from the small to the larger RNA.[57] In other experiments, hybridization of individual radiochemically pure poly(A)-containing and non-poly(A)-containing RNAs with restriction fragments has yielded a preliminary transcription map of *D. melanogaster* mtDNA, in which the sites of hybridization appear to be widely dispersed around the genome, saturating it almost completely.[59] Although no data are available on the mapping positions of the tRNA genes in this genome, the above hybridization data strongly suggest that, in mtDNA from *D. melanogaster* and from insects in general, the gene organization and mode of expression are probably very similar to those of mammalian mtDNA.

## IV. YEAST MITOCHONDRIAL DNA

### A. Sequence and Gene Organization

Yeast mtDNA varies considerably in size in different species, from the minimum of 18.9 kb found in *Torulopsis glabrata* to the maximum of 108 kb observed in *Brettanomyces custerii.*[62] In *Saccaromyces* strains, which have been the most extensively investigated, mtDNA has an average size of about 75 kb.[1,62] The limited evidence available suggests that in spite of variations of gene order the basic gene content and the essential features of gene organization are preserved in the mtDNAs from various yeasts, the size differences being mostly accounted for by the variable length of noncoding stretches and the variable number and size of intervening sequences in the mosaic genes.

In spite of the larger size, yeast mtDNA appears to have the same basic genetic information as mammalian mtDNA. However, its gene organization, as seen most typically in *Saccharomyces* strains, is strikingly different. In contrast to the mammalian mitochondrial genome, where genes are continuous and mostly butt-jointed to each other, with a nearly regular interspersion of structural and ribosomal RNA genes with tRNA genes, in yeast mtDNA the genes appear to be immersed in a background of apparently noncoding AT-rich sequences,[1] several of the genes are themselves fragmented,[1] and the tRNA genes are for the most part clustered[63-65] (Figure 10). The distribution of informational content between the two DNA strands is even more strongly asymmetric in yeast mtDNA than in mammalian mtDNA, all the genes being located on one strand with the exception of one[66] (see below).

One of the most striking features of yeast mtDNA is its very low GC content[67] (~18%). How such DNA can be functional has been clarified by the systematic investigations by Prunell and Bernardi[67,68] which have revealed the extreme compositional heterogeneity of this DNA. The evidence indicates that AT-rich stretches (GC <5%), averaging about 4 × $10^5$ daltons in size and accounting for up to about 50% of the genome, alternate with stretches with moderate GC content (average 26%) and of about the same average size, which also amount altogether to up to about 45% of the mtDNA length; furthermore there are, interspersed with the AT-rich and the moderately GC-rich stretches throughout the genome, short sequence elements with a very high GC content (45 to 62%) which account for about 6% of the genome. The conclusions by Prunell and Bernardi [67,68] about the general sequence organization of yeast mtDNA have been confirmed by more recent DNA sequence studies. However, the interpretation of the AT-rich stretches as "spacers" and of the moderately GC-rich stretches as "genes", although in general valid, has found some exceptions. Thus, the base composition of the structural gene for the var-1 protein is nearly 90% AT, with about one third of its GC content being present as a single 45-bp cluster.[69] Furthermore, DNA sequence and S1 mapping studies have revealed that the mRNAs so far analyzed have a long 5′-leader and a shorter 3′-tail very rich in AT; it seems likely that the 5′-leader and the 3′-tail have a function, though as yet undefined, and therefore cannot strictly be considered as noncoding spacers. The significance of the sequence elements with very high GC content is still unknown. It has been suggested that they are regulatory elements, like

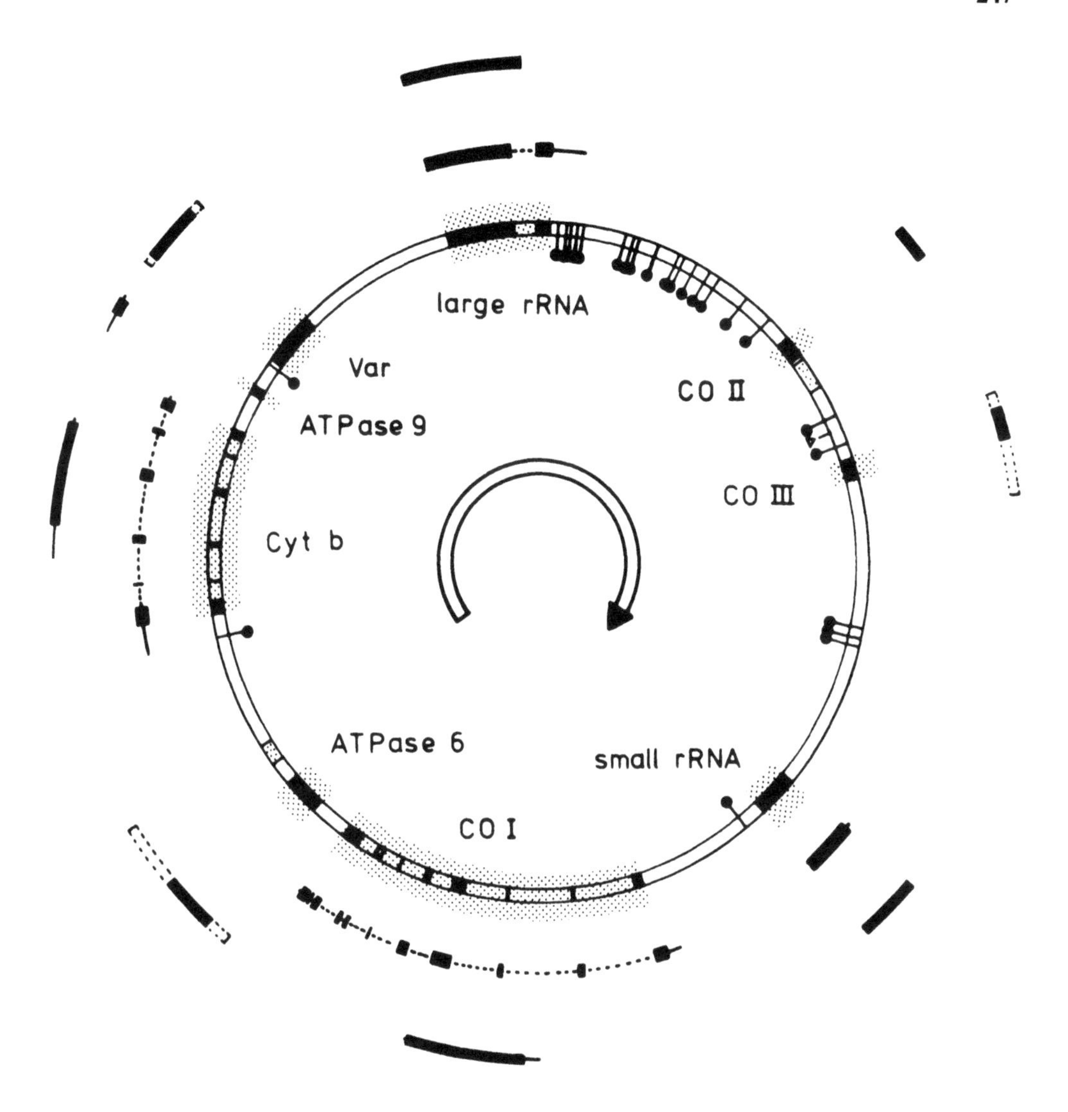


FIGURE 10. Genetic and transcription maps of the *S. cerevisiae* (strain KL14-4A) genome. The shaded areas correspond to the genes (with the black blocks representing the continuous or interrupted coding sequences), while the stippled blocks outside the genes represent long unidentified reading frames (URFs). The small filled circles represent the tRNA genes of the main coding strand, the open triangle (coding sequence) a tRNA gene ($tRNA^{Thr}_{GAU}$) on the complementary strand.[66] The putative mature mRNAs are shown in the outer ring as thick bars with 5′- and 3′-extensions; the open bar extensions indicate uncertainty or variability of limits. Between the outer ring and the DNA rings the putative precursors of the mRNAs are shown. The curved arrow in the center indicates the direction of transcription. (From Tabak, H. F., Grivell, L. A., and Borst, P., *CRC Crit. Rev. Biochem.*, With permission.)

promoters[68] or signals for RNA processing.[70] However, DNA sequence analysis has shown that such elements do not occur only between genes, but also within genes[69-72] and, in some cases, within introns.[73] They may be related to the GC-rich palindromic sequences containing double Pst I sites in *Neurospora crassa* mtDNA.

### B. Mitochondrial DNA Polymorphism

One of the most interesting aspects of gene organization in yeast mitochondria is represented by the gross variations in sequence which can be detected among different strains.

These variations have been best documented in *Saccharomyces* strains, where they occur in the form of major or minor additions or deletions of nucleotide stretches, which do not alter the overall gene order and do not have apparently any effect on gene expression. The major additions or deletions involve segments from a few hundred to several thousand nucleotide pairs.[74,75] Most of these have been shown to correspond to introns of mosaic genes, in particular, the 21S rRNA gene, the cytochrome *b (cob-box)* gene and the CO I*(oxi-3)* gene. Thus, the 21S rRNA gene exhibits a 1043-bp intron in the $\omega^+$ strains[76-78] (Figures 10 and 11), whereas it lacks it in the $\omega^-$ and $\omega^n$ strains.[74,75,78] The cytochrome *b* gene can exist in a "long" form with five introns as in *S. cerevisiae* KL14-4A[4,79,81] (Figures 10, 12, and 13a), or in a "short" form lacking introns bI1, bI2, and bI3 (for intron designation, the prefixes b and a will be used to indicate the cytochrome *b* and CO I gene introns, respectively), as in *S. cerevisiae* D273-10B[5] and in *S. carlsbergensis*[81] (Figures 12 and 13b). The CO I gene can likewise exist in a "long" form with 8 to 9 introns as in *S. cerevisiae* KL14-4[82,83] (Figures 10 and 14), or in a "middle-sized" form with 6 to 7 introns (where exons 5a, 5b, and 5c of the "long" form are fused, Figure 14), as in *S. cerevisiae* D273-10B,[6] or in a "short" form lacking aI1 and aI4 as in *S. carlsbergensis.*[81,83] Not all macroinserts found in the yeast mitochondrial genome appear to correspond to introns. The ~1.8 kb macroinsert, present in the AT-rich region downstream of the ATPase 6 *(oli-2)* gene in some *S. cerevisiae* strains (class I, like J69-1B), and absent in others (class II, like JM6), appears to be present in the tail of the putative ATPase 6 mRNA of class I strains.[84] A similar situation may exist in the *S. cerevisiae* strain LL20, where the CO III *(oxi-2)* specific mRNA has been found to be ~2 kb shorter at the 3′ end, as compared to the mRNA from the *S. cerevisiae* strain D273-10B.[85]

Besides the maxi-inserts described above, mini-inserts of fewer than 100 nucleotide pairs leading to polymorphisms have been described. Some of these appear to belong to the GC-rich clusters described by Prunell and Bernardi[68] which are widely distributed within the yeast mitochondrial genome and are located both between genes and within genes.[69-73] Thus, a 66-bp GC-rich cluster is present within the major exon of the 21S rRNA gene in the $\omega^+$ strains and absent in the $\omega^-$ and $\omega^n$ strains.[71] The 46-bp GC-rich cluster, which is present in all *var-1* alleles,[69] is duplicated in an inverted orientation, both in the gene and in its mRNA, in some strains of *S. cerevisiae,* but not in others (*a* element).[86] Both the 21S rRNA gene mini-insert and the *a* element of the *var-1* gene are flanked by a 2-bp terminal redundancy and, more significantly, have a palindromic sequence; this has led to the suggestion that they may represent transposable elements of the mitochondrial genome. An analogous type of GC-rich insert is present in the 15S rRNA gene in some yeast strains and absent in others.[72] In all cases so far investigated, these GC-rich sequences are found in the middle of high AT sequences (within a gene or between genes), suggesting that AT-rich sequences are preferred targets for this type of insertion.

Some of the mini-inserts which have been described in yeast mtDNA are AT-rich. To this class belong the *b* elements present in some strains within the *var-1* gene: these are represented by in-frame insertions of stretches of asparagine codons (AAT) which vary in number among different *var-1* alleles, resulting in the synthesis of larger *var-1* proteins.[69]

### C. Organization of Mitochondrial DNA in Multiple Transcription Units

As discussed above, in human mtDNA there appear to be one transcription unit for the L-strand and two overlapping transcription units for the H-strand, of which one includes the two rRNA genes and one or two of the adjacent tRNA genes, and the other comprises the other tRNA genes and the protein coding genes. In yeast mtDNA, on the contrary, transcription of the main coding strand is organized in multiple transcription units. This conclusion is based on the number of initiation sites for transcription, as identified by a mapping study of the mitochondrial RNA species which could be labeled in vitro at their 5′ end with

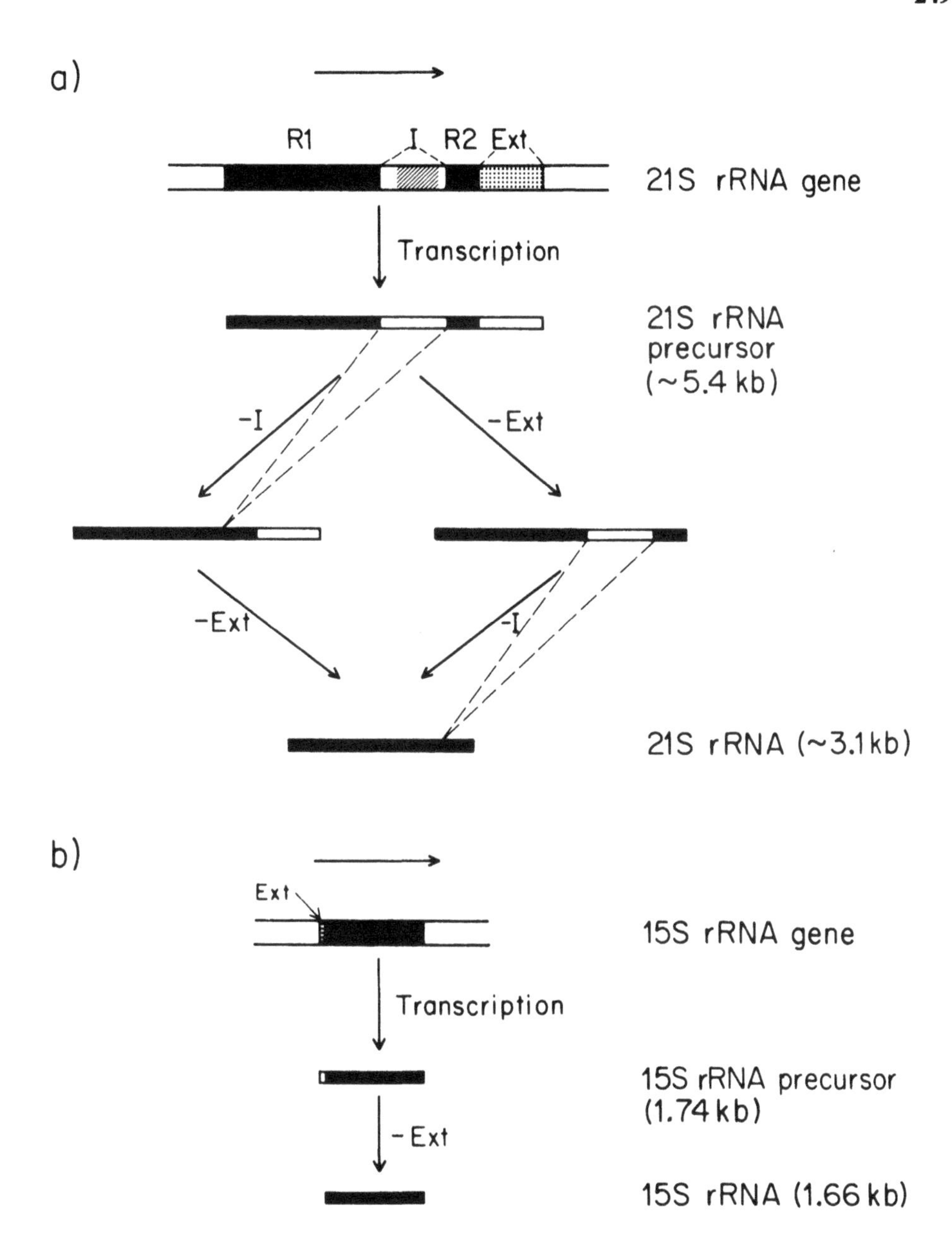


FIGURE 11. Synthesis and processing of the 21S rRNA (in an $\omega^+$ strain) (a) and the 15S rRNA (b) in *S. cerevisiae*. R1 and R2 in the 21S rRNA gene represent the two exons, and I, the intron, with the hatched area indicating an open reading frame; the stippled block in the 21S rRNA and 15S rRNA gene represents a 3′- and 5′-extension, respectively. The arrows indicate the direction of transcription.

[α-$^{32}$P]GTP and guanylyltransferase (“capping” enzyme) and which, therefore, had presumably preserved their 5′-terminal di- or triphosphate.[87] Seven to ten in vitro “capped” RNA species mapping at different positions have been identified by this approach in a wild-type strain (D273-10B).[88] Recent experiments indicate that the number of in vitro “capped” RNA species with different mapping specificities is at least 15.[89] Since single-deletion “petites” transcribe some of their RNAs in a manner which is identical to that observed in

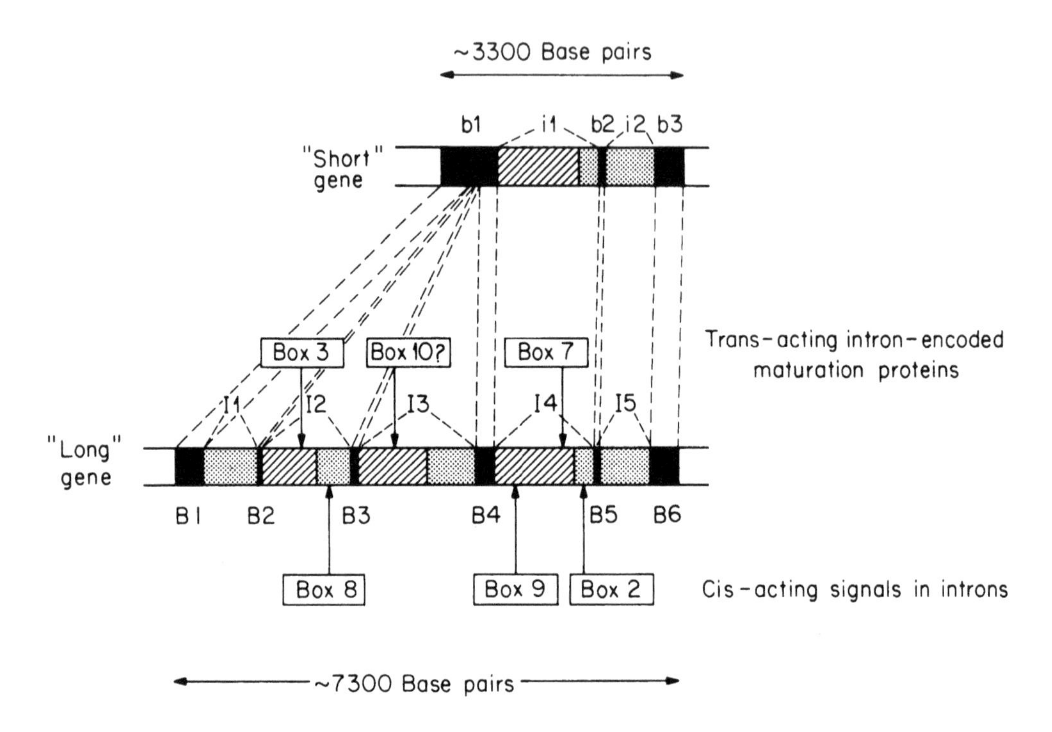


FIGURE 12. Structure of a "long" and a "short" cytochrome *b (cob-box)* gene. The "long" gene is from *S. cerevisiae* strain 777-3A,[8] the "short" gene from strain D273-10B.[5] Exons (B1 to B6 and b1 to b3) are represented as black blocks; in the introns (I1 to I5, i1 and i2), the opening reading frames in phase with upstream exons are shown as hatched areas, and the sequences blocked in the three registers, as stippled areas. Clusters of mutational sites within introns which block RNA splicing are indicated by boxed numbers. See text for details. (Modified from Jacq, C., Pajot, P., Lazowska, J., Dujardin, G., Claisse, M., Groudinski, O., de la Salle, H., Grandchamp, C., Labouesse, M., Gargouri, A., Guiard, B., Spyridakis, A., Dreyfus, M., and Slonimski, P. P., in *Mitochondrial Genes*, Slonimski, P., Borst, P., and Attardi, G., Eds., Cold Spring Harbor Laboratory, Cold Spring Harbor, New York, 1982, 155. With permission.)

the wild-type strain, it seems very unlikely that this apparent multiplicity of in vitro "capped" RNAs results from splicing of a single or a few shorter leader segments to different RNA species. The most plausible interpretation is that these distinct in vitro "capped" RNAs identify different transcription initiation sites on yeast mtDNA.

The most heavily "capped" transcript was found to be the 21S rRNA species, a result which shows that many of these molecules retain their initiating 5′-terminal ribonucleoside di- or triphosphate, and that processing at the 5′ end does not occur in this RNA species.[87] In contrast to the 21S rRNA, the 15S rRNA species was not "capped" in these in vitro experiments; however, an RNA species slightly larger (15.5S RNA) and mapping in the same region (Figures 10 and 11) was "capped", and was identified as the probable precursor of the 15S rRNA[87] (see below). In vitro "capping" of specific transcripts selected by hybridization with gene-specific cloned DNA fragments has allowed the identification and approximate localization of the initiation sites for transcription of the cytochrome *b* and CO I genes.[88] This powerful approach should allow in the near future the correlation with specific genes of the other transcriptional initiation sites detected in yeast mtDNA.

Another useful approach for the identification of promoter regions is provided by the study of transcription in *rho*$^-$ "petite" mutants. Thus, an analysis of the RNA synthesized in *rho*$^-$ mutants containing the CO III gene with varying lengths of 5′-flanking sequences has indicated that correct transcription and processing of the CO III mRNA depends on the

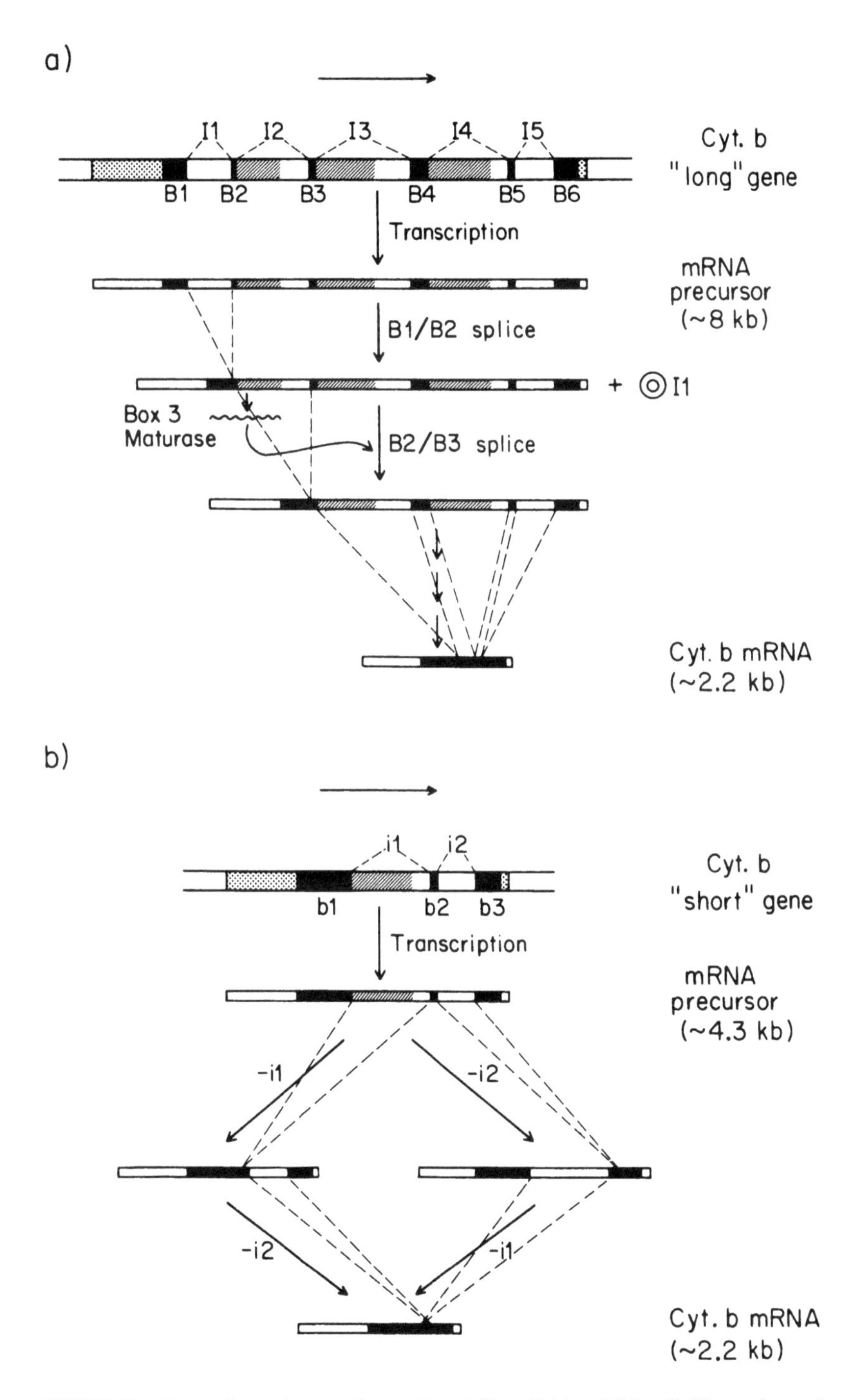


FIGURE 13. Processing pathways of transcripts of "long" (a) and "short" (b) cytochrome *b* gene. It is not certain whether the 8 kb and 4.3 kb mRNA precursors are the primary transcripts. See text for details.

presence of a 400 nt sequence located approximately 1.2 kb to the 5′ side of the initiator codon, upstream of the $tRNA^{Val}$ gene.[85] This critical sequence, which is also necessary for the transcription of the $tRNA^{Val}$ gene, may be the promoter of a primary transcript destined to be processed to the mature CO III mRNA by excision of the tRNA (see also below).

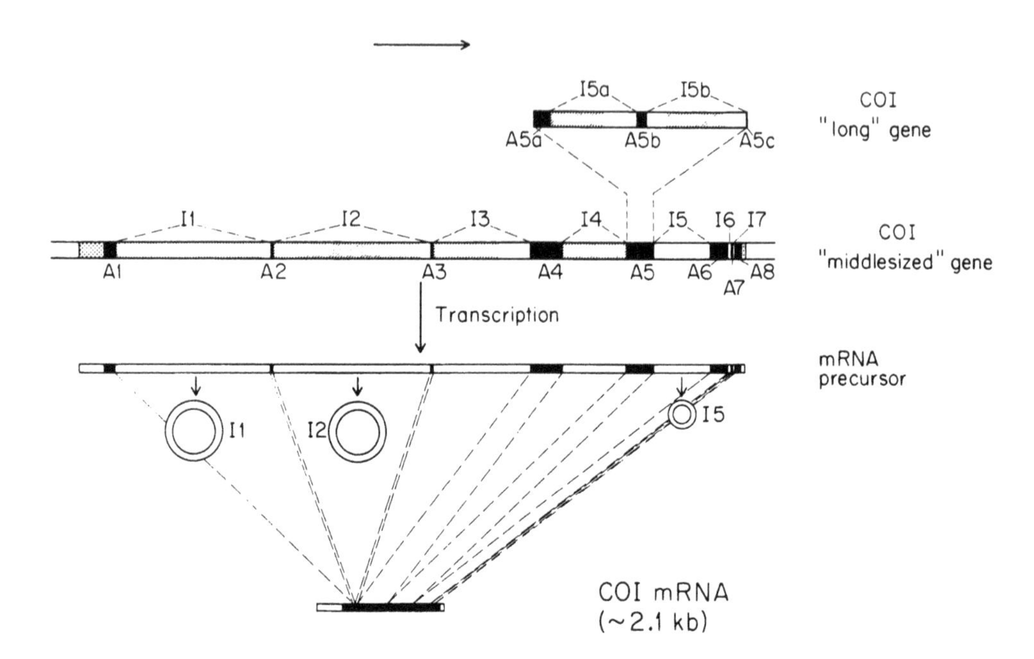


FIGURE 14. Structure of a "long" and a "middle-sized" CO I gene and synthesis of the CO I mRNA. The "long" gene is from *S. cerevisiae* strain KL14-4A,[82] the "middle-sized" gene from strain D237-10B.[6] Explanation of symbols as in Figure 12. See text for details.

The existence in the yeast mitochondrial genome of multiple transcription units raises the question of how a coordinate expression of different genes is achieved in this system. This question is especially pertinent for the coordination of expression of the two rRNA genes. In this case, the existence of a common sequence immediately preceding each gene, possibly representing an element of the rRNA promoter (see below), points to a control at the level of initiation of transcription.

## D. Strand Selection and Extent of Transcription

The discovery of complete symmetrical transcription of HeLa cell mtDNA[24,25] has raised the question of whether a similar situation applies to yeast mtDNA and has prompted several investigations aimed at answering this question. The distribution of genes in the two strands of yeast mtDNA appears to be almost totally asymmetric. With the exception of one, all the genes identified so far in *S. cerevisiae* are located on the same strand.[66] The single exception, the $tRNA^{Thr}_{CUN}$ gene, was assigned to the other strand on the basis of the DNA sequence of a *rho*$^-$ clone containing also a copy of the $tRNA^{Val}$ gene on the main coding strand (Figure 10). It seems unlikely that the unusual strand location of the $tRNA^{Thr}_{CUN}$ gene is the result of some rearrangement that has occurred during "petite" formation, since this location has been observed in several "petites".[91]

The first hint of the possible occurrence of some symmetrical transcription in yeast mtDNA came from experiments of saturation hybridization of mtDNA, in vitro labeled by nick-translation, with mitochondrial RNA.[92] In these experiments the saturation level was found to be reduced from 35 to 29% if the RNA was preannealed to a high Rot value, and to increase to 46% if the preannealed RNA was denatured prior to hybridization with DNA. The interpretation of these experiments is, however, difficult because they do not distinguish between formation and dissociation of intermolecular hybrids and effects of changes in the secondary structure of the RNA. Subsequent experiments, measuring the self-complementarity of mitochondrial RNA, pulse-labeled either in vivo[1] or in isolated organelles,[93] or the capacity of RNA synthesized in vitro by mtDNA transcription complexes to hybridize with excess mitochondrial RNA,[94] indicated that the transcripts complementary to the noncoding strand, if present at all, occur in low concentration. In agreement with this conclusion, experiments in which separate strands, radioactively labeled, of restriction fragments containing sequences of the CO II *(oxi-1)* gene were hybridized with total mitochondrial RNA transferred to DBM paper revealed discrete bands of hybridization only with the coding strand.[95,96] These experiments indicated that discrete antimessage transcripts corresponding to the CO II gene were not synthesized or were rapidly degraded, but did not exclude the presence of antimessage transcripts heterogeneous in size. In other experiments, in which separated strands of a fragment corresponding to the ATPase 6 gene were hybridized in solution with an excess of total mitochondrial RNA, the hybridization level of the coding strand was 95% while that of the complementary strand was only 5%.[92] In these experiments, however, self-annealing of any complementary RNA molecules would have reduced the extent of hybridization with the noncoding DNA strand.

Direct evidence for the presence of a small amount of self-complementary RNA in yeast mitochondrial RNA preparations has come from recent experiments in which a double-stranded RNA fraction has been isolated by a procedure involving the use of guanidine hydrochloride.[97] This RNA fraction, which was estimated to represent ~0.005% of total mitochondrial RNA, was characterized as being double stranded on the basis of its density in a $Cs_2SO_4$ gradient, its sensitivity to digestion by RNase III but not by RNase H or DNase I, and its sensitivity to RNases A and $T_1$ in low salt but not in high salt. From the capacity of this RNA fraction to hybridize to each of eight "petite" mtDNA probes containing sequences derived from widely different segments of mtDNA, it was deduced that the self-complementary RNA fraction corresponds to a major portion and possibly all of the length of the mtDNA molecule. Consistent with this conclusion are the results of an electron microscopic analysis of this RNA fraction, which showed double-stranded linear molecules resistant to DNase I but sensitive to RNase III, with a length distribution from 0.1 to 14 $\mu$, and with occasional longer molecules up to 23 $\mu$.

From the evidence discussed above it seems reasonable to conclude that some symmetrical transcription probably occurs in yeast mtDNA. This transcription may involve both genes and spacers and extend over a considerable portion of the mtDNA molecule, if not its totality. However, the results of the in vivo pulse-labeling and in vitro labeling experiments indicate that the rate of transcription of the noncoding strand is much lower than that of the coding strand, and that therefore, in contrast to the situation in HeLa cells, transcription in yeast mtDNA is strongly biased in favor of the coding strand. The significance of the low level of symmetrical transcription in yeast mtDNA remains to be determined.

The extent of transcription of the coding strand of *S. cerevisiae* mtDNA has been estimated to be at least 60% from the results of the saturation hybridization experiments mentioned above;[92] the estimate is close to those obtained from the sum of the lengths of the discrete transcripts of *S. cerevisiae* (56%) and *S. carlsbergensis* (50%).[81] There is clear evidence, derived from RNA mapping and DNA sequencing studies, that the transcribed portions of yeast mtDNA include both the AT-rich stretches and the stretches with moderate GC content, as well as the GC-rich clusters, identified by Prunell and Bernardi.[67,68]

### E. General Organization of Mitochondrial DNA Transcripts

A detailed analysis of the mapping properties of the yeast mtDNA transcripts provided the first insight into the complexity of transcription and RNA processing in this system[81,99] (Figure 10). Several conclusions emerged from these studies.

First, under conditions which were apt to reveal the presence of any transcript representing more than 0.01% of mitochondrial RNA, only about 50% of the mtDNA length appeared to be represented in these transcripts, in agreement with the results of the saturation experiments mentioned above. Most of the transcripts detected could be related to known genetic loci (Figure 10). No transcript larger than 4S RNA was found to hybridize with the segment of mtDNA corresponding to the major tRNA region, between the larger rRNA gene and the CO II gene. At each locus, two or more overlapping transcripts were observed, the size of these transcripts being in general much larger than required to specify the gene product corresponding to that locus (Figure 10). Finally, strain differences due to the presence or absence of macro-inserts in the 21S rRNA locus, the *cob-box* (cytochrome *b*) locus (Figure 13), and the *oxi-3* (CO I) locus (Figure 14) appeared to be reflected in the size of the larger transcripts corresponding to these loci.

The organization of mtDNA transcripts described above was consistent with a model of transcription from multiple promoters, possibly one for each locus. The presence at each locus of several overlapping transcripts of widely different size, but all larger than required for coding the product of that locus, strongly supported a complex pattern of processing from larger precursors to mature RNAs. Subsequent studies, involving a correlation of RNA mapping data with sophisticated genetic data, analysis of translation products in wild-type and mutant strains, and DNA sequencing, have confirmed the crucial role that RNA processing plays in yeast mitochondrial gene expression. In particular, it has been shown that several yeast mtDNA genes are discontinuous, with strain differences concerning the number or even the presence of discontinuities, and with a complex pattern of RNA splicing and trimming being involved in the maturation of the final RNA products.

## F. RNA Processing

### *1. General Features*

In no case has the largest transcript hybridizable with a given locus of yeast mtDNA been conclusively identified as the primary transcript of the corresponding gene. Two major types of RNA processing have been observed in yeast mitochondrial RNA: trimming of 5′- or 3′-extensions and RNA splicing. The 21S rRNA precursor exhibits a 3′-extension[100] and the 15.5S RNA[90,99,101] a 5′-extension, which are removed during maturation (Figure 11). The largest transcripts identified at the protein coding loci in general exhibit a 5′-leader sequence and a 3′-tailing sequence. These can be preserved intact or can be partly trimmed away during maturation of the precursors to mRNA. In every case analyzed in detail, the mature mRNAs have been shown to have a 5′-leader and a 3′-untranslated tail, conforming therefore to the general rule observed for eukaryotic and prokaryotic mRNAs and differing strikingly in this respect from mammalian mitochondrial mRNAs. The presence of short poly(A) segments (20 to 28 nt) at the 3′ ends of a small fraction (~5%) of yeast mitochondrial RNA which is retained on a poly(U)-Sepharose-column, has been reported,[102] but their presence has not been confirmed.[103] In any case, yeast mitochondrial mRNAs do not appear to be extensively polyadenylated at their 3′ ends as are their mammalian counterparts.

Like mammalian mitochondrial mRNAs, yeast mRNAs are not "capped" and there is no evidence of internal methylation. Secondary modifications of mitochondrial rRNA and tRNAs in yeast are much less extensive than in their cytoplasmic counterparts. Thus, only two methylated riboses and one pseudouridine residue have been found per 21S rRNA molecule, while no methyl groups and one pseudouridine have been detected per 15S rRNA molecule;[104] these levels have to be contrasted with the presence of ~43 methyl groups and ~25 pseudouridine residues per cytoplasmic 26S rRNA molecule, and of ~24 methyl groups and ~11 pseudourudine residues per cytoplasmic 17S rRNA molecule.[104] In yeast mitochondrial tRNA ~2.6 methyl groups per molecule have been found, compared to ~4.3 methyl groups per cytoplasmic tRNA molecule.[105]

### 2. *Processing of Transcripts from Mosaic Genes*

#### *a. 21S rRNA Gene*

Evidence that the 21S rRNA gene is split in *S. cerevisiae* $\omega^+$ strains, exhibiting an ~1160 bp intron at ~500 bp from the 3′ end, while it is continuous in *S. cerevisiae* $\omega^-$ and $\omega^n$ strains and in *S. carlsbergensis,* was originally provided by restriction mapping data[74,75] and electron microscopy of DNA-DNA or RNA-DNA hybrids.[75,76] Subsequently, a sequence analysis of mtDNA from three *S. cerevisiae* strains differing by their ω alleles ($\omega^+$, $\omega^-$, and $\omega^n$) has revealed that the gene of the $\omega^+$ strain differs from that of the $\omega^-$ and $\omega^n$ strains by the presence of an 1143-bp intron, as well as by a GC-rich mini-insert (66 bp) located 156 bp upstream within the exon.[71] By $S_1$ mapping, it has further been shown that the intervening sequence is at ~570 bp from the 3′ end of the gene, and that the sequences of the mature 21S rRNA transcribed from an interrupted *(S. cerevisiae* $\omega^+$ strain) and an uninterrupted gene *(S. carlsbergensis)* are identical in the region corresponding to that surrounding the intervening sequence in the gene.[106]

In the $\omega^+$ strains, a 21S rRNA putative precursor of ~5.4 kb has been identified which contains the sequences of the 1143-bp intron and of a ~1.1-kb 3′-extension[100] (Figure 11a). This precursor is processed by intron excision and 3′-extension removal. The two steps do not appear to occur in a rigorous sequence, since intermediates lacking either the intron or the 3′-extension have been found both in a "petite" derived from an $\omega^+$ strain and in an $\omega^+$ wild-type strain (Figure 11a). In the "petite" strain analyzed, intron excision seemed to precede in general, 3′-end removal.

An attempt to identify the 21S rRNA transcription unit has been made by DNA transfer hybridization experiments utilizing total mitochondrial RNA from a wild-type strain or from a "petite" (containing a region of mtDNA with the 21S rRNA gene and the five closest tRNA genes see [Figure 10]) and mtDNA restriction digests from the same "petite".[107] The results have indicated that the 21S rRNA gene region is transcribed, both in wild-type cells and in the "petite" analyzed, as a continuous 6 to 7 kb molecule encompassing the 21S rRNA and a large 3′-tail: this extends through the $tRNA^{Thr}_{ACN}$ gene and probably includes the $tRNA^{His}$ and $tRNA^{Cys}$ genes. These results would imply that the 5.4-kb 21S rRNA precursor referred to above is not the primary transcript, but is formed by excision from the polycistronic molecule of the $tRNA^{Thr}_{ACN}$ and possibly the $tRNA^{His}$ and $tRNA^{Cys}$.

Synthesis of the 21S rRNA putative precursor and its processing to mature 21S rRNA have also been observed in isolated mitochondria.[93,108,109] The efficiency of conversion of the precursor to mature 21S rRNA varied in the different systems used, and evidence has been obtained indicating the dependence of the process on ATP.[109]

#### *b. Cytochrome* b *(Cob-box) Gene*

The first evidence concerning the complex organization of the cytochrome *b* gene came from genetic studies. A detailed analysis of a large number of yeast mutants affected in the synthesis of cytochrome *b* or resistant to inhibitors of the $bc_1$ complex, by genetic and physical mapping, intragenic complementation, and characterization of abnormal protein products, suggested in fact that the structural gene for cytochrome *b* is discontinuous, and that some of its introns contain regulatory elements.[79,110-118] Subsequent electron microscopic analysis of hybrids formed between the smallest and most abundant RNA mapping in this region ($18S_E$, ~2.2 kb) and DNA segments from the same region provided direct evidence for the occurrence in the cytochrome *b* gene of at least five exons.[80] More recently, DNA sequence analysis of the "short" form of the cytochrome *b* gene as found in *S. cerevisiae* D273-10B[5] and in *S. carlsbergensis,* and of segments of the "long" form as found in *S. cerevisiae* 777-3A,[4] has led to the elucidation of the organization of the two forms of cytochrome *b* gene. This is shown in Figure 12. The "long" form of the cytochrome *b* gene covers a span of ~7.3 kb and is divided into six exons (B1 to B6) and five introns (bI1 to bI5). The "short" form of the gene (~3.3 kb) lacks the first three introns.

The longest transcript corresponding to the "long" gene is an ~8-kb RNA which covers the whole span of the gene and adjacent segments on the 5′ and 3′ sides;[80,81,119,120] the shortest and most abundant transcript, a $18S_E$ RNA, is probably the cytochrome *b* mRNA (Figure 13), as inferred from the results of in vitro translation experiments.[80,119] Its length (~2.2 kb) is about twice that required to code for cytochrome *b*. The bulk of the extra sequences are located at the 5′-end of the RNA and presumably form an untranslated leader. Between the 8 kb RNA precursor and the 2.2 kb putative mRNA there are several transcripts of intermediate size mapping in the *cob-box* locus (Figure 13a). These are splicing intermediates in the processing pathway leading to the formation of the mature mRNA, as shown by the analysis of their intron and exon composition. The details of this processing pathway have not all been worked out, but the available evidence indicates that each of the five intervening sequences is excised as a single event, and that in the wild-type strains there is not a strict order of splicing.[119,121-123] However, the excision of bI1 appears to be an early event[119-123] and the excision of bI2 a late event,[119,121-123] and the preferred order of excision events, as judged from the relative abundance of intermediates in wild-type cells, is bI1 → bI3 → bI4 → bI2,[122-123] with the removal of bI5 possibly occurring at any time.[119,121] This order, as observed in an asynchronously processed population of molecules in each cell, does not necessarily reflect the obligatory sequence of steps, as expected in a synchronously processed population of molecules and as can be reconstructed from the effects of intron mutations[122,123] (see below). An interesting observation has been that the excised intron bI1 accumulates as a circular RNA[120,121] ($10S_E$ RNA, Figure 15a) (see also below).

An analysis of the transcripts of the "short" cytochrome *b* gene has revealed (Figure 12) a 4.3-kb RNA containing both introns, two partially spliced intermediates — one (2.9 kb) lacking the first intron (i1), and the other (3.6 kb) lacking the second intron (i2) — and two major transcripts representing fully spliced products, one with an apparent size of ~2.2 kb (similar to that of the "long" gene mRNA) and the other with a size of ~2.0 kb; the two transcripts differ in the length of their 5′ nontranslated leaders.[124] The presence of two splicing intermediates lacking either the first or the second intron suggests that the cytochrome *b* mRNA(s) of the "short" gene can be formed by two alternate splicing pathways (Figure 13b). This possibility has been corroborated by an analysis of the transcripts in intron mutants defective in splicing.[124] Further insight into the pathway of splicing of the mRNA precursors in the "short" *cob-box* gene has come from an analysis of nuclear mutants defective in the transcription or processing of cytochrome *b* mRNA.[125] In particular, in seven of such mutants belonging to a single complementation group, a large number of novel cytochrome *b* transcripts have been observed. The size and hybridization pattern of two of these transcripts (2.6 and 1.9 kb) suggest that their composition is i1-b2-i2-b3 and i1-b2-b3, respectively; this would imply that the product of the mutated nuclear gene is required for the scission of the i1-b2 boundary. It seems likely that the i1 maturase (equivalent to the *box 7* maturase of the "long gene") is involved in the scission of the b1-i1 boundary. See below for further discussion of the nuclear mutants.

### *c. CO I* (Oxi-3) *Gene*

A DNA sequence analysis of the "middle-sized" form of the CO I gene from *S. cerevisiae* D273-10B and a comparison with the sequence of the human CO I gene[3] have shown that the CO I gene in this yeast strain is 9979 nt long and consists of 7 and possibly 8 exons (Figure 14), which account for only 16% of the gene sequences.[6] The "long" CO I gene of *S. cerevisiae* KL14-4A contains two additional introns which split A5 into three segments[82] (Figure 14). The "short" CO I gene of *S. carlsbergensis* differs from the *S. cerevisiae* D273-10B by the lack of introns aI1 and aI4.[83]

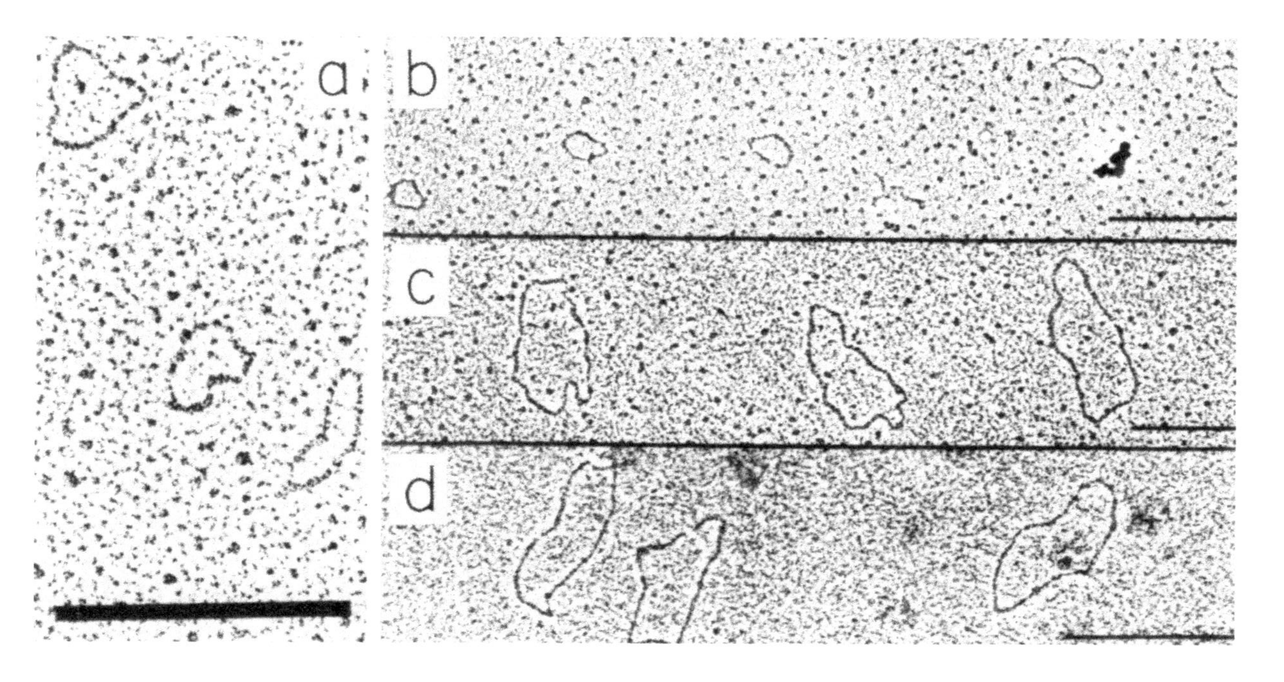


FIGURE 15. Electron micrographs of circular molecules present in mitochondrial RNA preparations from wild-type or *rho*⁻ *S. cerevisiae* cells. a): RNA from a *rho*⁻ derivative of *S. cerevisiae* KL14-4A retaining a segment of the cytochrome *b* gene, including the region B1 to B4; circular molecules of this size are present in the $10S_E$ fraction of RNA from wild-type cells[121] and show specific hybridization with the bI1 intron; bar = 0.25 μ. b), c) and d): Molecules of the $11S_E$ fraction (b), the $18S_E$-$19S_E$ fraction (c), and the $18S_E$-$19S_E$ fraction after treatment with glyoxal (d), from the RNA of *S. cerevisiae* KL14-4A;[126] $11S_E$ molecules hybridize with aI5, and the $18S_E$-$19S_E$ molecules, with aI1 and aI2 of the CO I gene;[82] bar = 0.3 μ. Taken from References 120 and 126.

Several overlapping transcripts have been identified which correspond to the *oxi-3* locus in *S. cerevisiae* D723-10B and KL14-4A.[81-83,119] Four abundant transcripts migrate at $11S_E$, $18S_E$, $19S_E$, and $19.2S_E$; of these, the $18S_E$ RNA species (~2.1 kb) which hybridizes with all exon probes is most likely the CO I mRNA.[6,83,119] The $11S_E$, $19S_E$, and $19.2S_E$ species hybridize, respectively, with probes derived from the aI5, aI2, and aI1 introns, and consist of circular molecules[82,83,119,126] (Figures 15b, c and d). There are also, mapping in the same region, larger molecules, which migrate in the range $21S_E$ to $32S_E$, and which show overlapping hybridization. These are presumably splicing intermediates of the large primary transcript. The processing scheme for this primary transcript has still to be worked out, although a tentative model concerning some of the possible steps has been presented.[82]

### 3. *Role of Introns in Controlling RNA Splicing*

#### *a. The "mRNA Maturase" Model*

The elegant genetic and biochemical analysis by several groups of investigators[79,111-118] of a large number of mutations located in the region of the cytochrome *b (cob-box)* "long" gene has revealed unexpected functions of the introns of this gene and unfolded an intricate network of intragenic and intergenic regulatory interactions. Mutations localized in the exons — B1 *(box 4/5)*, B3 *(box 8)*, B4 *(box 1)*, B5 *(box 2)* and B6 *(box 6)* — were found to have the effects expected of non-sense mutations (lack of cytochrome *b* activity and presence of shorter polypeptides related to apocytochrome *b*, resulting from premature termination of its synthesis) or mis-sense mutations (reduced cytochrome *b* activity, altered fingerprint of the apocytochrome *b* polypeptide).[111-118] Such mutations form a single complementation group.[79,127] Another class of mutations affecting the expression of the cytochrome *b* gene was localized in the introns — bI2 *(box 3, box 8)*, bI3 *(box 10)*, and bI4 *(box 2, box 7, box 9)* (Figure 12). These clusters of mutations constitute separate complementation groups, which are capable of complementing each other as well as mutations at *box 4/5, box 1* and *box 6*.[79,127] Many of these intron mutations are pleiotropic and affect, besides the synthesis of cytochrome *b*, also the synthesis of CO I (hence the name *box)*.[79,111,114,115,127,128]

The first insight concerning the role of introns in controlling cytochrome *b* and CO I gene expression was provided by the observation that cytochrome *b* intron mutations arrest the splicing of cytochrome *b* mRNA precursors[120,121,128] and cause the accumulation of novel polypeptides, sharing a portion of their sequence with apocytochrome *b*.[111-113,115-118,121,129,130] Furthermore, some of these mutations, in particular *box 3* and *box 7* mutations, can be complemented in *trans*, implying that a diffusible gene product provided by the intact intron can correct the splicing defect of the mutated intron.[79,127,131] The nature of this diffusible product in the case of the *box 3* locus has been revealed by an in-depth analysis by Slonimski and co-workers[4,132] of the structure of this locus and adjacent regions in wild-type and mutant strains. A DNA sequence analysis of bI1, B2, bI2, and contiguous intron-exon boundaries in a wild-type strain and in several *box 3* non-sense mutants did not provide any support for the earlier suggestion[128] that the intron RNA may be utilized for aligning sequences at the intron-exon boundaries. The striking observation was, instead, the presence within bI2, in the same strand containing the cytochrome *b* gene sequence, of a long open reading frame in phase with B2 and corresponding in location to the *box 3* genetic locus. By contrast, the sequence of bI1 was found to be blocked by multiple stop codons in the three registers of the coding strand. The presence in *box 3* mutants of novel polypeptides sharing a part of their sequence with apocytochrome *b*[111,117] could be accounted for if one assumed a translation of a hybrid exon-intron reading frame; furthermore, the size of these novel polypeptides could be fairly well correlated with the location in the DNA sequence of several *box 3* non-sense mutations, if one assumed that translation of the *box 3* reading frame started at the initiation codon for cytochrome *b* in B1 and proceeded through the spliced B1-B2 into bI2.[4,132] This idea was supported by evidence indicating that excision of bI1 precedes excision of bI2 (see above), and that the former results in the formation of a free circular RNA ($10S_E$ RNA) which accumulates (Figures 13a and 15a). On the basis of the above observations, it was proposed[4,132] that the diffusible product of the wild-type bI2 intron, which is able to correct in *trans* the splicing defects caused by *box 3* mutations, is a protein encoded in the intron. The same idea was put forward by Church and Gilbert[133] on the basis of the observation that "petite" mutants, which are incapable of mitochondrial protein synthesis, do not make mature cytochrome *b* mRNA.[133]

The concept that the diffusible products of bI2 is a protein ("mRNA maturase") has received strong support from a variety of observations like (1) the polar effect of nonsense mutations located in B1 on the in vivo complementation of *box 3* mutations, (2) the capacity of some mis-sense mutations in B1 to complement *box 3* mutants, (3) the fact that the $rho^-$ clones capable of complementing in *trans box 3* mutants are those which have retained, besides the *box 3* locus, a continuous mtDNA segment extending to the 5′ end of the cytochrome *b* gene, (4) the evidence that chloramphenicol blocks the processing of the cytochrome *b* mRNA precursors, in particular, the excision of bI2, bI3, and bI4, (5) the alleviation of some *box 3* non-sense mutations by nuclear and mitochondrial information suppressors,[134] and (6) the phenotypic cure of the same *box 3* mutations by paromomycin, presumably through antibiotic-induced ribosomal misreading.[135]

An important aspect of the maturase model is the autoregulatory property implied. In fact, the concentration of active maturase must be subject to a negative feedback, since the maturase activity destroys the RNA sequence of the mRNA precursor which encodes its own amino acid sequence ("splicing homeostasis").[4] It follows that in wild-type strains, only trace amounts of active maturase should be present, while an inactive protein or a protein fragment would accumulate in *box 3* mutants. The full-length maturase would be expected, according to the model, to have a size of 49 kilodaltons (kd).[4,136] A protein of ~42 kd has indeed been observed in some mis-sense *box 3* mutants:[111] considering the tendency of SDS-polyacrylamide gel electrophoresis to give underestimates of the true size of mitochondrially coded proteins,[6] the size of the observed protein is in reasonable agreement with the expectation.[4] Recently, a full length, possibly active wild-type maturase has been identified in a mutant affected in sequences downstream of *box 3 (box 8* mutant, see below).

Among the cytochrome *b* gene introns besides bI2, there are two others, bI3 and bI4, which contain open reading frames in the same strand containing the cytochrome *b* gene sequence[5,8] (Figure 12). By contrast, as mentioned above, the sequences of bI1, and bI5 are blocked in the three phases[4,5] and very probably do not specify any protein; in agreement with this idea, these introns can be excised from the mRNA precursors in the absence of mitochondrial protein synthesis.[120,133]

The intron bI3 contains a 1045-bp reading frame in phase with B3, followed by a 646-bp blocked frame, which contains an unusually GC-rich stretch of 150 bp[73] (Figure 12). It is interesting that the only intron of the *Aspergillus* cytochrome *b* gene is located at the same position as bI3 of the yeast "long" gene, and exhibits a significant sequence homology to it in its 5′-end and 3′-end proximal regions.[137] The bI3 reading frame could conceivably code for a maturase; however, only rare and leaky trans-recessive mutations forming a separate complementation group *(box 10)* have been localized in this intron.[8]

The intron bI4 which corresponds to the intron i1 of the "short" gene and is almost identical to it in sequence,[8,138] contains a 1152-bp reading frame in phase with B4, followed by a ~250-bp blocked frame (Figure 12). Several mutations have been located in bI4, and more precisely within the 3′-proximal half of the reading frame *(box 7)*.[8,138] These mutations: form a single complementation group, arrest the splicing of cytochrome *b* mRNA precursors at a later stage than *box 3* mutations, are recessive and can be complemented in *trans* by wild-type bI4 sequences; furthermore, they cause cessation of synthesis of CO I.

The polypeptide corresponding to the putative *box 7* mRNA maturase has not yet been unambiguously identified. Mutations located in the two domains flanking the *box 7* domain (*box 9 and box 2*, Figure 12) cause accumulation of three polypeptides with sizes of 23,000 (P23), 27,000 (P27), and 35,000 (P35) daltons.[130,136] Among these, P27 is the most likely candidate for being the *box 7* maturase, because it is the only one among the three polypeptides which responds specifically to *box 7* mutations; in particular, it is shortened by the expected amount in several *box 7* mutations creating stop codons in the bI4 reading frame.[138,139] The size of this polypeptide would imply that it is encoded completely within the bI4 reading frame, and probably within the 750 bp segment which includes all the *box 7* mutations so far identified. This possibility would be consistent with the lack of reactivity of P27 with anticytochrome *b* antibodies.[136] However, the lack of an AUG codon in the expected position or within a reasonable distance from it, and the evidence suggesting that the synthesis of the *box 7* maturase depends on the translation of upstream exons (B1, B2, B3, and B4)[127,138] and on the excision of bI2,[111,120,128] have suggested the possibility that this polypeptide results from the proteolytic cleavage of a larger precursor polypeptide, originating from fused sequences B1, B2, B3, B4, and the adjacent bI4 reading frame. According to this hypothesis, the P23 and P35 polypeptides, which by fingerprinting pattern and immunological cross reactivity appear to be related to cytochrome *b* and whose length is not modified by *box 7* non-sense mutations,[129,136] could be the fragments derived from the $NH_2$-proximal portion of the putative precursor. Recently, a study correlating the nucleotide sequence alterations detected in a set of *box 9* mutants with the changes in the size of the bI4 specific polypeptides and in the complementation behavior of these mutants has provided strong support for the hypothesis of the derivation of the P27 maturase from a precursor of an apparent molecular weight of 55,000 daltons, encoded in exon sequences B1 to B4 and in the intron bI4 open reading frame.[140] However, in apparent contradiction with the precursor model discussed above, are the observations indicating that non-sense mutations localized in *box 9* or in B4 do not or only partially abolish the expression of the putative P27 maturase.[138,140] A leakiness in chain termination at non-sense codons or a reinitiation of translation of P27 has been suggested as a plausible explanation of the above observations.[138,140] It should be mentioned that leakiness in frameshift mutations in the CO II gene of yeast mtDNA has been reported.[141]

Several of the introns of the CO I *(oxi-3)* gene also contain an open reading frame in continuity and in phase with the adjacent exon on the 5′ side[6] (Figure 14). This similarity in organization to the introns of the cytochrome *b* gene would suggest that the CO I gene introns may also code for maturases involved in the splicing of the CO I mRNA precursors. Consistent with this possibility is the observation that, with the exception of the excision of aI5, complete processing of the CO I gene transcripts required mitochondrial protein synthesis,[82,83] and the evidence suggesting that strains with intron mutations in this gene are affected in processing.[82,83,142,143] However, the fact that only a few mutations among several hundreds localized in these gene have been clearly assigned to introns,[142] that these mutations apparently form a single complementation group,[142,144] and that no trans-recessive mutations have been identified so far in this gene, raises the possibility that the reading frames of the CO I gene introns may not be normally translated into trans-active and diffusible mRNA maturases.[142] It is conceivable that the proteins encoded in the CO I gene introns have a structural role *in situ* as components of the splicing apparatus, as suggested by Slonimski and collaborators for the product of the fourth intron, aI4 (see below).[143]

The 21S rRNA gene intron also contains a 705-bp open reading frame[71] (Figure 11a): this could conceivably code for a 235-amino-acid-long polypeptide with a function dispensable for the mitochondria and specific for the $\omega^+$ strains, such as the removal of the intron. However, the observation that, in some *rho*$^-$ mutants derived from $\omega^+$ strains, processing of the 21S rRNA precursor to mature 21S rRNA occurs normally,[71,145,146] suggests that no mitochondrially synthesized proteins are required for the excision of the intron and splicing of the precursor. However, a possible role of this putative protein in controlling the polarity of recombination is still to be excluded.[71]

*b.* Cis-*Dominant Splicing Signals*

Some of the mutations localized in introns bI2 and bI4 of the cytochrome *b* gene disturb the splicing of the cytochrome *b* mRNA precursors without affecting the synthesis and activity of the maturase encoded in the mutated intron. These mutations are *cis*-dominant and can complement *box 3* and *box 7* mutations.[8,138] Some of these mutations are located outside the limits of the *box 3* and *box 7* regions, in the blocked frames of bI2 and bI4, respectively (Figure 12). Of such a nature are the *box 8* mutations, which affect the splicing of the cytochrome *b* mRNA precursor at an early stage,[8] and the *box 2* mutations which affect the splicing at a late stage.[138] Many others among the *cis*-dominant mutations affecting bI4 are clustered in a relatively small region *(box 9)* within the 5′-proximal half of the reading frame[138] (Figure 12). That the *box 9* mutations disturb splicing without affecting the maturase synthesis or activity is strongly suggested by their capacity to complement *box 7* mutations, by the fact that they accumulate the P27 polypeptide, and by the observation that one such mutation is a third base change.[138]

The cis-dominance of *box 8, box 2* and *box 9* mutations and the evidence indicating that they do not affect the synthesis or activity of the putative maturase encoded in the corresponding intron suggest that these mutations disturb splicing by modifying the RNA sequence which is the substrate of the splicing reaction. These critical sequences affected by the *cis*-dominant mutations may control splicing by interaction with proteins or other nucleotide sequences, as for example by creating secondary structures important for splicing. It may be significant, in this context, that, in eight of the *box 9* mutants studied, the sequence alterations are localized in a 12-nucleotide stretch exhibiting a perfect complementarity to a segment of the 15S rRNA located roughly in the middle of the molecule.[8,138,140] This observation has led to the suggestion that a ribosome-mRNA precursor interaction may be important for splicing, either by promoting the acquisition by the RNA of the proper secondary structure or by helping in the correct phasing of the splicing machinery.[8,138] A direct involvement of ribosomes in RNA processing has also been suggested on the basis of experiments showing the rapid inhibitory effects of chloramphenicol on the splicing reaction.[147]

The occurrence of cis-dominant mutations in the CO I gene will be discussed in the next section.

*c. Cytochrome* b *Gene—CO I Gene Interactions (*Box *Phenomenon)*

As mentioned above, a considerable amount of evidence indicates that intron bI4 of the "long" cytochrome b gene controls not only the expression of this gene, but also the expression of another mosaic gene, the CO I gene.[138,148,149] Thus, *box 7* mutations are pleiotropic in their effects, reducing or abolishing the synthesis of cytochrome *b* and abolishing completely the synthesis of CO I *(box* phenotype). The *box* phenomenon is not confined to strains with the "long" form of the cytochrome *b* gene, but is also observed in strains with the "short" form of the gene where intron i1 (homologous to intron bI4 of the "long" gene) is involved; therefore, this phemonenon is independent of the three intervening sequences in the upstream half of the cytochrome *b* gene.[149] The observation that *box 3* mutants also exhibit the *box* phenotype[111,120,127] is probably due to their inability to generate the trans-acting *box 7* product.

The finding that a block of CO I gene expression occurs also in mutants exhibiting a complete deletion of the *cob-box* gene indicates that the regulation of bI4 is a positive one. The fourth intron (aI4) of the "long" and "middle-sized" form of the CO I gene exhibits a striking sequence homology to bI4[6,138,142] whereas there is no homology between the exons of the two genes.[138] The homology of bI4 to aI4 probably represents the basis of the *box*

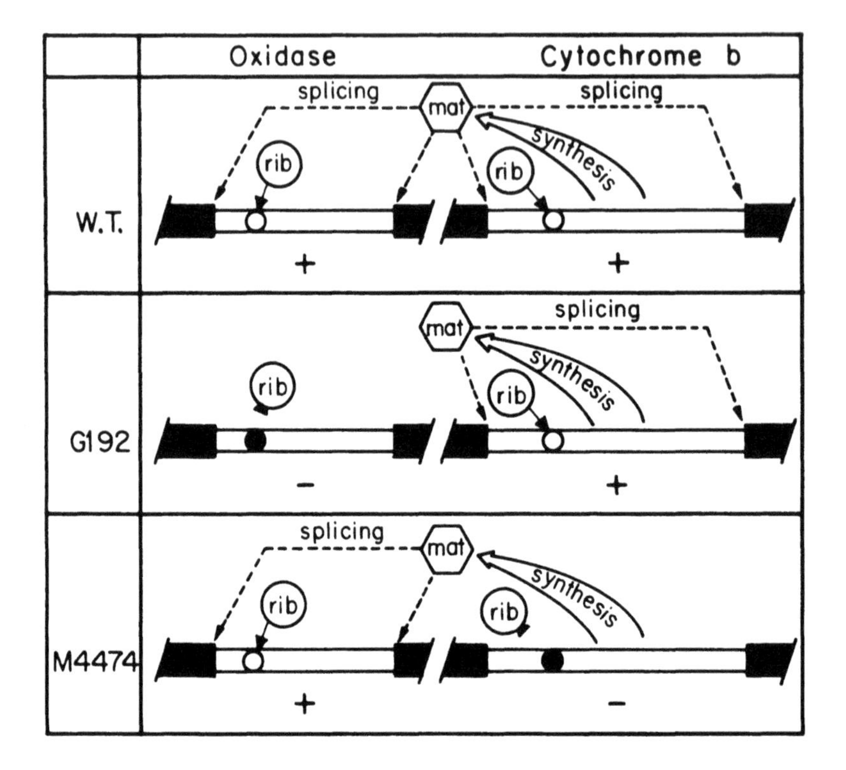


FIGURE 16. A model for the intergenic communication between the introns of the cytochrome *b (cob-box)* and CO I genes. Portions of the cytochrome *b* and CO I mRNA precursors are shown, in particular, intron bI4 and adjacent exons and intron aI4 and adjacent exons, respectively. In both introns, a circle indicates the position of the dodecanucleotide complementary to a sequence of the small rRNA[138,142] This sequence is assumed to be recognized by the mitochondrial ribosome (rib), possibly in conjunction with the maturase (mat). In the CO I G192 mutation, the signal (asterisked circle) of aI4 is no longer recognized, and, therefore, no aI4 excision and no CO I mRNA synthesis takes place. In the *box9* M4474 mutant, the bI4 signal (asterisked circle) is not recognized, and no cytochrome *b* mRNA is found. (From Netter, P. Jacq, C., Carignani, G., and Slonimski, P. P., *Cell*, 28, 733, 1982. Copyright by M.I.T. With permission.)

phenomenon by providing common recognition signals for the splicing machinery. In fact, several lines of evidence suggest that the *box 7* maturase encoded in part in bI4 is involved, not only in the excision of this intron from the cytochrome *b* mRNA precursors, but also in the excision of aI4 from the CO I mRNA precursors[8,138] (Figure 16).

In contrast to *box 7* mutations, *cis*-dominant mutations of bI4 which cause splicing deficiencies by modifying critical sequences of the substrate, but which accumulate active maturase like *box 9* and *box 2* mutations, allow synthesis of normal amounts of CO I.[111,123,138] Among the regions of highest homology between bI4 and aI4, besides the 3′-proximal two thirds of the reading frame (corresponding to *box 7* in bI4), there is a short segment upstream of the latter (corresponding to *box 9* in bI4), which exhibits an identical dodecanucleotide in the two introns, with base sequence complementarity to the 15S rRNA.[142] It is therefore particularly interesting that two *cis*-dominant mutations causing a complete deficiency of CO I have been localized in this dodecanucleotide in aI4 of the CO I gene[142] (Figure 16). These observations are consistent with the view that this dodecanucleotide represents a *cis*-acting signal sequence present in both cytochrome *b* and CO I mRNA precursors, which is recognized by the splicing machinery. The homology to the 15S rRNA sequence suggests

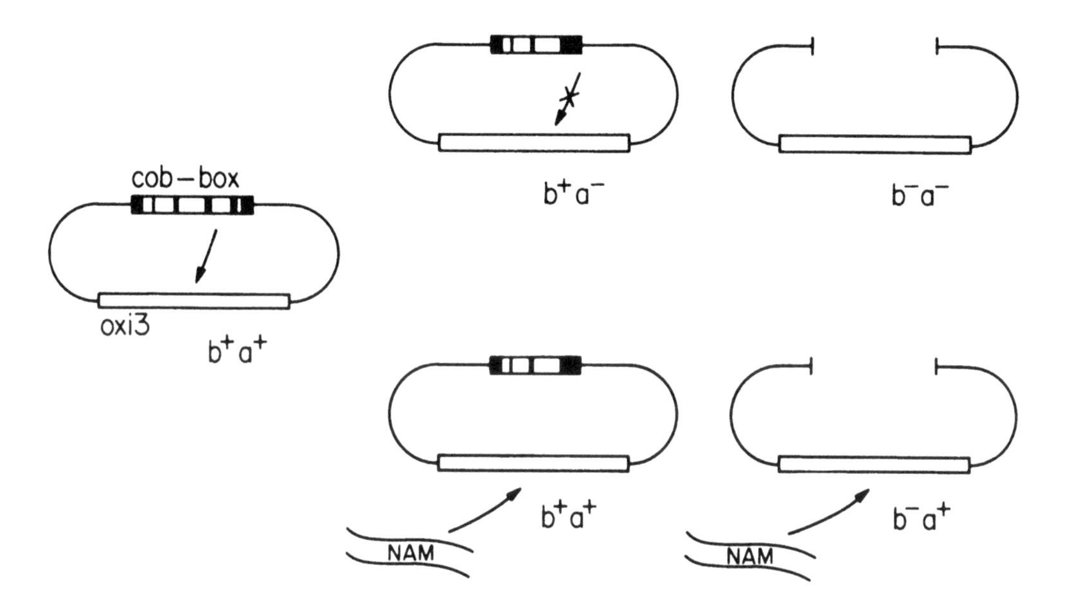


FIGURE 17. Interactions between the mitochondrial genes for cytochrome *b* and CO I and the nuclear gene NAM. The effects on the expression of the cytochrome *b* and CO I genes of the removal of $bI_4$ or of a complete deletion of the cytochrome *b* gene are relieved by the concomitant appearance of the NAM nuclear suppressor mutation. (From Jacq. C., Pajot, P., Lazowska, J., Dukardin, G., Claisse, M., Groudinski, O., de la Salle, H., Grandchamp, C., Labouesse, M., Gargouri, A., Guiard, B., Spyridakis, A., Dreyfus, M., and Slonimski, P., in *Mitochondrial Genes,*, Slonimsky, P., Borst, P., and Attardi, G., Eds., Cold Spring Harbor Laboratory, Cold Spring Harbor, New York, 1982, 155. With permission.)

that an interaction with the mitochondrial ribosome, in conjunction with *box 7* maturase, may be involved in this recognition[142] (Figure 16).

The cytochrome *b*-CO I gene interaction appears to be unidirectional. In fact, large deletions in the CO I gene appear to be without effect on cytochrome *b* synthesis.[128]

### *d.* Box 7 *Specific Suppressors*

One of the most intriguing observations concerning the role of introns in cytochrome *b* and CO I gene expression has been the discovery of specific suppressor mutations capable of alleviating the effects of *box 7* mutations or of the complete lack of bI4 (obtained by intron manipulation, see below) and of restoring cytochrome *b* and/or CO I synthesis.[8,150,151] Two such suppressor mutations, both dominant, have been found in nuclear genes NAM2-1 and NAMX, and one in mtDNA, mim2-1. In all three cases the suppression is partial; although there is restoration of the respiratory capacity, the content of cytochromes *b* and $aa_3$ never reaches that of wild-type cells.

It is noteworthy that, when NAM2-1 or mim2-1 is associated with a complete deletion of the cytochrome *b* gene, a deletion which would normally lead to a complete deficiency of both cytochrome *b* and CO I, one observes a partial restoration of CO I synthesis[8] (Figure 17). These results indicate that, in the presence of one of these suppressor mutations, the cytochrome *b* gene is no longer necessary for the expression of the CO I gene, and that in particular, the bI4 *(box 7)* maturase is no longer required for the splicing of the primary transcript of the latter gene. Either activation of a silent gene, nuclear (NAM2-1 or NAMX) or mitochondrial (mim2-1), coding for an active maturase which could function as a substitute for the *box 7* maturase, or a mutational change of a normally expressed gene with resulting acquisition of "maturase" activity has been suggested as an explanation of the above-described phenomenon (Figure 17). It is interesting that the mim2-1 mutation has been localized within the fourth intron of the CO I gene (aI4) and has recently been shown to be due to a single base substitution.[143] Furthermore, there is some evidence that a 56,000-dalton protein, exhibiting peptide homologies with CO I, and which could result from translation of the previously fused CO I exons A1, A2, A3, and A4 with the aI4 open reading frame, accumulates in *box 7* mutants and in *cis*-dominant aI4 mutants, as expected for a normally expressed translation product of aI4 having a role, still unknown, in aI4 excision.[143]

***e. Are the Mitochondrial Gene Introns Indispensable?***

The existence, mentioned above, of yeast strains lacking the single intron in the 21S rRNA gene, or some of the introns in the cytochrome *b* gene or in the CO I gene shows that the introns involved, in certain genetic backgrounds, are dispensable for the expression of the corresponding gene. In agreement with this conclusion are the results of experiments of intron manipulation, in which intron sequences have been subtracted *en bloc* from the cytochrome *b* gene.[8,139] These experiments have shown that any of the five introns of this gene can be removed, either alone or in combination, without affecting the synthesis of cytochrome *b*. The most striking observations are the reestablishment, after removal of bI4 and bI5, of cytochrome *b* formation in a mutant strain defective in cytochrome *b* as a result of a *box 2* mutation, and the synthesis of cytochrome *b* in a strain with a completely intronless gene.[8] It should be noted that, among all intronless derivatives that have been constructed, only the bI4$^{+}$ strains (including bI1$^{\Delta}$bI2$^{\Delta}$bI3$^{\Delta}$BI4$^{+}$bI5$^{\Delta}$) were able to synthesize CO I;[8,73] the only exception is a rare mutant (bI4$^{\Delta}$bI5$^{\Delta}$), in which the excision of bI4 and bI5 was selected concomitantly with a nuclear suppressor mutation (see above).[8,151]

The intron manipulation experiments mentioned above are interesting in two respects. First, they show that introns can be excised and exons ligated at the DNA level in such a way as to maintain the correct reading frame of the exons involved, or at least as to allow the correct removal of any residual intron sequences at the RNA level. Nothing is known about the mechanism involved, although it is formally analogous to a transposition phenomenon. Second, these observations indicate that, mechanistically, intron excision at the RNA level is not a cooperative event between various introns of the same gene. This does not imply that there is no *order* in the excision of introns. The intron manipulation data only show that removal of any given intron does not depend on the presence of the upstream or downstream intron sequences, nor, directly, on diffusible products encoded in these sequences. By contrast, the analysis of the intron and exon composition of the transcripts observed in different intron mutants of the cytochrome *b* gene has suggested that there is an obligatory order of excision, namely: bI1→ bI2→ bI3→ bI4, with bI5 possibly occurring at any time.[123] This polarity may be related to a cascade of synthesis of different maturases, the *box 3* (bI2) maturase synthesis depending on the splicing — under control of nuclear-coded enzyme(s) — of B1 and B2, then the *box 10* (bI3) maturase synthesis depending on the splicing of B1B2 to B3, and, finally, the *box 7* (bI4) maturase synthesis depending on the splicing of B1B2B3 to B4.[133] This order of intron excision may not be observed in a wild-type strain, since, in an asynchronously processed population of RNA molecules, the processing, in an obligatory order, of a few molecules may provide a sufficient amount of maturases to allow the out-of-sequence splicing out of the introns in the bulk of molecules in the population.[123] Indeed, the available data indicate the lack of a rigorous order of splicing events in the maturation of the cytochrome *b* mRNA in wild-type strains, although there is a preferred sequence of steps which presumably reflects the kinetics of the five cut-and-splice reactions involved in the whole process.[121,123]

The situation discussed above as concerns the sequence of events in the processing of the cytochrome *b* gene transcripts does not seem to apply to the CO I gene. In fact, studies carried out with mutants affected in the synthesis of CO I suggest that a fixed sequential order of splicing steps from the 5′ end to the 3′ end may not occur in the transcripts of this gene.[82,83] This situation may reflect a role of the reading frames of the introns of this gene other than that of coding for maturases, or may be due to the overlapping specificities of the putative maturases (as suggested, for example, by the striking sequence homology between aI1 and aI2[6]), or to an independent synthesis of the maturases encoded in the individual introns[82,83] (see also above).

If the intron of the 21S rRNA gene, those of the cytochrome *b* gene, and some at least of the introns of the CO I gene, as discussed above, do not play an indispensable role in the expression of the corresponding genes, one can ask what selective advantage is their presence conferring upon the cell whose genes harbor them. One can speculate that, as proposed for the introns of nuclear genes,[152] the intervening sequences of mitochondrial genes may accelerate evolution of these genes by facilitating recombination[153] and/or by reassortment.[153] However, any model concerning the possible significance of these introns must deal with the role of the open reading frames that most mitochondrial gene introns contain. A crucial question is how are these reading frames, some of which are very long — up to ~2350 nt (in aI2) — maintained, while the other two frames in each of the corresponding segments of the same strand exhibit the expected frequency of non-sense codons. The conclusion seems inescapable: these reading frames must at some time be expressed. A possible role in controlling the polarity of recombination of the putative protein encoded in the optional intron of the 21S rRNA gene[71] was mentioned above. The striking similarity in organization between bI2 or bI4 and other introns of the cytochrome *b* gene and of the CO I gene, each exhibiting a reading frame which is, in general, continuous with the upstream exon, cannot be fortuitous and points to a similar protein coding function for all these reading frames. The existence of distinct intron-coded proteins controlling the various splice steps of the same mRNA precursor, by creating multiple regulatory targets for nuclear gene products or environmental stimuli, would certainly increase the potential for selective gene regulation under different physiological circumstances in an eminently adaptable organism. Furthermore, the important discovery of the essential role of *box 7* maturase for the expression of the CO I gene suggests that other intronic reading frames as well may code for proteins which, apart from a maturase activity on the mRNA precursor(s) of the same gene, have a controlling function, directly or through the excised intron sequences, on the expression of different mitochondrial genes. The intron-coded proteins may thus provide the cell with the capacity of finely modulating and coordinating in a differential way the expression of the individual mitochondrial genes in response to different physiological or environmental conditions. The significance of the circular RNAs resulting from the processing of the mRNA precursors of the cytochrome *b* and CO I genes is not understood, but it is conceivable that they play an important role in the network of mitochondrial intergenic communications being considered here.

*4. Mechanism of RNA Splicing in Mitochondria*

In contrast to the considerable amount of evidence available concerning the role of introns in mRNA precursor splicing in yeast mitochondria, almost nothing is known about the mechanism involved in this splicing. Where sequence data for exon-intron junctions are available, *the splice points of the yeast mitochondrial mRNA precursors show no homology with the consensus sequence of the splice points of eukaryotic viral and nuclear RNAs:* AG ↓ GTAAGT . . . PyPyXPyAG ↓ .[154] In the case of the 21S rRNA precursor, several secondary structures have been proposed, involving base pairing of segments of the intron with exon segments near the splice points, [71,106] but all these structures have relatively low stabilities; therefore, there is no convincing evidence that the intron sequence may serve as a guide for the splicing. It has been suggested that splicing of the 21S rRNA precursor may occur after the assembly of the large ribosomal subunit, the ribosome itself providing the guide for splicing.[71] In other systems,[155,156] there is evidence that ribosomal proteins bind rapidly to ribosomal RNA precursors in vivo; therefore, a role of ribosomal proteins in the splicing of the 21S rRNA precursor appears to be plausible.

Nothing is known about the mechanism of action of the mRNA maturases, nor whether this action is enzymatic or structural. As mentioned before, the possible involvement of the ribosome in the splicing reaction has been suggested, based on the existence of a perfect base sequence complementarity between a dodecanucleotide in a critical region of the bI4 and aI4 introns and a stretch of the 15S rRNA,[8,138,142] and on the results of experiments showing the rapid inhibitory effects of chloramphenicol on the splicing.[147]

*5. Circular RNAs*

One of the most intriguing features of the processing of the transcripts of the mitochondrial mosaic genes in yeast is the accumulation of circular RNA species which accompanies this process[120,126] (Figure 15). As judged from the resistance of these circular molecules to pronase treatment and to stringent denaturing conditions, they are covalently closed. Four distinct circular RNA species have been identified and mapped on mtDNA. One of these species, $10S_E$ RNA (600 to 700 nt) (Figure 15a), appears to hybridize with sequences of the intron bI1 of the cytochrome *b* "long" gene,[120,121] while the three other species — $11S_E$, ~950 nt; (Figure 15b); $19S_E$, ~2450 nt; and $19.2S_E$, ~2540 nt (Figure 15c and d) — are complementary, respectively, to sequences of the introns aI5, aI2, and aI1 of the CO I "middle-sized" gene.[83,119,126] Length measurements of the RNAs and RNA-DNA hybridization mapping by blotting and electron microscopy have strongly suggested that the entire intron sequence is represented in the circle in each case, at least as far as the circular RNA species derived from CO I transcripts are concerned. The above findings are consistent with the derivation of each circular RNA species by precise excision of the corresponding intron from the mRNA precursor(s). The observation that not all excised introns give rise to circular molecules would tend to exclude the artifactual formation of these circles due to the presence of ligase activity at the splicing site, suggesting rather that their formation is intimately associated with the splicing process. An accumulation of the single intron of the cytoplasmic large rRNA precursor in *Tetrahymena* in the form of circular molecules (~0.4 kb) has been previously observed and related to the splicing event.[157] Why only some of the introns of yeast mtDNA transcripts accumulate in the form of circles is not known. Their stability suggests that they may have some as yet unknown function.

*6. Synthesis and Processing of Transcripts of Other Genes*

The other genes identified in yeast mtDNA, i.e., the 15S rRNA gene, the approximately 25 tRNA genes, and the structural genes coding for CO II *(oxi-1)*, CO III *(oxi-2)*, ATPase subunit 9 *(oli-1)* and subunit 6 *(oli-2)*, and *var-1* protein, appear to be continuous. With the exception of the 15S rRNA gene and the tRNA genes, very little is known about the transcription of these genes and the processing of the corresponding transcripts.

***a. 15S rRNA Gene***

An RNA species slightly larger than the mature 15S rRNA (15.5S rRNA) and mapping at the same site as the latter has been identified both in wild-type strains[87,101] and in "petite" mutants retaining the 15S rRNA gene.[90,99,101,158] In vitro capping experiments have shown that the 15.5S rRNA preserves its 5′-terminal di- or triphosphate, and therefore, its 5′ end presumably derives from initiation of transcription.[87,90] A sequencing analysis of the 5′-end proximal segment of 15.5S rRNA capped in vitro with [$\alpha$-$^{32}$P]GTP and guanylyltransferase and of the 5′-end proximal segment of mature 15S rRNA labeled with [$\gamma$-$^{32}$P]ATP and polynucleotide kinase has shown that the 15.5S rRNA has an ~80 nt 5′-extension, which must be removed during the maturation to 15S rRNA[90] (Figure 11b). The same conclusion has been reached on the basis of high-resolution S1 mapping experiments.[101] The mature

15S rRNA has staggered 5′ ends extending over five contiguous nucleotides, an observation which indicates a certain imprecision in the processing event[90] (Figure 18). A sequence analysis of the 3′ end proximal segment of the 15.5S and the 15S rRNA species, after labeling with [5′-$^{32}$P]pCp and RNA ligase, has further shown that the two species have the same 3′-terminus.[90] It is not known whether the 15.5S rRNA represents the primary transcript of 15S rRNA gene region, although this seems likely. A transcript of the same size as the 15.5S rRNA,[108,109] and containing sequences found in the 15S rRNA,[108] has been found to be synthesized in isolated mitochondria; however, processing of this putative precursor to 15S rRNA occurred only to a minor extent under the experimental conditions used. Processing the 15.5S rRNA to 15S rRNA can occur also in "petite" mutants,[90,99,101] an observation which indicates that the processing enzyme is not synthesized in the organelle, but is nuclear-coded and imported from the cytoplasm. It is interesting that a 16-nucleotide stretch very rich in AT (94%) immediately upstream of the 15.5S rRNA coding sequence has a 16/17 homology to the 17 nt stretch immediately upstream of the initiation site for transcription of the 21S rRNA gene[159] (Figure 18). It seems possible that this homologous sequence represents an element of the rRNA promoter.

***b. tRNA Genes***

Among the approximately 25 yeast mtDNA genes, 16 are located within 12% of the genome between the 21S rRNA gene and the CO II gene,[63,64] and are separated by moderately long to long stretches of AT-rich DNA, with the exception of the $tRNA^{Ser}_{AGY}$ and $tRNA^{Arg}_{CGY}$ genes which exist in tandem, separated by only 3 bp.[160,161] The other tRNA genes are scattered around the genome (Figure 10). As in mammalian mitochondrial tRNAs, the 3′-terminal -CCA of the yeast organelle tRNAs is not encoded in the DNA. The primary transcripts of the tRNA genes have not yet been identified. No consensus sequences identifiable as possible promoter or terminator signals for transcription can be recognized around the tRNA genes. RNA transfer hybridization experiments utilizing mitochondrial RNA from wild-type cells and tRNA gene-containing probes have detected only tRNA-size transcripts,[81,99] even in the case of the two tandemly arranged tRNA genes mentioned above[160] (for an exception, however, see below). The results are compatible with either the existence of separate promoters for the individual genes, situated very near to their 5′ ends or even within the genes themselves, or with the synthesis and very rapid processing of polycistronic transcripts containing the sequences of two or more tRNAs.

An insight into the mechanism of transcription of the tRNA genes has come from an analysis of the transcripts of "petite" mutants retaining tRNA genes. Early investigations had revealed that some "petites" synthesize mature tRNA capable of being aminoacylated, while others do not.[162,163] Furthermore, one "petite" was found to synthesize larger transcripts containing tRNA sequences:[163] it could not be decided, however, whether these larger transcripts reflected the transcription process as it occurs in wild-type cells or, rather, anomalous transcription events due to rearrangements in the "petite" genome. A correlation between the capacity of different "petite" mutants to make mature tRNAs and the presence in the "petite" genome of a region of the mitochondrial genome near the paromomycin locus led Morimoto et al.[99] to postulate the existence in that region of a locus necessary for the processing of all tRNAs, even those encoded in quite distant regions of the genome. It has recently been possible to restore the formation of mature $tRNA^{Ser}_{UCN}$ in a "petite" strain containing the $tRNA^{Ser}_{UCN}$ gene (originally capable of synthesizing large transcripts carrying its sequence, but not of making $tRNA^{Ser}_{UCN}$ in appreciable amounts) by mating it with another deletion mutant, which contained the $tRNA^{Trp}$, $tRNA^{Pro}$, and $tRNA^{fMet}$ genes and was capable of synthesizing these tRNAs because of the presence in the same "petite" genome of the "tRNA synthesis locus".[164] The available evidence suggests that the restoration was not due to recombination but, rather, to intergenic complementation, therefore pointing to an action in *trans* of the tRNA synthesis locus. This locus or its product could act at the level

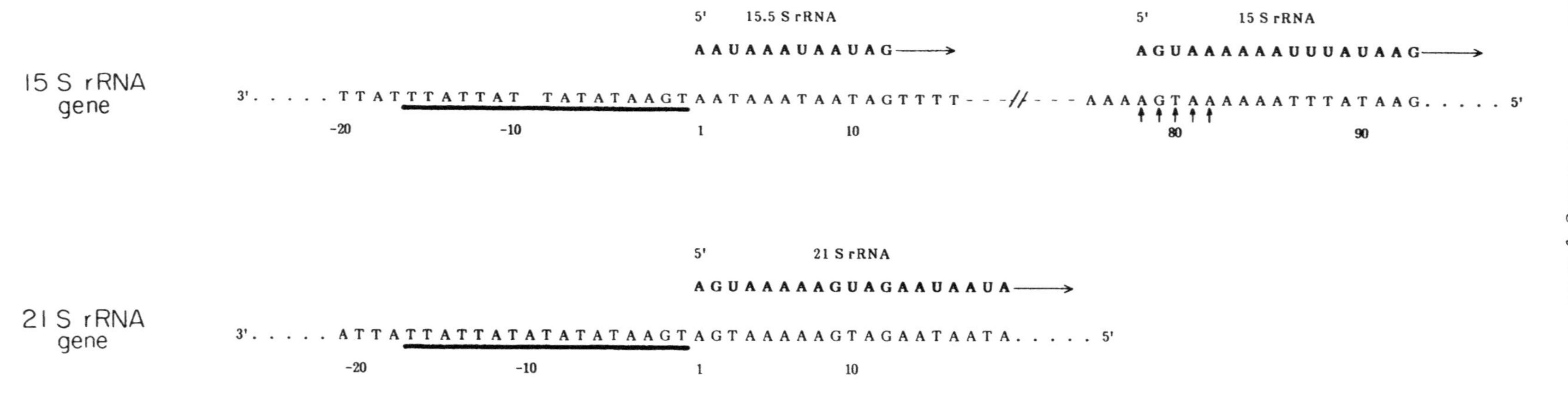


FIGURE 18. The sequences of the sense DNA strand in the regions around the 5′ ends of the 15.5S, 15S, and 21S rRNA coding segments[159] are aligned with the sequences of the 5′-end proximal segments of the 15.5S, 15S, and 21S rRNAs[87,90] The homologous nucleotide stretches immediately upstream of the 5′ ends of the 15.5S and 21S rRNAs[159] are underlined. The small arrows at positions 78 to 82 in the 15S rRNA gene sequence indicate the sites where the precursor is cleaved to generate the mature 15S rRNA.[90]

of transcription or posttranscriptional processing. An intriguing possibility is that the product of this locus is an RNA destined to be associated with a tRNA processing enzyme like the RNase P in *E. coli.*[164]

More recently,[165] an analysis, by RNA transfer hybridization and S1 mapping experiments, of the transcripts of "petite" mutants containing a group of tRNA genes from the large tRNA gene cluster, but defective in tRNA maturation, has revealed, besides high- molecular-weight transcripts containing tRNA sequences, low-molecular-weight RNA components (100 to 150 nt). Each of these components was shown to contain one tRNA sequence and the upstream nucleotide stretch up to the 3′ end of the preceding tRNA sequence. In the wild-type strain and in a "petite" mutant retaining the "tRNA synthesis locus", on the contrary, a family of S1-protected fragments ranging from the expected length for a mature tRNA to a size longer by 1 to 5 nt has been observed with each specific probe. While it is not clear whether this 5′-end heterogeneity is real or artifactual the above observations have suggested a model whereby polycistronic transcripts carrying several tRNA sequences are destined to be processed to mature tRNAs in two steps. In the first step, the precursor is cleaved at the 3′ end of each tRNA sequence; in the second step, the 5′ leader sequence is removed: it is apparently this second step that requires the presence of the "tRNA synthesis locus". It is interesting that in the work described above, using a high-specific-activity probe, the presence of a small amount of a high-molecular-weight transcript has been detected in the wild-type RNA.

Mention was made above of the evidence indicating that the $tRNA^{Thr}_{ACN}$ gene, and possibly also the $tRNA^{His}$ and $tRNA^{Cys}$ genes, may be cotranscribed in a common precursor with the 21S rRNA gene[107] and that the $tRNA^{Val}$ gene may be cotranscribed with the CO III gene.[85]

***c. Other Genes***

It seems to be a general rule that the yeast mitochondrial mRNAs have a 5′ leader and a 3′ tail; however, the size of these extensions varies considerably from mRNA to mRNA. Thus, the apocytochrome *b* mRNA has a leader of ~940 nt and an ~100-nt 3′ tail for an 1155 nt coding sequence;[65] the ATPase 9 *(oli-1)* mRNA has an ~500-nt 5′ leader and a 70-nt 3′ tail for a 225 nt coding sequence;[65,98] the CO I *(oxi-3)* mRNA has an ~400-nt long 5′ leader and a 30 to 100-nt 3′-extension for a 1530 nt coding sequence;[83] the CO II *(oxi-1)* mRNA has a 54-nt 5′ leader and a 3′ tail of >75 nt for a 756 nt coding sequence.[95,96] A 3.6-kb transcript, which is the major transcript mapping in the CO III *(oxi-2)* locus in mtDNA from *S. cerevisiae* D273-10B and other strains of yeast, and which is presumably the CO III mRNA, has a 490-nt 5′ leader and an ~2450-nt 3′ extension for an 807 nt coding sequence.[85] The reason for this variability in size of the 5′ and 3′ extensions among different mRNAs is not known, but may be related to possible regulatory functions of these extensions. It should be mentioned, however, that in the strain *S. cerevisiae* LL20, the 3′ extension of the putative CO III mRNA is shorter by ~2 kb, arguing against a specific role of the long 3′ tail in the expression of the gene.[85] A similar strain-dependent variability in the length of the 3′ extension has been observed in the ATPase 6 *(oli-2)* putative mRNA. In fact, a difference in the length of the 3′ tail of this mRNA has been reported between *S. cerevisiae* strains of class I (like J69-1B), whose mtDNA contains an ~1.8-kb macroinsert downstream of the ATPase 6 gene, and strains of class II (like JM6) whose mtDNA lacks this insert.[84] In particular, the putative ATPase 6 mRNA of class I strain (~4500 nt) has a 2300-nt-long 3′ tail, while the mRNA of class II has a 200-nt-long tail. The 5′ and 3′ ends of the two types of mRNA appear to coincide, the difference between class I and class II mRNAs being an internal stretch of ~2000 nt which is missing from the 3′ tail of class II mRNA, and which corresponds in position to the optional mtDNA macroinsert.

### G. Nuclear Control of DNA Transcription and RNA Processing in Mitochondria

The observation that petite mutants are capable of transcribing and processing correctly many mitochondrial RNA species (21S rRNA, 15S rRNA, tRNAs, etc.), and partially some others, indicates that the major part of the components of the mitochondrial transcription apparatus and of the RNA processing machinery is specified by nuclear genes. The complexity of the functions involved is suggested by the evidence that has already become available concerning the role of nuclear genes in the transcription and processing of cytochrome *b* mRNA. Among 130 complementation groups of nuclear mutants selected for their inability to grow on nonfermentable substrates, 5 have been found to be affected in the proper initiation of transcription of the cytochrome *b* "short" gene or in the excision of intervening sequences from precursor transcripts.[166] Of these, three complementation groups are affected in genes coding for products needed for the excision of the first intron of the "short" gene (i1, equivalent to I4 of the "long" gene; Figure 12); all the nuclear mutants defective in the excision of i1 are also unable to process the transcripts of the CO I gene. In another complementation group, as discussed in a previous section, novel cytochrome *b* transcripts have been observed[125] which consist of the first intron sequences linked to the downstream exon and intron sequences, reflecting the inability of the mutants to cut and splice at the i1-b2 boundary.

Mention was made above of the discovery of nuclear suppressor mutations (NAM2-1 and NAMX), which alleviate the effects of *box 7* mutations, either by activating a silent gene coding for a maturase or by changing the activity of a normal product.[8,150,151]

Although the general rule is that the protein components of the mitochondrial replication, transcription, and translation machineries are structurally and genetically distinct from the homologous proteins of the corresponding nuclear-cytoplasmic machineries, evidence for possible exceptions is starting to emerge. Thus, it has recently been shown that, in two nuclear tRNA modification mutants, a single point mutation affects the methylation of guanosine to $N^2$,$N^2$-dimethylguanosine, or the methylation of uridine to 5′-methyluridine, in both mitochondrial and cytoplasmic tRNAs.[167] Similar observations have been made for a nuclear point mutation affecting the addition of the isopentenyl side chain to adenosine.[168] It is not known yet whether each of the mutations discussed above involves a structural or a regulatory gene, nor whether, in the former case, the mitochondrial and cytoplasmic enzymes are identical, or two related proteins derived from a single gene as a result of transcriptional, translational or post-translational variation, or two different multimeric proteins sharing one subunit affected by the mutation.

## V. MITOCHONDRIAL DNA OF FILAMENTOUS FUNGI

### A. *Neurospora crassa*

#### 1. *Organization of rRNA and tRNA Genes*

*Neurospora crassa* mapping studies have shown that the two rRNA genes and most of the tRNA genes are concentrated in one third of the mitochondrial genome (~20 kb)[169-173] (Figure 19). The genes for the two rRNA species (19S and 25S) are separated by about 5 kb, and the large rRNA gene contains an intervening sequence of about 2.3 kb.[170,172,174] The tRNA genes located in this region are divided into two major clusters, one between the two rRNA genes, and the other on the 3′-side of the large rRNA gene. The mtDNA segment encompassing the first tRNA gene cluster also contains the CO III gene.[175] Hybridization studies with separated strands of cloned mtDNA fragments and DNA sequence analysis have shown that the two rRNA genes, the CO III gene, and all the tRNA genes are encoded in the same DNA strand.[170] Furthermore, from the DNA sequence it has been inferred that the direction of transcription is from the small to the large rRNA gene,[176] as is true for the rRNA genes of almost all systems (see below). The DNA sequence studies, besides providing

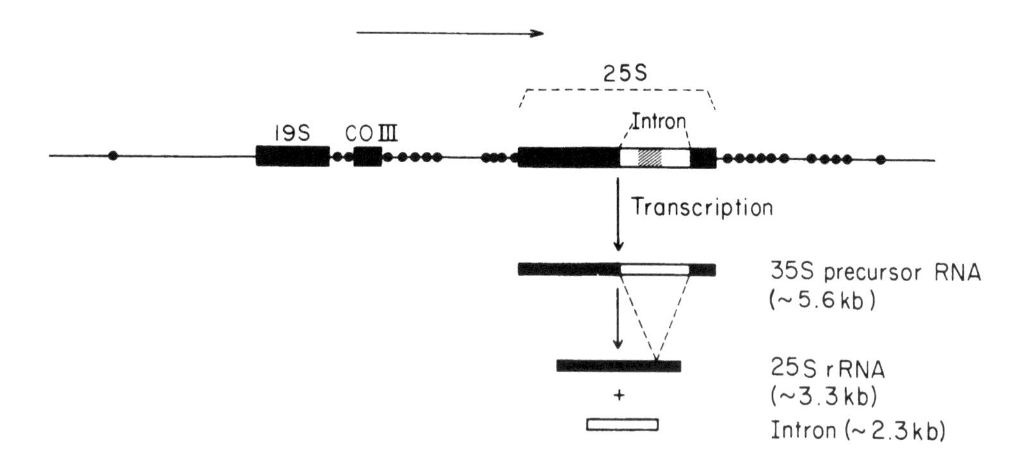


FIGURE 19. Gene organization in the rRNA and tRNA region of *Neurospora crassa* mtDNA, and synthesis of 25S rRNA. The filled circles on the DNA strand represent tRNA genes; the hatched area in 25S rRNA gene intron represents an open reading frame. It is not known whether the 35S precursor RNA is the primary transcript. The arrow indicates the direction of transcription of all the genes shown. Redrawn from data presented in References 175 and 176.

information on the location and the structure of the intervening sequence of the large rRNA gene, have revealed the primary structure of 23 tRNA genes.[175] None of these exhibits intervening sequences; in all cases, the -CCA end of the tRNAs is not encoded in the DNA and must be added posttranscriptionally. The tRNA genes are separated by stretches of variable lengths (from a few nucleotides to several hundred nucleotides).[173]

The DNA sequence analysis has also shown the peculiar distribution of the PstI restriction sites and the organization of the sequences which encompass them.[173,177] PstI cleaves *N. crassa* mtDNA into 50 to 100 fragments. The majority of the PstI sites are found in the region of the rRNA and tRNA genes. With one exception, the PstI sites in this region occur in doublets, the two homologous sequences being separated by the dinucleotide TA. Each doublet of PstI sites is preceded by a stretch of pyrimidine and followed by a stretch of purines, the whole region forming a palindromic structure which can be folded in a hairpin-like fashion; examples of these potential secondary structures are given in Figure 20. The GC content of these hairpin structures is extremely high (~70% on the average, to be compared with the overall GC content of *N. crassa* mtDNA which is ~40%). The GC-rich palindromic sequences exhibit a striking conservation of an 18-bp "core sequence", which has been found in 22 out of 23 such palindromic sequences, with a substantial homology also in the region surrounding this 18-bp fully conserved sequence.

The distribution of the highly conserved GC-rich palindromic sequences containing double PstI sites suggests that they may play a role in RNA synthesis and/or processing. In fact, the GC-rich palindromic sequences present in the region of the rRNA and tRNA genes appear to be distributed with a certain regularity, flanking either single tRNA genes or small groups of them[173,177] (Figure 21). The distance between the hairpin structures containing double PstI sites and the tRNA genes varies between 3 and more than 100 nucleotides. Furthermore, one such structure is found near each edge of the intervening sequence in the large rRNA gene.[177] The latter observation makes it unlikely that these sequences are involved, at least exclusively, in initiation or termination of transcription, but rather points to a role in RNA processing. This processing may conceivably occur through the activity of an enzyme like RNase III of *E. coli;* this could recognize the stems of the individual palindromic regions or, possibly, stems formed by base pairing of two adjacent or more

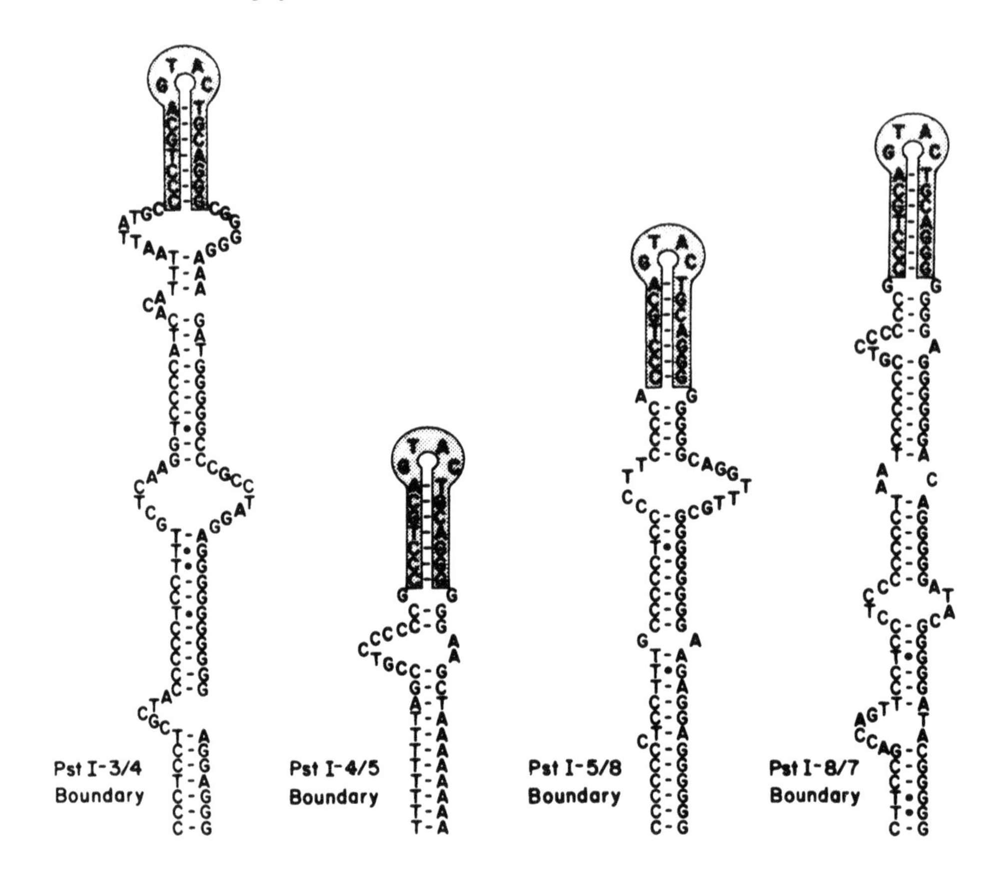


FIGURE 20. Potential secondary structures formed by the sequences (noncoding strand) containing the tandem Pst I sites. The shaded segments represent the 18-base conserved "core sequence" including the Pst I sites. (From Yin, S., Burke, J., Chang, D. D., Browning, K. S., Heckman, J. E., Alzner-De Weerd, B., Potter, M. J., and RajBhandary, U. L., in *Mitochondrial Genes,* Slonimski, P., Borst, P., and Attardi, G., Eds., Cold Spring Harbor Laboratory, Cold Spring Harbor, New York, 1982, 361. With permission.)

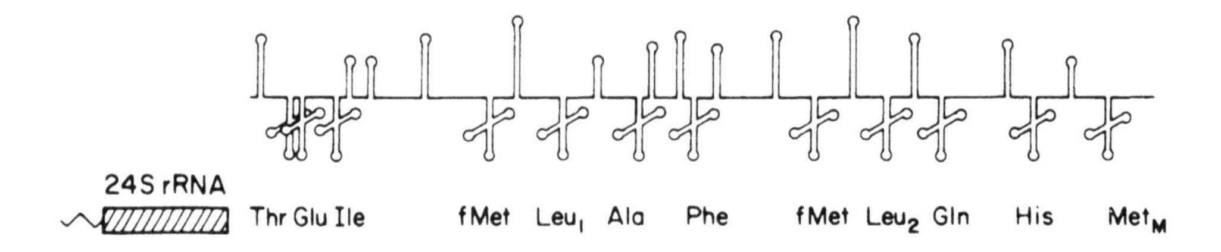


FIGURE 21. Distribution of the palindromic sequences containing double Pst I sites in the tRNA gene cluster downstream of the 25S rRNA gene. (From Yin, S., Burke, J., Chang, D. D., Browning, K. S., Heckman, J. E., Alzner-DeWeerd, B., Potter, M. J., and RajBhandary, U. L., in *Mitochondrial Genes,* Slonimski, P., Borst, P., and Attardi, G., Eds., Cold Spring Harbor Laboratory, Cold Spring Harbor, New York, 1982, 361. With permission.)

distantly located such regions, with the resulting looping out of the RNA in between, as is proposed to occur in the processing of the 16S and 23S rRNA of *E. coli* (see Chapters 1 and 2 in this volume). Other interpretations of the significance of the GC-rich palindromic DNA sequences have been proposed, such as their having a role in site-specific recombination or DNA replication, or their being equivalent to translocatable elements that can mediate DNA rearrangements, or their having no role at all, as "selfish" DNA.[177]

*2. Transcription and RNA Processing in the rRNA and tRNA Region*

***a. General Features***

No information is available concerning the location of promoters in the tRNA and rRNA region. RNA transfer hybridization experiments utilizing DNA probes corresponding to different portions of this region have revealed high-molecular-weight RNA species which span the entire segment including the 19S and the 25S rRNA genes.[176] The data suggest that there may be a small number of transcription units, and possibly only one, and that most of the discrete RNA species mapping in this region may result from processing of larger precursors. The processing of the 25S rRNA from a 35S RNA precursor, which is the best investigated among such phenomena, will be discussed below. A 5.0-kb and a 3.8-kb RNA mapping in the 19S rRNA region may be precursors of the 19S rRNA, and the 5.0-kb, 2.8-kb, 2.4-kb, and 1.8-kb RNAs, which map in the middle of the DNA segment between the two rRNA genes, may be precursors of multiple tRNA species. Several alterations in the latter transcripts, including a striking accumulation of a 1.8-kb RNA species, have been detected at 37°C in a temperature-sensitive mutant affected in the splicing of the 35S precursor of 25S rRNA (see below), pointing to either an alteration of the splicing of transcripts in this region or to a dependence of the processing of these transcripts on the splicing of the 35S RNA.

***b. Synthesis of the Large rRNA***

As mentioned above, the gene for the large rRNA (25S rRNA) contains an intervening sequence of about 2.3 kb near its 3′ end, in a region which shows 76% homology with the *E. coli* large rRNA and 81% homology with the yeast large rRNA.[173] The location of the intervening sequence in the 25S rRNA gene is the same as in the yeast 21S rRNA gene. Like the yeast 21S rRNA intron, the *N. crassa* 25S rRNA intron contains an open reading frame capable of coding for a protein more than 258 amino acids long, while the other two reading frames of the same strand are blocked by stop codons with the expected frequency.[173] There is no evidence so far that this reading frame is translated in *N. crassa*.

Several nuclear mutants of *N. crassa*, some of which are temperature sensitive, have been identified.[178-182] These show a decreased ratio of 25S to 19S rRNA and accumulate a 35S RNA species (5.2 to 5.6 kb), containing the 25S rRNA coding sequence plus the intervening sequence, and in some cases, unusual intron- or exon-derived RNA species.[182] S1 protection experiments and R-loop analysis have shown that the 35S RNA is a collinear transcript of the 25S rRNA gene, and that the 35S RNA does not possess substantial 5′- or 3′-terminal extensions as compared to 25S rRNA, although extension of less than 300 nt could not be excluded.[176] RNA transfer hybridization experiments utilizing RNA from wild-type cells and probes corresponding to different segments of the 25S rRNA gene have revealed the presence of a low concentration of an RNA ~5.2 kb in size giving a positive signal with both intron and exon probes, and of a relatively abundant RNA 2.3 kb in size showing hybridization only with the intron-specific probes and presumably representing the excised intron.[176] On the basis of these observations, it has been proposed that the 25S rRNA gene is transcribed in the form of a molecule identical or similar to the 35S RNA; this RNA, in the wild-type cells, is spliced in a single step leading to the excision of a 2.3-kb intron RNA, which is moderately stable within the mitochondria.

In mutants defective in splicing, the 35S RNA precursor accumulates.[178-182] The mutants isolated so far identify three different nuclear genes, whose products are required for splicing the large mitochondrial rRNA.[182] Since all the mutants accumulate the 35S RNA precursor, the mutation in every case must affect the endonuclease activity necessary for splicing. The primary defect, however, could be either in a structural gene coding for the RNA splicing enzyme or in a ribosomal protein or some other component which must bind and/or modify the RNA for the proper splicing to occur, or in a regulatory gene controlling the synthesis or activity of one or more of these components. It is interesting that mutations at one locus, *cyt 4*, are subject to partial phenotypic suppression, including a significant decrease in the ratio of 35S RNA to mature 25S rRNA, by the electron transport inhibitor, antimycin.[181,182] This finding has been interpreted to suggest that the synthesis or activity of at least one component required for the large rRNA splicing is increased in response to a deficiency of electron transport.

The accumulation of the 35S RNA precursor in *N. crassa* splicing mutants has offered the opportunity to investigate the secondary structure of this RNA by either RNase III digestion or electron microscopy of RNA spread under partially denaturing conditions.[183] RNase III was found to cleave the 35S RNA into two molecules of roughly equal size (2.5 to 3kb) by acting at one or more sites near the 5′-intron-exon boundary. Electron microscopy of 35S RNA showed a relatively large hairpin (~180 nt) near the putative RNase III sensitive site(s); such a structure was absent in 25S rRNA. The 35S RNA which accumulates in the splicing mutants can be isolated in the form of a ribonucleoprotein particle containing an almost complete set of proteins of the large ribosomal subunit;[184] in such particles, the hairpin structure mentioned above is still accessible to RNase III, a finding that is consistent with a role of such a structure in the in vivo splicing, which may occur on RNA already complexed with ribosomal proteins. The 35S RNA precursor which accumulates in the splicing mutants will prove to be very valuable for in vitro processing studies.

**B. *Aspergillus nidulans***

The mitochondrial genome of *Aspergillus nidulans*, though only about half the size of the *Neurospora crassa* mtDNA, has a very similar gene organization, but in a more compact form.[185-188] Thus, most of the tRNA genes (i.e., 20) and the two rRNA genes are located in one third of the molecule, the small and large rRNA genes are separated by a long spacer (2.8 kb) containing a cluster of tRNA genes and the CO III gene, and the large rRNA gene is flanked on its 3′-side by another cluster of tRNA genes. The large rRNA gene contains a 1.8-kb-long intron which exhibits, here too, an open reading frame. As in *N. crassa*, all the genes mentioned above are on the same strand and are transcribed in the direction from the small to the large rRNA gene. In the rest of the genome are the cytochrome *b* gene, with one intron corresponding exactly in position to bI3 of the "long" gene of *S. cerevisiae*,[137] the CO I gene with three introns, the continuous genes for CO II and ATPase 6, and several URFs.[188-191] The intron of the cytochrome *b* gene and two of the introns of the CO I gene exhibit long open reading frames contiguous and in phase with the upstream exons, suggesting that they may code for mRNA maturases.[190,191]

Very little is known about DNA transcription and RNA processing in *A. nidulans* mitochondria. Discrete transcripts corresponding to the main genes have been mapped by electron microscopy of RNA-DNA hybrids between cloned restriction fragments of mtDNA and unfractionated mitochondrial RNA.[187,188] The tRNA genes in each of the two main clusters in the tRNA-rRNA region are separated from each other by AT-rich spacer sequences, usually consisting of only a few nucleotides and without any significant palindromic structures; two tRNA genes are joined end to end.[186] Such arrangements suggest a transcription of the tRNA genes in the form of polycistronic molecules, with a processing of the transcripts by an enzyme(s) recognizing the cloverleaf structure of the tRNA, like the *E. coli* RNase P.[186] If the GC-rich palindromic structures containing double Pst I sites in *N. crassa* mtDNA function as signals for RNA processing, one must conclude that, in spite of the similarity in overall mitochondrial tRNA and rRNA gene organization between *N. crassa* and *A. nidulans*, which extends even to the order of tRNA genes, different mechanisms of RNA processing operate in the two organisms.[177]

# VI. MITOCHONDRIAL DNA OF PROTOZOA

## A. Kinetoplast DNA

The kinetoplast DNA represents the mtDNA of trypanosomatids.[192-194] It has a remarkable structure, consisting of thousands of DNA circles interlocked to form a large network situated within the single mitochondrion of the cell. The circles are of two types: minicircles, present in 4000 to 10,000 copies per cell and of sizes varying between 0.8 kb and 2.5 kb in different species, which constitute 95% of the mass of the network and are usually heterogeneous in sequence; and maxicircles, present in 25 to 100 copies per cell, which make up only 5% of the mass of the kinetoplast and range in size between 20 kb and 38 kb, depending upon the species. While the minicircles are apparently not transcribed and have not been associated, as yet, with any function, the maxicircles (~20% GC) are transcribed, and are the functional equivalent of the mtDNA of other organisms.

Among the maxicircle genes, the two rRNA genes were the first to be identified and mapped by hybridization of maxicircle DNA with the two rRNA species[195-199] (see below). Recently, low-stringency hybridization experiments utilizing yeast mtDNA probes corresponding to different mitochondrial genes have allowed a tentative identification and approximate localization in the *Leishmania tarentolae* maxicircle of the genes for CO I, CO II, and cytochrome *b*.[201,202] The CO III and ATPase probes hybridized to several fragments, preventing an even tentative localization of the corresponding genes, while no hybridization was observed with an ATPase 9 probe.

Information about the transcription patterns in the maxicircles comes mostly from studies carried out on *Trypanosoma brucei*,[198-200,203] *L. tarentolae*,[195-197,201,204] and *Cristidia* species.[198,205,206] In these organisms, the most striking features of the transcription of the maxicircles so far detected have been the very small size of the two rRNA species, 9S RNA (apparent size of ~520 to 590 nt) and 12S RNA (~1020 to 1080 nt),[195,203] which are the smallest rRNAs known, and the unusual pattern of their transcription (see below). No ribosomal particles containing these RNA species have been identified as yet. However, evidence pointing to the rRNA nature of these species comes from their relative abundance, from their 1:1 stoichiometry, from the observations that their size and sequence are highly conserved in several trypanosomatids,[203,205,207] and from the existence of a small but significant homology of their sequence with that of the rRNAs of *E. coli* and human mitochondria.[199,208] An analysis of the DNA sequence of the 12S RNA gene has revealed that this gene lacks one of the two highly conserved sequences that appear to be involved in the chloramphenicol sensitivity of prokaryotic or organelle ribosomes, while the other one is very different;[208] this finding may account for the lack of effect of chloramphenicol on cellular protein synthesis in trypanosomatids.[209]

In *T. brucei*[199,203] and *L. tarentolae*,[201,204] the 9S and 12S rRNA genes map close to each other, with a spacer which in *T. brucei* appears to be less than 350 bp long;[203] the hybridization data indicate that they are continuous transcripts, are encoded in the same strand and, most unusually, are transcribed in the direction from the large to the small rRNA.[199,201,203,204] It is not known whether the two rRNA species are synthesized from a common precursor.

In addition to the two rRNA species, several low abundance transcripts sufficient in number and size to specify the known mtDNA coded products (varying in size between 0.25 kb and 1.8 kb in *L. tarentolae)* have been detected both in *T. brucei*[203,209] and *L. tarentolae*.[201,204] Most of these transcripts are retained on oligo(dT)-cellulose, suggesting that they are polyadenylated.[203,204] Maps of these transcripts have been constructed which show that a large portion of the maxicircle is transcriptionally active.[203,204] These maps have revealed several overlapping polyadenylated transcripts, an observation which points to the possible occurrence of processing events.

In the *L. tarentolae* map, most of the transcripts map in a 6.6-kb fragment containing the 9S and 12S rRNA genes, and have been shown to be encoded in the same DNA strand as the two rRNA species.[204] A puzzling fact is that so far no tRNA has been detected among the transcription products of maxicircles.[203] This failure, however, may be due to technical problems related to the high AT content of the tRNA genes.

### B. *Tetrahymena pyriformis*

Relatively little is known about the gene organization and expression of the linear mtDNA from *Tetrahymena pyriformis* (17.5 μ, ~46 kb, in strain ST). The only aspects which have been investigated are the rRNA gene organization and the origin of the mitochondrial tRNAs. The genes for the large (21S) and small (14S) rRNAs are widely separated in *Tetrahymena* mtDNA.[210,211] Furthermore, while there is only one copy of the small rRNA gene, the large rRNA gene is, most unusually, present in two or three copies depending upon the strain, the extra copy (copies) being located in a terminal or subterminal duplication-inversion. The reason for this inequality of genetic information for the two rRNA species is unknown. The 14S rRNA gene has the same polarity as the nearest 21S rRNA gene (situated at ~8 or 10 kb depending upon the strain). No intervening sequence appears to exist in either gene.

Suyama and Hamada[212] have obtained evidence suggesting that only a fraction of the tRNAs functioning inside mitochondria in *Tetrahymena* are endogenous, with the remainder being nuclear DNA coded and imported. In recent experiments,[213] up to 36 tRNA species from *Tetrahymena* mitochondria have been resolved by two-dimensional gel electrophoresis. Hybridization of $^{32}$P-labeled mitochondrial tRNAs to mtDNA immobilized on a nitrocellulose filter, followed by elution and two-dimensional gel electrophoresis, has revealed that only 10 of the 36 tRNA species had hybridized to mtDNA. Apart from the report of the apparent import into yeast mitochondria from the cytoplasmic-nuclear compartment of a lysine-accepting tRNA species (CUU), which does not seem to be involved in mitochondrial translation,[214] Suyama's findings on *Tetrahymena* represent the only exception, so far, to the general rule that all the RNAs functioning in mitochondria are endogenous. If confirmed by DNA sequencing data, these observations would provide still another example of the multiplicity of pathways that evolution of mitochondria has followed in different organisms.

### C. *Paramecium*

The mtDNA of *Paramecium* represents another form of linear mtDNA (14 μ, ~40 kb). Its mode of replication has been the object of investigation for some time.[215] By contrast, only recently has its gene organization and mode of expression started being explored. The evidence obtained so far indicates that, as in *Tetrahymena* mtDNA, the two rRNA (20S and 14S) genes are far apart in this genome (separated by 10 and 12 kb in two different species):[216] furthermore, by DNA sequencing, two tRNA genes have been located in the genome, one ($tRNA^{Tyr}_{UAY}$) fairly close (127 bp on the 3′-side) to the large rRNA gene,[217] and the other ($tRNA^{Trp}_{UGR}$) at some distance (~1500 bp on the 3′-side) from the small rRNA gene.[218] The tRNA genes appear to be flanked by short AT-rich stretches. All four genes are located on the same strand. No intervening sequences have been detected by R-loop analysis in the 20S rRNA gene,[219] nor do any appear to exist in the two tRNA genes sequenced so far.[217,218] These glimpses of gene organization indicate that the *Paramecium* mtDNA does not have the compactness of gene arrangement found in mammalian mtDNA.

## VII. MITOCHONDRIAL DNA OF PLANTS

The mtDNA of higher plants appears to be much larger and more heterogeneous than that of other organisms.[220,221] In spite of the considerable number of investigations, a coherent picture has not yet emerged on the organization of the plant mitochondrial genome, in particular concerning the native molecular form of mtDNA in these organisms, the size of the genome and its kinetic complexity. This is in part due to the variations that appear to exist among species and in part to the different analytical procedures used. In separate electron microscopic studies carried out on mtDNA from different plants, a homogeneous population of circular molecules of ~30-μ contour length has been observed in DNA released from osmotically shocked mitochondria (pea),[222] a heterogeneous population of circular molecules from 0.6 to 40 μ in length has been found in a closed circular fraction of maize mtDNA,[223] and a heterogeneous set of linear molecules 1 to 62 μ, or up to 28 μ, in length, has been seen in total mtDNA extracted from *Phaseolus vulgaris*[224] or from potato tubers,[225] respectively. From renaturation studies, a kinetic complexity varying between 220 and 1600 $\times$ $10^6$ daltons has been calculated,[226] with only minor repetitive components being detected. The published restriction enzyme digestion patterns of plant mtDNA appear to be very complex, with a large number of fragments in differing molar amounts, and a minimum molecular weight estimate varying between 150 and 500 $\times$ $10^6$ daltons.[220] This complexity in restriction patterns is probably a reflection of heterogeneity resulting from large scale or small scale rearrangements, like deletions, duplications, inversions, insertions, etc. or from microheterogeneity.[221]

Plant mitochondria have distinctive ribosomes, which contain 18S and 26S rRNA, differing from their cytoplasmic counterparts in nucleotide sequence,[227-229] and for their lower level of methylation and lower pseudouridine content.[230] In addition, unique so far among the mitochondrial ribosomes, plant mitochondrial ribosomes contain a 5S RNA,[228,231] which also differs from its cytoplasmic counterpart in nucleotide sequence.[232] The plant mitochondrial 5S RNA has structural characteristics of both prokaryotic and eukaryotic 5S RNAs, as well as unique features.[233]

DNA transfer hybridization experiments have shown the presence in wheat mtDNA restriction digests of fragments hybridizing specifically to mitochondrial 26S, 18S, and 5S rRNAs, as well as to mitochondrial tRNAs.[229] The hybridization patterns obtained with these RNA species has allowed some preliminary conclusions about their organization in the wheat germ mitochondrial genome. Thus, the observation that the 18S and 26S rRNAs always hybridized to different fragments in digests obtained with eight different endonucleases suggests that the genes for these two species are far apart; conversely, the observation that 5S rRNA consistently hybridized to the same restriction fragment as 18S rRNA points to a close linkage of the 5S rRNA and 18S rRNA genes (probably $<$500 bp apart). The above-described arrangement of the rRNA genes is clearly different from that known for prokaryotic cells and chloroplasts, where the order is 16S, 23S, and 5S rRNA genes, with the 23S and 5S rRNA genes being closely linked.

Another interesting observation was that each rRNA species (26S, 18S, and 5S) hybridized to several fragments in a given restriction digest, suggesting the possibility of multiple genes for each rRNA species or of a discontinuity of these genes. Finally, wheat mitochondrial 4S RNA was found to hybridize to many large restriction fragments and several small ones. This observation suggests that the tRNA genes are widely distributed in the wheat mitochondrial genome.

Heterologous DNA probes have also been utilized for the identification of plant mitochondrial genes. Thus, by using a specific yeast mtDNA probe, the maize mitochondrial gene coding for CO II has been identified in DNA transfer hybridization experiments and subsequently cloned.[234] DNA sequence analysis has revealed that in contrast to the mammalian and yeast homologous genes, the maize CO II gene is discontinuous, exhibiting two exons separated by a centrally located intron of 794 bp.[234] There are no long open reading frames in this intron. By using probes derived from this gene in RNA transfer hybridization experiments, several overlapping transcripts complementary to the probes have been identified in mitochondrial RNA. In particular, the results obtained with exon and intron probes have indicated the presence of abundant RNA species, with a size of 2.6 kb and smaller, and of larger transcripts from which the smaller species presumably derive by splicing; it is not clear whether the multiplicity of the larger RNAs and of the abundant RNAs reflects multiple initiation or stop points for transcription or a multistep processing pathway.[234]

## VIII. OVERVIEW AND PERSPECTIVES

The survey made in this article has revealed an extraordinary variety of transcription and RNA processing patterns in the mitochondria from different organisms. These range from the extreme simplicity and economy of the mammalian systems, where the entire genome is transcribed in the form of a polycistronic molecule destined to be cut by precise endonucleolytic cleavages, directed by tRNA sequences, into near-to-mature products, to the high degree of sophistication exhibited by the yeast system, with its intricate network of intragenic and intergenic interactions at the RNA processing level.

Extensive symmetrical transcription still remains unique to the mammalian, and possibly, in general animal cell systems, as is the organization of the protein coding genes and most tRNA genes of the H-strand in one single transcription unit and that of the rRNA genes and adjacent tRNA genes in another, overlapping, transcription unit. This pattern of transcription of the H-strand genes, while providing the cell with the capacity of independent transcriptional control of the rRNA and protein coding genes, places the main burden of regulation of expression of the genes within the same transcription unit on a differential stability of their transcripts. Also unique to the mammalian systems is the extraordinary device of using polyadenylation for creating termination codons at the 3′ end of most reading frames.

The yeast mitochondrial system, with its multiple transcription units, conforms more closely to the general prokaryotic and eukaryotic models, which exhibit independent transcriptional control for rRNA, mRNA, and tRNA synthesis. How a coordinate control of the synthesis of the two mitochondrial rRNA species is achieved in yeast is not known, although the existence of a homologous stretch of 16 nt immediately upstream of the two rRNA genes points to a common recognition signal for the RNA polymerase. The structural features of the yeast mitochondrial mRNAs, apparently all exhibiting a 5′-leader and a 3′-nontranslated extension, also conform to the general pattern of the prokaryotic and eukaryotic messengers. The evolutionary emergence of the complex processing pathways of the transcripts of mosaic genes, in particular, of the cytochrome *b* and CO I genes, requiring the involvement of transacting intron-coded "maturases", cis-acting critical intron sequences, and nuclear-coded products, implies a selective advantage for the organisms that display them. These sophisticated maturation schemes suggest some regulatory functions of the introns, as exemplified by the proposed control by the fourth intron of the cytochrome *b* gene on the expression of the CO I gene. Whatsoever these regulatory functions may be, they must not be indispensable in all genetic backgrounds, since strains lacking some or all of these introns exist in nature or can be constructed in the laboratory.

The organization and mode of expression of the two rRNA genes in different mitochondrial systems provides a striking example of the structural and functional diversity of these systems. Thus, one finds rRNA species varying in size from somewhat larger than that of the bacterial rRNAs, as in the filamentous fungi, to one third of that, as in *Trypanosomatids;* the two rRNA genes can be separated by a spacer barely long enough to accommodate one tRNA gene, as in mammalian mtDNA, or by a DNA stretch as long as 25 to 30 kb, like in yeast mtDNA; furthermore, the large rRNA gene can be continuous or interrupted. Finally, transcription can proceed from the small to a large rRNA, as in almost all mitochondrial systems, or from the large to the small gene, as in *Trypanosomatids.*

From the survey made above, it appears that a considerable amount of detailed information is already available, or is rapidly accumulating, concerning RNA synthesis and processing in human mitochondria, as a paradigm of mammalian systems, and in yeast and *Neurospora crassa* mitochondria, as examples of the fungal systems; various models of transcription and RNA processing pathways in different systems have been proposed. It is clear, however, that further decisive progress in the understanding of the mechanisms and regulatory circuits involved will depend to a great extent on the development of suitable in vitro transcription or RNA processing systems. These could be at the level of intact or partially disrupted organelles, or at the level of reconstituted systems utilizing mtDNA, the specific mitochondrial RNA polymerase(s) and other mitochondrial components, or suitable natural or in vitro-synthesized precursors and processing enzyme preparations. The preliminary results obtained on RNA synthesis and processing in isolated yeast mitochondria and the demonstration of correct initiation of transcription on mtDNA fragments by purified yeast mitochondrial RNA polymerase[235] have already provided an indication of the power of these approaches.

It is clear from what has been said above that the transcription and RNA processing patterns exhibited by the mitochondrial systems do not conform in a simple or univocal way to a prokaryotic or eukaryotic model. One finds in these patterns prokaryotic traits, like the absence of "cap" structures in the mRNAs and the existence of polycistronic transcripts as well as, often in the same organism, typically eukaryotic features, like the polyadenylation of the mitochondrial mRNAs and the existence of RNA splicing mechanisms. The mitochondrial systems also display unique characteristics, like the symmetrical transcription and the existence of mRNAs without 5′ leaders or 3′ tails and even with incomplete termination codons, as observed in the mammalian systems, and the coupling of translation and RNA processing, as found in yeast. The presence of intervening sequences in mitochondrial genes of more primitive eukaryotic cells and the absence of such sequences in the homologous genes of more evolved eukaryotes emphasizes the danger of deriving evolutionary implications from the absence or presence in mitochondria of a certain eukaryotic or prokaryotic trait. It should be mentioned, in this connection, that even in the chloroplast genome, which has certainly more prokaryotic features than the mitochondrial genome, some of the genes have intervening sequences, like the large rRNA gene in *Chlamydomonas*[236] and several tRNA genes in maize.[237,238]

## NOTE ADDED IN PROOF

Recent sequence studies on *Drosophilia yakuba* mtDNA (Clary, D. O., Goddard, J. M., Martin, S. C., Fauron, C. M.-R., and Wolstenholme, D. R., *Nucleic Acids Res.*, 10, 6619, 1982) have revealed that, although the genes so far found in this mtDNA correspond to mammalian mitochondrial genes, the order in which they are arranged is different. In particular, in contrast to the situation in mammalian mtDNA, the two rRNA genes, the $tRNA^{Val}$ gene and URF1 are located, on the same strand, on the side of the replication origin which is replicated first, while the $tRNA^{Ile}$, $tRNA^{Gln}$ and $tRNA^{F\text{-}Met}$ genes and URF2 lie, in the order given, on the side of the origin which is replicated last, and, except for the $tRNA^{Gln}$, are contained in the opposite strand relative to the rRNA genes. This difference in gene order is probably the result of a major translocation and inversion of a segment of an ancestral mtDNA molecule.

A detailed analysis of the structure of the intervening sequences of many mitrochrondrial mRNA and rRNA precursors from *S.cerevisiae*, *N. crassa* and *A. nidulans* has revealed the presence of several conserved sequences potentially able to interact in a pair-wise way to form an almost identical "core" secondary structure, that is presumably important for splicing (Davies, R. W., Waring, R. B., Ray, J. A., Brown, T. A., and Scazzocchio, C., Making ends meet: a model for RNA splicing in fungal mitochondria, *Nature,* 300, 719, 1982; Michel, F., Jacquier, A., and Dujon, B., Comparison of fungal mitochondria introns reveals extensive homologies in RNA secondary structure, *Biochimie,* 64, 867, 1982; Hensgens, L. A. M., Bonen, L., de Haan, M.,van der Horst, G., and Grivell, L. A., Two intron sequences in yeast mitochondrial COX1 gene: homology among URF-containing introns and strain-dependent variation on flanking exons, *Cell*, 32, 379, 1983). This "core" secondary structure is able to bring the ends of each intron into close proximity and, in many cases, to allow an internal RNA sequence to pair with exon bases adjacent to the splice junctions. Most of the conserved sequences and the consensus secondary structure have also been found, suprisingly, in the intron of the nuclear large rRNA precursors of *Tetrahymena* and of *Physarum polycephalum*. These observations point to an evolutionary relationship between these nuclear introns and the mitochondrial introns, and to a similarity in their splicing mechanism and in the proteins involved.

## ACKNOWLEDGMENTS

I am very grateful to several colleagues who have sent me reprints and preprints of their work and some illustration material. The investigations from my laboratory mentioned in this chapter were supported by N.I.H. grant GM-11726.

## REFERENCES

1. **Borst, P. and Grivall, L. A.,** The mitochondrial genome of yeast, *Cell*, 15, 705, 1978.
2. **Attardi, G.,** Organization and expression of the mammalian mitochondrial genome, *TIBS*, 6, 86, 1981; *TIBS*, 6, 100, 1981.
3. **Anderson, S., Bankier, A. T., Barrell, B. G., de Bruijn, M. H. L., Coulson, A. R., Drouin, J., Eperon, I. C., Nierlich, D. P., Roe, B. A., Sanger, F., Schreier, P. H., Smith, A. J. H., Staden, R., and Young, I. G.,** Sequence and organization of the human mitochondrial genome, *Nature (London)*, 290, 457, 1981.
4. **Lazowska, J., Jacq, C., and Slonimski, P. P.,** Sequence of introns and flanking exons in wild-type and *box 3* mutants of cytochrome *b* reveals an interlaced splicing protein coded by an intron, *Cell*, 22, 333, 1980.
5. **Nobrega, F. G., and Tzagoloff, A.,** DNA sequence and organization of the cytochrome *b* gene in *S. cerevisiae* D273-10B, *J. Biol. Chem.*, 255, 9828, 1980.
6. **Bonitz, G. S., Coruzzi, G., Thalenfeld, E., Tzagoloff, A., and Macino, G.,** Structure and sequence of the gene coding for subunit 1 of yeast cytochrome oxidase, *J. Biol. Chem.*, 255, 11927, 1980.
7. **Lewin, B.,** Alternative for splicing: An intron coded protein, *Cell*, 22, 645, 1980.
8. **Jacq, C., Pajot, P., Lazowska, J., Dujardin, G., Claisse, M., Groudinski, O., de la Salle, H., Grandchamp, C., Labouesse, M., Gargouri, A., Guiard, B., Spyridakis, A., Dreyfus, M., and Slonimski, P. P.,** Role of introns in the yeast cytochrome *b* gene: cis- and trans-acting signals, intron manipulation, expression, and intergenic communications, in *Mitochondrial Genes*, Slonimski, P., Borst, P., and Attardi, G., Eds., Cold Spring Harbor Laboratory, Cold Spring Harbor, New York, 1980, 155.
9. **Kuntzel, H., Köchel, H. G., Lazarus, C. M., and Lunsdorf, H.,** Mitochondrial genes in *Aspergillus*, in *Mitochondrial Genes*, Slonimski, P., Burst, P., and Attardi, G., Eds., Cold Spring Harbor Laboratory, Cold Spring Harbor, New York, 1982, 391.
10. **Davies, R. W., Scazzocchio, C., Waring, R. B., Lee, S., Grisi, E., McPhail Berks, M., and Brown, T. A.,** Mosaic genes and unidentified reading frames that have homology with human mitochondrial sequences are found in the mitochondrial genome of *Aspergillus nidulans*, in *Mitochondrial Genes*, Slonimski, P., Borst, P., and Attardi, G., Eds., Cold Spring Harbor Laboratory, Cold Spring Harbor, New York, 1982, 405.
11. **Ephrussi, B., Hottinguer, H., and Tarlitzki, J.,** Action de l'acriflavine sur les levures. II. Étude génétique du mutant "petite colonie", *Ann. Inst. Pasteur*, 76, 419, 1949.
12. **Mitchell, M. B. and Mitchell, H. K.,.** A case of "maternal" inheritance in *Neurospora crassa*, *Proc. Natl. Acad. Sci. U.S.A.*, 38, 442, 1952.

13. **Ojala, D., Merkel, C., Gelfand, R., and Attardi, G.,** The tRNA genes punctuate the reading of genetic information in human mitochondrial DNA, *Cell,* 22, 393, 1980.
14. **Crews, S. and Attardi, G.,** The sequences of the small ribosomal RNA gene and the phenylalanine tRNA gene are joined end to end in human mitochondrial DNA, *Cell,* 19, 775, 1980.
15. **Dubin, D. T., Montoya, J., Timko, K. D., and Attardi, G.,** Sequence analysis and precise mapping of the 3′ ends of HeLa cell mitochondrial ribosomal RNAs, *J. Mol. Biol.,* 157, 1, 1982.
16. **Montoya, J., Ojala, D., and Attardi, G.,** Distinctive features of the 5′-terminal sequences of the human mitochondrial mRNAs, *Nature (London),* 290, 465, 1981.
17. **Ojala, D., Montoya, J., and Attardi, G.,** The tRNA punctuation model of RNA processing in human mitochondria, *Nature (London),* 290, 470, 1981.
18. **Chomyn, A., Hunkapiller, M. W., and Attardi, G.,** Alignment of the amino terminal amino acid sequence of human cytochrome *c* oxidase subunits I and II with the sequence of their putative mRNAs, *Nucleic Acids Res.,* 9, 867, 1981.
19. **Anderson, S., de Bruijn, M. H. L., Coulson, A. R., Eperon, I. C., Sanger, F., and Young, I. G.,** The complete sequence of bovine mitochondrial DNA: conserved features of the mammalian mitochondrial genome, *J. Mol. Biol.,* 156, 683, 1982.
20. **Bibb, M. J., Van Etten, R. A., Wright, C. T., Walberg, M. W., and Clayton, D. A.,** Sequence and gene organization of mouse mitochondrial DNA, *Cell,* 26, 167, 1981.
21. **Grosskopf, R. and Feldmann, H.,** Analysis of a DNA segment from rat liver mitochondria containing the genes for the cytochrome oxidase subunits I, II and III, ATPase subunit 6, and several tRNA genes, *Curr. Genet.,* 4, 151, 1981.
22. **Quagliariello, C., De Benedetto, C., Fanizzi, F. P., Gallerani, R., Iuele, R., Mazzula, S., Spena, A., and Saccone, C.,** The nucleotide sequence of URF2 and adjacent tRNA genes in rat liver mitochondria, *Italian J. Biochem.,* in press.
23. **Aloni, Y. and Attardi, G.,** Expression of the mitochondrial genome in HeLa cells. II. Evidence for complete transcription of mitochondrial DNA, *J. Mol. Biol.,* 55, 251, 1971.
24. **Aloni, Y. and Attardi, G.,** Symmetric *in vivo* transcription of mitochondrial DNA in HeLa cells, *Proc. Natl. Acad. Sci. U.S.A.,* 68, 1957, 1971.
25. **Murphy, W. I., Attardi, B., Tu, C., and Attardi, G.,** Evidence for complete symmetrical transcription *in vivo* of mitochondrial DNA in HeLa cells, *J. Mol. Biol.,* 99, 809, 1975.
26. **Cantatore, P. and Attardi, G.,** Mapping of nascent light and heavy strand transcripts on the physical map of HeLa cell chondrial DNA, *Nucleic Acids Res.* 8, 2605, 1980.
27. **Amalric, F., Merkel, C., Gelfand, R., and Attardi, G.,** Fractionation of mitochondrial RNA from HeLa cells by high-resolution electrophoresis under strongly denaturing conditions, *J. Mol. Biol.,* 118, 1, 1978.
28. **Hirsch, M. and Penman, S.,** Mitochondrial polyadenylic acid-containing RNA: Localization and characterization, *J. Mol. Biol.,* 80, 379, 1973.
29. **Ojala, D. and Attardi, G.,** Expression of the mitochondrial genome in HeLa cells. XIX. Occurrence in mitochondria of polyadenylic acid sequences, "free" and covalently linked to mitochondrial DNA-coded RNA, *J. Mol. Biol.,* 82, 151, 1974.
30. **Gelfand, R. and Attardi, G.,** Synthesis and turnover of mitochondrial ribonucleic acid in HeLa cells: The mature ribosomal and messenger ribonucleic acid species are metabolically unstable, *Molec. Cell. Biol.,* 1, 497, 1981.
31. **Grohmann, K., Amalric, F., Crews, S., and Attardi, G.,** Failure to detect "cap" structures in mitochondrial DNA-coded poly(A)-containing RNA from HeLa cells, *Nucleic Acids Res.,* 5, 637, 1978.
32. **Ojala, D., Crews, S., Montoya, J., Gelfand, R., and Attardi, G.,** A small polyadenylated RNA (7S RNA), containing a putative ribosome attachment site, maps near the origin of human mitochondrial DNA replication, *J. Mol. Biol.,* 150, 303, 1981.
33. **Sekiya, T., Kobayashi, M., Seki, T., and Koike, K.,** Nucleotide sequence of a cloned fragment of rat mitochondrial DNA containing the replication origin, *Gene,* 11, 53, 1980.
34. **Steitz, J. A. and Jakes, K.,** How ribosomes select initiator regions in mRNA: Base pair formation between the 3′-terminus of 16S rRNA and the mRNA during initiation of protein synthesis in *Escherichia coli, Proc. Natl. Acad. Sci. U.S.A.,* 72, 4734, 1975.
35. **Attardi, B. and Attardi, G.,** Expression of the mitochondrial genome in HeLa cells. I. Properties of the discrete RNA components from the mitochondrial fraction, *J. Mol. Biol.,* 55, 231, 1971.
36. **Dubin, D. T and Taylor, R. H.,** Modification of mitochondrial ribosomal RNA from hamster cells: The presence of GmG and late-methylated UmGmU in the large subunit (17S) RNA, *J. Mol. Biol.,* 121, 523, 1978.
37. **Dubin, D. T., Taylor, R. H., and Davenport, L. W.,** Methylation status of 13S ribosomal RNA from hamster mitochondria: The presence of a novel riboside, $N^4$-methylcytidine, *Nucleic Acids Res.,* 5, 4385, 1978.

38. **Nierlich, D. P.**, Fragmentary 5S rRNA gene in the human mitochondrial genome, *Molec. Cell. Biol.*, 2, 207, 1982.
39. **Lynch, D. C. and Attardi, G.**, Amino acid specificity of the transfer RNA species coded for by HeLa cell mitocondrial DNA, *J. Mol. Biol.*, 102, 125, 1976.
40. **Davenport, L. W. Taylor, R. H., and Dubin, D. T.**, Comparison of human and hamster mitochondrial transfer RNA. Physical properties and methylation status, *Biochim. Biophys. Acta,* 447, 285, 1976.
41. **Kastelein, R. A., Remaut, E., Fiers, W., and van Duin, J.**, Lysis gene expression of RNA phage MS2 depends on a frameshift during translation of the overlapping coat protein gene, *Nature (London),* 295, 35, 1982.
42. **Montoya, J., Christianson, T., Levens, D., Rabinowitz, M., and Attardi, G.**, Identification of initiation sites for heavy strand and light strand transcription in human mitochondrial DNA, *Proc. Natl. Acad. Sci. U.S.A.*, 79, 7195, 1982.
43. **Robertson, H. D., Altman, S., and Smith, J.**, Purification and properties of a specific *E. coli* ribonuclease which cleaves a tyrosine transfer ribonucleic acid precursor, *J. Biol. Chem.*, 247, 5243, 1972.
44. **Bothwell, A. L. M., Stark, B. C., and Altman, S.**, Ribonuclease P substrate specificity: Cleavage of a bacteriophage ø80-induced RNA, *Proc. Natl. Acad. Sci. U.S.A.*, 73, 1912, 1976.
45. **Attardi, G., Cantatore, P., Chomyn, A., Crews, S., Gelfand, R., Merkel, C., Montoya, J., and Ojala, D.**, A comprehensive view of mitochondrial gene expression in human cells, in *Mitochondrial Genes,* Slonimski, P., Borst, P., and Attardi, G., Eds., Cold Spring Harbor Laboratory, Cold Spring Harbor, New York, 1982, 51.
46. **Attardi, G., Cantatore, P., Ching, E., Crews, S., Gelfand, R., Merkel, C., Montoya, J., and Ojala, D.**, The remarkable features of gene organization and expression of human mitochondrial DNA, in *The Organization and Expression of the Mitochondrial Genome,* Kroon, A. M. and Saccone, C., Eds., Elsevier/North-Holland, Amsterdam, 1980, 103.
47. **Rosenberg, M. and Court, D.**, Regulatory sequences involved in the promotion and termination of RNA transcription, *Ann. Rev. Genet.*, 13, 319, 1979.
48. **Clayton, D. A.**, Replication of animal mitochondrial DNA, *Cell,* 28, 693, 1982.
49. **Kobayashi, M., Yaginuma, K., Seki, T., and Koike, K.**, Nucleotide sequences of the cloned EcoA fragment of rat mitochondrial DNA, in *The Organization and Expression of the Mitochondrial Genome,* Kroon, A. M. and Saccone, C., Eds., Elsevier/North-Holland, Amsterdam, 1980, 221.
50. **Saccone, C., Cantatore, P., Gadaleta, G., Gallerani, R., Lanave, C., Pepe, G., and Kroon, A. M.**, The nucleotide sequence of the large ribosomal RNA gene and the adjacent tRNA genes from rat mitochondria, *Nucleic Acids Res.*, 9, 4139, 1981.
51. **Van Etten, R. A., Michael, N. L., Bibb, M. J., Brennicke, A., and Clayton, D. A.**, Expression of the mouse mitochondrial DNA genome, in *Mitochondrial Genes,* Slonimski, P., Borst, P., and Attardi, G., Eds., Cold Spring Harbor Laboratory, Cold Spring Harbor, New York, 1982, 73.
52. **Greco, M., Pepe, G., Bakker, H., Kroon, A. M., and Saccone, C.**, The genetic localization of presumptive mitochondrial messenger RNAs on rat-liver mitochondrial DNA, *Biochem. Biophys. Res. Commun.*, 88, 199, 1979.
53. **Ohi, S., Ramirez, J. L., Upholt, W. B., and Dawid, I. B.**, Mapping of mitochondrial 4S RNA genes in *Xenopus laevis* by electron microscopy, *J. Mol. Biol.*, 121, 299, 1978.
54. **Ramirez, J. L. and Dawid, I. B.**, Mapping of mitochondrial DNA in *Xenopus laevis* and *X. borealis:* The positions of ribosomal genes and D-loops, *J. Mol. Biol.*, 119, 133, 1978.
55. **Rastl, E. and Dawid, I. B.**, Expression of the mitochondrial genome in *Xenopus laevis:* a map of transcripts, *Cell,* 18, 501, 1979.
56. **Wolstenholme, D. R., Fauron, C. M.-R., and Goddard, J.**, The adenine and thymine-rich region of *Drosophila* mitochondrial DNA molecules, in *The Organization and Expression of the Mitochondrial Genome,* Kroon, A. M. and Saccone, C., Eds., Elsevier/North-Holland, Amsterdam, 1980, 241.
57. **Klukas, C. K. and Dawid, I. B.**, Characterization and mapping of mitochondrial ribosomal RNA and mitochondrial DNA in *Drosophilia melanogaster, Cell,* 9, 615, 1976.
58. **Spradling, A., Pardue, M. L., and Penman, S.**, Messenger RNA in heat-shocked *Drosophila* cells, *J. Mol. Biol.*, 109, 559, 1977.
59. **Bonner, J. J., Berninger, M., and Pardue, M. L.**, Transcription of polytene chromosomes and of the mitochondrial genome in *Drosophila melanogaster,* in *Chromatin,* Cold Spring Harbor Symp., Vol. 42, Cold Spring Harbor, New York, 1978, 803.
60. **Eaton, B. T. and Randlett, D. J.**, Origin of the actinomycin D insensitive RNA species in *Aedes albopictus* cells, *Nucleic Acids Res.*, 5, 1301, 1978.
61. **Dubin, D. T., HsuChen, C. C., Timko, K. D., Azzolina, T. M., Prince, D. L., and Ranzini, J. L.**, 3′ Termini of mammalian and insect mitochondrial rRNAs, in *Mitochondrial Genes,* Slonimski, P., Borst, P., and Attardi, G., Eds., Cold Spring Harbor Laboratory, Cold Spring Harbor, New York, 1982, 89.

62. **Clark-Walker, G. Dand Sriprakash, K. S.,** Size diversity and sequence rearrangements in mitochondrial DNAs from yeasts, in *Mitochondrial Genes,* Slonimski, P., Borst, P., and Attardi, G., Eds., Cold Spring Harbor Laboratory, Cold Spring Harbor, New York, 1982, 349.
63. **Van Ommen, G.-J. B., Groot, G. S. P., and Borst, P.,** Fine structure physical mapping of 4S RNA genes on mitochondrial DNA of *Saccharomyces cerevisiae, Mol. Gen. Genet.,* 154, 255, 1977.
64. **Wesolowski, M. and Fukuhara, H.,** The genetic map of transfer RNA genes of yeast mitochondria: Correction and extension, *Mol. Gen. Genet.,* 170, 261, 1979.
65. **Tabak, H. F., Grivell, L. A., and Borst, P.,** Transcription of mitochondrial DNA, *CRC Crit. Rev. Biochem.,* in press.
66. **Li, M. and Tzagoloff, A.,** Assembly of the mitochondrial membrane system: Sequences of yeast mitochondrial valine and an unusual threonine tRNA gene, *Cell,* 18, 47, 1979.
67. **Prunell, A. and Bernardi, G.,** The mitochondrial genome of wild-type yeast cells. IV. Genes and spacers, *J. Mol. Biol.,* 86, 825, 1974.
68. **Prunell, A. and Bernardi, G.,** The mitochondrial genome of wild-type yeast cells. VI. Genome organization, *J. Mol. Biol.,* 110, 53, 1977.
69. **Hudspeth, M. E. S., Ainley, W. M., Shumard, D. S., Butow, R. A., and Grossman, L. I.,** Location and structure of the *var 1* gene on yeast mitochondrial DNA: Nucleotide sequence of the 40.0 allele, *Cell,* 30, 617, 1982.
70. **Tzagoloff, A., Nobrega, M., Akai, A., and Macino, G.,** Assembly of the mitochondrial membrane system. Organization of yeast mitochondrial DNA in the *oli 1* region, *Curr. Genet.,* 2, 149, 1980.
71. **Dujon, B.,** Sequence of the intron and flanking exons of the mitochondrial 21S rRNA gene of yeast strains having different alleles at the ω and *rib-1* loci, *Cell,* 20, 185, 1980.
72. **Sor, F. and Fukuhara, H.,** Nucleotide sequence of the small ribosomal RNA gene from the mitochondria of *Saccharomyces cerevisiae,* in *Mitochondrial Genes,* Slonimski, P., Borst, P., and Attardi, G., Eds., Cold Spring Harbor Laboratory, Cold Spring Harbor, New York, 1980, 255.
73. **Slonimski, P.,** personal communication, 1982.
74. **Sanders, J. P. M., Heyting, C., Verbeet, M. Ph., Meijlink, F. C. P. W., and Borst, P.,** The organization of genes in yeast mitochondrial DNA. III. Comparison of the physical maps of the mitochondrial DNAs from three wild-type *Saccharomyces* strains, *Mol. Gen. Genet.,* 157, 239, 1977.
75. **Jacq, C., Kujawa, C., Grandchamp, C., and Netter, P.,** Physical characterization of the difference between yeast mitDNA alleles $\omega^+$ and $\omega^-$, in *Mitochondria 1977. Genetics and Biogenesis of Mitochondria,* Bandlow, W., Schweyen, R. J., Wolf, K., and Kaudewitz, F., Eds., Walter de Gruyter, New York, 1977, 255.
76. **Bos, J. L., Heyting, C., Borst, P., Arnberg, A. C., and van Bruggen, E. F. J.,** An insert in the single gene for the large ribosomal RNA in yeast mitochondrial DNA, *Nature (London),* 275, 336, 1978.
77. **Faye, G., Dennebouy, N., Kujawa, C., and Jacq, C.,** Inserted sequence in the mitochondrial 23S ribosomal RNA gene of the yeast *Saccharomyces cerevisiae, Mol. Gen. Genet.,* 168, 101, 1979.
78. **Heyting, C., Meijlink, F. C. P. W., Verbeet, M. Ph., Sanders, J. P. M., Bos, J. L., and Borst, P.,** Fine structure of the 21S ribosomal RNA region in yeast mitochondrial DNA. I. Construction of the physical map and localization of the cistron for 21S mitochondrial ribosomal RNA, *Mol. Gen. Genet.,* 168, 231, 1979.
79. **Slonimski, P. P., Pajot, P., Jacq, C., Foucher, M., Perrodin, G., Kochko, A., and Lamouroux, A.,** Mosaic organization and expression of the mitochondrial DNA region controlling cytochrome *c* reductase and oxidase. I. Genetic, physical and complementation maps of the *box* region, in *Biochemistry and Genetics of Yeast,* Bacila, M., Horecker, B. L., and Stoppani, A. O. M., Eds., Academic Press, New York, 1978, 339.
80. **Grivell, L. A., Arnberg, A. C., Boer, P. H., Borst, P., Bos, J. L., van Bruggen, E. F. J., Groot, G. S. P., Hecht, N. B., Hensgens, L. A. M., van Ommen, G.-J. B., and Tabak, H. F.,** Transcripts of yeast mitochondrial DNA and their processing in *Extrachromosomal DNA,* Cummings, D. J., Borst, P., Dawid, I. B., and Weissman, S. M., Eds., *ICN-UCLA Symp. Molec. Cell. Biology,* Vol. 15, Academic Press, New York, 1979, 305.
81. **Van Ommen, G.-J. B., Groot, G. S. P., and Grivell, L. A.,** Transcription maps of mtDNAs of two strains of *Saccharomyces:* Transcription of strain-specific insertions; complex RNA maturation and splicing, *Cell,* 18, 511, 1979.
82. **Grivell, L. A., Hensgens, L. A. M., Osinga, K. A., Tabak, H. F., Boer, P. H., Crusius, J. B. A., van der Laan, J. C., de Haan, M., van der Horst, G., Evers, R. F., and Arnberg, A. C.,** RNA processing in yeast mitochondria, in *Mitochondrial Genes,* Slonimski, P., Borst, P., and Attardi, G., Eds., Cold Spring Harbor Laboratory, Cold Spring Harbor, New York, 1982, 225.
83. **Hensgens, L. A. M., Arnberg, A. C., Roosendaal, E., van der Horst, G., van der Veen, van Ommen, G.-J. B., and Grivell, L. A.,** Variation transcription and circular RNAs of the mitochondrial gene for subunit I of cytochrome C oxidase, *J. Molec. Biol.,* 164, 35, 1983.

84. **Cobon, G. S., Beiharz, M. W., Linnane, A. W., and Nagley, P.**, Biogenesis of mitochondria: mapping of transcripts from the *oli-2* region of mitochondrial DNA in two grande strains of *Saccharomyces cerevisiae*, *Curr. Genet.*, 5, 97, 1982.
85. **Thalenfeld, B. E., Hill, J., and Tzagoloff, A.**, Assembly of the mitochondrial membrane system. Characterization of the *oxi2* transcript and localization of its promoter in *S. cerevisiae* D273-10B, *J. Biol. Chem.*, 258, 610, 1983.
86. **Butow, R. A.**, personal communication, 1982.
87. **Levens, D., Tichs, B., Ackerman, E., and Rabinowitz, M.**, Transcriptional initiation and 5′ termini of yeast mitochondrial RNA, *J. Biol. Chem.*, 256, 5226, 1981.
88. **Levens, D., Christianson, T., Edwards, J., Lustig, A., Ticho, B., Locker, J., and Rabinowitz, M.**, Transcriptional initiation of yeast mitochondrial RNA and characterization and synthesis of mitochondrial RNA polymerase, in *Mitochondrial Genes*, Slonimski, P., Borst, P., and Attardi, G., Eds., Cold Spring Harbor Laboratory, Cold Spring Harbor, New York, 1980, 295.
89. **Christianson, T. and Rabinowitz, M.**, personal communication, 1982.
90. **Christianson, T., Edwards, J., Levens, D., Locker, J., and Rabinowitz, M.**, Transcriptional initiation and processing of the small ribosomal RNA of yeast mitochondria, *J. Biol. Chem.*, 257, 6494, 1982.
91. **Tzagoloff, A.**, personal communication, 1982.
92. **Jakovcic, S., Hendler, F., Halbreick, A., and Rabinowitz, M.**, Transcription of yeast mitochondrial deoxyribonucleic Acid, *Biochemistry*, 18, 3200, 1979.
93. **Groot, G. S. P., Van Harten-Loosbroeck, N., van Ommen, G.-J. B., and Pijst, H. L. A.**, RNA synthesis in isolated yeast mitochondria, *Nucleic Acids Res.*, 9, 6369, 1981.
94. **Lewin, A., Morimoto, R., Merten, S., Martin, N., Berg, P., Christianson, T., Levens, D., and Rabinowitz, M.**, Physical mapping of mitochondrial genes and transcripts in *Saccharomyces cerevisiae*, in *Mitochondria 1977. Genetics and Biogenesis of Mitochondria*, Bandlow, W., Schweyen, R. J., Wolf, K., and Kaudwitz, F., Eds., Walter de Gruyter, New York, 1977, 271.
95. **Fox, T. D., and Boerner, P.**, Transcripts of the *oxi-1* locus are asymmetric and may be spliced, in *The Organization and Expression of the Mitochondrial Genome*, Kroon, A. M. and Saccone, C., Eds., Elsevier/North-Holland, Amsterdam, 1980, 191.
96. **Coruzzi, G., Bonitz, S. G., Thalenfeld, B. E., and Tzagoloff, A.**, Assembly of the mitochondrial membrane system. Analysis of the nucleotide sequence and transcripts in the *oxi-1* region of yeast mitochondrial DNA, *J. Biol. Chem.*, 256, 12780, 1981.
97. **Beilharz, M. W., Cobon, G. S., and Nagley, P.**, A novel species of double stranded RNA in mitochondria of *Saccharomyces cerevisiae*, *Nucleic Acids Res.*, 10, 1051, 1982.
98. **Hensgens, L. A. M., Grivell, L. A., Borst, P., and Bos, J. L.**, Nucleotide sequence of the mitochondrial structural gene for subunit 9 of yeast ATPase complex, *Proc. Natl. Acad. Sci. U.S.A.*, 76, 1663, 1979.
99. **Morimoto, R., Locker, J., Synenki, R. M., and Rabinowitz, M.**, Transcription, processing and mapping of mitochondrial RNA from grande and petite yeast, *J. Biol. Chem.*, 254, 12461, 1979.
100. **Merten, S., Synenki, R. M., Locker, J., Christianson, T., and Rabinowitz, M.**, Processing of precursors of 21S ribosomal RNA from yeast mitochondria, *Proc. Natl. Acad. Sci. U.SA.*, 77, 1417, 1980.
101. **Osinga, K. A., Evers, R. F., van der Laan, J. C., and Tabak, H. F.**, A putative precursor for the small ribosomal RNA from mitochondria of *Saccharomyces cerevisiae*, *Nucleic Acids Res.*, 9, 1351, 1981.
102. **Hendler, F. J., Padmanaban, G., Patzer, J., Ryan, R., and Rabinowitz, M.**, Yeast mitochondrial RNA contains a short polyadenylic acid segment, *Nature*, 258, 357, 1975.
103. **Moorman, A. F. M., van Ommen, G.-J. B., and Grivell, L. A.**, Transcription in yeast mitochondria: Isolation and physical mapping of messenger RNAs for subunits of cytochrome *c* oxidase and ATPase, *Mol. Gen. Genet.* 160, 13, 1978.
104. **Klootwijk, J., Klein, I., and Grivell, L. A.**, Minimal post-transcriptional modification of yeast mitochondrial ribosomal RNA, *J. Mol. Biol.*, 97, 337, 1975.
105. **Martin, R., Schneller, J. M., Stahl, A. J. C., and Dirheimer, G.**, Studies of odd bases in yeast mitochondrial tRNA: II. Characterization of rare nucleosides, *Biochem. Biophys. Res. Commun.*, 70, 997, 1976.
106. **Bos, J. L., Osinga, K. A., van der Horst, G., Hecht, N. B., Tabak, H. F., van Ommen, G.-J. B., and Borst, P.**, Splice point sequence and transcripts of the intervening sequence in the mitochondrial 21S ribosomal RNA gene of yeast, *Cell*, 20, 207, 1980.
107. **Locker, J. and Rabinowitz, M.**, Transcription in yeast mitochondria: Analysis of the 21S rRNA region and its transcripts, *Plasmid*, 6, 302, 1981.
108. **Newman, D. and Martin, N.**, Synthesis of RNA in isolated mitochondria from *Saccharomyces cerevisiae*, *Plasmid*, 7, 66, 1982.
109. **Boerner, P., Mason, T. L., and Fox, T. D.**, Synthesis and processing of ribosomal RNA in isolated yeast mitochondria, *Nucleic Acids Res.*, 9, 6379, 1981.

110. **Kotylak, Z. and Slonimski, P. P.,** Joint control of cytochromes *a* and *b* by a unique mitochondrial DNA region comprising four genetic loci, in *The Genetic Function of Mitochondrial DNA*, Saccone, C. and Kroon, A. M., Eds., North-Holland, New York, 1976, 143.
111. **Claisse, M. L., Spyridakis, A., Wambier-Kluppel, M. L., Pajot, P., and Slonimski, P. P.,** Mosaic organization and expression of the mitochondrial DNA region controlling cytochrome *c* reductase and oxidase. II. Analysis of proteins translated from the *box* region, in *Biochemistry and Genetics of Yeast*, Bacila, M., Horecker, B. L., and Stoppani, A. D. M., Eds., Academic Press, New York, 1978, 364.
112. **Slonimski, P. P., Claisse, M. L., Foucher, M., Jacq, C., Kochko, A., Lamouroux, A., Pajot, P., Perrodin, G., Spyridakis, A., and Wambier-Kluppel, M. L.,** Mosaic organization and expression of the mitochondrial DNA region controlling cytochrome *c* reductase and oxidase. III. A model of structure and function, in *Biochemistry and Genetics of Yeast*, Bacila, M., Horecker, B. L., and Stoppani, A. O. M., Eds., Academic Press, New York, 1978, 391.
113. **Mahler, H. R., Hanson, D., Miller, D., Lin, D. D., Alexander, N. J., Vincent, R. D., and Perlman, P. S.,** Regulatory aspects of mitochondrial biogenesis, in *Biochemistry and Genetics of Yeast*, Bacila, M., Horecker, B. L., and Stoppani, A. O. M., Eds., Academic Press, New York, 1978, 513.
114. **Alexander, N. J., Vincent, R. D., Perlman, P. S., Miller, D. H., Hanson, D. K., and Mahler, H. R.,** Regulatory interactions between mitochondrial genes. I. Genetic and biochemical characterization of some mutant types affecting apocytochrome *b* and cytochrome oxidase, *J. Biol. Chem.*, 254, 2471, 1979.
115. **Haid, A., Schweyen, R. J., Bechmann, H., Kaudewitz, F., Solioz, M., and Schatz, G.,** The mitochondrial COB region in yeast codes for apocytochrome *b* and is mosaic, *Eur. J. Biochem.*, 94, 451, 1979.
116. **Kreike, J., Bechmann, H., Van Emert, F. J., Schweyen, R. J., Boer, P. H., Kaudewitz, F., and Groot, G. S. P.,** The identification of apocytochrome *b* as a mitochondrial gene product and immunological evidence for altered apocytochrome *b* in yeast strains having mutations in the COB region of mitochondrial DNA, *Eur. J. Biochem.*, 101, 607, 1979.
117. **Solioz, M. and Schatz, G.,** Mutations in putative intervening sequences of the mitochondrial cytochrome *b* gene of yeast produce abnormal cytochrome *b* polypeptides, *J. Biol. Chem.*, 254, 9331, 1979.
118. **Hanson, D. K., Miller, D. H., Mahler, H. R., Alexander, N. J., and Perlman, P. S.,** Regulatory interactions between mitochondrial genes. II. Detailed characterization of novel mutants mapping within one cluster in the COB 2 region, *J. Biol. Chem.*, 254, 2480, 1979.
119. **Grivell, L. A., Arnberg, A. C., Hensgens, L. A. M., Roosendaal, E., van Ommen, G.-J. B., and van Bruggen, E. F. J.,** Split genes on yeast mitochondrial DNA: Organization and expression, in *The Organization and Expression of the Mitochondrial Genome*, Kroon, A. M. and Saccone, C., Eds., Elsevier/North-Holland, Amsterdam, 1980, 37.
120. **Halbreich, A., Pajot, P., Foucher, M., Grandchamp, C., and Slonimski, P.,** A pathway of cytochrome *b* mRNA processing in yeast mitochondria: Specific splicing steps and an intron-derived circular RNA, *Cell*, 19, 321, 1980.
121. **Van Ommen, G.-J. B., Boer, P. H., Groot, G. S. P., de Haan, M., Roosendaal, E., Grivell, L. A., Haid, A., and Schweyen, R. J.,** Mutations affecting RNA splicing and the interaction of gene expression of the yeast mitochondrial loci *cob* and *oxi-3*, *Cell*, 20, 173, 1980.
122. **Schweyen, R. J., Francisci, S., Haid, A., Ostermayr, R., Rödel, G., Schmelzer, C., Schroeder, R., Weiss-Brummer, B., and Kaudewitz, F.,** Transcripts of yeast mitochondrial DNA: Processing of a split-gene transcript and expression of RNA species during adaptation and differentiation processes, in *Mitochondrial Genes*, Slonimski, P., Borst, P., and Attardi, G., Eds., Cold Spring Harbor Laboratory, Cold Spring Harbor, New York, 1982, 201.
123. **Schmelzer, C., Haid, A., Grosch, G., Schweyen, R. J., and Kaudewitz, F.,** Pathways of transcript splicing in yeast mitochondria. Mutations in intervening sequences of the split gene COB reveal a requirement for intervening sequence-encoded products, *J. Biol. Chem.*, 256, 7610, 1981.
124. **Bonitz, S. G., Homison, G., Thalenfeld, B. E., Tzagoloff, A., and Nobrega, F. G.,** Assembly of the mitochondrial membrane system. Processing of the apocytochrome *b* precursor RNAs in *Saccharomyces cerevisiae* D273-10B, *J. Biol. Chem.*, 257, 6268, 1982.
125. **Dieckmann, C. L., Pape, L. K., and Tzagoloff, A.,** Identification and cloning of a yeast nuclear gene (CBP1) involved in expression of mitochondrial cytochrome *b*, *Proc. Natl. Acad. Sci. U.S.A.*, 79, 1805, 1982.
126. **Arnberg, A. C., van Ommen, G.-J. B., Grivell, L. A., van Bruggen, E. F. J., and Borst, P.,** Some yeast mitochondrial RNAs are circular, *Cell*, 18, 313, 1980.
127. **Lamouroux, A., Pajot, P., Kochko, A., Halbreich, A., and Slonimski, P. P.,** Cytochrome *b* messenger RNA maturase encoded in an intron regulates the expression of the split gene: II. *Trans*- and *cis*-acting mechanisms of mRNA splicing, in *The Organization and Expression of the Mitochondrial Genome*, Kroon, A. and Saccone, C., Eds., Elsevier/North-Holland, Amsterdam, 1980, 153.
128. **Church, G. M., Slonimski, P. P., and Gilbert, W.,** Pleiotropic mutations within two yeast mitochondrial cytochrome genes block mRNA processing, *Cell*, 18, 1209, 1979.

129. **Alexander, N. J., Perlman, P. S., Hanson, D. K., and Mahler, H. R.,** Mosaic organization of a mitochondrial gene: Evidence from double mutants in the cytochrome *b* region of *Saccharomyces cerevisiae, Cell,* 20, 199, 1980.
130. **Claisse, M., Slonimski, P. P., Johnson, J., and Mahler, H. R.,** Mutations within an intron and its flanking sites: Pattern of novel polypeptides generated by mutants in one segment of the *cob-box* region of yeast mitochondrial DNA, *Mol. Gen. Genet.,* 177, 375, 1980.
131. **Kochko, A., Colson, A. M., and Slonimski, P. P.,** Expression en *cis* lors de la complémentation des exons du gène mosaïque mitochondrial controlant le cytochrome *b* chez *Saccharomyces cerevisiae, Arch. Int. Physiol. Biochem.,* 87, 619, 1979.
132. **Jacq, C., Lazowska, J., and Slonimski, P. P.,** Sur un nouveau méchanisme de la régulation de l'expression génétique, *C. R. Acad. Sci. Paris, Ser. D.,* 290, 89, 1980.
133. **Church, G. M. and Gilbert, W.,** Yeast mitochondrial intron products required in *trans* for RNA splicing, in *Mobilization and Reassembly of Genetic Information,* Vol. 17, Scott, W. A., Werner, R., Joseph, D. R., and Schultz, J., Eds., Academic Press, New York, 1980, 379.
134. **Kruszewska, A.,** Nuclear and mitochondrial informational suppressor of *box 3* intron mutations in *Saccharomyces cerevisiae,* in *Mitochondrial Genes,* Slonimski, P., Borst, P., and Attardi, G., Eds., Cold Spring Harbor Laboratory, Cold Spring Harbor, New York, 1982, 323.
135. **Dujardin, G., Groudinsky, O., Kruszewska, A., Pajot, P., and Slonimski, P. P.,** Cytochrome b messenger RNA maturase encoded in an intron regulates the expression of the split gene: III. Genetic and phenotypic suppression of intron mutations, in *The Organization and Expression of the Mitochondrial Genome,* Kroon, A. M. and Saccone, C., Eds., Elsevier/North-Holland, Amsterdam, 1980, 157.
136. **Bechmann, H., Haid, A., Schweyen, R. J., Mathews, S., and Kaudewitz, F.,** Expression of the "split gene" COB in yeast mtDNA. Translation of intervening sequences in mutant strains, *J. Biol. Chem.,* 256, 3525, 1981.
137. **Lazowska, J., Jacq, C., and Slonimski, P. P.,** Splice points of third intron in the yeast mitochondrial cytochrome *b* gene, *Cell,* 27, 12, 1981.
138. **De La Salle, H., Jacq, C., and Slonimski, P. P.,** Critical sequences within mitochondrial introns: Pleiotropic mRNA maturase and cis-dominant signals of the *box* intron controlling reductase and oxidase, *Cell,* 28, 721, 1982.
139. **Mahler, H. R., Hanson, D. K., Lamb, M. R., Perlman, P. S., Anziano, P. Q., Glaus, K. R., and Haldi, M. L.,** Regulatory interactions between mitochondrial genes: Expressed introns. Their function and regulation, in *Mitochondrial Genes,* Slonimski, P., Borst, P., and Attardi, G., Eds., Cold Spring Harbor Laboratory, Cold Spring Harbor, New York, 1982, 185.
140. **Weiss-Brummer, B., Rödel, G., Schweyen, R. J., and Kaudewitz, F.,** Expression of the split gene *cob* in yeast: Evidence for a precursor of a "maturase" protein translated from intron 4 and preceding exons, *Cell,* 29, 527, 1982.
141. **Fox, T. D. and Weiss-Brummer, B.,** Leaky +1 and −1 frameshift mutations at the same site in a yeast mitochondrial gene, *Nature (London),* 288, 60, 1980.
142. **Netter, P., Jacq, C., Carignani, G., and Slonimski, P. P.,** Critical sequences within mitochondrial introns: cis-dominant mutations of the "cytochrome-*b*-like" intron of the oxidase gene, *Cell,* 28, 733, 1982.
143. **Dujardin, G., Jacq, C., and Slonimski, P. P.,** Single base substitution in an intron of oxidase gene compensates splicing defects of the cytochrome *b* gene, *Nature (London),* 298, 628, 1982.
144. **Foury, F. and Tzagoloff, A.,** Assembly of the mitochondrial membrane system. Genetic complementation of mit$^-$ mutations in mitochondrial DNA of *Saccharomyces cerevisiae, J. Biol. Chem.,* 253, 3792, 1978.
145. **Faye, G., Kujawa, C., and Fukuhara, H.,** Physical and genetic organization of *petite* and *grande* yeast mitochondrial DNA. IV. *In vivo* transcription products of mitochondrial DNA and localization of 23S ribosomal RNA in *petite* mutants of *Saccharomyces cerevisiae, J. Mol. Biol.,* 88, 185, 1974.
146. **Tabak, H. F., van der Laan, J., Osinga, K. A., Schouten, J. P., van Boom, J. H., and Veeneman, G. H.,** Use of a synthetic DNA oligonucleotide to probe the precision of RNA splicing in a yeast mitochondrial petite mutant, *Nucleic Acids Res.,* 9, 4475, 1981.
147. **Schmelzer, C. and Schweyen, R. J.,** Evidence for ribosomes involved in splicing of yeast mitochondrial transcripts, *Nucleic Acids Res.,* 10, 513, 1982.
148. **Pajot, P., Wambier-Kluppel, M. L., and Slonimski, P. P.,** Cytochrome *c* reductase and cytochrome oxidase formation in mutants and revertants in the "box" region of mitochondrial DNA, in *Mitochondria 1977. Genetics and Biogenesis of Mitochondria,* Bandlow, W., Schweyen, R. J., Wolf, K., and Kaudewitz, F., Eds., Walter de Gruyter, New York, 1977, 173.
149. **Dhawale, S., Hanson, D. L., Alexander, N. J., Perlman, P. S., and Mahler, H. R.,** Regulatory interactions between mitochondrial genes: Interactions between two mosaic genes, *Proc. Natl. Acad. Sci. U.S.A.,* 78, 1778, 1981.

150. **Dujardin, G., Pajot, P., Groudinsky, O., and Slonimski, P. P.,** Long range control circuits within mitochondria and between nucleus and mitochondria. I. Methodology and phenomenology of suppressors, *Mol. Gen. Genet.*, 179, 469, 1980.
151. **Groudinsky, O., Dujardin, G., and Slonimski, P. P.,** Long range control circuits within mitochondria and between nucleus and mitochondria. II. Genetic and biochemical analyses of suppressors which selectively alleviate the mitochondrial intron mutations, *Mol. Gen. Genet.*, 184, 493, 1981.
152. **Gilbert, W.,** Why genes in pieces? *Nature (London)*, 271, 501, 1978.
153. **Borst, P.,** The biogenesis of mitochondria in yeast and other primitive eukaryotes, in *International Cell Biology 1980—1981*, Schweiger, H. G., Ed., Springer-Verlag, Berlin, 1981, 239.
154. **Seif, I., Khoury, G., and Dhar, R.,** BKV splice sequences based on analysis of preferred donor and acceptor sites, *Nucleic Acids Res.*, 6, 3387, 1979.
155. **Warner, J. R. and Soeiro, R.,** Nascent ribosomes from HeLa cells, *Proc. Natl. Acad. Sci. U.S.A.*, 58, 1984, 1967.
156. **Liau, M. C. and Perry, R. P.,** Ribosome precursor particles in nucleoli, *J. Cell Biol.*, 42, 272, 1969.
157. **Grabowski, P. J., Zaug, A. J., and Cech, T. R.,** The intervening sequence of the ribosomal RNA precursor is converted to a circular RNA in isolated nuclei of Tetrahymena, *Cell*, 23, 467, 1981.
158. **Faye, G., Kujawa, C., Dujon, B., Bolotin-Fukuhara, M., Wolf, K., Fukuhara, H., and Slonimski, P. P.,** Localization of the gene coding for the mitochondrial 16S ribosomal RNA using rho$^-$ mutants of *Saccharomyces cerevisiae*, *J. Mol. Biol.*, 99, 203, 1975.
159. **Osinga, K. A. and Tabak, H. F.,** Initiation of transcription of genes for mitochondrial ribosomal RNA in yeast: Comparison of the nucleotide sequence around the 5′-ends of both genes reveals a homologous stretch of 17 nucleotides, *Nucleic Acids Res.*, 10, 3617, 1982.
160. **Miller, D. L. and Martin, N. C.,** Organization and expression of a tRNA gene cluster in *Saccharomyces cerevisiae* mitochondrial DNA, *Curr. Genet.*, 4, 135, 1981.
161. **Martin, N. C., Miller, D., Hartley, J., Moynihan, P., and Donelson, J. E.,** The $tRNA^{Ser}_{AGY}$ and $tRNA^{Arg}_{CGY}$ genes form a gene cluster in yeast mitochondrial DNA, *Cell*, 19, 339, 1980.
162. **Casey, J. W., Hsu, H.-J., Rabinowitz, M., Getz, G. S., and Fukuhara, H.,** Transfer RNA genes in the mitochondrial DNA of cytoplasmic *petite* mutants of *Saccharomyces cerevisiae*, *J. Mol. Biol.*, 88, 717, 1974.
163. **Martin, N., Rabinowitz, M., and Fukuhara, H.,** Isoaccepting mitochondrial glutamyl-tRNA species transcribed from different regions of the mitochondrial genome of *Saccharomyces cerevisiae*, *J. Mol. Biol.*, 101, 285, 1976.
164. **Martin, N. C. and Underbrink-Lyon, K.,** A mitochondrial locus is necessary for the synthesis of mitochondrial tRNA in the yeast, *Saccharomyces cerevisiae*, *Proc. Natl. Acad. Sci. U.S.A.*, 78, 4743, 1981.
165. **Frontali, L., Palleschi, C., and Francisci, S.,** Transcripts of mitochondrial tRNA genes in *Saccharomyces cerevisiae*, *Nucleic Acids Res.*, 10, 7283, 1982.
166. **Dieckmann, C. L., Bonitz, S. G., Hill, J., Homison, G., McGraw, P., Pape, L., Thalenfeld, B. E., and Tzagoloff, A.,** Structure of the apocytochrome-*b* gene and processing of apocytochrome-*b* transcripts in *Saccharomyces cerevisiae*, in *Mitochondrial Genes*, Slonimski, P., Borst, P., and Attardi, G., Eds., Cold Spring Harbor Laboratory, Cold Spring Harbor, New York, 1982, 213.
167. **Hopper, A. K., Furukawa, A. H., Pham, H. D., and Martin, N. C.,** Defects in modification of cytoplasmic and mitochondrial transfer RNAs are caused by single nuclear mutations, *Cell*, 28, 543, 1982.
168. **Martin, N. C. and Hopper, A. K,.** Isopentenylation of both cytoplasmic and mitochondrial tRNAs is affected by a single nuclear mutation, *J. Biol. Chem.*, 257, 10562, 1982.
169. **Terpstra, P., de Vries, H., and Kroon, A. M.,** Properties and genetic localization of mitochondrial transfer RNAs of *Neurospora crassa*, in *Mitochondria 1977. Genetics and Biogenesis of Mitochondria*, Bandlow, W., Schweyen, R. J., Wolf, K., and Kaudewitz, F., Walter de Gruyter, New York, 1977, 291.
170. **Heckman, J. E. and RajBhandary, U. L.,** Organization of tRNA and rRNA genes in *N. crassa* mitochondria: intervening sequence in the large rRNA gene and strand distribution of the RNA genes, *Cell*, 17, 583, 1979.
171. **Heckman, J. E., Yin, S., Alzner-De Weerd, B., and RajBhandary, U. L.,** Mapping and cloning of *Neurospora crassa* mitochondrial transfer RNA genes, *J. Biol. Chem.*, 254, 12694, 1979.
172. **de Vries, H., de Jonge, J. C., Bakker, H., Meurs, H., and Kroon, A.,** The anatomy of the tRNA-rRNA region of the *Neurospora crassa* mitochondrial DNA, *Nucleic Acids Res.*, 6, 1791, 1979.
173. **Yin, S., Burke, J., Chang, D. D., Browning, K. S., Heckman, J. E., Alzner-De Weerd, B., Potter, M. J., and RajBhandary, U. L.,** *Neurospora crassa* mitochondrial tRNAs and rRNAs: Structure, gene organization, and DNA sequences, in *Mitochondrial Genes*, Slonimski, P., Borst, P., and Attardi, G., Eds., Cold Spring Harbor Laboratory, Cold Spring Harbor, New York, 1982, 361.
174. **Hahn, U., Lazarus, C. M., Lünsdorf, H., and Küntzel, H.,** Split gene for mitochondrial 24S ribosomal RNA of *Neurospora crassa*, *Cell*, 17, 191, 1979.

175. **Browning, K. S. and RajBhandary, U. L.**, Cytochrome oxidase subunit III gene in *Neurospora crassa* mitochondria, *J. Biol. Chem.*, 257, 5253, 1982.
176. **Green, M. R., Grimm, M. F., Goewert, R. R., Collins, R. A., Dole, M. D., Lambowitz, A. M., Heckman, J. E., Yin, S., and RajBhandary, U. L.**, Transcripts and processing patterns for the ribosomal RNA and transfer RNA region of *Neurospora crassa* mitochondrial DNA, *J. Biol. Chem.*, 256, 2027, 1981.
177. **Yin, S., Heckman, J., and RajBhandary, U. L.**, Highly conserved GC-rich palindromic DNA sequences flank tRNA genes in *Neurospora crassa* mitochondria, *Cell*, 26, 325, 1981.
178. **Mannella, C.A., Collins, R. A., Green, M. R., and Lambowitz, A. M.**, Defective splicing of mitochondrial rRNA in cytochrome-deficient nuclear mutants of *Neurospora crassa*, *Proc. Natl. Acad. Sci. U.S.A.*, 76, 2635, 1979.
179. **Lambowitz, A. M.**, Mitochondrial ribosome assembly and RNA splicing in *Neurospora crassa*, in *The Organization and Expression of the Mitochondrial Genome*, Kroon, A. M. and Saccone, C., Eds., Elsevier/North-Holland, Amsterdam, 1980, 291.
180. **Grant, D. M. and Lambowitz, A. M.**, Mitochondrial ribosomal RNA genes, in *The Cell Nucleus*, Vol. 10, Busch, H. and Rothblum, L., Eds., Academic Press, New York, 1982, 387.
181. **Garriga, G., Collins, R. A., Grant, D. M., Lambowitz, A. M., and Bertrand, H.**, Mitochondrial RNA splicing in *Neurospora crassa*, in *Mitochondrial Genes*, Slonimski, P., Borst, P., and Attardi, G., Eds., Cold Spring Harbor Laboratory, Cold Spring Harbor, New York, 1982, 381.
182. **Bertrand, H., Bridge, P., Collins, R. A., Garriga, G., and Lambowitz, A. M.**, RNA splicing in *Neurospora* mitochondria. Characterization of new nuclear mutants with defects in splicing the mitochondrial large rRNA, *Cell*, 29, 517, 1982.
183. **Grimm, M. F., Cole, M. D., and Lambowitz, A. M.**, Ribonucleic acid splicing in *Neurospora* mitochondria: Secondary structure of the 35S ribosomal precursor ribonucleic acid investigated by digestion with ribonuclease III and by electron microscopy, *Biochemistry*, 20, 2836, 1981.
184. **LaPolla, R. J. and Lambowtiz, A. M.**, Binding of mitochondrial ribosomal proteins to a mitochondrial ribosomal precursor RNA containing a 2.3-kilobase intron, *J. Biol. Chem.*, 254, 11746, 1979.
185. **Lazarus, C. M., Lünsdorf, H., Hahn, U., Stepień, P. P., and Küntzel, H.**, Physical map of *Aspergillus nidulans* mitochondrial genes coding for ribosomal RNA: An intervening sequence in the large rRNA cistron, *Mol. Gen. Genet.*, 177, 389, 1980.
186. **Köchel, H. G., Lazarus, C. M., Basak, N., and Küntzel, H.**, Mitochondrial tRNA gene clusters in *Aspergillus nidulans*: Organization and nucleotide sequence, *Cell*, 23, 625, 1981.
187. **Küntzel, H., Basak, N., Imam, G., Köchel, H., Lazarus, C. M., and Lünsdorf, H.**, The mitochondrial genome of *Aspergillus nidulans*, in *The Organization and Expression of the Mitochondrial Genome*, Kroon, A. M. and Saccone, C., Eds., Elsevier/North-Holland, Amsterdam, 1980, 79.
188. **Küntzel, H., Köchel, H. G., Lazarus, C. M., and Lünsdorf, H.**, Mitochondrial genes in *Aspergillus*, in *Mitochondrial Genes*, Slonimski, P., Borst, P., and Attardi, G., Eds., Cold Spring Harbor Laboratory, Cold Spring Harbor, New York, 1980, 391.
189. **Macino, G., Scazzocchio, C., Waring, R. B., McPhail Berks, M., and Davies, R. W.**, Conservation and rearrangement of mitochondrial structural gene sequences, *Nature (London)*, 288, 404, 1980.
190. **Waring, R. B., Davies, R. W., Lee, S., Grisi, E., McPhail Berks, M., and Scazzocchio, C.**, The mosaic organization of the apocytochrome *b* gene of *Aspergillus nidulans* revealed by DNA sequencing, *Cell*, 27, 4, 1981.
191. **Davies, R. W., Scazzocchio, C., Waring, R. B., Lee, S., Grisi, E., McPhail Berks, M., and Brown, T. A.**, Mosaic genes and unidentified reading frames that have homology with human mitochondrial sequences are found in the mitochondrial genome of *Aspergillus nidulans*, in *Mitochondrial Genes*, Slonimski, P., Borst, P., and Attardi, G., Eds., Cold Spring Harbor Laboratory, Cold Spring Harbor, New York, 1982, 405.
192. **Borst, P. and Hoeijmakers, J. H. J.**, Kinetoplast DNA, *Plasmid*, 2, 20, 1979.
193. **Simpson, L., Simpson, A. M., Kidane, G., Livingston, L., and Spithill, T. W.**, The kinetoplast DNA of the hemoflagellate protozoa, *Am. J. Trop. Med. Hyg.*, 29, 1053, 1980.
194. **Englund, P. T., Hajduk, S. L., and Marini, J. C.**, The molecular biology of Trypanosomes, *Ann. Rev. Biochem.*, 51, 1982, 695.
195. **Simpson, L. and Simpson, A.**, Kinetoplast RNA of *L. tarentolae*, *Cell*, 14, 169, 1978.
196. **Masuda, H., Simpson, L., Rosenblatt, H., and Simpson, A. M.**, Restriction map, partial cloning and localization of 9S and 12S kinetoplast RNA genes on the maxicircle component of the kinetoplast DNA of *Leishmania tarentolae*, *Gene*, 6, 51, 1979.
197. **Simpson, L., Simpson, A. M., Masuda, H., Rosenblatt, H., Michael, N., and Kidane, G.**, Replication and transcription of kinetoplast DNA, in *Extrachromosomal DNA*, Cummings, D. J., Borst, P., Dawid, I. B., and Weissman, S. M., Eds., ICN-UCLA Symp. *Molec. Cell. Biol.*, Vol. 15, Academic Press, New York, 1979, 533.
198. **Borst, P., and Hoeijmakers, J. H. J.**, Structure and function of kinetoplast DNA of the African trypanosomes, in *Extrachromosomal DNA*, Cummings, D. J., Borst, P., Dawid, I. B., and Weissman, S. M., Eds., Academic Press, ICN-UCLA Symp. *Molec. Cell. Biol.*, Vol. 15, New York, 1979, 515.

199. **Borst, P., Hoeijmakers, J. H. J., Frasch, A. C. C., Snijders, A., Janssen, J. W. G., and Fase-Fowler, F.,** The kinetoplast DNA of *Trypanosoma brucei:* Structure, evolution, transcription, mutants, in *The Organization and Expression of the Mitochondrial Genome,* Kroon, A. M. and Saccone, C., Eds., Elsevier/North-Holland, Amsterdam, 1980, 7.
200. **Simpson, A. M. and Simpson, L.,** Kinetoplast DNA and RNA of *Trypanosoma brucei, Mol. Biochem. Parasitol.,* 2, 93, 1980.
201. **Simpson, L., Simpson, A. M., Spithill, T. W., and Livingston, L.,** Sequence organization of maxicircle kinetoplast DNA from *Leishmania tarentolae,* in *Mitochondrial Genes,* Slonimski, P., Borst, P., and Attardi, G., Eds., Cold Spring Harbor Laboratory, Cold Spring Harbor, New York, 1980, 435.
202. **Simpson, L., Spithill, T. W., and Simpson, A. M.,** Identification of maxicircle DNA sequences in *Leishmania tarentolae* homologous to sequences of specific yeast mitochondrial structural genes, *Mol. Biochem. Parasitol.,* 6, 253, 1982.
203. **Hoeijmakers, J. H. J., Snijders, A., Janssen, J. W. G., and Borst, P.,** Transcription of kinetoplast DNA in *Trypanosoma brucei* bloodstream and culture forms, *Plasmid,* 5, 329, 1981.
204. **Simpson, A. M., Simpson, L., and Livingston, L.,** Transcription of the maxicircle kinetoplast DNA of *Leishmania tarentolae, Mol. Biochem. Parasitol.,* 6, 237, 1982.
205. **Hoeijmakers, J. H. J., Schoutsen, B., and Borst, P.,** Kinetoplast DNA in the insect trypanosomes *Crithidia luciliae* and *Crithidia fasciculata, Plasmid,* 7, 199, 1982.
206. **Hoeijmakers, J. H. J. and Borst, P.,** RNA from the insect trypanosome *Crithidia luciliae* contains transcripts of the maxicircle and not of the minicircle component of kinetoplast DNA, *Biochim. Biophys. Acta,* 521, 407, 1978.
207. **Cheng, D. and Simpson, L.,** Isolation and characterization of kinetoplast DNA and RNA of *Phytomonas davidi, Plasmid,* 1, 297, 1978.
208. **Eperon, I. C., Janssen, J. W. G., Hoeijmakers, J. H. J., and Borst, P.,** The major transcripts of the kinetoplast DNA of *Trypanosoma brucei* are very small ribosomal RNAs, *Nucleic Acids Res.,* 11, 105, 1983.
209. **Borst, P.,** personal communication, 1982.
210. **Goldbach, R. W., Borst, P., Bollen-De Boer, J. E., and van Bruggen, E. F. J.,** The organization of ribosomal RNA genes in the mitochondrial DNA of *Tetrahymena pyriformis* strain ST, *Biochim. Biophys. Acta,* 521, 169, 1978.
211. **Goldbach, R. W., Bollen-De Boer, J. E., van Bruggen, E. F. J., and Borst, P.,** Conservation of the sequence and position of the ribosomal RNA genes in *Tetrahymena pyriformis* mitochondrial DNA, *Biochim. Biophys. Acta,* 521, 187, 1978.
212. **Suyama, Y. and Hamada, J.,** Imported tRNA: Its synthetase as a probable transport protein, in *Genetics and Biogenesis of Chloroplasts and Mitochondria,* Bücher, Th., Neupert, W., Sebald, W., and Werner, S., Eds., North-Holland, Amsterdam, 1976, 763.
213. **Suyama, Y.,** Native and imported tRNAs in *Tetrahymena* mitochondria: Evidence for their involvement in intramitochondrial translation, in *Mitochondrial Genes,* Slonimski, P., Borst, P., and Attardi, G., Eds., Cold Spring Harbor Laboratory, Cold Spring Harbor, New York, 1980, 449.
214. **Martin, R. P., Schneller, J. M., Stahl, A. J. C., and Dirheimer, D.,** Import of nuclear deoxyribonucleic acid coded lysine-accepting transfer ribonucleic acid (anticodon CUU) into yeast mitochondria, *Biochemistry,* 18, 4600, 1979.
215. **Goddard, J. M. and Cummings, D. J.,** Mitochondrial DNA replication in *Paramecium aurelia.* Cross-linking of the initiation end, *J. Mol. Biol.,* 109, 327, 1977.
216. **Cummings, D. J. and Laping, J. L.,** Organization and cloning of mitochondrial deoxyribonucleic acid from *Paramecium tetraaurelia* and *Paramecium primaurelia, Molec. Cell. Biol.,* 1, 972, 1981.
217. **Seilhamer, J. J. and Cummings, D. J.,** Structure and sequence of the mitochondrial 20S rRNA and tRNA tyr gene of *Paramecium primaurelia, Nucleic Acids Res.,* 9, 6391, 1981.
218. **Seilhamer, J. J. and Cummings, D. J.,** personal communication, 1982.
219. **Cummings, D. J., Maki, R. A., Conlon, P. J., and Laping, J.,** Anatomy of mitochondrial DNA from *Paramecium aurelia, Mol. Gen. Genet.,* 178, 499, 1980.
220. **Leaver, C. J., Forde, B. G., Dixon, L. K., and Fox, T. D.,** Mitochondrial genes and cytoplasmically inherited variation in higher plants, in *Mitochondrial Genes,* Slonimski, P., Borst, P., and Attardi, G., Eds., Cold Spring Harbor Laboratory, Cold Spring Harbor, New York, 1982, 457.
221. **Leaver, C. J. and Gray, M. W.,** Mitochondrial genome organization and expression in higher plants, *Ann. Rev. Plant Physiol.,* 33, 373, 1982.
222. **Kolodner, R. and Tewari, K. K.,** Physicochemical characterization of mitochondrial DNA from pea leaves, *Proc. Natl. Acad. Sci. U.S.A.,* 69, 1830, 1972.
223. **Levings, C. S., III, Shah, D. M., Hu, W. W. L., Pring, D. R., and Timothy, D. H.,** Molecular heterogeneity among mitochondrial DNAs from different maize cytoplasms, in *Extrachromosomal DNA,* Cummings, D. J., Borst, P., Dawid, I. B., and Weissman, S. M., Eds., ICN-UCLA Symp. *Molec. Cell Biol.,* Vol. 15, Academic Press, New York, 1979, 63.

224. **Wolsterholme, D. R. and Gross, N. J.,** The form and size of mitochondrial DNA of the red bean, *Phaseolus vulgaris, Proc. Natl. Acad. Sci. U.S.A.*, 61, 245, 1968.
225. **Vedel, F. and Quetier, F.,** Physico-chemical characterization of mitochondrial DNA from potato tubers, *Biochim. Biophys. Acta,* 340, 374, 1974.
226. **Ward, B. L., Anderson, R. S., and Bendich, A. J.,** The mitochondrial genome is large and variable in a family of plants, *(Cucurbitaceae), Cell,* 25, 793, 1981.
227. **Gray, M. W., Bonen, L., Falconet, D., Huh, T. Y., Schnare, M. N., and Spencer, D. F.,** Mitochondrial ribosomal RNAs of *Triticum aestivum* (wheat): Sequence analysis and gene organization, in *Mitochondrial Genes,* Slonimski, P., Borst, P., and Attardi, G., Eds., Cold Spring Harbor Laboratory, Cold Spring Harbor, New York, 1982, 483.
228. **Cunningham, R. S., Bonen, L., Doolittle, W. F., and Gray, M. W.,** Unique species of 5S, 18S and 26S ribosomal RNA in wheat mitochondria, *FEBS Lett.*, 69, 116, 1976.
229. **Bonen, L. and Gray, M. W.,** The genes for wheat mitochondrial ribosomal and transfer RNA: Evidence for an unusual arrangement, *Nucleic Acids Res.*, 8, 319, 1980.
230. **Cunningham, R. S. and Gray, M. W.,** Isolation and characterization of $^{32}$P-labeled mitochondrial and cytosol ribosomal RNA from germinating wheat embryos, *Biochim. Biophys, Acta,* 475, 476, 1977.
231. **Leaver, C. J. and Harmey, M. A.,** Higher plant mitochondrial ribosomes contain a 5S ribosomal RNA component, *Biochem. J.*, 157, 275, 1976.
232. **Spencer, D. F., Bonen, L., and Gray, M. W.,** Primary sequence of wheat mitochondrial 5S ribosomal ribonucleic acid: Functional and evolutionary implications, *Biochemistry,* 20, 4022, 1981.
233. **Gray, M. W. and Spencer, D. F.,** Is wheat mitochondrial 5S ribosomal RNA prokaryotic in nature? *Nucleic Acids Res.*, 9, 3523, 1981.
234. **Fox, T. D. and Leaver, C. J.,** The *Zea mays* mitochondrial gene coding cytochrome oxidase subunit II has an intervening sequence and does not contain TGA codons, *Cell,* 26, 315, 1981.
235. **Edwards, J. C., Levens D., and Rabinowitz, M.,** Analysis of transcriptional initiation of yeast mitochondrial DNA in a homologous in vitro transcription system, *Cell,* 31, 337, 1982.
236. **Rochaix, J. D. and Malnoe, P.,** Anatomy of the chloroplast ribosomal DNA of *Chlamydomonas reinhardii, Cell,* 15, 661, 1978.
237. **Koch, W., Edwards, K., and Kössel, H.,** Sequencing of the 16S-23S spacer in a ribosomal RNA operon of *Zea mays* chloroplast DNA reveals two split tRNA genes, *Cell,* 25, 203, 1981.
238. **Steinmetz, A., Gubbins, E. J., and Bogorad, L.,** The anticodon of the maize chloroplast gene for $tRNA^{Leu}_{UAA}$ is split by a large intron, *Nucleic Acids Res.*, 10, 3027, 1982.

Chapter 11

# MODIFIED NUCLEOSIDES IN RNA — THEIR FORMATION AND FUNCTION

**Glenn R. Björk**

## TABLE OF CONTENTS

# I. INTRODUCTION

Both ribosomal RNA (rRNA) and transfer RNA (tRNA) are fundamental to protein synthesis. It is therefore not surprising that the synthesis of these RNA species constitutes a major biosynthetic pathway in the cell. Both rRNA and tRNA contain modified nucleosides, which are derivatives of the ordinary nucleosides adenosine (A), guanosine (G), cytidine (C) and uridine (U) (Figure 1). It was earlier believed that this was a unique feature of all stable RNAs, but it has now been established that the stable 5S RNA does not contain any modified nucleosides.[1] Although mRNA from *Escherichia coli* does not contain methylated nucleosides,[2] the mRNA from eukaryotic organisms does (see chapters by Moss and Flint, this volume). Except for 5S RNA all other rRNA and tRNA species so far investigated contain modified nucleosides but to varying degrees. Both rRNA as well as tRNA are initially made as larger transcripts than their mature products. Several endo- and exonucleases operate during the stepwise maturation of RNA (reviewed in this book by other authors). However, during maturation several of the four ordinary nucleosides, A, G, C, and U are enzymatically modified in a specific manner. Such modifications occur concomitantly to processing of the primary transcription products and thus are an integrated part of the maturation process. Therefore, all modification reactions except one (see below) occur at the polynucleotide level, i.e., after the polymerization by the DNA-dependent-RNA polymerase. Modified nucleosides are derivatives of the four ordinary nucleosides, and Figure 1 shows some structures of modified nucleosides (e.g., 1-methylguanosine($m^1G$) contains a methyl group in position 1 of guanosine, etc.). A more complete catalog of structures of modified nucleosides has recently been published.[3] This review concentrates on recent results of the dynamic process of modification, i.e., its genetics, its regulation of formation, and its functional aspects. Earlier results and some other aspects of modification which are not discussed here are excellently dealt with elsewhere.[4-16]

# II. PRESENCE OF MODIFIED NUCLEOSIDES

## A. In rRNA

Early reports revealed that the pattern of rRNA modification in eubacterial and eukaryotic organisms differs both qualitatively and quantitatively.[17-25] All base modifications present in eukaryotic rRNA, except for amψ, are also present in eubacterial rRNA (Table 1). However, some base-methylated nucleosides, like $m^1G$, $m^2G$, $m^4Cm$, are specific for eubacteria. There is also a difference between different rRNA species, e.g., $m^2G$ are found in both 16S rRNA and 23S rRNA from *E. coli* while others, like $m^1G$ and $m^5U$, are present only in 23S rRNA. The most striking difference between modification of rRNA in eubacterial and eukaryotic organisms is the extent of ribose methylation. Whereas *E. coli* has only one ribose methylated nucleoside (Gm) in 16S rRNA and three (Gm, Cm, Um) in 23S rRNA, the 18S rRNA and 28S rRNA from *Xenopus leavis* contain 33 and 62—63 ribose methylated nucleosides, respectively. Ribose methylation may be a late evolutionary event: perhaps it has been selected in order to protect the larger eukaryotic ribosomes from endonucleolytic cleavage.[18]

The locations of the modified nucleosides are rather specific, e.g., in 18S rRNA from *X. leavis* the 2′O-ribose methylations are concentrated at the 5′-end while base methylations are clustered toward the 3′-end.[26-28] In 28S rRNA most methyl groups are located at the 3′-end. Furthermore, all methylated nucleosides are present in regions of rRNA, which are preserved during the maturation process.[18,29] The 3′-end of 18S rRNA is evolutionary conserved and this sequence contains the two copies of $m_2^6A$ which are present in the small

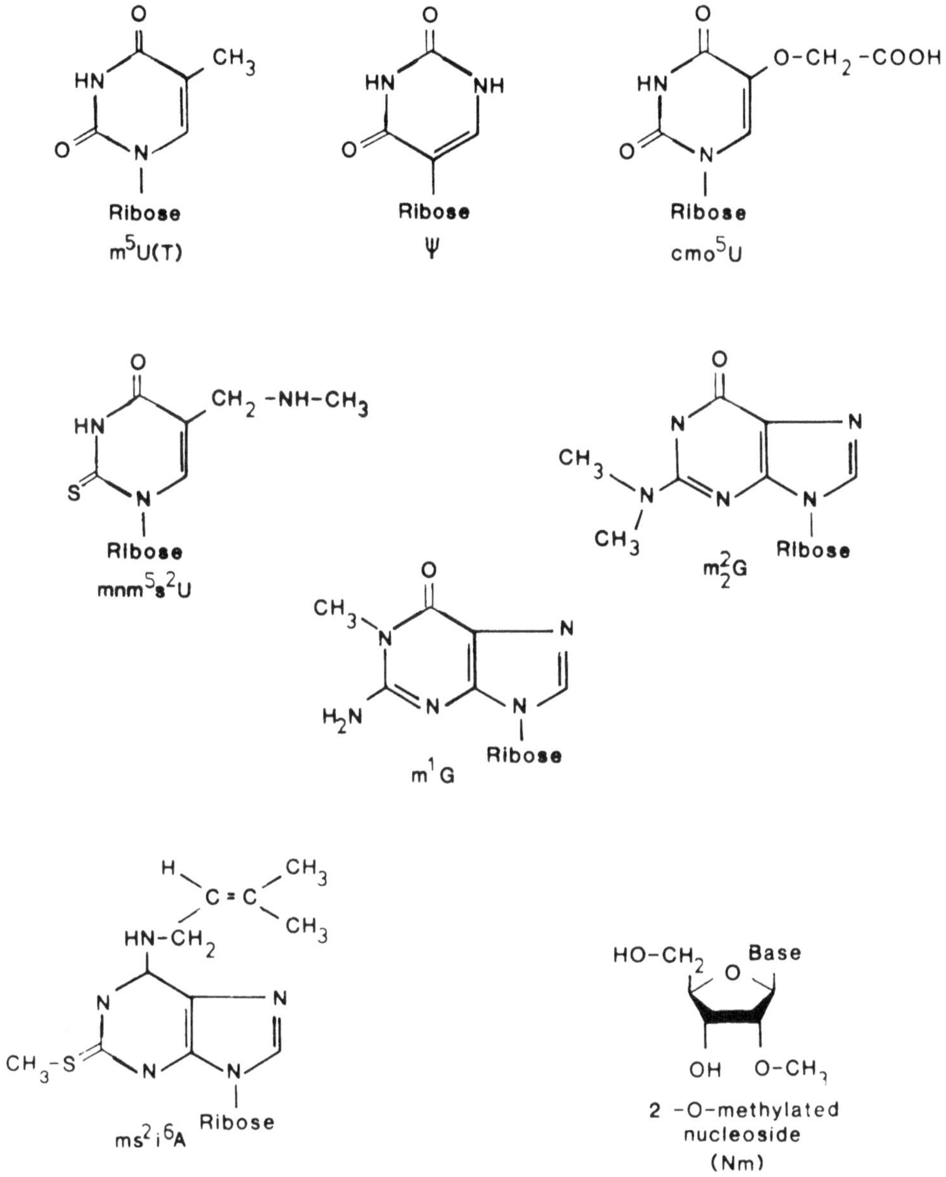

FIGURE 1. Structure of some modified nucleosides and their abbreviations.

**Abbreviations** — ψ, pseudouridine; m$^5$U, 5-methyluridine; s$^2$m$^5$U, 2-thio-5-methyluridine; mo$^5$U, 5-methoxuyridine; cmo$^5$U, uridine-5-oxyacetic acid; mcmo$^5$U, methylester of cmo$^5$U; s$^2$U, 2-thiouridine; mnm$^5$s$^2$U, 5-methylamino-methyl-2-thiouridine; cmnm$^5$s$^2$U, 5-carboxymethylaminomethyl-2-thiouridine; mcm$^5$U, 5-methoxycarbonylmethyluridine; mcm$^5$s$^2$U, 5-methoxycarbonylmethyl-2-thiouridine; s$^4$U, 4-thiouridine; amψ, 3-(3-amino-3-carboxypropyl)-1-methyl pseudouridine; acp$^3$U, 3-(3-amino-3-carboxypropyl)uridine; D, dihydrouridine; m$^3$C, 3-methylcytidine; m$^5$C, 5-methylcytidine; ac$^4$C, N$^4$-acetylcytidine; s$^2$C, 2-thiocytidine; I, inosine; m$^1$I, 1-methylinosine; m$^1$G, 1-methylguanosine; m$^2$G, 2-methylguanosine; m$^2$G, 2-methylguanosine; m$^7$G, 7-methylguanosine; m$^1$A, 1-methyladenosine; m$^2$A, 2-methyladenosine; m$^6$A 6-methyladenosine; m$^6_2$A, 6-dimethyladenosine; i$^6$A, *N*-6-($\Delta^2$-isopentyl)adenosine; ms$^2$i$^6$A$^2$; *N*-6-($\Delta^2$-isopentenyl)-2-methylthioadenosine, t$^6$A, *N*-9-(β-D-ribofuranosyl)purin-6-ylcarbamoyl threonine, mt$^6$A, N- 9(β-D-ribofuranosyl)purin-6-yl-*N*-methylcarbamoyl threonine; ms$^2$t$^6$A; N-(β-D-ribofuranosyl-2-methyltinopurin-6-yl)carbamoyl threonine; yW, wybutosine; o$_2$yW, wybutoxosine; Q, 7-(4,5-cisdihydroxy-1-cyclopenten-3-yl aminomethyl)-7-deazaguanosine Am, Gm, Cm, Um, ψm m$^5$Um 2′-*O*-ribosemethylated derivative of the corresponding nucleosides. Transfer RNA-modifying enzymes catalyzing the formation of m$^5$U, ψ, etc., are denoted tRNA(m$^5$U)methyltransferase, tRNA(ψ) synthetase, etc.

**Table 1**
**MODIFIED NUCLEOSIDES IN rRNA FROM DIFFERENT ORGANISMS (MOL/MOL RNA.[a])**

| | RNA from small subunit | | | | | | | | rRNA from large subunit | | | | | | |
|---|---|---|---|---|---|---|---|---|---|---|---|---|---|---|---|
| | 16S | | | 17S | | 18S | | | 23S | | | 26S | | 28S | |
| Mod nucl[b] | Ec | Mc | Al | Sc | Ap | Xl | HeLa | Rat liver | Ec | Mc | Al | Sc | Ap | Xl | HeLa |
| $m^1A$ | | | | | | | | | | | | 2 | 2 | 1 | |
| $M^6A$ | 2 | | | | 1 | 1 | 1(mA) | 1 | 2,5 | 0,4 | 0,5 | | | 1 | 2(mA) |
| $m^6_2A$ | 2 | 2,2 | 2,0 | 2 | 2 | 2 | 2 | 2 | | | | | | | |
| $m^1G$ | | | | | | | | | 0,8 | 0,4 | 0,5 | | | | |
| $m^2G$ | 2 | | 2,8 | | | | | | 2,4 | | | | | | |
| $m^2_2G$ | | 1.2 | | | | | | | | | | | | | |
| $m^7G$ | 1 | 1,0 | 1,4 | 1 | 1 | 1 | 1 | 1 | 0,8 | | | | | | |
| $m^5C$ | 2 | 1,4 | | | 0—1 | | | | 1,3 | | 1,4 | 2 | 3 | | 2(mC) |
| $m^4Cm$ | 1 | 1,0 | 1,0 | | | | | | | | | | | | |
| $m^3U$ | | | 1,2 | | | | | | | 0,5 | | 2 | 2 | 1 | |
| $m^5U$ | | | 0,5 | | | | | | 3,0 | | 6,3 | | | | 1(mU) |
| Am | | | | 8 | 15 | 12 | 13 | 12 | | | | 12 | 23 | 17,5 | 19—20 |
| Gm | 1 | | | 5 | 7 | 7 | 6—7 | 8 | 0,9 | 2.8 | 1,8 | 10 | 22 | 20—21 | 21—22 |
| Cm | | | | 3 | 6 | 5 | 7 | 8 | 0,7 | 1.2 | 2,7 | 7 | 19 | 15 | 15 |
| Um | | | | 2 | 12 | 9 | 12—13 | 9 | 1,0 | | | 8 | 20 | 8 | 8 |
| ψ | | | | 12—14 | | | | 38 | | | | 24—25 | | | |
| ψm | | | | | 1 | | | | | | | | | | 1 |
| amψ | | | | 1 | 1 | 1 | 1 | | | | | | | | |
| $ac^4C$ | | | | | | | | 1 | | | | | | | |

[a] Ec, *Escherichia coli*,[19,24-25]; Mc, *Mycoplasma capricolum*[24]; Al, *Acholeplasma laidlawii*[24]; Sc, *Sacharomyces carlsbergensis*[17]; Ap, *Acer pseudoplatanus*[22]; Xl, *Xenopus leavis*[20]; HeLa, Hela cells[18,30]; rat liver.[21]

[b] For abbreviations, see text to Figure 2.

subunit rRNA in all organisms studied so far (Table 1). Since this evolutionary conserved sequence $m^6_2A$-$m^6_2A$-C appears to be exposed on the surface of the ribosome, it is likely to be of significant functional importance.[35,36]

## A. In tRNA

Transfer RNA is heavily modified and a number of different modified nucleosides have been identified in the 177 tRNAs sequenced so far[37] (Figure 2). It is clear that tRNAs from eukaryotes are more modified than those from eubacteria. Thus, transfer $RNA^{Ser}_1$ from *E. coli* contains 9 modified nucleosides, while $tRNA^{Ser}_1$ from rat liver contains 14 modified nucleosides.

Some modified nucleosides ($m^2G$, $m^2_2G$, $m^1I$, yW, $o_2yW$, $i^6A$, $m^3C$, $m^5C$, $mcm^5U$, $mcm^5s^2U$ and $m^5Um$) occur only in eukaryotic tRNAs, whereas others ($m^2A$, $m^6A$, $ms^2i^6A$, $s^2C$, $s^4U$, $mnm^5s^2U$, $mo^5U$, $cmo^5U$, $mcmo^5U$, and $cmnm^5s^2U$) are specific for eubacterial tRNA. Several modified nucleosides, like $m^5U$ and ψ are (on the other hand) found in both types of organisms. Figure 2 shows that only specific positions in the tRNA are prone to modification. Thus, more than 80% of the different tRNA species from eukaryotes or eubacteria have ψ in position 55 and more than 75% of eukaryotic tRNA have $m^5U$ in position 54.[37] All eubacterial tRNAs, except tRNA species from *Mycoplasma* (kid) and $tRNA^{Gly}_{1A,B}$ from *Staphylococcus epidermidis,* contain $m^5U$ in position 54. Furthermore, some regions of the tRNA molecules, i.e., the anticodon region, are more heavily modified, while others, i.e., the amino acid stem, are almost devoid of modified nucleosides. A remarkable feature is the large variety of modified nucleosides which occur at position 34 (wobble position) and position 37 (3′ side of the anticodon). The latter modification is correlated to the codons read by the different tRNAs[16] (Table 2). Thus, tRNA species reading codons starting with U (Phe, Ser, Tyr, Cys, Trp) have a large hydrophobic residue, like $ms^2i^6A$ (eubacteria) or $i^6A$ (eukaryotes), in this position, whereas tRNAs reading codons starting with A have hydrophilic residues, like $t^6A$ (eubacteria and eukaryotes) or $mt^6A$ (eubacteria). Transfer RNA reading codewords starting with C or G may be unmodified or may contain methylated derivatives like $m^1G$, $m^1I$, $m^2A$ or $m^6A$ in position 37.

Transfer tRNAs encoding the valine, serine, threonine, and alanine codon families, i.e., four codons having their first two nucleotides in common but coding for the same amino acid, contain at position 34 modified nucleosides which have been suggested to extend the wobble characteristic of the corresponding tRNAs.[39-42] Thus, tRNA, specific for valine, serine, threonine, and alanine from *Bacillus subtilis*, contains $mo^5U$, from *E. coli* $cmo^5U$ or $mcmo^5U$, and from yeast and mammals inosine (I). Those tRNAs having I in this position read according to Crick's wobble hypothesis[43] while those having $mo^5U$ and $cmo^5U$ do not (Table 3). These modifications have therefore been suggested to extend the wobble characteristics of U. Alternatively, the 2 out of 3 reading has been suggested by Lagerkvist and collaborators to explain the decoding of these codon families.[44] In any case organisms can be grouped as having either $mo^5U$, $cmo^5U$, or I at position 34 in the tRNA species specific for valine, serine, and alanine.[16]

All tRNAs both from eubacteria and eukaryotes that read codons ending with U or C (nonfamily codons) have Q or its derivatives in position 34, while tRNAs which read codons ending with A or G have $s^2U$-derivatives in position 34 (the wobble position; Table 4). In tRNA from eubacteria, such $s^2U$-derivatives are $mnm^5s^2U$ or $cmnm^5s^2U$ (*B. subtilis)*, while in tRNA from eukaryotes the $s^2U$-derivative is $mcm^5s^2U$.

Figure 2 shows that some modified nucleosides, like $\psi$, are present in seven positions in tRNA from eubacteria, while as many as 13 positions are represented in eukaryotic tRNAs. Other modified nucleosides may be present in different positions although not to the same extent as $\psi$. The methylated nucleoside $m^1G$ is present in two positions (9 and 37) in eukaryotic tRNA, Cm in three positions (positions 4, 32, 34) in eukaryotic tRNAs, and Gm in four positions (positions 18, 19, 34, 38) in eukaryotic tRNAs. Some modifications like the ribose-methylated nucleosides usually occur within loop-structures although exceptions exist (Cm and Gm at position 4).

Recently it has been suggested that "archaebacteria" is distinct from "eubacteria" (true bacteria) and the eukaryotes.[45] Transfer RNA from "archaebacteria" has an unusual modification pattern since all archaebacterial species examined lacked $m^5U$ and $m^7G$ in their tRNA. Furthermore, tRNA from all except one species lacked D in their tRNAs.[46,47] However, $m^1\psi$ replaces $m^5U$ in the $m^5U$-$\psi$-C sequence of tRNA and this modified nucleoside is unique to archaebacterial tRNAs.[48] Furthermore, such modified nucleosides as $m^2G$ and $m^2_2G$ which are only found in eukaryotic tRNAs were present in tRNA from most "archaebacteria" species examined. Therefore, the pattern of modification of tRNA places the "archaebacteria" as a distinct group. Its tRNA modification pattern resembles more that of the eukaryotic than that of eubacterial kingdom.

## III. PROPERTIES OF RNA-MODIFYING ENZYMES

### A. rRNA-Modifying Enzymes

Assuming that one enzyme is needed for the formation of each modified nucleoside at a specific site, one can estimate that *E. coli* needs nine modifying enzymes for 16S rRNA.

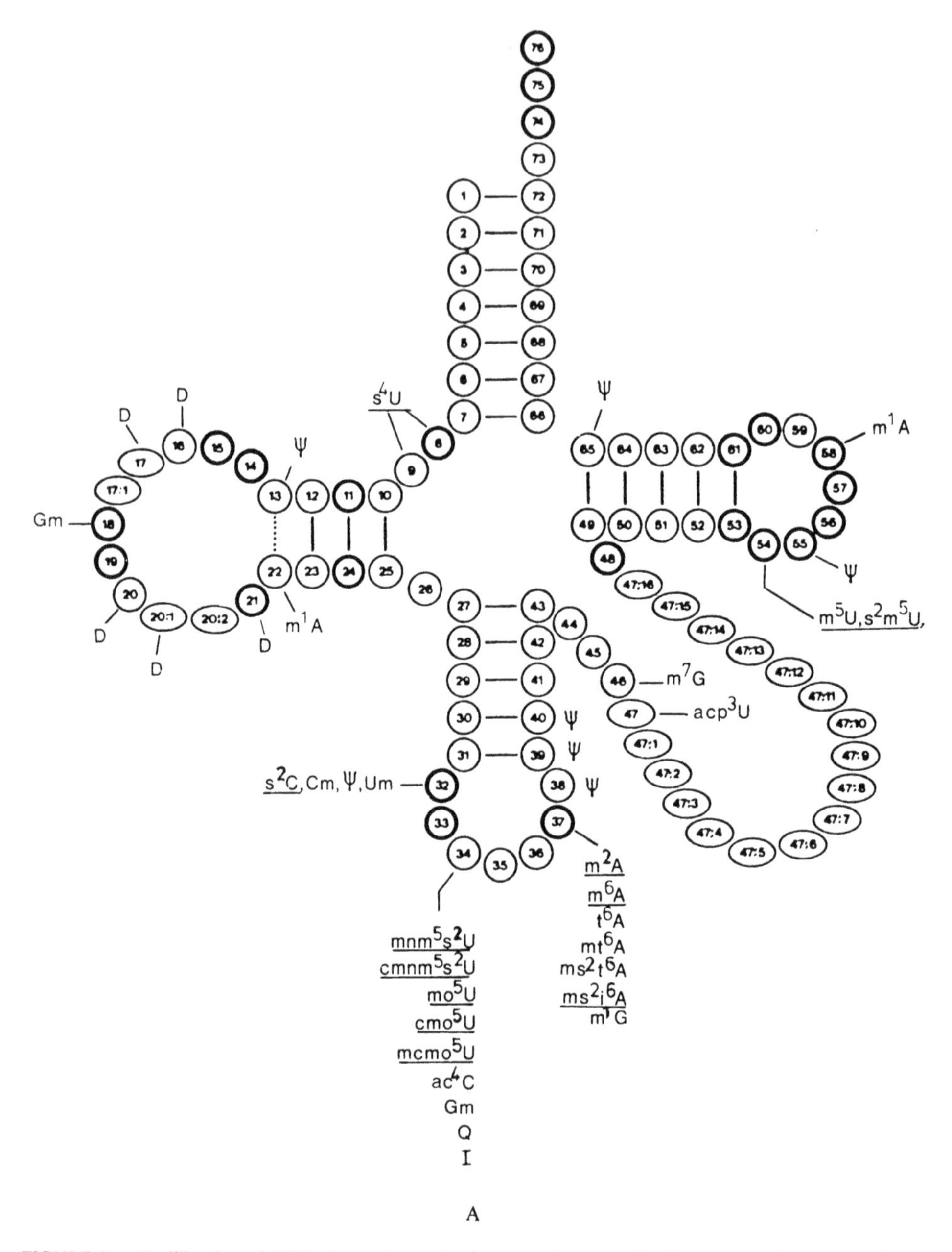


FIGURE 2. Modification of tRNA from eubacteria (A) and eukaryotes (B). Data were collected from Gauss and Sprinzl.[37] (modified from Kersten.[38]) The underlined modified nucleosides are found only in the group of organisms where they are listed.

Moreover, if different enzymes are required for 23S rRNA, an additional 13 enzymes may be needed. Thus, about 20 different rRNA-modifying enzymes may be required for the proper maturation of rRNA in *E. coli.* At present only a few enzymes have been purifed and partially characterized (Table 5). Only the rRNA ($m_2^6A$)methyltransferase from *E. coli* has been purified to homogeneity.[49] Most of these small enzymes (Mr of about 30,000) require magnesium ions for activity. The rRNA ($m_2^6A$)methyltransferase activity is inhibited by ribosomal protein S21, most probably by displacement of the enzyme from the substrate, the 30S ribosomal subunit. The rRNA($m^1G$)methyltransferase activity is also influenced by proteins, one stimulatory and one which degrades the substrate *S*-adenosyl-L-methionine

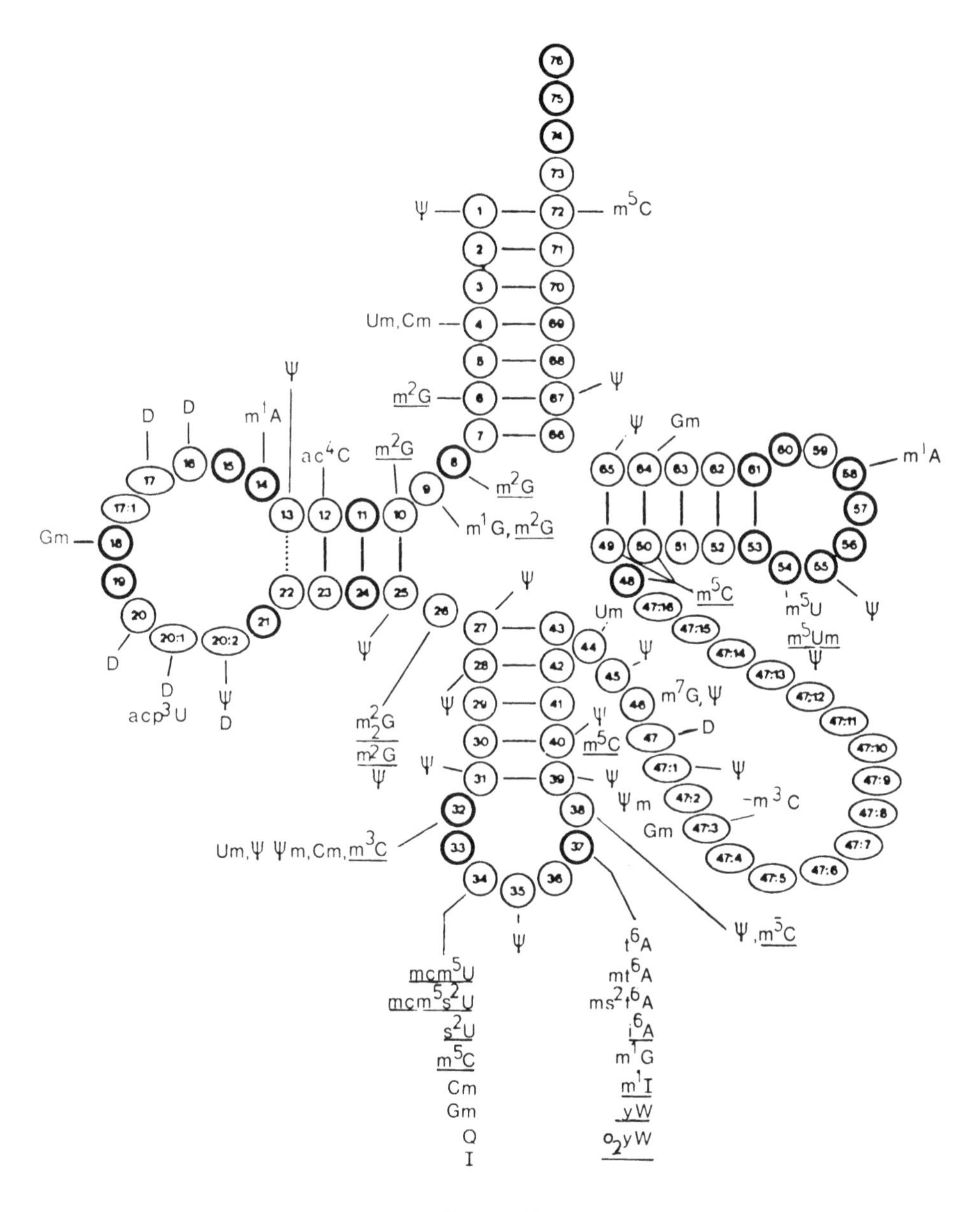


FIGURE 2B

(AdoMet). Table 5 shows that while some rRNA-methylating enzymes, like the rRNA($m^6_2$A)-methyltransfease, require a ribonucleoprotein particle (30S) as substrate, others, like the rRNA($m^1$G)methyltransferase, use 23S rRNA as well as 50S ribosomal subunit as substrates. The rRNA($m^6$A)methyltransferase is remarkable since it can use not only the high molecular weight rRNA as a substrate but also the low molecular weight compound β-9-ribosyl-2-6-diaminopurine.[51,52] So far this is the only example of a rRNA methyltransferase which is also able to use a low molecular precursor analog to rRNA as a substrate.

The ability of the rRNA($m^6_2$A)-and rRNA($m^1$G)methyltransferase to use 30S and 50S particles as substrates, respectively, implies that these modifications may occur late in ribosome assembly and also that the sequence to be methylated is exposed. Other studies which have revealed that the small rRNA sequence $m^6_2$A-$m^6_2$A-C is sensitive to S1-nuclease digestion also suggest that this part of the small rRNA in the small ribosomal subunit is exposed.[35-36]

**Table 2**
**MODIFICATION AT POSITION 37 (3′ SIDE OF THE ANTICODON) IN DIFFERENT tRNA SPECIES AND THEIR ENCODING CAPACITY**

| Codon[a] (nucleotide) | | | Amino acid inserted | Position 37 modification in tRNA from | |
|---|---|---|---|---|---|
| 1st, | 2nd, | 3rd | | Eubacteria | Eukaryotes |
| U | N | U, C | Phe, Ser<br>Tyr, Cys | ms$^2$i$^6$A,m$^1$G | yW, i$^6$A, m$^1$G |
| | | A, G | Leu, Ser, Trp | ms$^2$i$^6$A | i$^6$A, m$^1$G |
| C | N | U, C | Leu, Pro<br>His, Arg | m$^1$G, m$^2$A, | m$^1$G |
| | | A, G | Leu, Pro<br>Gln, Arg | m$^1$G, m$^2$A | m$^1$G |
| A | N | U, C | Ile, Thr<br>Asn, Ser | t$^6$A, mt$^6$A | t$^6$A |
| | | A, G | Ile, Met<br>Thr, Lys, Arg | t$^6$A | t$^6$A |
| G | N | U, C | Val, Ala<br>Asp, Gly | m$^2$A | m$^1$G, m$^1$I |
| | | A, G | Val, Ala<br>Gln, Gly | m$^2$A, m$^6$A | m$^1$I |

[a] N, any of the four ordinary nucleotides

## B. tRNA-Modifying Enzymes

Only five tRNA-modifying enzymes have been isolated to more than 50% purity (Table 6). Since they are present in low amounts in the cell (see below), their purification has been difficult. Furthermore, lack of specific tRNA substrates for the individual enzymes as well as their instability during the enrichment has hampered the biochemical characterization of the enzymes. Nevertheless, as shown in Table 6, most of the enzymes are small, acidic proteins showing several features in common; e.g., tRNA methyltransferases have similar Km for AdoMet and they are all inhibited by *S*-adenosyl-L-homocysteine (AdoHcy), and most of them require an intact SH-group for activity. Those enzymes which have been purified most extensively — the tRNA methyl-transferases producing m$^1$G *(E. coli)*, m$^5$U *(E. coli)*, m$^5$U *(Streptococcus faecalis)*, mcmo$^5$U and ψ (*Salmonella typhimurium*), m$^1$A (rat liver), and the G-inserting enzyme *(E. coli)* — all seem to be composed of one polypeptide in the active state, except tRNA(m$^5$U)methyltransferase from *S. faecalis,* which is composed of two identical subunits.[61] Furthermore, the tRNA(s$^4$U)synthetase is also a more complexed enzyme since it is also composed of at least two different subunits.[63]

**Table 3**
**MODIFICATION AT POSITION 34 (THE WOBBLE POSITION) IN tRNA ENCODING THE SERINE, ALANINE, OR VALINE CODON FAMILIES[a]**

| Codons (nucleotides) 1st + 2nd | 3rd | Amino acid | Wobble nucleosides in tRNA from *B. subtilis* | *E. coli* | yeast | mammals |
|---|---|---|---|---|---|---|
| GU<br>UC<br>GC | U | Val<br>Ser<br>Ala | $mo^5U$ | $cmo^5U$<br>or<br>$mcmo^5U$ | I | I |
| GU<br>UC<br>GG | C | Val<br>Ser<br>Ala | | | I | I |
| GU<br>UC<br>GC | A | Val<br>Ser<br>Ala | $mo^5U$ | $cmo^5U$<br>or<br>$mcmo^5U$ | I | I |
| GU<br>UC | G | Val<br>Ser<br>Ala | $mo^5U$ | $cmo^5U$<br>or<br>$mcmo^5U$ | | |

[a] In Gauss and Sprinzl[37] only $cmo^5U$ is suggested to be present in Val1, Ala1, Ser1 tRNA. However, the methylester $mcmo^5U$ is labile in base, but it cannot be excluded that in the cell some of the tRNAs reported to have $cmo^5U$ may contain instead $mcmo^5U$, i.e., the methylesther of $cmo^5U$.

The tRNA-modifying enzymes are highly specific. Table 7 shows how different mutations influence the modifications of rRNA and tRNA. Not only are the tRNA-modifying enzymes specific for tRNA *(trmA/trmD/rrmA)* but they are also specific for a specific position in the tRNA *(hisT)*. Furthermore, Smolar et al.[67] have shown that in yeast there exist two tRNA($m^1G$)methyltransferases with different target specificity — one enzyme catalyzing the formation of $m^1G$ at position 37 and the other at position 9. Kraus and Staehelin[68-69] have also demonstrated two distinct enzymes catalyzing the formation of $m^2G$ at two different sites in the tRNA. Table 2 shows that tRNAs reading codon starting with U have $ms^2i^6A$ at position 37, while tRNAs reading codons starting with A have $mt^6A/t^6A$ at this position. The modifying enzymes for these tRNAs, therefore, seem to have a recognition specificity for position 36. This has been verified by analysis of a set of missense and nonsense suppressors. Transfer $RNA_2^{Gly}$ which has the anticodon region sequence $5'U_{33}$-$U^+$–C–C–$A_{37}$–A–3′ is changed in the *glyTsu36*$^+$ (HA) mutant to 5′–$U_{33}$–$U^+$–C–*U*–$t^6A_{37}$–A–3′.[70] Another suppressor derivative of $tRNA_2^{Gly}$, which reads UGA, is changed not only in position 36 in the anticodon (C to A) but also in the modification of $A_{37}$ which results in an anticodon sequence of 5′–$U_{33}$ $U^+$CA $ms^2i^6A_{37}$–A–3′.[71] Furthermore, the $tRNA_3^{Gly}$ with the anticodon sequence: 5′–$U_{33}$–G–C–*C*–$A_{37}$–A–3′ is changed to 5′–$U_{33}$–G–C–*A*–$ms^2i^6$–$A_{37}$–A–3′ in the mutant *glyWsu* 78.[72] These three examples clearly show that the $ms^2i^6A$- and $t^6A$-producing enzymes have the nucleoside at the 5′-side of the target (i.e., nucleoside at position 36) as part of their recognition sequence.

Although the primary structure surrounding the modification site at least in these cases constitutes part of the recognition signal for tRNA-modifying enzymes, the tertiary structure of the tRNA is also important. When tRNA was fragmented, tRNA methyltransferases were unable to catalyze the methylation reaction although the intact tRNA molecules acted as

**Table 4**
**MODIFICATION IN POSITION 34 (WOBBLE POSITION) OF tRNAs ENCODING THE NONFAMILY CODONS**

| Codon (nucleotide) | | Amino acid | Wobble nucleosides in tRNA from | |
|---|---|---|---|---|
| 1st + 2nd | 3rd | inserted | Eubacteria | Eukaryotes |
| UA | U | Tyr | Q | galQ |
| | C | Tyr | | |
| | A | och | | |
| | G | amb | | |
| CA | U | His | Q | Q |
| | C | His | | |
| | A | Gln | $mnm^5s^2U$ | |
| | G | Gln | | |
| AA | U | Asn | Q | Q |
| | C | Asn | | |
| | A | Lys | $mnm^5s^2U$ | $mcm^5s^2U$ |
| | G | Lys | $cmnm^5s^2U$ | |
| GA | U | Asp | Q | manQ |
| | C | Asp | | |
| | A | Glu | $mnm^5s^2U$ | $mcm^5s^2U$ |
| | G | Glu | | |

substrates for the enzymes.[73-75] Furthermore, tRNA($m^1A$)methyltransferase from *B. subtilis* not only requires a specific sequence but also a specific tertiary structure.[76] Base substitutions in the primary sequence of phage T4-specific dimeric tRNA precursors, which influence the three-dimensional structure, abolish or diminish the modification of only the tRNA derived from the percursor half in which the mutation occurs.[77] Thus, although sequences were presented for the modifying enzymes, the reaction does not occur if the spatial organization of the target is changed. Transfer RNA methyltransferases also bind to immobilized tRNA irrespective of whether or not the tRNA is a substrate.[78] Thus, these enzymes have an affinity for the three dimensional structure of tRNA. Probably, the recognition signal for the tRNA-modifying enzymes may be attributed both to an overall specific conformation of the structure of the tRNA and in addition a polynucleotide sequence. The importance of these two contributions of the recognition signal may be different among the various tRNA-modifying enzymes since, for instance, the rate of transfer of the isopentenyl group to fragmented tRNA was 20-fold higher than compared to using intact tRNA as an acceptor in vitro.[79] More extensive genetical and biochemical studies using purified enzymes, however, are necessary before we understand the high specificity of the tRNA-modifying enzymes.

All known tRNA modifications, except one, occur on the polynucleotide level. The exception is the formation of the Q base which occurs on the monomeric level.[80-81] The Q base is inserted into the tRNA by a tRNA transglycosylase, which replaces a guanine for a quenine (base of Q).[82] In most RNA transmethylation reactions AdoMet is the methyl donor. So far only tRNA($m^5U$)methyltransferase from Gram-positive organisms uses 5-, 10-methylene tetrahydrofolate as the methyl donor.[83-84] However, transfer RNA from the Gram-

**Table 5**
**CHARACTERISTICS OF RIBOSOMAL RNA METHYLTRANSFERASES**

| rRNA methyl-transferase | Source | Substrates | Mr[a] ($10^{-3}$) | Divalent[b] ions | Monovalent ions[a] | Poly-amines[b] | Protein influencing activity | Km (n*M*) | Km (AdoMet) | Mutants available | Ref. |
|---|---|---|---|---|---|---|---|---|---|---|---|
| $m^6_2A$ | *E. coli* | 30S | 30(D) | Mg(+) | K(+) $NH_4(+)$ | | S21(−) IF3(−) | | | Yes (ksgA) | 49 |
| $m^1G$ | *E. coli* | 23S, rRNA 50S, 70S | | Mg(−) | K(+) $NH_4^+$ (±) | Sperm(−) Purtr(±) | Fact Al(−) B(+) | 53(23SrRNA) | 4.6 | Yes (rrmAl) | 50 |
| $m^2G$ | *E. coli* | 23S rRNA | | Mg(±) | — | — | A2(−) | | | Yes | 50 |
| $m^6A$ | *E. coli* | 23S rRNA, DAPR[c] | 31.6(N) | Mg(+) | K(+) | | | | | No | 51—52 |
| $m^6_2A$ | Plasmid born | 50S | 29(D) | | | | | | | Yes | 53—55 |

[a] Molecular weight determinations in the presence (D) or absence (N) of denaturing agent
[b] Signs following ions, polyamines, spermidine, and putrescine, or protein factors indicate stimulation (+) or inhibition (−) by the presence of respective agent
[c] DAPR = β-9-ribosyl-2-6-diaminopurine, an adenosine analog

**Table 6**
**CHARACTERISTICS OF tRNA MODIFYING ENZYMES PURIFIED TO NEAR HOMOGENEITY**

| tRNA Modifying enzyme producing | Source | Mr ($10^{-3}$) | AdoMet (Km[μ*M*]) | AdoHcy (Ki[μ*M*]) | pH opt | Divalent ion | Monovalent ions | Purity (%) | Infl. by SH-reagent[a] | pI | Km tRNA (μ*M*) | Poly-amine | Ref. |
|---|---|---|---|---|---|---|---|---|---|---|---|---|---|
| $m^1G$ | *E. coli* | 46(N) 32(D) | 5.0 | 6.0 | 8—8.5 | Mg(−) | K,Na(−) $NH_4$(−) | >95 | pCMB(−) | 5.2 | 20 | (+) | 56 |
| $m^5U$ | *E. coli* | 42(D) 56(N) | 17 | Inhib | 8.0 | Mg(+) | K, Na, $NH_4$ (+) | ~25 | pCMB(−) | 4.7 | 0.08 | (+) | 57 |
| | | 38(D) 42(N) | 12.5 | Inhib. | 8.4 | Mg(+) | $NH_4$ (+) | >95 | | 4.8 | 1.1 | (+) | 58 |
| G-insert | *E. coli* | 58(N) 46(d) | | | 7.0 | Mg(+) | K,Na(−) | >95 | | 4.6—4.8 | | | 59 |
| $mcmo^5U$[b] | *S. typhimurium* | 50(D) (N) | 2.0 | Inhib. | 7.5 | Mg(+) | K,Na (+) | ~50 | pCMB | — | 0.35 | | 60 |
| ψ | *S. typhimurium* | 50(D) 58(D) | | | | | | ~90 | IAcA(−) | 4.1 | | | 60a |
| $m^5U$ | *S. faecalis* | 115(N) | 1000[c] | | | | | >95 | | | 2.5 | | 61 |
| $m^1A$ | Rat-liver | 95(N) | 0.3 | 0.85 | 8.0 | Mg(+) | K,Na,$NH_4$(+) | >95 | | | .012-.033 | (+) | 62 |

*Note:* See explanations for signs used in footnote b, table 5.

[a] pCMB and IAcA denote *p*-chloromercuribenzoate and iodoacetamide, respectively.
[b] The identification of the product formed is tentative.
[c] This indicates the Km for the one carbon donor, 5, 10-methylenetetrahydrofolate.

## Table 7
## SPECIFICITY OF tRNA-MODIFYING ENZYME

| | rRNA | | | tRNA | | | |
|---|---|---|---|---|---|---|---|
| **Genotype** | **$m^5U$** | **$m^1G$** | **ψ** | **$m^5U$** | **$m^1G$** | **ψ38/39** | **ψ55** |
| wt | + | + | + | + | + | + | + |
| *trmA5* | + | + | + | − | + | + | + |
| *rrmA1* | + | − | + | + | + | + | + |
| *trmD1* | + | + | + | + | − | + | + |
| *hisT* | nd | nd | nd | + | + | − | + |

+ , normal level of the indicated nucleosides
− , absent or deficient in the indicated nucleoside
nd, not determined

Data from References 64 (*trmA, rrmA*), 65(*trmD*), and 66 (*hisT*).

positive organism, *B. subtilis* deficient in $m^5U$, still accepts methyl groups from tRNA($m^5U$)methyltransferase from *E. coli* using AdoMet as the methyl donor.[85] Furthermore, *E. coli* tRNA lacking $m^5U$ is a substrate for the enzyme from *B. subtilis.*[83] This suggests that both types of enzymes have the same specificity although they use different cofactors. A biochemical comparison of the two enzymes may reveal some common features which are important in the tRNA-protein interaction and may illustrate the differences in the cofactor binding sites.

# IV. GENETICS OF RNA MODIFICATION

## A. Genetics of rRNA-Modifying Enzymes

Björk and Isaksson[64] isolated the first mutants with aberrant modification of rRNA. Ribosomal RNA from one of these *E. coli* mutants completely lacked $m^1G$ in the 23S rRNA, whereas another was partly deficient in $m^2G$ in rRNA. Since these mutants are viable, the complete absence of $m^1G$ in rRNA is not obviously deleterious to cell growth. Furthermore, since the 50S subunit is a substrate for the enzyme in vitro, the $m^1G$ is located in an exposed area of the ribosome (see above). Ribosomal RNA from these mutants was used as specific substrate for the biochemical characterization of the corresponding enzymes.[50,86] Furthermore, the mutants have facilitated the identification of a dimeric precursor tRNA from T4 infected cells.[86] Resistance in *E. coli* to kasugamycin is due to the complete lack of $m^6_2A$ at the 3′-end of 16S rRNA.[88-89] The structural gene for the rRNA($m^6_2A$)methyltransferase *(ksgA)* is located close to min 1 on the *E. coli* chromosomal map.[88] In the biosynthesis of $m^6_2A$, $m^6A$ may be an intermediate. Since the level of $m^6A$ in 23S rRNA is normal in the *ksgA* mutant, it is likely that two different enzymes exist which catalyze the formation of $m^6_2A$ and $m^6A$.[88] As discussed above, there are probably about 20 rRNA modifying enzymes in *E. coli*. Mutants affecting only three of these enzymes have so far been characterized. Furthermore, no mutants defective in modification of rRNA from other organisms than *E. coli* have hitherto been characterized. Therefore many more mutants defective in other modifying enzymes are necessary before we can obtain a thorough understanding of the synthesis and function of the modified nucleosides in the ribosome.

## B. Genetics of tRNA-Modifying Enzymes

To date about 37 tRNA species from *E. coli, Salmonella,* or phage T4 have been sequenced and among them 25 different modified nucleosides have been identified. Some, like ψ, are found in different positions of the tRNA. Therefore, most likely more than one enzyme is involved in its biosynthesis (Figure 2). Other modified nucleosides, like $ms^2i^6A$, $cmo^5U$, and $mnm^5s^2U$, have a complicated structure and therefore more than one enzyme is probably involved in their synthesis. From such considerations one can infer that at least 40 different tRNA-modifying enzymes are present in *E. coli* or *Salmonella.* Assuming an average gene size of one kilobase of DNA (see below) for a modifying enzyme, at least 40 kb of DNA would be required to govern the synthesis of tRNA-modifying enzymes. This constitutes as much as 1% of the total genetic information in bacteria. Another set of about 20 different enzymes is involved in bacterial rRNA modification. Thus, a substantial amount of genetic information (1.5% of the total DNA content) is required to ensure a proper modification of RNA. However, each gene seems to be expressed at a low level (see below), thus, the energetic load is not as large as one could assume from the amount of genes involved.

In order to understand not only the function of the modified nucleosides in tRNA, but also the regulation of the corresponding modifying enzymes and the gene organization, it is necessary to obtain mutants defective in the modification of tRNA. Björk and Isaksson[64] devised a screening procedure which requires mutants able to grow while producing undermethylated tRNA. One gene, *trmA,* was shown to be the structural gene for the tRNA($m^5U$)-methyltransferase. This screening procedure was later used by Marinus and Morris[90] to isolate mutants defective in DNA methylation as well as in tRNA methylation. Using an improved screening method, two additional mutants, *trmC,* which governs the synthesis of $mnm^5s^2U$ and *trmD,* which governs the synthesis of $m^1G$, were characterized.[65,91] The three genes *trmA, trmC* and *trmD* have been mapped and the corresponding wild type genes have been cloned on different plasmids[92-94] (Figure 3). The gene *trmA* is 70% cotransducible with *argH* (88 min) and is located between *argH* and *bfe.*[95] The gene *trmC* has recently been located close to *aroC* (50 min) on the chromosomal map.[94] The *trmD* gene is located close to *tyrA* (56 min).[93] The gene *trmB* affects the formation of $m^7G$ in the tRNA and the gene is located in a region indicated in Figure 3.[96] The recessive UGA suppressor, *supK*, is located between *lys* and *serA* at 62 min.[97] The *supK* mutants are most probably deficient in the tRNA($mcmo^5U$)methyltransferase.[98] In a recently developed method for screening of modification mutants, transposons (Tn5 or Tn10) were employed as mutagens and a novel link between the biosynthesis of aromatic acids and modification of tRNA was discovered (see below).[99]

The regulatory behavior of the histidine and tryptophan operons has been worked out in detail. Mutants *hisT* and *miaA* (earlier called *trpX)* have been isolated which derepress the respective operons.[100,101] The *hisT* gene governs the synthesis of ψ in the anticodon stem of several tRNA species including $tRNA^{His}$, and the *miaA* is responsible for the formation of $ms^2i^6A$ in several tRNA species including $tRNA^{Trp}$.[66,102-105] Two other genes, *nuvA* and *nuvB,* are involved in the formation of $s^4U$ in tRNA.[102,103] The gene *trt* is responsible for the formation of the Q base in the wobble position of tRNA specific for tyrosine, histidine, asparagine, aspartic acid, and tryptophane (Table 4). The *trt*-gene has been cloned and it is likely to be the structural gene for the G-inserting enzyme.[105a] Thus, only seven out of at least 40 different modifying enzymes have been genetically characterized, and it is obvious that more mutants defective in tRNA modification must be isolated. The mutants already characterized, however, clearly show that the tRNA modification genes are not clustered but scattered all around the chromosome (Figure 3).

Phillips and Kjellin-Stråby[108] discovered the first mutant with aberrant RNA modification. The origanism they used was *Saccharomyces cerevisiae*, and tRNA from the mutant is completely deficient in $m^2_2G$. This nucleoside is present at position 26 in about 40% of all tRNA chains in yeast (Figure 2). Laten et al.[109] and Janner et al.[110] have isolated antisuppressor mutations in yeast which are deficient in $i^6A$. Recently yeast mutants deficient in $m^5U$ or D formation in tRNA have been characterized.[111-112]

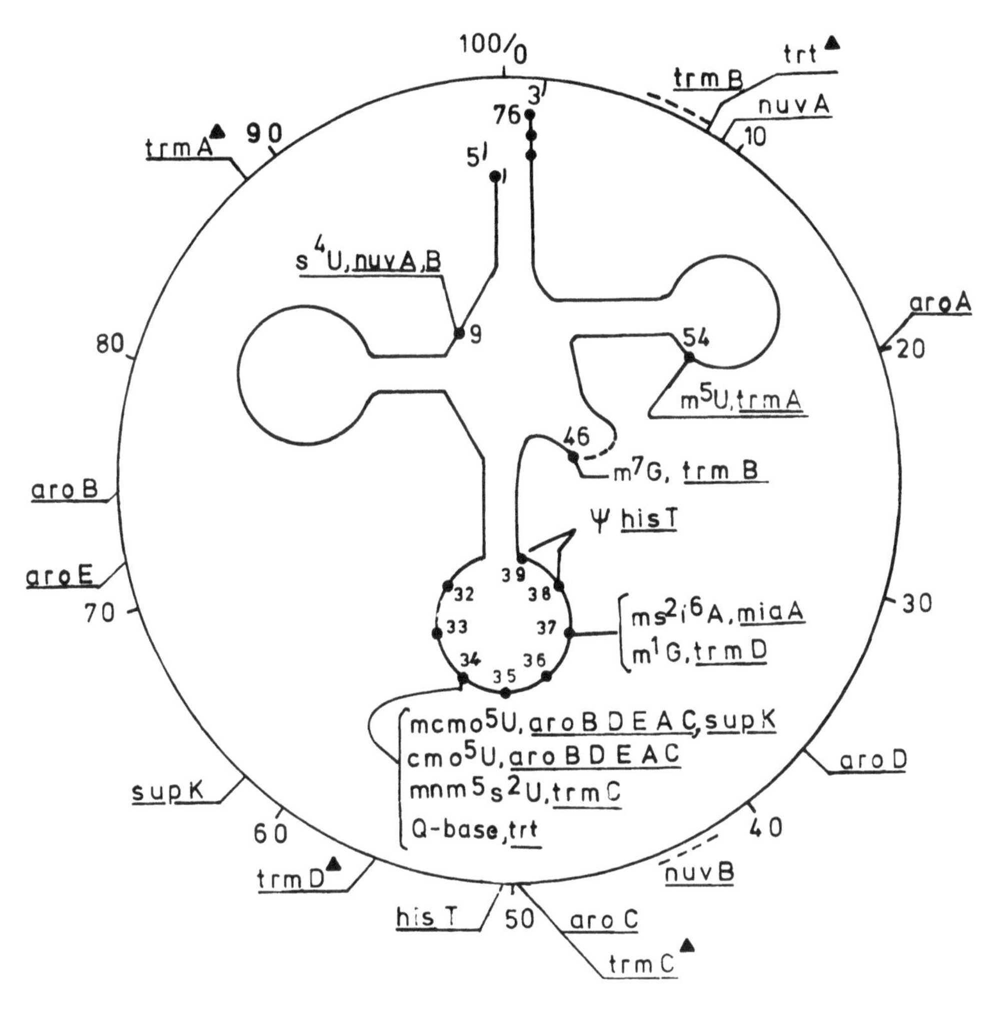


FIGURE 3. Known genes involved in the biosynthesis of modified nucleosides in bacterial tRNA. The position of the different modified nucleosides in the tRNA are indicated as well as the corresponding genes which are involved in their synthesis. The location of *trmC* in relation to *aroC* is uncertain.[94] The genes with a ▲ have all been cloned.

# V. FORMATION OF MODIFIED NUCLEOSIDES

## A. Regulation and Gene Organization

### *1. rRNA Modification*

Both in eubacteria and eukaryotes the modification of rRNA occurs stepwise and in a highly ordered manner.[113-116] The modification reaction is regulated in a concerted way with the processing of the primary transcript. In eukaryotes, ribose methylation and formation of ψ occur early, probably during transcription (Figure 4).[23,117] In *Saccharomyces carlsbergensis* the 37S rRNA precursor contains all ribose methylated nucleosides except Gm, in the conserved sequence UmGmψ.[117] Furthermore, all ψ present in the mature 26S rRNA and 17S rRNA are also found at the 37S rRNA precursor level.[23] Most base methylation occurs late and formation of $m_2^6A$ in yeast rRNA occurs after the 18S rRNA precursor has been transported out to the cytoplasm[118,119] (Figure 4). The sequential order of modification of HeLa cell rRNA is similar.[18,120-121] The formation of $m_2^6A$ in a conserved region of 16S rRNA from *E. coli* as well as in *Drosophila* and in HeLa cells is also a late event.[122-125] Furthermore one 2′-*O*-ribose methylation, Gm in the highly conserved sequence UmGmψ,

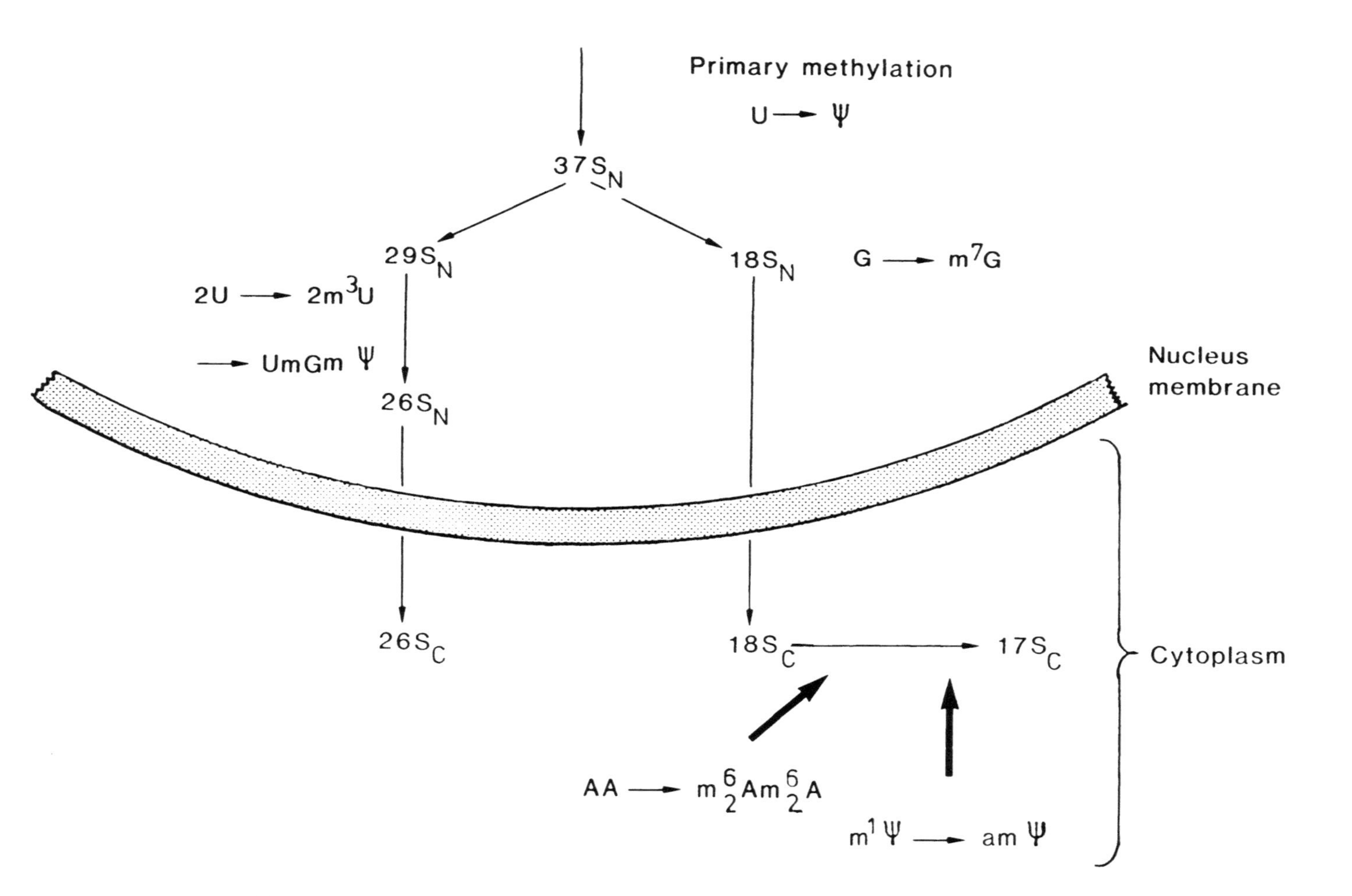


FIGURE 4. Modification of and processing of rRNA from yeast. Modified from Planta et al.[117]

is also formed late in *Drosophila,* mammal and yeast cells.[18,118,124] The highly modified nucleoside amψ, present in eukaryotic 17—18S rRNA, is formed in a stepwise manner. Following the formation of ψ, one methyl group is added at position 1 ($m^1ψ$). These two modification steps occur in the nucleus, while the last step, the introduction of the α-amino-α-carboxylpropyl side chain, occurs in the cytoplasm.[117] Thus some modifying enzymes must be present in the nucleus, like most of the ribose methylating enzymes, while others, like rRNA($m^6_2A$)methyltransferase and the last enzyme catalyzing the formation of amψ, are located in the cytoplasm.[117] Little is known of the regulation of rRNA methyltransferases. As discussed above, however, the activity of some of the *E. coli* enzymes seem to be sensitive to different proteins and factors.

Plasmid specific resistance towards erythromycin is associated with the introduction of $m^6_2A$ in the 23S rRNA in some Gram-positive organisms.[126] The presence of $m^6_2A$ in the 50S subunit reduces the binding of the antibiotic.[127] Erythromycin can, at low concentration ($10^{-8}$ to $10^{-7}M$), induce resistance with a concomitant introduction of methyl groups into the ribosome. Accordingly, a specific induction of a rRNA($m^6_2A$)methyltransferase occurs.[55] The transcriptional rate for its mRNA was not altered upon induction but its functional half-life increased.[127] Thus, the regulation of the rRNA($m^6_2A$)methyltransferase is posttranscriptional. The sequence of the entire gene is known.[54,128] Between the promoter and the start of translation of the tRNA($m^6_2A$)methyltransferase there exists an open reading frame of 19 amino acids, a region called the "leader region". Such a feature is reminiscent of the attenuator mechanism for the regulation of many amino acid biosynthetic operons. An attenuator-like mechanism has been proposed for the regulation of the rRNA($m^6_2A$)methylatransferase which would operate entirely at the translation level and not at the level of transcriptional termination.[54-55,128] Erythromycin is suggested to stall the ribosome during the translation of the 19-amino acid leader peptide which results in an unfolding of some of the potential stem and loop structures in such a way that the Shine and Dalgarno sequence for the structural gene for the rRNA($m^6_2A$)methyltransferase becomes exposed and available for translation. There is always a basal level of the enzyme present in the cell, resulting in a few methylated ribosomes which are resistant to erythromycin. Thus, erythromycin stalls only nonmethylated ribosomes. The few methylated ribosomes available may attach at the translational initiation site on the mRNA for the rRNA($m^6_2A$)methyltransferase and the enzyme will be produced. A slow exponential escape from the inhibition of high concentration of erythromycin will then occur.

*2. tRNA Modification*

Maturation of tRNA is a highly ordered process.[129] In prokaryotes, the modification of some nucleotides takes place at distinct stages of the maturation process.[130] However, the different precursors seem to be differentially modified, indicating that the modification occurs stepwise.[131] For example, the multimeric precursor for $tRNA^{Leu}_I$ contains $m^5U$, ψ and D, but not $m^1G37$. The latter methylated nucleoside is, however, present in the monomeric precursor, which is still lacking Gm18.[131] These results are, however, based on analysis of mutants defective in tRNA maturation. Since such mutants accumulate precursors, it is not clear what is the preferred substrate for the tRNA modifying enzymes. In fact, results have been obtained suggesting that the preferred substrates for tRNA methyltransferases are RNA molecules of molecular weight similar to that of a mature tRNA.[132-133] Thus, some tRNA modification reactions can occur at the precursor level, but the rate may be low compared to that of the precursor trimming reaction. However, some modifications, like ribose methylation, require an almost mature tRNA since these nucleosides have never been found in multimeric precursors.

In eukaryotes, base modifications, like $m^5C$, $m^1A$, and ψ, occur in the nucleus in an obligatory order.[134] Formation of ψ in the TψC-loop preceeds the formation of ψ in the anticodon stem. The latter modification does not occur until the 5′-leader sequence is removed

**Table 8**
**REGULATION OF tRNA BIOSYNTHETIC ENZYMES**

| tRNA methyl-transferase producing | Response of spec. act. as function of growth rate | Response to shift-up | Response to shift-down | Response to amino acid limitation in *relA*⁺/*relA1* strains |
|---|---|---|---|---|
| $m^5U$ | Increases | As stable RNA[a] | As stable RNA[a] | Stringent[b] |
| $m^1G$ | Slight increase | — | — | Nonstringent |
| $mnm^5s^2U$ | Constant | — | — | Nonstringent |

[a] Synthesis of stable RNA (rRNA and tRNA) increases and decreases immediately following a shift to a richer or poorer medium, respectively.

[b] Stringent regulation indicates increase synthesis in relAl strain and decrease in *relA*$^+$ strain. This behavior is similar to stable RNA and ribosomal proteins.[138] The *relAl* gene governs the synthesis of ppGpp under amino acid limitation resulting in high level of ppGpp in *relA*$^+$ strain and low in *relAl* strain.

Data from References 139, 140, and 141.

and the 3′-CCA terminal is added. Furthermore, for those tRNA precursors containing an intervening sequence the modification around the anticodon occurs after the removal of the intervening sequence.[135-136] Therefore, during the maturation of tRNA precursors, size reduction and modification are intimately related processes and occur in concert.

Modified nucleosides confer on the tRNA the ability to read efficiently different messenger RNAs (see below). It is therefore important for the cell to synthesize the tRNA-modifying enzymes in amounts adequate for optimal modification of the tRNA. It is known that the mechanism regulating tRNA is similar to that controlling the synthesis of rRNA and proteins involved in translation.[137-138] However, little is known about the regulation of the tRNA-modifying enzymes. Such enzymes can be regarded as tRNA biosynthetic enzymes. In order better to understand the dynamic aspects of tRNA, it is necessary not only to understand the regulation of synthesis and maturation of tRNA, but also the regulation of the synthesis of the tRNA-modifying enzymes.

Using tRNA from mutants defective in the formation of $m^5U$*(trmA)* $mnm^5s^2U$-*(trmC)* and $m^1G$*(trmD)*, it was possible to measure the activity of the corresponding enzymes in crude extracts. The level of tRNA($m^5U$)methyltransferase increases with increasing growth rate while the level of the other two enzymes only slightly increases or stays constant (Table 8). The expression of the three genes also shows different response to the accumulation of ppGpp in the cell (Table 8). Furthermore, the tRNA($m^5U$)methyltransferase was not only regulated as ribosomal RNA, ribosomal protein, and other proteins involved in translation with regard to growth rate and ppGpp but also to shift-up and shift-down conditions (Table 8). Thus, the tRNA($m^5U$)methyltransferase is regulated like proteins normally found in large quantities in the cell, although the tRNA($m^5U$)methyltransferase is present in small amounts. The other two enzymes, tRNA($mnm^5s^2U$) and tRNA($m^1G$)methyltransferase are regulated more like other proteins which are produced in low amounts in the cell. Thus, although the enzymes are all involved in the modification of tRNA, they are regulated differently. It should be pointed out that the regulation is not 'constitutive', because if so a lower specific activity would be observed as growth rate increases.

As described by Lemaux et al.,[142] temperature can specifically influence the expression of genes. This may be one reason why the formation of Gm in *Bacillus thermophilus* is induced by high temperature.[143] Thiolation during the formation of $m^5s^2U$ in tRNA from *Hermus thermophilus* is also stimulated by high temperature.[144] Thus, several interesting regulatory features may be unravelled as more thorough understanding of the regulatory mechanism which governs the synthesis of the tRNA biosynthetic enzyme is achieved.

The gene organization of the *trmA* and *trmD* gene has already been established. The *trmA* gene has been cloned.[92] Subcloning and Tn5 insertion in the relevant chromosomal area indicated that the transcriptional unit for the expression of the tRNA($m^5U$)methyltransferase does not include any other proteins.[144a] The transcription of the *trmA* gene which is located close to the origin of replication is counterclockwise and opposite the replication of the chromosome.[145] Other genes in this region with higher expression than the *trmA* gene are normally transcribed in the same direction as that of DNA replication. Whether this is relevant to the regulation of *trmA* expression awaits further studies. Thus, the organization of *trmA*, which is regulated in a similar way to stable RNA and protein made in high amounts, is simple.

The *trmD* gene maps close to *tyrA* and has been cloned (Figure 3). The gene organization was elucidated using subcloning, deletion mapping, and gene inactivation by Tn5 transposons.[146] The organization for this gene is more complex than that of *trmA*. The *trmD* gene is the third promoter distal gene of a polycistronic operon of four genes (Figure 5). This has been confirmed by DNA sequencing and the establishment of the N-terminal of the purified tRNA-($m^1G$)methyltransferase. Surprisingly, the first and the last gene in the *trmD* operon codes for ribosomal protein S16 and L19, respectively.[147] These two ribosomal proteins are made in 100-fold larger amounts than the tRNA ($m^1G$)-methyltransferase. Thus, it is obvious that the regulatory mechanism governing the production of the tRNA($m^1G$)methyltransferase is complex and perhaps operates both at the transcriptional as well as the translational levels.

**B. Metabolic Aspects of the Formation of Modified Nucleosides**

There are many modifying enzymes which use amino acid derivatives as cofactors. It is well established that the formation of methylated nucleosides involves transfer of the methyl group from methionine via the methyl donor *S*-adenosyl-L-methionine.[4] This is true for all tRNA methylation reactions in most organisms except for some Gram-positive bacterial species which utilize tetrahydrofolate as methyl donor only in the case when $m^5U$ is formed.[83-84] The formation of $t^6A$ occurs by the enzymatic incorporation of threonine into tRNA.[148-150] One additional C atom must, however, be introduced prior to the incorporation of threonine to produce $t^6A$. This C atom originates from bicarbonate.[151] Many modified uridines, like $s^4U$ and $mnm^5s^2U$, and adenosines, like $ms^2i^6A$, contain a sulfur atom. In all cases, L-cysteine serves as the sulfur donor, since in vivo labeling using $^{35}S$-cysteine resulted in radioactivity in $ms^2i^6A$, $mnm^5s^2U$, two identified nucleosides as well as in $s^4U$.[105] Thus, methionine, threonine, or cysteine are direct precursors to these groups of modified nucleosides, and a deficiency of these amino acids will lead to undermodified tRNA. Depending on the severity of the amino acid limitation, the effect on the synthesis of different modified nucleosides is probably not the same, since the amount of enzymes available in the cell and their Km for the corresponding substrates may be different. Such changes in the pattern of modification of tRNA also result in an altered pattern of isoaccepting tRNAs.

Other changes in the metabolism of the cell which bear no apparent relationship to substrates of modifying enzymes can still induce change in the modification pattern of tRNA. (For earlier results, see Reference 152.) Starvation for leucine or arginine results in a severe unbalanced growth which induces specific undermodification in $tRNA^{Phe}$ and $tRNA^{Leu}$.[153] Mainly D16 and $ms^2i^6A37$ in $tRNA^{Phe}$ and ψ38 and D17 in $tRNA^{Leu}$ were lacking.[154] Other deficiencies were also observed. Modified nucleosides like $m^5U$, $s^4U$, Gm, $m^7G$, ψ32, and D20 in $tRNA^{Phe}$ and ψ40 in $tRNA^{Leu}$ were, however, found in normal amounts. Thus, upon starvation for leucine, there was a differential effect on the pattern of modification which has no apparent relation to the amino acid withdrawn. Imposing unbalanced growth by other means than removal of a required amino acid, e.g., by the addition of antibiotics to growing cells, also results in the appearance of undermodified tRNA species suggesting that unbalanced growth as such may result in the synthesis of undermodified tRNA.[155-156] Threonine

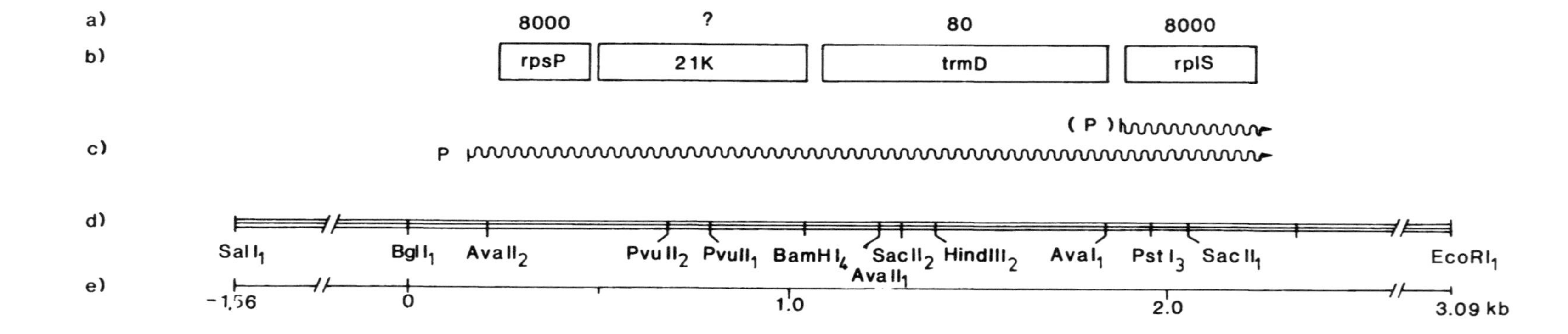


FIGURE 5. Gene organization of the *trmD* operon. Line a, number of molecules present/genome at a specific growth rate of 1.0. Line b, structural genes in the operon: *rpsP*, ribosomal protein S16; 22K, unknown protein; *trmD* tRNA($m^1$G)methyltransferase and *rplS*, ribosomal protein L19. Line c, transcription units from the *trmD* operon. The internal promoter (p) is only tentative. Line d, different restriction sites. Line e, relative physical distance (kilobase, kb). O has been arbitrarily set at the $Bgl_1$ restriction site.

or leucine limitation during steady state of growth leads not only to specific changes in the concentration of cognate isoaccepting tRNAs, but also in other tRNA species.[157] The changes in the tRNA profiles were similar to the one observed under starvation conditions (see above). Therefore, it is likely that the same undermodification occurred in cells growing in balanced growth in amino acid-limiting media. Furthermore, in mutants which overproduce threonyl-tRNA ligase, concentrations of isoaccepting species for $tRNA^{Thr}$ are also changed.[158] A relationship has been suggested between mutations in structural genes for amino acid biosynthetic enzymes and alterations in concentrations of isoaccepting tRNAs.[158] A mutation in *ilvU* changes the concentration of isoaccepting $tRNA^{Ile}$ and $tRNA^{Val}$, and also influences the regulation of isoleucyl-tRNA ligase.[159] The authors suggest that the *ilvU*$^+$ gene product not only allows derepression of isoleucyl-tRNA ligase but also retards the conversion of isoaccepting species, most likely by controlling the modification of these tRNA species. Thus, unbalanced growth, amino acid metabolism, and the availability of amino acid in the cell play an important role in the regulation of tRNA modification.

Enterochelin is an iron chelating compound, which is secreted upon iron limitation by *Escherichia, Salmonella,* and *Klebsiella*[160-161] (Figure 6). This compound is able to aquire $Fe^{3+}$ from iron-binding proteins like transferrin and lactoferrin present in different body fluids. *E. coli* growing under iron restriction conditions produces a receptor for the $Fe^{3+}$-enterochelin complex. Furthermore, such cells also contain undermodified tRNA.[156,162-163] This undermodification is specific — only the thiomethylation reaction in the formation of $ms^2i^6A$ is inhibited.[163] The same change in tRNA modification occurs when *E. coli* is growing in body fluids where iron-binding protein is present.[164] It has been suggested that this tRNA alteration is connected to the adaptation of *E. coli* to growth in body fluids and with bacterial pathogenecity.[164] Cogenate tRNAs are involved in the transport of the aromatic amino acids.[165-166]. Deficiency of the thiomethyl-group of $ms^2i^6A$ in tRNA stimulates the transport of the three aromatic amino acids[167] (Figure 6). Thus, synthesis of aromatic amino acids and enterochelin is intimately coupled to the formation of $ms^2i^6A$ in tRNA and its function in sensing the level of the three aromatic amino acids in the cell.

Another unexpected link between biosynthesis of aromatic amino acids and tRNA modification was recently reported.[168] All mutants defective in the common pathway of aromatic amino acids (Figure 6, *aroA, D, E, B,* or *C)* are deficient in $cmo^5U$ and $mcmo^5U$ in their tRNAs. Presence of shickimic acid in the growth medium leads to the resumption of synthesis of $cmo^5U/mcmo^5U$ by an *aroD* mutant, while this effect is absent in an *aroC* mutant. No metabolites in the specific pathway leading to tyrosine, tryptophane, phenylalanine and the four vitamins known to branch out from chorismic acid, are involved in the modification of tRNA (reference 168 and Björk, G. R., unpublished observation). Therefore chorismic acid itself or some unknown metabolite thereof must play a key role in the formation of $cmo^5U/mcmo^5U$ in the tRNA (Figure 6).

One possible way to modulate the translation efficiency could be to alter the modification pattern of the tRNA. During the development of the slime mold *Dictostelium discoideum* and the amphibian *Rana catesbeiana,* changes occur in the modification pattern of tRNA.[170-172] In *D. discoideum* the contents of $m^5U$, $\psi$ and $m^5C$ decrease during development from vegetative cells to spore formation.[172] Furthermore, $m^5U$ containing tRNA species are preferentially used in protein synthesis.[173] Since eukaryotic tRNA containing $m^5U54$ is more efficient in translation than tRNA containing U54, changes in $m^5U$ during development might regulate translation in a subtle but important way.[174] Furthermore, soon after the onset of development one of the two major $tRNA^{Asp}$ is completely deacylated. Such deacylated tRNA contains guanosine in the wobble position instead of quenine (Q).[172] No change occurs in the transcription of tRNA genes during the development of *D. discoideum* although several posttranscriptional modifications changes occur especially in the formation of the Q base.[170] Transfer RNA from tumor tissue has exclusively guanine instead of Q in the wobble position

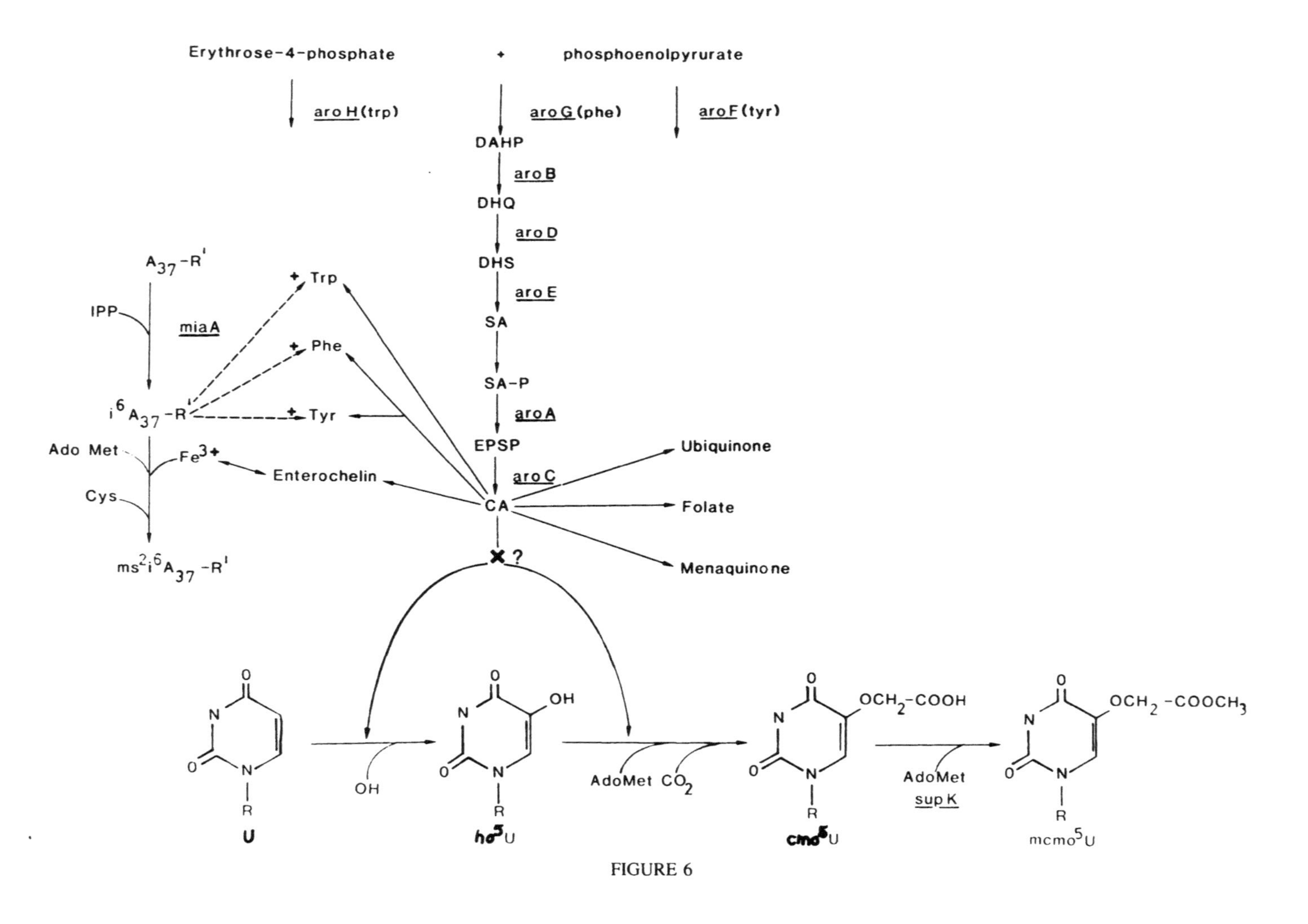

Erythrose-4-phosphate
+
phosphoenolpyrurate
aroH(trp)
aroG(phe)
aroF(tyr)
DAHP
aroB
DHQ
aroD
DHS
aroE
SA
SA-P
aroA
EPSP
aroC
CA
X ?
Ubiquinone
Folate
Menaquinone
+ Trp
+ Phe
+ Tyr
Enterochelin
Fe3+
A37-R'
IPP
miaA
i6A37-R'
Ado Met
Cys
ms2i6A37-R'
OH
U
ho5U
AdoMet CO2
OCH2-COOH
cmo5U
AdoMet
supK
OCH2-COOCH3
mcmo5U


FIGURE 6

of tRNAs specific for tyrosine, histidine, asparagine, and aspartic acid.[176-177] During the development of *Drosophila* the content of Q base changes dramatically, but this effect may also be influenced by the age of the growth medium and the temperature of cultivation.[178-181] Thus the synthesis of certain modified nucleosides, like $m^5U$ and Q, changes during the developmental process in eukaryotes. It has been proposed that translation of termination codons may play a regulatory role.[182-184] Since Q may be present in $tRNA^{Tyr}$ and is important for possible read-through,[185] the developmental changes in Q content may reflect translational regulation during this process. (See Section VI. B.3.)

*B. subtilis* is used as a simple model system for developmental studies. Temporal changes in relative amounts of tRNA isoaccepting species for several amino acids occur during the development of *B. subtilis*. In some cases this altered tRNA profile is due to a change in the modification pattern.[186] In vegetative cells the predominant species of $tRNA^{Tyr}$ contains $i^6A37$, while in stationary phase $ms^2i^6A37$ is present.[187-188] Similar changes also occurs during modification of $tRNA^{Trp}$. Changes in the aeration also influence the formation of $ms^2i^6A37$ in $tRNA^{Phe}$ as well as the formation of Gm34.[189] Furthermore, brain contains $tRNA^{Ser}$ which is lacking Gm but is otherwise identical to liver $tRNA^{Ser}$.[190] Therefore developmental changes as well as changes in growth conditions strongly influence the modification of tRNA. Such modifications may provide the tRNA with a novel capacity to translate more efficiently the mRNA pools characteristic of the particular stages of development and may thus be of selective advantage to the organism.

## VI. FUNCTIONAL ASPECTS

Figures 2 illustrates that certain positions in the tRNA are more prone to modification than others. In the three-dimensional structure of yeast tRNA, none of the modified nucleosides present appeared to be essential for maintaining the structure.[191] The presence of the modified groups, however, increases the acceptable surface area by 20%, suggesting that the modified nucleosides are there to be recognized by various proteins and/or nucleic acids. All mutants defective in either rRNA or tRNA modification, except *supK*[97] and a mutant deficient in $mnm^5s^2U$ formation,[192] appear to be quite viable. Furthermore, some species of *Mycoplasma* do not contain many of the modified nucleosides. This suggests that most of the modified nucleosides in rRNA/tRNA are nonessential. The function of modified nucleosides might be manifold. The presence of a modified nucleoside may *increase* or *decrease* various interactions. To make the genetic code less sensitive to codon context, the introduction of modified nucleosides next to the anticodon may either increase or decrease the interaction of the tRNA with the adjacent nucleotide. Each modification therefore may have its unique function and may have evolved as a consequence of changes in the tRNA molecule. Feldman[13] has proposed that some modified nucleosides are intended for covalent fixation of the tRNA to the ribosome. After this fixation, the tRNA-ribosome binding is split and the modification group is lost and recycled outside the ribosome. Thus, demodification should occur, and actually one such enzyme, which removes the isopentenyl group of $i^6A$, has been characterized.[193] Therefore at least some tRNA modifications may be reversible. To summarize, it is likely that the modified nucleosides in both rRNA and tRNA

FIGURE 6. Interrelationship between biosynthesis of aromatic amino acids and tRNA modification. Arrows ($\rightarrow^+$) denote the stimulation of the transport of the corresponding amino acids by the cognate tRNAs containing an unmodified A at position 37.[167] The biosynthetic block in the hypothetical biosynthetic pathway of $cmo^5U$ in tRNA induced by the lack of chorismic acid (CA) is tentative.[168] See discussion in references 168 and 169 about possible biosynthetic pathways for $cmo^5U$/$mcmo^5U$. The genetic symbols corresponding to each of the known structural genes of the common aromatic amino acid pathway are indicated as well as *miaA*, and *supK*, which are specific for the synthesis of $ms^2i^6A$ of some tRNAs and $mcmo^5U$ in position 37 and in position 34, respectively, of some tRNAs.

are serving as a fine tuning device of some of the functions of these nucleic acids. In this review I will concentrate on recent results. Earlier studies regarding the function of modified nucleosides are found in several older reviews.[4-16]

**A. Function of Modified Nucleosides in rRNA**

Methylation of rRNA is not an absolute prerequisite for normal maturation of rRNA in *E. coli*.[194] Subtle effects are, however, not excluded. Methylation of rRNA in an eukaryotic cell is on the other hand essential for the maturation process.[195] Since the eukaryotic rRNA is degraded when methylation is inhibited, the large amount of 2′-*O*-ribose-methyl groups normally present in the precursor may protect the eukaryotic rRNA from degradation. The different effects of methylation on the maturation process in eubacteria and eukaryotic rRNA may thus reflect vast differences in the degree of ribose methylation present in the two groups of organisms.

Only three mutants which are defective in rRNA modification have so far been characterized. Two mutants deficient in the single $m^1G$ or the single $m^5C$ in 23S RNA are still able to grow reasonably well and therefore these methylated nucleosides in the 23S rRNA are not essential.[64,91] No detailed functional analysis has, however, been performed to reveal any possible subtle changes in the activity of the ribosome. Bacteria which are resistant to kasugamycin are lacking $m_2^6A$ in the 3′ part of the 16S rRNA, due to a mutation in the structural gene, *(ksgA)*, for the rRNA($m_2^6A$)methyltransferase.[88-89] Growth characteristics of such mutants lacking $m_2^6A$ in their 16S rRNA are almost similar to the wild type.[196] Thus the lack of $m_2^6A$, which is evolutionary conserved and present in the small subunits of the ribosome from all organisms (Table 1), is not essential for growth. Careful physical characterization, however, has revealed that the presence of $m_2^6A$ influences the formation of a hairpin structure in which the $m_2^6A$ is part of the loop.[197] The Shine and Dalgarno sequence (position 3 to 9 from the 3′-end in 16S rRNA) is important in the initiation of translation and kasugamycin is known to inhibit this part of the translation process.[198-201] The two $m_2^6A$ nucleosides are present at position 24 and 25 starting from the 3′-end of the 16S rRNA, i.e., rather close to the Shine and Dalgarno sequence. Ribosome 30S subunits lacking $m_2^6A$ require more IF3 to initiate protein synthesis.[202] The IF3 is also known to bind close to the $m_2^6A$ sequence.[203] Furthermore, the presence of the $m_2^6A$ in the 16S rRNA increases the rate of polyphenylalanine synthesis in vitro by the accelerated binding of ribosomal protein S1 and an unidentified protein.[196] Thus, the $m_2^6A$ conserved sequence which is exposed on the 30S subunit[204] is involved in initiation of translation. The function of this region in other organisms may well be similar to the one established for *E. coli*.

The mechanism behind resistance to erythromycin is different from kasugamycin resistance since in this case a novel $m_2^6A$ in the 23S rRNA is introduced.[126] Resistance to thiostrepton in *Streptomyces azureus* is also due to an acquisition of a new methylated group but the methylated product has not been yet identified.[205] The novel methylated nucleosides discussed above are known to inhibit the binding of the corresponding antiobiotics to the ribosome. Other physiological consequences for the cell due to the presence of a novel methylgroup in the ribosome have not yet been established.

**B. Function of Modified Nucleosides in tRNA**

*1. Function of Modified Nucleosides in Position 37*

Figure 2 shows that position 37 is highly prone to be modified. As mentioned in Section II. B, there is a correlation between the modification at position 37 and the coding specificity (Table 3). Thus, tRNAs reading UXX codons usually have a bulky hydrophobic residue, like isopentenyl, at position 37, while tRNAs reading AXX codons have a hydrophilic residue, like threonine, at position 37. It has therefore been suggested that these modifications may strengthen the rather weak A-U/U-A basepairing next to position 37 in these anticodon-codon interactions.[7] In fact, measuring the anticodon-codon interaction in dimers of different tRNA species the stability is about the same for tRNA pairs irrespectively of the GC-content of the anticodon.[206] This has been attributed to stacking interactions between bases within and adjacent to the anticodon. *E. coli* initiatior $tRNA_f^{Met}$, which has an unmodified A37, is able to read not only AUG (methionine) but also GUG (valine) and UUG (leucine).[207,208] However, yeast initiator $tRNA_f^{Met}$, which contains a $t^6A37$, is unable to read GUG in vivo.[209] Furthermore, *E. coli* $tRNA_m^{Met}$, which also has a $t^6A37$, reads only AUG (methionine). Therefore, it has been proposed that the function of the $t^6A$ might be to prevent such wobbling in the *E. coli* elongator $tRNA_m^{Met}$ and yeast initiator $tRNA_f^{Met}$.[210] However, this latter $tRNA_f^{Met}$, which does contain $t^6A37$, binds both AUG and GUG triplets in vitro,[211] indicating that the different coding properties of the initiator and elongator $tRNA^{Met}$ in *E. coli* cannot entirely be due to presence or absence of $t^6A$ at position 37. The presence of $t^6A$ at position 37 in yeast $tRNA_{III}^{Arg}$ was found to stabilize U-A as well as U-G base pairing adjacent to the 5′ side of the $t^6A$, most probably by stacking.[212] Furthermore, $t^6A$ deficient $tRNA^{Ile}$ binds less efficiently to poly (AUC) programmed ribosomes.[213] Transfer RNAs reading AXX have as third letter of the anticodon and its 3′-adjacent nucleotide $Upt^6A$. Physical measurements showed that the presence of $t^6A$ stabilize the stacking interaction relative to UpA.[214] In addition to increased stacking, the presence of $t^6A$ may also prevent wobble on the 3′ side of the anticodon.[214] Thus, the function of $t^6A$ is to stabilize the anticodon-codon interaction, making it more efficient. Its role in the specificity of the anticodon-codon interaction remains to be elucidated. There are data to suggest that mispairing in the third nucleotide of the anticodon is not an intrinsic property of $t^6A$ alone.[211-212]

Transfer $RNA^{Tyr}$ from *E. coli* deficient in $ms^2i^6A$ binds less efficiently to the ribosome.[215] Yeast antisuppressor mutations, *sinl* and *mod5-1*, reduce the activity of the serine-inserting UGA suppressor and the tyrosine-inserting UAA suppressor, respectively.[109-110] Both *sinl* and *mod5-1* strains lack completely the $i^6A$ normally present in position 37 of the suppressor tRNAs. However both mutants grow relatively well, showing only a 10% reduction in growth rate in the case of the *sinl* mutation.[110] Furthermore, *E. coli* $tRNA^{Phe}$ carrying $i^6A37$ instead of $ms^2i^6A37$, is less active in ribosomal binding.[216] Yeast $tRNA^{Phe}$ contains the yW-base in position 37.[37] Removal of this hypermodified base results in inability to bind to ribosomes and to form polyphenylalanine in a polyU programmed translation system.[217] Furthermore, yeast $tRNA^{Phe}$ lacking the yW-base changes its coding properties as well as its ability to bind oligonucleotides in vitro.[218-219] This modified base has been suggested to intercalate between the two codon-anticodon triplet duplexes present in the A- and P-sites on the ribosome.[220] The model suggests that both in the A- and P-site the yW-base is in contact with the mRNA and may stabilize the codon-anticodon interactions by stacking.[220] Transfer RNA from *Lactobacillus acidophilus,* deficient in $i^6A$, and tRNA from *Mycoplasma,* which normally lacks $ms^2i^6A$, are both fully active in polyphenylalanine synthesis in vitro.[221-222] An *E. coli* mutant *miaA* (formerly called *trpX)* lacks $ms^2i^6A$.[104-105,223] This mutant has a derepressed tryptophan operon, probably caused by an inability of the $tRNA^{Trp}$, which normally contains $ms^2i^6A$, to read efficiently the two tryptophan codons present in the tryptophan-leader sequence.[104,223] The mutation does not, however, influence the activity of a tryptophan-inserting UGA suppressor.[104] Furthermore, conformational studies of the $i^6A$ and $t^6A$ monomers have shown that the side chains might block the two N-sites present in the adenine base to form hydrogen bonds, i.e., to prevent Watson-Crick base pairing.[224-225] These results are consistent with the hypothesis that these kinds of modifications at position 37 promote a single-stranded structure of the anticodon loop and stabilize the codon-anticodon interaction, probably by stacking. Furthermore, the presence of $i^6A$ is not essential for all tRNA species and its function is sensitive to the reading context, the particular tRNA, and the conformation of the anticodon.

Transfer RNA reading codons starting with G or C usually have unmodified purine or a simple modification (like $m^1G$, $m^2A$, etc.) present at position 37 (Figure 2). The energetically more stable GC pairs should have a lower demand for stabilization of the anticodon-codon interaction. At present no direct experimental evidence exists for the function of these kinds of modified nucleosides. However, some of them, like $m^1G$ and $m^6A$, should prevent normal Watson-Crick base pairing. Thus, the function might be to prevent hydrogen bonding to a nucleotide on the 5′-side of the codon in the mRNA, to a nucleotide in the 16S rRNA or to a ribosomal protein which might be close to the anticodon. If so, such modification would be to *prevent* too strong binding of the tRNA to the mRNA-ribosome complex. This may contribute to an equal "resting time" for tRNAs on the ribosome, resulting in a smooth translation. If the interaction is at the mRNA level, mutants defective in these kinds of modifications should show strong context effects during translation. Only one mutant *(trmD)* defective in such modification has been characterized.[65] This mutant is deficient in $m^1G$ at a high temperature (43.5°) However, the level of $m^1G$ in the tRNA is only reduced to 40% of the normal level. Still, the growth rate of the mutant is reduced.[229a] This would suggest that the $m^1G$ represents an essential modification — and in fact no unmodified G is present in the tRNAs so far sequenced.[37] Even tRNAs from a *Mycoplasma* strain, which contains few modified nucleosides, contain $m^1G$ at an even higher level than found in tRNA from *E. coli.*[24] It seems, therefore, that it is highly unfavorable for tRNA and thus for protein synthesis to have an unmodified G at position 37 of the tRNA.

*2. Function of Pseudouridine in the Anticodon Stem*

Mutation in the *hisT* gene results in a ψ deficiency in the anticodon stem (positions 38 and 39) in many tRNA species, among them $tRNA^{His}$.[66,101-103] Such mutants were isolated by their ability to derepress the histidine operon.[101] Recently it has been revealed that $tRNA^{His}$ lacking ψ reads the seven histidine codons present in a row in the histidine leader mRNA inefficiently.[226] This very special reading context thus requires a modified $tRNA^{His}$. However, the reduction in growth rate by a *hisT* mutation is moderate, which suggest that in most reading contexts the ψ-deficient tRNA is able to function fairly well.[227-228] Also $tRNA^{Gln}$ in a *hisT* mutant lacks ψ in the anticodon stem. Bossi and Roth[229] could show that a certain *hisD* amber mutation, *hisD6404,* is well suppressed by the *supE* suppressor($tRNA^{Gln}_{am}$) mutation. However, in a *hisT* background, this *hisD6404* mutation is not suppressed at all and the bacterial strain is phenotypically $His^-$ (Table 9). In an elegant approach they next selected for $His^+$ mutants and could show that some of them were in fact suppressed by a second mutation next to the primary amber (UAG) codon in the mRNA (Table 9). This second mutation involved always a change from a C to an A at the 3′-side of the UAG codon. This change in the reading context in the mRNA makes the defective $tRNA^{Gln}_{am}$ able to read this UAG codon despite the lack of ψ38 and ψ39. However, the ψ-deficiency as such does not increase the sensitivity to reading context for the $tRNA^{Gln}_{am}$, since the efficiency of the *supE* even when tRNA is fully modified depends on codon context (Table 9). Therefore, the lack of ψ reduces the efficiency of the $tRNA^{Gln}_{am}$ at both these sites. If this is true at all codon contexts the function of ψ in the anticodon stem may be to adjust the anticodon to an optimal conformation for an efficient reading.

*3. Function of Modified Nucleosides in Position 34 (Wobble Position)*

Figure 2 shows that there is a large variety of modified nucleosides at position 34. Some of these, $mo^5U$, $cmo^5U$ and I (Table 3) cause an increased wobble capacity. These modified nucleosides are only found in tRNAs reading a family of four codons for one and the same amino acid. Thus, $mo^5U$ and $cmo^5U$ are able to read not only codons ending with A or G, according to the wobble hypothesis, but also codons ending with U (Table 3).[39-40,230] However, they are less able to read codons ending with C. It has been suggested that these kinds

**Table 9**
**INFLUENCE OF CODON CONTEXT AND *hisT* ON EFFICIENCY OF SUPPRESSION**

| Relevant genetic background | HisD activity resulting from suppression of indicated site | | |
|---|---|---|---|
| | wild types (–CAGC–) | *hisD6404* (–UAGC–) | *hisD6404-Cl* (–UAGA–) |
| wild type | 107.3 ($His^+$) | <0.2 ($His^-$) | <0.2 ($His^-$) |
| *supE20* | 100.0 ($His^+$) | 2.1 ($His^+$) | 21.2 ($His^+$) |
| *supE20, hisT1504* | 93.3 ($His^+$) | <0.2 ($His^-$) | 1.0 ($His^+$) |

*Note:* The value of the specific activity in strain (*hisD*$^+$, supE20) was taken as 100; all other activities are expressed relative to this value. The His phenotype of the different strains are indicated within parentheses. (Data from Bossi and Roth.[232])

of modification of uridine do not interact with the anticodon loop backbone but shift the keto-enol tautomerism of these modified uridines.[231-233] This may result in a less specific hydrogen bonding which would explain their extended wobble capacities. However, it does not explain why these modifications are present, since all these codon families are also read by at least one other tRNA species which reads codons ending with U/C, according to the wobble hypothesis. In the valine codon family the major $tRNA_1^{Val}$ contains cmo$^5$U in position 34 and reads GUG, GUA, GUU, and to some extent also GUC.[40,230] The minor $tRNA_{2A+B}^{Val}$, which contain a G in position 34, reads GUU and GUC according to the wobble hypothesis. Therefore it is not obvious why cmo$^5$U is present in $tRNA_1^{Val}$, if the function of cmo$^5$U is only to extend the wobble. However, $Aro^-$ mutants which lack cmo$^5$U in their tRNAs grow at a 20% reduced growth rate in rich medium.[233a] This large growth rate difference would indicate that the presence of cmo$^5$U is important. The GUU codon is used almost as much as GUA and GUG together for proteins, such as ribosomal proteins, made in large amounts.[234] Therefore, the presence of cmo$^5$U might be important in certain reading contexts like the one present in mRNA for ribosomal proteins. This wobble base is within four Å distance from a pyrimidine in 16S rRNA when the tRNA is in the P-site on the ribosome.[235-238] This may extend the anticodon stack into the 16S rRNA and stabilize the tRNA ribosome interaction. The importance of the modification as such in these interactions has not been elucidated.

A recessive UGA suppressor *(supK)* was isolated which also suppresses some frameshift mutations.[97,239] The *supK* mutants are deficient in a tRNA methyltransferase which most likely catalyzes the formation of mcmo$^5$U in some tRNA species[98] (see Figure 6). These results suggest that a tRNA having cmo$^5$U instead of mcmo$^5$U is able to cause suppressions of UGA or framshift mutations. However, a tRNA species has so far not been identified as the suppressing agent The mechanism for this suppression is still unknown.

Other substitutions in the 5-position of the uridine, like the mcm- and mnm- restrict the wobble interaction of U.[240-241] Such uridine derivatives, which often also contain a sulfur in the 2-position of U, bind to codons ending with A but not to those ending with G. The presence of a thio group in the 2-position of U is not involved in this restriction, but has been implicated in increasing the stacking interaction within the anticodon.[242-244] These bulky 5-substitutes of U bind to the anticodon backbone and for geometric reasons restrict the binding to G.[245-247] The possible presence of these types of modified nucleosides in eukaryotic UAA specific suppressors would explain the inability of such suppressors to read UAG codons.[248] Waldron et al.[249] have shown that the yeast UAA specific *suq5-01* suppressor is changed in the anticodon from –U*GA–to–T*UA–. The U* and T* were suggested to be a $mcm^5U$ and $mcm^5s^2U$, respectively. This suppressor reads only UAA and not UAG; the presence of these modified nucleosides would explain the restricted wobble. One temperature-sensitive mutant of *E. coli* has been isolated which is deficient in $mnm^5s^2U$ normally present in $tRNA_2^{Glu}$.[192] If this is the only genetic lesion in this mutant, its temperature-sensitive effect would suggest that this nucleoside may be essential for cell growth. Therefore a more profound function of $mnm^5s^2U$ than only restriction of the wobble against G should be suspected. The presence of $mnm^5s^2U$ in the tRNAs reading split codon families might prevent a low level of misreading of codons ending with C or U (Table 4). Other *E. coli* mutants, *trmC*, are lacking $mnm^5s^2U$.[65,96] Most probably only one methyl group is absent, which results in the presence of $nm^5s^2U$ instead of $mnm^5s^2U$. The *trmC* mutants are viable, but it is not known whether or not the lack of the methyl group influences the wobble characteristics or misreading within the non codon families.

Elongator $tRNA_m^{Met}$ of *E. coli* contains $ac^4C$ in position 34 while the initiator $tRNA_f^{Met}$ contains only C.[37] Incubation of $tRNA_m^{Met}$ with bisulfite specifically removes the acetyl group, leaving an unmodified C in the wobble position.[250] The presence of $ac^4C$ has no effect on the rate or yield of amino-acylation of $tRNA_m^{Met}$ in vitro, while the $tRNA_m^{Met}$ lacking $ac^4C$ bind to ribosome almost twice as well as the $ac^4C$ containing $tRNA_m^{Met}$.[250] However, the presence of $ac^4C$ decreases the misreading in vivo of AUA, which normally codes for isoleucine. Thus, the primary function of $ac^4C$ appears to be to reduce the misreading of the AUA isoleucine codon. This is achieved by a somewhat reduced efficiency in reading the AUG methionine codon.[250]

The hypermodified nucleoside Q is found in the wobble position of tRNAs specific for tyrosine, histidine, asparagine, and aspartic acid from all organisms so far analyzed (Table 4).[37] Transfer RNAs containing Q or its derivatives all recognize codons $NA_C^U$.[251] The wobble characteristics of Q are not significantly different from those of G.[251] However, measured as anticodon-anticodon binding in tRNA dimer formation, the Q-U wobble pair was about 3 times stronger than the G-U wobble pair.[206] Mutant *trt* lacking Q dies more rapidly in stationary phase.[251a] Transfer $RNA^{Tyr}$ from *Drosophila melanogaster* containing G in position 34 reads a terminator codon (most probably UAG), while Q containing $tRNA^{Tyr}$ does not.[185] It has been suggested that partial suppression for termination codons might be a normal regulatory device.[182-184] Thus, the modification of G to Q may be involved in the regulation of gene expression. The content of Q in tRNA changes during development of *Drosophila*[178] and Q is completely lacking in tumor cells,[176-177] which might be a reflection of a translational regulatory mechanism.

*4. Function of Modified Nucleoside in Other Places Than in the Anticodon Region*

The $m_2^2G$ is found in position 26 in about 40% of the tRNA species in yeast. A yeast mutant *(trm1)* deficient in $m_2^2G$ shows a slight reduction in growth rate.[252] Furthermore, the in vitro charging capacity of $tRNA^{Ser}$ is reduced 20%, suggesting that some tRNA species are in a nonchargeable conformation.[252] The results suggest that the presence of $m_2^2G$ in the tRNA might stabilize the three-dimensional structures of tRNA. Transfer RNA, deficient in D supports synthesis of polyphenylalanine in vitro as well as tRNA containing this modified nucleosides.[111] Transfer $RNA^{Phe}$ contains $acp^3U$ in position 47. Chemical derivatization of $acp^3U$ did not influence the tRNA in aminoacylation or polyphenylalanine synthesis in vitro.[253] Specific chemical reduction of $m^7G46$ disrupts the C13-G22-$m^7G46$ base triple which leads to a slightly less ordered tRNA structure.[254] Furthermore the finding that a mutant, *trmB*, defective in $m^7G$ formation, might also grow slower than a *trmB*$^+$ cell further supports the idea that $m^7G$ stabilizes the structure of tRNA.[96] Heterologous in vitro methylation producing $m^7G$ or $m^2G$ results in an altered kinetics of aminoacylation.[216,255] A mutant completely lacking $s^4U$ in its tRNA shows identical growth characteristics to the wild type cell.[264-265] However, $s^4U$ is involved in the photoprotection phenomenon. Thus, modification in other part of the tRNA seems to be involved in the stabilization of tRNA ($m^7G$, $m_2^2G$), but perhaps most of them may participate in specific interaction with different macromolecules. Therefore much more specific assays might be necessary to reveal the function of these modified nucleoside.

Ribothymidine ($m^5U$) was originally thought to be present in all tRNA species. However, bulk tRNA from wheat germ is a specific substrate for tRNA($m^5U$)methyltransferase from yeast.[259] In fact, wheat germ contains a subset of tRNA species completely lacking $m^5U$ in its tRNA.[260-261] Since $m^5U$ is one of the most abundant modified nucleosides and is present in tRNA from most organisms, it was thought to be essential. However, *E. coli* mutant *(trmA5),* completely lacking $m^5U$ in its tRNA, is viable but is outgrown by wild type cells in a mixed population experiment.[262] The difference in growth rate between wild type and *trmA5* mutant is 4%.[262a] A yeast mutant completely lacking $m^5U$ in its tRNA grows normally.[112] Furthermore, some species of microorganisms normally lack $m^5U$ in the tRNA.[263-265] Lack of $m^5U$ facilitates initiation of protein synthesis in *Streptococcus faecalis* with unformylated initiator $tRNA_f^{Met}$ and a similar mechanism may be operating in *E. coli.*[266-268] All these results obtained in vivo show that $m^5U$ is not essential for cell growth but may enhance slightly tRNA function. Experiments in vitro using eukaryotic tRNAs lacking $m^5U$ demonstrated that depending on the class of tRNA used the presence of $m^5U$ may either increase or decrease the activity of the tRNA in protein synthesis.[174,269] Thus, the level of $m^5U$ in some eukaryotic tRNAs may regulate protein synthesis.[174]

The content of $m^5U$ seems to influence the elongation factor directed A-site binding.[270] Kersten and collaborators[270] showed that there is misincorporation of leucine during polyU-directed polyphenylalanine synthesis. It was further shown that the lack of $m^5U$ augmented the intrinsic misreading capacity of $tRNA_4^{Leu}$ and the lack of $m^5U$ does not induce some other $tRNA^{Leu}$ to misread.[38] The stability of the tertiary structure of $tRNA^{Met}$ is increased when $m^5U54$ is present as compared to a molecule which contains only U54.[271] This stability is further increased upon introduction of $s^2m^5U54$, which is normally present in *Hermus thermophilus,* and this may partially explain the extreme heat stability of tRNA from this organism. Therefore, $m^5U$ and $m^5s^2U$ stabilize the tRNA structure.[271] The functional differences demonstrated in vitro in the presence or absence of $m^5U$ may explain the small but significant growth rate differences of the *E. coli trmA* mutant obtained in vivo.

## VII. OUTLOOK

Several new techniques, such as introduction of transposon in bacterial genetics, cloning of specific DNA fragments, methods to sequence both DNA and RNA, and more sensitive in vitro systems for transcription and translation, have been introduced during recent years. This has made it possible to better analyze the formation and function of modified nucleosides in RNA. The importance of well-defined mutants defective in RNA modification will even be more apparent when these new techniques are more extensively applied to different aspects of RNA modification. The formation of modified nucleosides is complex and is interlinked on different levels with several aspects of cell metabolism. The function of the modified nucleosides seems to be primarily in fine tuning of the function of tRNA and rRNA. However, this does not rule out specific functions for some of the modified nucleosides, especially those found in a few tRNA species, to interact with certain proteins or RNA species. Thus a more thorough analysis of existing RNA modification mutants and characterization of new mutants may reveal more versatile functions of modified nucleosides and also contribute to a better understanding of the function of rRNA and tRNA.

## ACKNOWLEDGMENTS

This work was supported by the Swedish Cancer Society (project no. 680), the Swedish Natural Science Research Council (project B-Bu 2930-102), and the Swedish board for technical development. The stimulating discussion and the critical reading of the manuscript by my colleagues A. Byström (Umeå), P. Gustafsson (Umeå), T. Hagervall (Umeå), K. Hjalmarsson (Umeå), L. Isaksson (Uppsala), K. Kjellin-Stråby, and P. Lindström (Umeå), I. Tittawella (Umeå), and M. Wikström (Umeå) are greatfully acknowledged. I also wish to thank K. Lönneborg and M. Rodriguez for their patience and expert help in typing the manuscript.


## REFERENCES


1. **Brownlee, G. G., Sanger, F., and Barrel, B. G.,** The sequence of 5S ribosomal ribonucleic acid, *J. Mol. Biol.*, 34, 379, 1968.
2. **Moore, P. B.,** Methylation of messenger RNA in *Escherichia coli, J. Mol. Biol.*, 18, 38, 1966.
3. **Nishimura, S.,** Structures of modified nucleosides found in tRNA, in *Transfer RNA: Structure, Properties and Recognition*, Schimmel, P. R., Söll, D. and Abelson, J. N., Eds., Cold Spring Harbor Laboratory, Cold Spring Harbor, New York, 1979, 547.
4. **Borek, E., and Srinivasau, P. R.,** The methylation of nucleic acids, *Annu. Rev. Biochem.*, 35, 275, 1966.
5. **Starr, J. L. and Sells, B. H.,** Methylated ribonucleic acids *Physiol. Rev.*, 49, 623, 1969.
6. **Hall, R. H.,** *The modified nucleosides in Nucleic Acid*, Columbia University Press, New York, 1971.
7. **Nishimura, S.,** Minor components in transfer RNA: Their characterization, location, and function, *Prog. Nucleic Acid Res. Mol. Biol.*, 12, 49, 1972.
8. **Kerr, S. J. and Borek, E.,** The tRNA methyltransferases, *Adv. Enzymol.*, 36, 1, 1972.
9. **Söll, D.,** Enzymatic modification of transfer RNA: Modified nucleosides form at the polynucleotide level, but their function is not established, *Science*, 173, 293, 1973.
10. **Schäfer, K. P. and Söll, D.,** New aspects in tRNA biosynthesis, *Biochimie*, 56, 795, 1974.
11. **Nau, F.,** The methylation of tRNA, *Biochimie*, 58, 629, 1976.
12. **Agris, P. F. and Söll, D.,** The modified nucleosides in transfer RNA, in *Nucleic Acid and Protein Recognition*, Vogel, H., Ed., Academic Press, New York, 1977, 321.
13. **Feldman, M. Ya,** Minor components in transfer RNA. The location-function relationships, *Prog. Biophys. Mol. Biol.*, 32, 83, 1977.
14. **McCloskey, J. A. and Nishimura, S.,** Modified nucleosides in transfer RNA, *Acc. Chem. Res.*, 10, 403, 1977.
15. **Nishimura, S.,** Modified nucleosides and isoaccepting tRNA, in *Transfer RNA*, Altman, S., Ed., Cambridge, MIT Press, 1978, 168.
16. **Nishimura, S.,** Modified nucleosides in tRNA, in *Transfer RNA: Structure, Properties, and Recognition*, Schimmel, P. R., Söll, D., and Abelson, J. N., Eds., Cold Spring Harbor Laboratory, Cold Spring Harbor, New York, 1979, 59.
17. **Klootwijk, J. and Planta, R. J.,** Analysis of the methylation sites in yeast ribosomal RNA, *Eur. J. Biochem.*, 39, 325, 1973.
18. **Maden, B. E. H. and Salim, M.,** The methylated nucleotide sequences in HeLa cell ribosomal RNA and its precursors, *J. Mol. Biol.*, 88, 133, 1974.
19. **Carbon, P., Ehresmann, C., Ehresmann, B., and Ebel, J. P.,** The sequence of *Escherichia coli* ribosomal 16SRNA determined by new rapid gel methods, *FEBS Lett.*, 94, 152, 1978.
20. **Khan, M. S. N., Salim, M., and Maden, B. E. H.,** Extensive homologies between the methylated nucleotide sequences in several vertebrate ribosomal ribonucleic acids, *Biochem. J.*, 169, 531, 1978.
21. **Thomas, G., Gordon, J., and Rogg, H.,** $N^4$-acetylcytidine. A previously unidentified labile component of the small subunit of eukaryotic ribosomes, *J. Biol. Chem.*, 253, 1101, 1978.
22. **Cecchini, J.-P. and Miassod, R.,** Studies on the methylation of cytoplasmic ribosomal RNA from cultured higher plant cells, *Eur. J. Biochem.*, 98, 203, 1979.
23. **Brand, R. C., Klootwijk, J., Sibum, C. P., and Planta, R. J.,** Pseudouridylation of yeast ribosomal precursor RNA, *Nucl. Acids Res.*, 7, 121, 1979.

24. **Hsuchen, C.-C. and Dubin, D. T.**, Methylation patterns of *Mycoplasma* transfer and ribosomal ribonucleic acid, *J. Bacteriol.*, 144, 991, 1980.
25. **Branlant, C., Krol, A., Machatt, M. A., Ponyet, J., and Ebel, J.-P.**, Primary and secondary structures of *Escherichia coli* MRE600 23S ribosomal RNA. Comparison with models of secondary structure for maize chloroplast 23SrRNA and large portions of mouse and human 16S mitochondrial rRNAs, *Nucl. Acids Res.*, 9, 4303, 1981.
26. **Maden, B. E. H. and Reeder, R. H.**, Partial mapping of methylated sequences in *Xenopus leavis* ribosomal RNA by preparative hybridization to cloned fragments of ribosomal DNA, *Nucleic Acids Res.*, 6, 817, 1979.
27. **Brand, R. C. and Gerbi, S. A.**, Fine structure of ribosomal RNA II. Distribution of methylated sequences within *Xenopus leavis* rRNA, *Nucleic Acids Res.*, 7, 1497, 1979.
28. **Maden, B. E. H.**, Methylation map of *Xenopus leavis* ribosomal RNA, *Nature*, 288, 293, 1980.
29. **Grouse, R. L. and Gerbi, S. A.**, Fine structure of ribosomal RNA III. Location of evolutionarily conserved regions within ribosomal DNA, *J. Mol. Biol.*, 140, 321, 1980.
30. **Khan, M. S. N. and Maden B. E. H.**, Nucleotide sequences within the ribosomal ribonucleic acids of HeLa cells, *Xenopus leavis* and chick embryo fibroblasts, *J. Mol. Biol.*, 101, 235, 1976.
31. **Alberty, H., Raba, M., and Gross, H. J.**, Isolation from rat liver and sequence of a RNA fragment containing 32 nucleotides from position 5 to 36 from the 3′ end of ribosomal 18S RNA, *Nucleic Acids Res.*, 5, 425, 1978.
32. **Choi, Y. C. and Busch, H.**, Modified nucleotides in $T_1$RNase oligonucleotides of 18S ribosomal RNA of the Novikoff Hepatoma, *Biochemistry*, 17, 2551, 1978.
33. **Azad, A. A. and Deacon, N. J.**, The 3′-terminal primary structure of five eukaryotic 18S RNAs determined by the direct chemical method of sequencing. The highly conserved sequences include an invariant region complementary to eukaryotic 5S rRNA, *Nucleic Acids Res.*, 8, 4365, 1980.
34. **Van Charldrop, R. and Van Knippenberg, P. H.**, Sequence modified nucleotides and secondary structure at the 3′-end of small ribosomal subunit RNA, *Nucleic Acids Res.*, 10, 1149, 1982.
35. **Khan, M. S. N. and Maden, B. E. H.**, Conformation of methylated sequences in HeLa cell 18-S ribosomal RNA. Nuclease $S_1$ as a probe, *Eur. J. Biochem.*, 84, 241, 1978.
36. **Vass, J. K. and Maden, B. E. H.**, Studies on the conformation of the 3′ terminus of 18-SrRNA, *Eur. J. Biochem.*, 85, 241, 1978.
37. **Gauss, D. H. and Sprinzl, M.**, Compilation of tRNA sequences, *Nucleic Acids Res.*, 9, r1, 1981.
38. **Kersten, H.**, tRNA methylation: On the role of modified nucleosides in transfer RNA, in *Biochemistry of S-Adenosylmethionine and Related Compounds*, Usdin, E., Borchardt, R. T., and Creveling, C. R., Eds., McMillan Publishers, New York, 1982, 357.
39. **Ishikura, H., Yamada, Y., and Nishimura, S.**, Structure of serine tRNA from *Escherichia coli*. I. Purification of serine tRNAs with different codon response, *Biochim. Biophys. Acta*, 228, 471, 1971.
40. **Takemoto, T., Takeishi, K., Nishimura, S., and Ukita, T.**, Transfer of valine into rabbit haemoglobin from various isoaccepting species of valyl-tRNA differing in codon recognition, *Eur. J. Biochem.*, 38, 489, 1973.
41. **Albani, M., Schmidt, W., Kersten, H., Geibel, K., and Lüderwald, I.**, 5-methyloxyuridine, a new modified constituent in tRNAs of *Bacillaceae*, *FEBS Lett.*, 70, 37, 1976.
42. **Murao, K., Hasegawa, T., and Ishikura, H.**, 5-methoxyuridine: A new minor constituent located in the first position of the anticodon of $tRNA^{Ala}$, $tRNA^{Thr}$ and $tRNA^{Val}$ from *Bacillus subtilis*, *Nucleic Acids Res.*, 3, 2851, 1976.
43. **Crick, E. H. C.**, Codon-anticodon pairing: The wobble hypothesis, *J. Mol. Biol.*, 19, 548, 1966.
44. **Lagerkvist, U.**, "Two out of three": An alternative method for codon reading, *Proc. Natl. Acad. Sci. U.S.A.*, 75, 1759, 1978.
45. **Woese, C. R. and Fox, G. E.**, Phylogenetic structure of the prokaryotic domain: The primary kingdoms, *Proc. Natl. Acad. Sci. U.S.A.*, 74, 5088, 1977.
46. **Best, A. N.**, Composition and characterization of tRNA from *Methonococcus vanniellii*, *J. Bacteriol.*, 133, 240, 1978.
47. **Gupta, R. and Woese, C. R.**, Unusual modification patterns in the transfer ribonucleic acids of *Archaebacteria*, *Current Microbiology*, 4, 245, 1980.
48. **Pang, H., Ihara, M., Kuchino, Y., Nishimura, S., Gupta, R., Woese, C. R., and McCloskey, J. A.**, Structure of a modified nucleoside in archaebacterial tRNA which replaces ribosylthymine. 1-methylpseudouridine, *J. Biol. Chem.*, 257, 3589, 1982.
49. **Poldermans, B., Roza, L., and Van Knippenberg, P. H.**, Studies on the function of two adjacent $N^6$,$N^6$-dimethyladenosines near the 3′ end of 16S ribosomal RNA of *Escherichia coli*. III. Purification and properties of the methylating enzyme and methylase-30S interactions, *J. Biol. Chem.*, 254, 9094, 1979.
50. **Isaksson, L. A.**, Partial purification of ribosomal RNA($m^1G$)- and rRNA($m^2G$)methylase from *Escherichia coli* and demonstration of some proteins affecting their apparent activity, *Biochim. Biophys. Acta*, 312, 122, 1973.

51. **Sipe, J. E., Anderson, W. M., Remy, C. N., and Love, S. H.,** Characterization of *S*-adenosylmethionine: ribosomal ribonucleic acid-adenine($N^6$)methyltransferase of *Escherichia coli* strain B, *J. Bacteriol.,* 110, 81, 1972.
52. **Anderson, W. M., Jr., Remy, C. N., and Sipe, J. E.,** Ribosomal ribonucleic acid-adenine ($N^6$-)methylase of *Escherichia coli* strain B: Ionic and substrate site requirements, *J. Bacteriol.,* 114, 988, 1973.
53. **Shivakumar, A. G., Hahn, J., and Dubnau, D.,** Studies on the synthesis of plasmid-coded proteins and their control in *Bacillus subtilis* minicells, *Plasmid,* 2, 279, 1979.
54. **Gryczan, T. J., Grandi, G., Hahn, J., Grandi, R., and Dubnau, D.,** Conformational alteration of mRNA structure and the post-transcriptional regulation of erythromycin-induced drug resistance, *Nucleic Acids Res.,* 8, 6081, 1980.
55. **Dubnau, D., Grandi, G., Grandi, R., Gryczan, T. J., Hahn, J., Kozloff, Y., and Shivakumar, A. G.,** Regulation of plasmid-specified MLS-resistance in *B. subtilis* by conformation alteration of mRNA structure, in *Molecular Biology, Pathogenecity and Ecology of Bacterial Plasmids,* Levy, S. B., Clowes, R. C., and Koenig, E. L., Eds., Plenum Press, New York, 1981, 157.
56. **Hjalmarsson, K. J., Byström, A. S., and Björk, G. R.,** Purification and characterization of tRNA($m^1G$)methyltransferase from *Escherichia coli,* strain K12, *J. Biol. Chem.,* 258, 1343, 1983.
57. **Ny, T.,** Regulation and purification of tRNA($m^5U$)methyltransferase from *Escherichia coli,* Thesis, Umeå University, Sweden, 1979.
58. **Greenberg, R. and Dudock, B.,** Isolation and characterization of $m^5U$-methyltransferase from *Escherichi coli, J. Biol. Chem.,* 255, 8296, 1980.
59. **Okada, N. and Nishimura, S.,** Isolation and characterization of a guanine insertion enzyme, a specific tRNA transglycosylase from *Escherichia coli, J. Biol. Chem.,* 254, 3061, 1979.
60. **Pope, W. T. and Reeves, R. H.** Purification and characterization of a tRNA methylase from *Salmonella typhimurium, J. Bacteriol.,* 136, 191, 1978.
60a. **Arena, F., Ciliberto, G., Ciampi, S., and Cortese, R.,** Purification of pseudouridylate synthetase I from *Salmonella typhimurium, Nucleic Acid Res.,* 5, 4523, 1978.
61. **Delk, A. S., Nagle, D. P., Jr., and Rabinowitz, J. C.,** The methylene-tetrahydrofolate-dependent biosynthesis of ribothymidine in the transfer-RNA of *Streptococcus faecalis,* in *Development of Biochemistry,* Vol. 4, *Chemistry and Biology of Pteridines,* Kislink, R. L., and Brown, G. N., Eds., 1979, 389.
62. **Glick, J. M. and Leboy, P. S.,** Purification and properties of tRNA(adenine-1)methyltransferase from rat liver, *J. Biol. Chem.,* 252, 4790, 1977.
63. **Abrell, J. W., Kaufman, E. E., and Lipsett, M. N.,** The biosynthesis of 4-thiouridylate. Separation and purification of two enzymes in the transfer ribonucleic acid-sulfurtransferase system, *J. Biol. Chem.,* 246, 294, 1971.
64. **Björk, G. R. and Isaksson, L. A.,** Isolation of mutants of *Escherichia coli* lacking 5-methyluracil in transfer ribonucleic acid or 1-methylguanine in ribosomal RNA, *J. Mol. Biol.,* 51, 83, 1970.
65. **Björk, G. R., and Kjellin-Stråby, K.,** *Escherichia coli* mutants with defects in the biosynthesis of 5-methylaminomethyl-2-thiouridine or 1-methylguanosine in their tRNA, *J. Bacteriol.,* 133, 508, 1978.
66. **Singer, C. E., Smith, G. R., Cortese, R., and Ames, B. N.,** Mutant $tRNA^{His}$ ineffective in repression and lacking two pseudouridine modifications, *Nature (London) New Biol.,* 238, 76, 1972.
67. **Smolar, N., Hellman, U., and Svensson, I.,** Two transfer RNA (1-methylguanine)methylases from yeasts, *Nucleic Acids Res.,* 2, 993, 1975.
68. **Kraus, J. and Staehelin, M.,** $N^2$-Guanine specific transfer RNA methyltransferase I from rat liver and leukemic rat spleen, *Nucleic Acids Res.,* 1, 1455, 1974.
69. **Kraus, J. and Staehelin, M.,** $N^2$-Guanine specific transfer RNA methyltransferase II from rat liver, *Nucleic Acids Res.,* 1, 1479, 1974.
70. **Roberts, J. W. and Carbon, J.,** Nucleotide sequence studies of normal and genetically altered glycine-tRNAs from *Eschericia coli, Nature,* 250, 412, 1974.
71. **Prather, N. E., Murgola, E. J., and Mims, B. H.,** Primary structure of an unsual glycine tRNA UGA suppressor, *Nucleic Acids Res.,* 9, 6421, 1981.
72. **Carbon, J. and Fleck, E. W.,** Genetic alteration of structure and function in glycine transfer RNA of *Escherichia coli:* Mechanism of suppression of the tryptophan synthetase A78 mutation, *J. Mol. Biol.,* 85, 371, 1974.
73. **Kuchino, Y., Seno, T., and Nishimura, S.,** Fragmented *E. coli* methionine tRNA as methyl acceptor for rat liver tRNA methylase: Alteration of the site of methylation by conformational change of tRNA structure resulting from fragmentation, *Biochem. Biophys. Res. Comm.,* 43, 476, 1971.
74. **Shersneva, L. P., Venkstern, T. V., and Bayev, A. A.,** A study of tRNA methylases by the dissected molecule method, *FEBS Lett.,* 14, 279, 1971.
75. **Shersneva, L. P., Venkstern, T. V., and Bayev, A. A.,** A study of transfer RNA methylation, *Biochim. Biophys. Acta,* 294, 250, 1973.

76. **Kersten, H., Raettig, R., Weissenbach, J., and Dirheimer, G.,** Recognition of individual procaryotic and eucaryotic transfer-ribonucleic acids by *B. subtilis* adenine-1-methyl-transferase specific for the dihydrouridine loop, *Nucleic Acids Res.,* 5, 3033, 1978.
77. **McClain, W. H. and Seidman, J. G.,** Genetic perturbations that reveal tertiary conformation of tRNA precursor molecules, *Nature,* 257, 106, 1975.
78. **Gambaryan, A. S., Morozov, I. A., Venkstern, T. V., and Bayev, A. A.,** The mechanism of action of tRNA methylases studied with immobilized tRNAs, *Nucleic Acids Res.,* 6, 1001, 1979.
79. **Vickers, J. D., and Logan, D. M.,** Isopentenyl adenosine: Synthesis in *E. coli* tRNA, *Biophys. Soc. Abstr.,* 10, 166a, 1970.
80. **Farkas, W. R. and Singh, R. D.,** Guanylation of transfer ribonucleic acid by a cell-free lysate of rabbit reticulocytes, *J. Biol. Chem.,* 248, 7780, 1973.
81. **Okada, N., Harada, F., and Nishimura, S.,** Specific replacement of Q base in the anticodon of tRNA by guanine catalyzed by a cell-free extract of rabbit reticulocyte, *Nucleic Acids Res.,* 3, 2593, 1976.
82. **Okada, N., Noguchi, S., Kasai, H., Shindo-Okada, N., Ohgi, T., Goto, T., and Nishimura, S.,** Novel mechanisms of post-transcriptional modification of tRNA. Insertion of bases of Q precursors into tRNA by a specific tRNA transglycosylase reaction, *J. Biol. Chem.,* 254, 3067, 1979.
83. **Kersten H., Sandig, L., and Arnold, H. H.,** Tetrahydrofolate-dependent 5-methyluracil tRNA transferase activity in *B. subtilis, FEBS Lett.,* 55, 57, 1975.
84. **Delk, A. S. and Rabinowitz, J. C.** Biosynthesis of ribosylthymine in tRNA of *Streptococcus faecalis:* A folate-dependent methylation not involving *S*-adenosyl-L-methionine, *Proc. Natl. Acad. Sci. U.S.A.,* 72, 528, 1975.
85. **Arnold, H. H., Schmidt, W., Raettig, R., Sandig, L., Domdey, H., and Kersten, H.,** *S*-adenosylmethionine and tetrahydrofolate dependent methylation of tRNA in *Bacillus subtilis.* Incomplete methylations caused by trimethoprim, pactamycin, or chloramphenicol, *Arch. Biochem. Biophys.,* 176, 12, 1976.
86. **Isaksson, L. A.,** Formation *in vitro* of 1-methylguanine in 23-S RNA from *Escherichia coli.* The effects of spermidine and two proteins, *Biochim. Biophys. Acta,* 312, 134, 1973.
87. **Björk, G. R.,** Identification of bacteriophage T4-specific precursor tRNA by using a host mutant defective in the methylation of tRNA, *J. Virology,* 16, 741, 1975.
88. **Helser, T. L., Davis, J. E., and Dahlberg, J. E.,** Change in methylation of 16S ribosomal RNA associated with mutation to kasugamycin resistance in *Escherichia coli, Nature (London) New Biol.,* 233, 12, 1971.
89. **Helser, T. L., Davis, J. E., and Dahlberg, J. E.,** Mechanism of kasugamycin resistance in *Escherichia coli, Nature (London) New Biol.,* 235, 6, 1972.
90. **Marinus, M. G. and Morris, N. R.,** Isolation of deoxyribonucleic acid methylase mutants of *Eschericha coli* K-12, *J. Bacteriol.,* 114, 1143, 1973.
91. **Björk, G. R. and Kjellin-Stråby, K.,** General screening procedure for RNA modificationless mutants: Isolation of *Escherichia coli* strains with specific defects in RNA methylation, *J. Bacteriol.,* 133, 499, 1978.
92. **Ny, T. and Björk, G. R.,** Cloning and restriction mapping of the *trm*A gene coding for transfer ribonucleic acid (5-methyluridine)methyltransferase in *Escherichia coli* K-12, *J. Bacteriol.,* 142, 371, 1980.
93. **Byström, A. S. and Björk, G. R.,** Chromosomal location and cloning from the (*trmD*) gene responsible for the synthesis of tRNA-($m^1G$)methyltransferase in *Escherichia coli* K-12, *Mol. Gen. Genetics,* 188, 440, 1982.
94. **Hagervall, T. and Björk, G. R.,** unpublished results, 1982.
95. **Björk, G. R.,** Transductional mapping of gene *trm*A responsible for the production of 5-methyluridine in transfer ribonucleic acid of *Escherichia coli, J. Bacteriol.,* 124, 92, 1975.
96. **Marinus, M. G., Morris, N. R., Söll, D., and Kwong, T. C.,** Isolation and partial characterization of three *Escherichia coli* mutants with altered transfer ribonucleic acid methylases, *J. Bacteriol.,* 122, 257, 1975.
97. **Reeves, R. H., and Roth, J. R.,** A recessive UGA suppressor, *J. Mol. Biol.,* 56, 523, 1971.
98. **Reeves, R. H. and Roth, J. R.,** Transfer ribonucleic acid methylase deficiency found in UGA suppressor strains, *J. Bacteriol.,* 124, 332, 1975.
99. **Björk, G. R. and Olsén, A.,** A method for isolation of *Escherichia coli* mutants with aberrant RNA methylation using translocatable drug resistance elements, *Acta Chem. Scand.,* ser. B, 33, 591, 1979.
100. **Yanofsky, C. and Soll, L.,** Mutations affecting $tRNA^{Trp}$ and its charging and their effect on regulation of transcription termination at the attenuator of the tryptophan operon, *J. Mol. Biol.,* 113, 663, 1977.
101. **Roth, J. R., Anton, D. N., and Hartman, P.,** Histidine regultory mutants in *Salmonella typhimurium.* I. Isolation and general properties, *J. Mol. Biol.,* 22, 305, 1966.
102. **Turnbough, C. L., Jr., Neill, R. J., Landsberg, R., and Ames, B. N.,** Pseudouridylation of tRNAs and its role in regulation in *Salmonella typhimurium, J. Biol. Chem.,* 254, 5111, 1979.
103. **Cortese, R., Landsberg, R., von der Haar, R. A., Umbarger, H. E., and Ames, B. N.,** Pleitropy of *his*T mutants blocked in pseudouridine synthesis in tRNA; Leucine and isoleucine-valine operons, *Proc. Natl. Acad. Sci. U.S.A.,* 71, 1857, 1974.

104. **Eisenberg, S. P., Yarus, M., and Soll, L.,** The effect of an *Escherichia coli* regulatory mutation on transfer RNA structure, *J. Mol. Biol.*, 135, 111, 1979.
105. **Vold, B. S., Lazar, J. M. and Gray, A. M.,** Characterization of a deficiency of $N^6$-($\Delta^2$-isopentenyl)-2-methyltioadenosine in the *Escherichia coli* mutant *trp*X by use of antibodies to $N^6$-($\Delta^2$-isopentenyl)-adenosine, *J. Biol. Chem.*, 254, 7362, 1979.
105a. **Nishimura, S.,** personal communication.
106. **Jagger, J.,** Growth delay and photoprotection induced by near-ultraviolet light, *Res. Prog. Org.-Biol. Med. Chem.*, 21, 383, 1972.
107. **Lipsett, M. N.** Enzymes producing 4-thiouridine in *Escherichia coli* tRNA: approximate chromosomal locations of the genes and enzyme activities in a 4-thiouridine-deficient mutant, *J. Bacteriol.*, 135, 993, 1978.
108. **Phillips, J. H. and Kjellin-Stråby, K.,** Studies on microbial ribonucleic acid IV. Two mutants of *Saccharomyces cerevisiae* lacking $N^2$-dimethylguanine in soluble ribonucleic acid,*J. Mol. Biol.*, 26, 509, 1967.
109. **Laten, H., Gorman, J., and Bock, R. M.,** Isopentenyladenosine deficient tRNA from an antisuppressor mutant of *Saccharomyces cerevisiae, Nucleic Acids Res.*, 5, 4329, 1978.
110. **Janner, F., Vögeli, G., and Fluri, R.,** The antisuppressor strain *sin1* of *Schizosaccharomyces pombe* lacks the modification isopentenyladenosine in transfer RNA, *J. Mol. Biol.*, 139, 207, 1980.
111. **Lo, R. Y. C., Bell, J. B., and Roy, K. L.,** Dihydrouridine-deficient tRNAs in *Saccharomyces cerevisiae, Nucleic Acids Res.*, 10, 889, 1982.
112. **Hopper, A. K., Furukawa, A. H., Pham, H. D., and Martin, N. C.,** Defects in modification of cytoplasmic and mitochondrial transfer RNAs are caused by single nuclear mutations, *Cell*, 28, 543, 1982.
113. **Gegenheimer, P. and Apirion, D.,** Processing of procaryotic ribonucleic acid, *Microbiol Rev.*, 45, 502, 1981.
114. **Abelson, J.,** RNA processing and the intervening sequence problem, *Annu. Rev. Biochem.*, 48, 1035, 1979.
115. **Perry, R. P.,** Processing of RNA, *Annu. Rev. Biochem.*, 45, 605, 1976.
116. **Hadjiolov, A. A. and Nikolaev, N.,** Maturation of ribosomal ribonucleic acids and the biogenesis of ribosomes, *Prog. Biophys. Molec. Biol.*, 31, 95, 1976.
117. **Planta, R. J., Klootwijk, J., Rané, H. A., Brand, R. C., Veldman, G. M., Stiekema, W. J., and de Jonge, P.,** Transcription and processing of ribosomal RNA in yeast and bacilli, in *RNA Polymerase, tRNA and Ribosomes*, Osawa, S., Ozeki, H., Uchida, H., and Yura, T., Eds., 1979, 451.
118. **Brand, R. C., Klootwijk, J., Van Steenbergen, T. J. M., de Kok, A. J., and Planta, R. J.,** Secondary methylation of yeast ribosomal precursor RNA, *Eur. J. Biochem.*, 75, 311, 1977.
119. **Klootwijk, J., van den Bos, R. C., and Planta, R. J.,** Secondary methylation of yeast ribosomal RNA, *FEBS Lett.*, 27, 102, 1972.
120. **Zimmerman, E. F.,** Secondary methylation of ribosomal ribonucleic acid in HeLa cells, *Biochemistry*, 7, 3156, 1968.
121. **Salim, M. and Maden, B. E. H.,** Early and late methylation in HeLa cell ribosome maturation, *Nature*, 244, 334, 1973.
122. **Lowry, C. V. and Dahlberg, J. E.,** Structural differences between the 16S ribosomal RNA of *E. coli* and its precursor, *Nature (London) New Biol.*, 232, 52, 1972.
123. **Feunteun, J., Rosset, R., Ehresmann, C., Stiegler, P., and Fellner, P.,** Abnormal maturation of precursor 16S RNA in a ribosomal assembly defective mutant of *E. coli, Nucleic Acids Res.*, 1, 141, 1974.
124. **Levis, R. and Penman, S.,** Processing steps and methylation in the formation of the ribosomal RNA of cultured *Drosophila* cells, *J. Mol. Biol.*, 121, 219, 1978.
125. **Maden, B. E. H., Salim, M., and Robertson, J. S.,** Progress in the structural analysis of mammalian 45S and ribosomal RNA, in *Ribosomes*, Nomura, M., Tissieres, A., and Lengyel, P., Eds., Cold Spring Harbor Laboratory, Cold Spring Harbor, New York, 1974, 829.
126. **Lai, C.-J. and Weisblum, B.,** Altered methylation of ribosomal RNA in an erythromycin-resistant strain of *Staphylococcus aureus, Proc. Natl. Acad. Sci. U.S.A.*, 68, 856, 1971.
127. **Shivakumar, A. G., Hahn, J., Grandi, G., Kozlov, Y., and Dubnau, D.,** Posttranscriptional regulation of an erythromycin resistance protein specified by plasmid pE194, *Proc. Natl. Acad. Sci. U.S.A.*, 77, 3903, 1980.
128. **Horinouchi, S., and Weisblum, B.,** Posttranscriptional modification of mRNA conformation: Mechanism that regulates erythromycin-induced resistance, *Proc. Natl. Acad. Sci., U.S.A.*, 77, 7079, 1980.
129. **Mazzara, G. P. and McClain, W. H.,** tRNA synthesis, in *Transfer RNA: Biological Aspects*, Söll, D., Abelson, J. N., and Schimmel, P. R., Cold Spring Harbor Laboratory, Cold Spring Harbor, New York, 1980.
130. **Sakano, H., Shimura, Y., and Ozeki, H.,** Selective modification of nucleosides of tRNA precursors accumulated in a temperature-sensitive mutant of *Escherichia coli, FEBS Lett.*, 48, 117, 1974.

131. **Shimura, Y., Sakano, H., Kubokawa, S., Nagawa, F., and Ozeki, H.**, tRNA precursors in RNaseP mutants, in *Transfer RNA: Biological Aspects*, Söll, D., Abelson, J. N., and Schimmel, P. R., Eds., Cold Spring Harbor Laboratory, Cold Spring Harbor, New York, 1980, 43.
132. **Davis, A. R. and Nierlich, D. P.**, The methylation of transfer RNA in *Escherichia coli*, *Biochim. Biophys. Acta*, 374, 23, 1974.
133. **Schaefer, K. P., Altman, S., and Söll, D.**, Nucleotide modification *in vitro* of the precursor of $tRNA^{Tyr}$ of *Escherichia coli*, *Proc. Natl. Acad. Sci. U.S.A.*, 70, 3626, 1973.
134. **Melton, D. A., De Robertis, E. M., and Cortese, R.**, Order and intracellular location of the events involved in the mutaration of a spliced tRNA, *Nature*, 284, 143, 1980.
135. **Ogden, R. C., Knapp, G., Peebles, C. L., Kang, H. S., Beckmann, J. S., Johnson, P. F., Fuhrman, S. A., and Abelson, J. N.**, Enzymatic removal of intervening sequences in the synthesis of yeast tRNAs, in *Transfer RNA: Biological Aspects*, Söll, D., Abelson, J. N., and Schimmel, P. R., Eds., Cold Spring Harbor Laboratory, Cold Spring Harbor, New York, 1980, 173.
136. **Valenzuela, P., O'Farrel, P. Z., Cordell, B., Maynard, T., Goodman, H. M. and Rutter, W. J.**, Yeast tRNA precursors: structure and removal of intervening sequences by an excision-ligase activity, in *Transfer RNA: Biological Aspects*, Söll, D., Abelson, J. N., and Schimmel, P. R., Eds., Cold Spring Harbor Laboratory, Cold Spring Harbor, New York, 1980, 191.
137. **Maaløe, O.**, Regulation of the protein-synthesizing machinery-ribosomes, tRNA, factors and so on, in *Biological Regulation and development*, Vol. 1, Gene Expression, Goldberg, R. F., Ed., Plenum Press, New York, 1979, 487.
138. **Nierlich, D. P.**, Regulation of bacterial growth, RNA, and protein synthesis, *Annu. Rev. Microbiol.*, 32, 393, 1978.
139. **Ny, T. and Björk, G. R.**, Stringent regulation of the synthesis of a transfer ribonucleic acid biosynthetic enzyme: Transfer ribonucleic acid ($m^5U$)methyltransferase from *Escherichia coli*, *J. Bacteriol.*, 130, 635, 1977.
140. **Ny, T. and Björk, G. R.**, Growth-rate-dependent regulation of transfer ribonucleic acid(5-methyluridine)methyltransferase in *Escherichia coli* B/r. *J. Bacteriol.*, 141, 67, 1980.
141. **Ny, T., Thomale, J., Hjalmarsson, K., Nass, G., and Björk, G. R.**, Non-coordinate regulation of enzyme involved in transfer RNA metabolism in *Escherichia coli*, *Biochim. Biophys. Acta.*, 607, 277, 1980.
142. **Lemaux, P. G., Herendeen, S. L., Bloch, P. L., and Neidhardt, F. C.**, Transient rates of synthesis of individual polypeptides in *E. coli* following temperature shifts, *Cell*, 13, 427, 1978.
143. **Agris, P. F., Koh, H., and Söll, D.**, The effect of growth temperatures on the *in vivo* ribose methylation of *Bacillus stearothermophilus* transfer RNA, *Arch. Biochem. Biophys.*, 154, 277, 1973.
144. **Watanabe, K., Shinma, M., Oshima, T., and Nishimura, S.**, Heat-induced stability of tRNA from an extreme thermophile, *Thermus thermophilus*, *Biochem. Biophys. Res. Comm.*, 72, 1137, 1976.
144a. **Lindström, P., and Björk, G. R.**, unpublished results.
145. **Lindström, P. and Stüber, D.**, unpublished results.
146. **Byström, A. S. and Björk, G. R.**, The structural gene (*trm*D) for the tRNA($m^1G$)methyltransferase in part of a four polypeptide operon in *Escherichia coli* K-12, *Mol. Gen. Genetics*, 188, 447, 1982.
147. **Byström, A. S., Hjalmarsson, K., Wikström, M., and Björk, G. R.**, The nucleotide sequence of an operon encoding the tRNA($m^1G$)-methyltransferase and two ribosomal proteins, S16 and L19, in *Escherichia coli* K-12, in Hjalmarsson, K. Ph.D. thesis, Umeå University, Umeå, Sweden, 1982.
148. **Chheda, G. B., Hong, C. I., Piskorz, C. F., and Harmon, G. A.**, Biosynthesis of *N*-(Purine-6-ylcarbamoyl)-L-threonine riboside. Incorporation of L-threonine *in vivo* into modified nucleoside of transfer ribonucleic acid, *Biochem. J.*, 127, 515, 1972.
149. **Powers, D. M. and Peterkofsky, A.**, Biosynthesis and specific labeling of *N*-(purin-6-ylcarbamoyl)threonine of *Escherichia coli* transfer RNA, *Biochem. Biophys. Res. Comm.*, 46, 831, 1972.
150. **Körner, A. and Söll, D**, *N*-(Purin-6-ylcarbamoyl)threonine: Biosynthesis *in vitro* in transfer RNA by an enzyme purified from *Escherichia coli*, *FEBS Lett.*, 39, 301, 1974.
151. **Elkins, B. N. and Kelter, E. B.**, The enzymatic synthesis of *N*-(Purin-6-ylcarbamoyl)threonine, an anticodon-adjacent base in transfer ribonucleic acid, *Biochemistry*, 13, 4622, 1974.
152. **Littauer, U. Z. and Inouye, H.**, Regulation of tRNA, *Annu. Rev. Biochem.*, 42, 439, 1973.
153. **Kitchingman, G. R. and Fournier, M. J.**, Unbalanced growth and the production of unique transfer ribonucleic acids in relaxed-control *Escherichia coli*, *J. Bacteriol.*, 124, 1382, 1975.
154. **Kitchingman, G. R. and Fournier, M. J.**, Modification-deficient transfer ribonucleic acids from relaxed control *Escherichia coli:* Structures of the major undermodified phenylalanine and leucine transfer RNAs produced during leucine starvation, *Biochemistry*, 16, 2213, 1977.
155. **Huang, P. C. and Mann, M. B.**, Comparative fingerprint and composition analysis of the three forms of $^{32}P$-labeled phenylalanine tRNA from chloramphenicol-treated *Escherichia coli*, *Biochemistry*, 13, 4704, 1974.

156. **Juares, H., Skjöld, A. C. and Hedgcoth, C.**, Precursor relationship of phenylalanine transfer riibonucleic acid from *Eschericia coli* treated with chloramphenicol or starved for iron, methionine, or cysteine, *J. Bacteriol.*, 121, 44, 1975.
157. **Thomale, J. and Nass, G.**, Alteration of the intracellular concentration of aminoacyl-tRNA synthetases and isoaccepting tRNAs during amino-acid-limited growth in *Escherichia coli, Eur. J. Biochem.*, 85, 407, 1978.
158. **Thomale, J. and Nass, G.**, Genetically determined differences in concentrations of isoaccepting tRNAs in *Escherichia coli Nucleic Acids Res.*, 4, 4313, 1977.
159. **Fayerman, J. T., Coker Vann, M., Williams, L. S., and Umbarger, H. E.**, *ilvU*, a locus in *Escherichia coli* affecting the derepression of isoleucyl-tRNA synthetase and the RPC-5 chromatographic profiles of $tRNA^{Ile}$ and $tRNA^{Val}$ *J. Biol. Chem.*, 254, 9429, 1979.
160. **Bullen, J. J., Rogers, H. J., and Griffiths, E.**, Role of iron in bacterial infection, *Curr. Top. Microbiol. Immun.*, 80, 1, 1978.
161. **Weinberg, E. D.**, Iron and infection *Microbiol. Rev.*, 42, 45, 1978.
162. **Wettstein, F. O. and Stent, G. S.**, Physiologically induced changes in the property of phenylalanine tRNA in *Escherichia coli, J. Mol. Biol.*, 38, 25, 1968.
163. **Rosenberg, A. H. and Gefter, M. L.**, An iron-dependent modification of several transfer RNA species in *Escherichia coli, J. Mol. Biol.*, 46, 581, 1969.
164. **Griffiths, E. and Humphreys, J.**, Alterations in tRNAs containing 2-methyl-$N^6$-($\Delta^2$-isopentenyl)adenosine during growth of enteropathogenic *Escherichia coli* in the presence of iron-binding proteins, *Eur. J. Biochem.*, 82, 503, 1978.
165. **Yoo, S. H., Pratt, M. L., and Shive, W.**, Evidence for a direct role of tRNA in an amino acid transport system, *J. Biol. Chem.*, 254, 1013, 1979.
166. **Yoo, S. H. and Shive, W.**, Evidence for a role of specific isoacceptor species of tRNA in amino acid transport, *Biochem. Biophys. Res. Comm.*, 88, 552, 1979.
167. **Buck, M. and Griffiths, E.**, Regulation of aromatic amino acid transport by tRNA: role of 2-methyl-$N^6$-($\Delta^2$-isopentenyl)-adenosine, *Nucleic Acids Res.*, 9, 401, 1981.
168. **Björk, G. R.**, A novel link between the biosynthesis of aromatic amino acids and transfer RNA modification in *Escherichia coli, J. Mol. Biol.*, 140, 391, 1980.
169. **Murao, K., Ishikura, H., Albani, M., and Kersten, H.**, On the biosynthesis of 5-methoxyuridine and uridine-5-oxyacetic acid in specific procaryotic transfer RNAs, *Nucleic Acids Res.*, 5, 1273, 1978.
170. **Palatnik, C. M., Katz, E. R., and Brenner, M.**, Isolation and characterization of transfer RNAs from *Dictyostelium discoideum* during growth and development, *J. Biol. Chem.*, 252, 694, 1977.
171. **Klee, H. J., DiPietro, D., Fournier, M. J., and Fischer, M. S.**, Characterization of transfer RNA from liver of the developing amphibian, *Rana catesbeiana, J. Biol. Chem.*, 253, 8074, 1978.
172. **Dingermann, T., Schmidt, W., and Kersten., H.**, Modified bases in tRNA of *Dictyostelium discoideum:* alterations in the ribothymidine content during development, *FEBS Lett.*, 80, 205, 1977.
173. **Dingermann, T., Pistel, F., and Kersten, H.**, Functional role of ribosylthymine in transfer RNA. Preferential utilization of tRNAs containing robosylthymine instead of uridine at position 54 in protein synthesis of *Dictyostelium discoideum, Eur. J. Biochem.*, 104 33, 1980.
174. **Roe, B. A. and Tsen, H-Y.**, Role of ribothymidine in mammalian $tRNA^{Phe}$. *Proc Natl. Acad. Sci. U.S.A.*, 74, 3696, 1977.
175. **Dingerman, T., Ogilvie, A., Pistel, F., Muhlhofer, W., and Kersten, H.**, Reduced aminoacylation of asparagine-transfer RNA early in the developmental cycle of *Dictyostelium discoideum:* Modification pattern and possible significance of the unchanged isoacceptor $tRNA^{Asn}{}_3$, *Hoppe-Seyler's Z. Physiol. Chem. Bd.*, 363, 763, 1981.
176. **Roe, B. A., Stankierwicz, A. F., Rizi, H. L., Weiss, C., DiLanro, M. N., Pike, D., Chen, C. Y. and Chen, E. Y.**, Comparison of rat liver and Walker 256 carcinosarcoma tRNAs, *Nucleic Acids Res.*, 6, 673, 1979.
177. **Okada, N., Noguchi, S., Nishimura, S., Ohgi, T., Goto, T., Crain, P. F., and McCloskey, J. A.**, Detection of unique tRNA species in tumor tissues by *Escherichia coli* guanine insertion enzyme, *Proc. Natl. Acad. Sci. U.S.A.*, 75, 4247, 1978.
178. **White, B. N., Tener, G. M., Holder, J., and Suzuki, D. T.**, Activity of a transfer RNA modifying enzyme during the development of *Drosophila* and its relationship to the su(s) locus, *J. Mol. Biol.*, 74, 635, 1973.
179. **Hosbach, H. A. and Kubli, E.**, Transfer RNA in aging *Drosophila.* II. Isoacceptor patterns, *Mech. Aging Dev.*, 10, 141, 1979.
180. **Owenby, R. K., Stulberg, M. P., and Jacobson, K. B.**, Alteration of the Q family of transfer RNAs in adult *Drosophila melanogaster* as a function of age, nutrition, and genotype, *Mech. Aging Dev.*, 11, 91, 1979.

181. **Wosnick, M. A., and White, B. N.,** A doubtful relationship between tyrosine tRNA and suppression of the vermilion mutant in *Drosophila, Nucleic Acids Res.*, 4, 3919, 1977.
182. **Philipson, L.,** Can suppressor tRNA control translation in mammalian cells?, in *Nonsense Mutations and tRNA Suppressors,* Celis, J. E. and Smith, J. D., Eds., Academic Press, New York, 1979, 313.
183. **Philipson, L., Andersson, P., Olshevsky, U., Weinberg, R., Baltimore, D., and Gesteland, R.,** Translation of MuLV and MSV RNAs in nuclease-treated reticulocyte extracts. Enhancement of the gag-pol polypeptide with yeast suppressor tRNA, *Cell,* 13, 189, 1978.
184. **Geller, A. I. and Rich, A.,** A UGA termination suppression $tRNA^{Trp}$ active in rabbit reticulocytes, *Nature,* 283, 41, 1980.
185. **Bienz, M. and Kubli, E.,** Wild-type $tRNA^{Tyr}_{G}$ reads the TMV RNA stop codon, but Q base-modified $tRNA^{Tyr}_{Q}$ does not, *Nature,* 294, 188, 1981.
186. **Singhal, R. P. and Vold, B.,** Changes in transfer ribonucleic acids of *Bacillus subtilis* during different growth phases, *Nucleic Acids Res.*, 3, 1249, 1976.
187. **Vold, BS:** Post-transcriptional modifications of the anti-codon loop region: Alterations in isoaccepting species of tRNA's during development in *Bacillus subtilis, J. Bacteriol.*, 135, 124, 1978.
188. **Keith, G., Rogg, H., Dirheimer, G., Menichi, B., and Heyman, T.,** Post-transcritpional modification of tyrosine tRNA as a function of growth in *Bacillus subtilis, FEBS Lett.*, 61, 120, 1976.
189. **Arnold, H. H., Raettig, R., and Keith, G.,** Isoaccepting phenylalanine tRNAs from *Bacillus subtilis* as a function of growth conditions. Differences in the content of modified nucleosides, *FEBS Lett.*, 73, 210, 1977.
190. **Rogg, H., Müller, P., Keith, G., and Staehelin, M.,** Chemical basis for brain-specific serine transfer RNAs, *Proc. Natl. Acad. Sci.*, 74, 4243, 1977.
191. **Kim, S. H.,** Crystal structure of yeast $tRNA^{Phe}$ and general structural features of other tRNAs, in *Transfer RNA: Structure, Properties and Recognition,* Schimmel, P. R., Söll, D., and Abelson, J. N., Cold Spring Harbor Laboratory, Cold Spring Harbor, New York, 1979, 83.
192. **Colby, D. S., Schedl, P., and Guthrie, G.,** A functional requirement for modification of the wobble nucleotide in the anticodon of a T4 suppressor tRNA, *Cell,* 9, 449, 1976.
193. **McLennan, B. D.,** Enzymatic demodification of transfer RNA species containing $N^6$-($\Delta^2$-isopentenyl)adenosine, *Biochem. Biophys. Res. Comm.*, 65, 345, 1975.
194. **Chelbi-Alix, M. K., Expert-Bezancou, A., Hayes, F., Alix, J.-H., and Branlant, C.,** Properties of ribosomes and ribosomal RNAs synthesized by *Escherichia coli* grown in the presence of ethionine. Normal maturation of ribosomal RNA in the absence of methylation, *Eur. J. Biochem.*, 115, 627, 1981.
195. **Vaughan, M. H., Jr., Soeiro, R., Warner, J. R., and Darnell, J. E.,** The effects of methionine deprevation on ribosome synthesis in HeLa cells, *Proc. Natl. Acad. Sci., U.S.A.*, 58, 1527, 1967.
196. **Igarashi, K., Kishida, K., Kashiwagi, K., Tatokoro, I., Kakegawa, T., and Hirose, S.,** Relationship between methylation of adenine near the 3′ end of 16-S ribosomal RNA and the activity of 30-S ribosomal subunits, *Eur. J. Biochem.*, 113, 587, 1981.
197. **Van Charldorp, R., Hens, H. A., and Van Knippenberg, P. H.,** Adenosine dimethylation of 16S ribosomal RNA: effect of the methylgroups on local conformational stability as deduced from electrophoretic mobility of RNA fragmnts in denaturing polyacrylamide gels, *Nucleic Acids Res.*, 9, 267, 1981.
198. **Shine, J. and Dalgarno, L.,** The 3′-terminal sequence of *Escherichia coli* 16S ribosomal RNA: Complementarity to nonsense triplets and ribosome binding sites. *Proc. Natl. Acad. Sci. U.S.A.*, 71, 1342, 1974.
199. **Okuyama, A., Machiyama, N., Kinoshita, T., and Tanaka, N.,** Inhibition by kasugamycin of initiation complex formation on 30S ribosomes, *Biochem. Biophys. Res. Commun.*, 43, 196, 1971.
200. **Okuyama, A. and Tanaka, N.,** Differential effects of aminoglycosides on cistron-specific initiation of protein synthesis, *Biochem. Biophys. Res. Commun.*, 49, 951, 1972.
201. **Polderman, B., Goosen, N., and Van Knippenberg, P. H.,** Studies on the function of two adjacent $N^6N^6$-dimethyladenosines near the 3′ end of 16S ribosomal RNA of *Escherichia coli.* I. The effect of kasugamycin on initiation of protein synthesis, *J. Biol. Chem.*, 254, 9085, 1979.
202. **Polderman, B., Van Buul, C. P. J. J., and Van Knippenberg, P. H.,** Studies on the function of two adjacent $N^6N^6$-dimethyladenosines near the 3′ end of 16S ribosomal RNA of *Escherichia coli.* II. The effect of the absence of the methyl groups on initiation of protein biosynthesis. *J. Biol. Chem.*, 254, 9090, 1979.
203. **Langhrea, M., Dondon, J., and Grunberg-Monago, M.,** The relationship between the 3′-end of 16S RNA and the binding of initiation factor IF-3 to the 30S subunit of *E. coli, FEBS Lett.*, 91, 265, 1978.
204. **Politz, S. M. and Glitz, D. G.,** Ribosome structure. Localization of $N^6$,$N^6$-dimethyladenosine by electron microscopy of a ribosome-antibody complex, *Proc. Natl. Acad. Sci. U.S.A.*, 74, 1468, 1977.
205. **Clundliffe, E.,** Mechanism of resistance to thiostrepton in the producing-organism *Streptomyces azureus, Nature,* 272, 792, 1978.
206. **Grosjean, H. J., deHenau, S., and Crothers, D. M.,** On the physical basis for ambiguity in genetic cooding interactions, *Proc. Natl. Acad. Sci. U.S.A.*, 75, 610, 1978.
207. **Miller, J. H.,** GUG and UUG are initiation codons *in vivo, Cell,* 1, 73, 1974.

208. **Steitz, J. A.,** Genetic signals and nucleotide sequences in messenger RNA, in *Biological Regulation and Development,* Vol. 1, *Gene Expression,* Goldberg, R. F. Ed., Plenum Press, New York, 1979, 349.
209. **Stewart, J. W., Sherman, F., Shipman, N. A., and Jackson, M.,** Identification and mutation relocation of the AUG codon initiating translation of iso-1-cytochrome c in yeast, *J. Biol. Chem.,* 246, 7429, 1971.
210. **Dube, S. K., Marker, K. A., Clark, B. F. C., and Cory, S.,** Nucleotide sequence of *N*-formylmethionyl-transfer RNA, *Nature,* 218, 232, 1968.
211. **Freier, S. M. and Tinoco, I., Jr.,** The binding of complementary oligonucleotides to yeast initiator transfer RNA, *Biochemistry,* 14, 3310, 1975.
212. **Weissenbach, J. and Grosjean, H. J.,** Effect of threonylcarbamoyl modification ($t^6A$) in yeast $tRNA^{Arg}_{III}$ on codon-anticodon and anticodon-anticodon interactions. A theromodynamic and kinetic evaluation, *Eur. J. Biochem.,* 116, 207, 1981.
213. **Miller, J. P., Hussain, Z., and Schweizer, M. P.,** The involvement of the anticodon adjacent modified nucleoside *N*-9-(β-D-ribofuranosyl-purin-6-ylcarbamoyl)-threonine in the biological function of *E. coli* $tRNA^{Ile}$, *Nucleic Acids Res.,* 3, 1185, 1976.
214. **Watts, M. T. and Tinoco, I., Jr.,** Role of hypermodified bases in transfer RNA. Solution properties of dinucleoside monophosphates, *Biochemistry,* 17, 2455, 1978.
215. **Gefter, M. L. and Russel, R. L.,** Role of modification in tyrosine transfer RNA. A modified base affecting ribosome binding, *J. Mol. Biol.,* 39, 145, 1969.
216. **Hoburg, A., Aschhoff, H. J., Kersten, H., Manderschied, U., and Gassen, H. G.,** Function of modified nucleosides 7-methylguanosine, ribothymidine, and 2-thiomethyl-$N^6$-(isopentenyl)adenosine in procaryotic transfer ribonucleic acid, *J. Bacteriol.,* 140, 408, 1979.
217. **Thiebe, R. and Zachau, H. G.,** A specific modification next to the anticodon of phenylalanine transfer ribonucleic acid, *Eur. J. Biochem.,* 5, 546, 1968.
218. **Ghosh, K. and Ghosh, H. P.,** Role of modified nucleosides in transfer ribonucleic acid. Effect of removal of the modified base adjacent to 3′ end of the anticodon in codon-anticodon interaction, *J. Biol. Chem.,* 247, 3369, 1972.
219. **Pongs, O. and Reinwald, E.,** Function of Y in codon-anticodon interaction of $tRNA^{Phe}$, *Biochem. Biophys. Res. Commun.,* 50, 357, 1973.
220. **Fairclough, R. H. and Cantor, C. R.,** An energy transfer equilibrium between two identical copies of a ribosome-bound fluorescent transfer RNA analogue: Implications for the possible structure of codon-anticodon complexes, *J. Mol. Biol.,* 132, 587, 1979.
221. **Litwack, M. D. and Peterkofsky, A.,** Transfer ribonucleic acid deficient in $N^6$-($\Delta^2$-isopentenyl)adenosine due to mevalonic acid limitation, *Biochemistry,* 10, 994, 1971.
222. **Kimball, M. E. and Söll, D.,** The phenylalanine tRNA from *Mycoplasma* sp. (Kid): a tRNA lacking hypermodified nucleosides functional in protein synthesis, *Nucleic Acids Res.,* 1, 1713, 1974.
223. **Yanofksy, C., and Soll, L.,** Mutations affecting $tRNA^{Trp}$ and its charging and their effect on regulation of transcription termination at the attenuator of the tryptophan operon, *J. Mol. Biol.,* 113, 663, 1977.
224. **Bugg, C. E. and Thewalt, U.,** Crystal structure of $N^6$-($\Delta^2$-isopentenyl)adenine, a base in the anticodon loop of some tRNAs, *Biochem. Biophys. Res. Commun.,* 46, 779, 1972.
225. **Parthasarathy, R., Ohrt, J. M., and Chheda, G. B.,** Conformation and possible role of hypermodified nucleosides adjacent to 3′-end of anticodon in tRNA: *N*-(purin-6-ylcarbamoyl)-L-theonine riboside, *Biochem. Biophys. Res. Commun.,* 60, 211, 1974.
226. **Johnston, M. H., Barnes, W. M., Chumley, F. G., Bossi, L., and Roth, J. R.,** Model for regulation of the histidine operon of *Salmonella, Proc. Natl. Acad. Sci. U.S.A.,* 77, 508, 1980.
227. **Lewis, J. A. and Ames, B. N.,** Histidine regulation in *Salmonella typhimurium* XI. The percentage of transfer $RNA^{His}$ charged *in vivo,* and its relation to the repression of the histidine operon, *J. Mol. Biol.,* 66, 131, 1972.
228. **Bruni, C. B., Colantuoni, V., Spordone, L., Cortese, R., and Blasi, L.,** Biochemical and regulatory properties of *Escherichia coli* K-12 *hisT* mutants, *J. Bacteriol.,* 130, 4, 1977.
229. **Bossi, L. and Roth, J. R.,** The influence of codon context on genetic code translation, *Nature,* 286, 123, 1980.
229a. **Hjalmarsson, K. J.,** unpublished results.
230. **Mitra, S. K., Lustig, F., Åkesson, B., Lagerkvist, U., and Strid, L.,** Codon-anticodon recognition in the valine codon family, *J. Biol. Chem.,* 252, 471, 1977.
231. **Morikawa, K., Torii, K., Iitaka, Y., Tsuboi, M., and Nishimura, S.,** Crystal and molecular structure of the methyl ester of uridine-5-oxyacetic acid: a minor constituent of *Escherichia coli* tRNAs, *FEBS Lett.,* 48, 279, 1974.
232. **Hillen, W., Egert, E., Lindner, H. J., and Gassen, H. G.,** Restriction of amplification of wobble recognition. The structure of 2-thio-5-methylaminomethyluridine and the interaction of odd uridines with the anticodon loop backbone, *FEBS Lett.,* 94, 361, 1978.

233. **Hillen, W., Egert, E., Lindner, H. J., Gassen, H. G., and Vorbrüggen, H.,** 5-methoxyuridine: The influence of 5-substituents on the keto-enol tautomerism of the 4-carbonyl group, *J. Carbohydrates, Nucleasides, Nucleotides,* 5, 23, 1978.
233a. **Björk, G. R.,** unpublished results.
234. **Grosjean, H.,** Codon usage in several organisms, in *Transfer RNA: Biological Aspects,* Söll, D., Abelson, J. N., and Schimmel, P. R., Eds., Cold Spring Harbor Laboratory, Cold Spring Harbor, New York, 1980, 565.
235. **Ofengand, J., Liou, R., Kohut, J., III, Schwartz, I., and Zimmermann, R. A.,** Covalent cross-linking of transfer ribonucleic acid to the ribosomal P site. Mechanism and site of reaction in transfer ribonucleic acid, *Biochemistry,* 18, 4322, 1979.
236. **Zimmermann, R. A., Gates, S. M., Swartz, I., and Ofengand, J.,** Covalent cross-linking of transfer ribonucleic acid to the ribosomal P site. Site of reaction in 16S ribonucleic acid *Biochemistry,* 18, 4333, 1979.
237. **Ofengand, J. and Liou, R.,** Correct codon-anticodon base pairing at the 5′-anticodon position blocks covalent cross-linking between transfer ribonucleic acid and 16S RNA at the ribosomal P site, *Biochemistry,* 20, 552, 1981.
238. **Taylor, B. H., Prince, J. B., Ofengand, J., and Zimmermann, R. A.,** Nonanucleotide sequence from 16S ribonucleic acid binding site of the *Escherichia coli* ribosome, *Biochemistry,* 20, 7581, 1981.
239. **Atkins, J. F. and Ryce, S.,** UGA and non-triplet suppressor reading of the genetic code, *Nature,* 249, 527, 1974.
240. **Sekiya, T., Takeishi, K., and Ukita, T.,** Specificity of yeast glutamic acid transfer RNA for codon recognition, *Biochim. Biophys. Acta,* 182, 411, 1969.
241. **Yoshida, M., Takeishi, K., and Ukita, T.,** Structural studies on a yeast glutamic acid tRNA specific to GAA codon, *Biochim Biophys Acta,* 228, 153, 1971.
242. **Sen, G. C. and Ghosh, H. P.,** Role of modified nucleosides in tRNA: Effect of modification of the 2-thiouridine derivative located at the 5′-end of the anticodon of yeast transfer $RNA_2^{Lys}$, *Nucleic Acids Res.,* 3, 523, 1976.
243. **Weissenbach, J. and Dirheimer, G.,** Pairing properties of the methylester of 5-carboxymethyl uridine in the wobble position of yeast $tRNA_3^{Arg}$, *Biochim. Biophys. Acta,* 518, 530, 1978.
244. **Mazumdar, S. K., Saenger, W., and Scheit, K. H.,** Molecular structure of poly-2-thiouridylic acid, a double helix with non-equivalent polynucleotide chains, *J. Mol. Biol.,* 85, 213, 1974.
245. **Berman, H. M., Marcu, D., Narayanan, P., Fissekis, J. D., and Lipnick, R. L.,** Modified bases in tRNA: the structure of 5-carbamoylmethyl- and 5-carboxymethyl uridine, *Nucleic Acids Res.,* 5, 893, 1978.
246. **Hillen, W., Egert, E., Lindner, H. J., and Gassen, H. G.,** Crystal and molecular structure of 2-thio-5-carboxymethyluridine and its methyl ester: Helix terminator nucleosides in the first position of some anticodons, *Biochemistry,* 17, 5314, 1978.
247. **Kasai, H., Nishimura, S., Vorbrüggen, H., and Iitaka, Y.,** Crystal and molecular structure of the acetonide of 5-methylaminomethyl-2-thiouridine. A minor constituent of *Escherichia coli* tRNAs, *FEBS Lett.* 103, 270, 1979.
248. **Sherman, R., Ono, B., and Stewart, J. R.,** Use of the iso-l-cytochrome C system for investigating nonsense mutants and suppressors in yeast, in *Nonsense Mutations and tRNA Suppressors,* Celis, J. E. and Smith, J. D., Eds., 1979, 133.
249. **Waldron, C., Cox, B. S., Wills, N., Gesteland, R. F., Piper, P. W., Colby, D., and Guthrie, C.,** Yeast ochre suppressor SUQ5-ol is an altered $tRNA_{UCA}^{Ser}$, *Nucleic Acids Res.,* 9, 3077, 1981.
250. **Stern, L. and Schulman, L. H.,** The role of the minor base $N^4$-acetylcytidine in the function of the *Escherichia coli* noninitiator methionine transfer RNA, *J. Biol. Chem.,* 253, 6132, 1978.
251. **Harada, F. and Nishismura, S.,** Possible anticodon sequences of $tRNA^{His}$, $tRNA^{Asn}$, and $tRNA^{Asp}$ from *Escherichia coli* B. Universal presence of nucleoside Q in the first position of the anticodons of these transfer ribonucleic acids, *Biochemistry,* 11, 301, 1972.
251a. **Noguchi S., Nishimura, Y., Hirota, Y., and Nishimura, S.,** Isolation and characterization of an *Escherichia coli* mutant, lacking tRNA-guanine transalycosylase. Function and biosynthesis of quesine in tRNA, *J. Biol. Chem.,* 257, 6544, 1982.
252. **Björk, G. R. and Kjellin-Stråby, K.,** Mutants of bacteria and yeast with specific defects in transfer ribonucleic acid methylation, in *The Biochemistry of Adenosylmethionine,* Salvatore, F., Borek, E., Zappia, V., Williams-Ashman, H. G., and Schlenk, F., Eds., 1977, 216.
253. **Friedman, S.,** The effect of chemical modification of 3-(3-amino-3-carboxypropyl)uridine on tRNA function, *J. Biol. Chem.,* 254, 7111, 1979.
254. **Arcari, P. and Hecht, S. M.,** Isoenergetic hydride transfer. A reversible tRNA modification with concomitant alteration of biochemical properties, *J. Biol. Chem.,* 253, 8278, 1978.
255. **Roe, B., Michael, M., and Dudock, B.,** Function of $N^2$-methylguanine in phenylalanine transfer RNA, *Nature (London) New Biol.,* 246, 135, 1973.
256. **Ramabhadran, T. V., Fossum, T., and Jagger, J.,** *Escherichia coli* mutant lacking 4-thiouridine in its transfer ribonucleic acid, *J. Bacteriol.,* 128, 671, 1976.

257. **Thomas, G. and Favre, A.,** 4-thiouridine triggers both growth delay induced by near-ultraviolet light and photoprotection, *Eur. J. Biochem.*, 113, 67, 1980.
258. **Ramabhadran, T. V. and Jagger, J.,** Mechanism of growth delay induced in *Escherichia coli* by near ultraviolet radiation, *Proc. Natl. Acad. Sci. U.S.A.*, 73, 59, 1976.
259. **Svensson, I., Björk, G. R., and Lundahl, P.,** Studies on microbial RNA. Properties of tRNA methylases from *Saccharomyces cerevisiae, Eur. J. Biochem.*, 9, 216, 1969.
260. **Marcu, K., Mignery, R., Reszelbach, R., Roe, B., Sirover, M., and Dodock, B.,** The absence of ribothymidine in specific eukaryotic tRNAs: I. Glycine and threonine tRNAs of wheat embryo, *Biochem. Biophys. Res. Commun.*, 55, 477, 1973.
261. **Marcu, K., Marcu, D., and Dudock, B.,** Wheat germ tRNAs containing uridine in place of ribothymidine: a characterization of an unusual class of eukaryotic tRNAs, *Nucleic Acids Res.*, 5, 1075, 1978.
262. **Björk, G. R. and Neidhardt, F. C.,** Physiological and biochemical studies on the function of 5-methyluridine in the transfer ribonucleic acid of *Escherichia coli, J. Bacteriol.*, 124, 99, 1975.
262a. **Björk, G. R.,** unpublished.
263. **Johnson, L., Hayashi, K., and Söll, D.,** Isolation and properties of a transfer ribonucleic acid deficient in ribothymidine, *Biochemistry*, 9, 2823, 1970.
264. **Delk, A. N., Romeo, J. M., Nagle, D. P., Jr., and Rabinowitz, J. C.,** Biosynthesis of ribothymidine in the transfer RNA of *Streptococcus faecalis* and *Bacillus subtilis, J. Biol. Chem.*, 251, 7649, 1976.
265. **Vani, B. R., Ramakrishnan, T., Taya, Y., Noguchi, S., Yamaizumi, Z., and Nishimura, S.,** Occurence of 1-methyladenosine and absence of ribothymidine in transfer ribonucleic acid in *Mycobacterium smegmatis, J. Bacteriol.*, 137, 1084, 1979.
266. **Samuel, C. E. and Rabinowitz, J. C.,** Initiation of protein synthesis by folate-sufficient and folate-deficient *Streptococcus fecalis* R. Biochemical and biophysical properties of methionine transfer ribonucleic acid, *J. Biol. Chem.*, 249, 1198, 1974.
267. **Delk, A. S. and Rabinowitz, J. C.,** Partial nucleotide sequence of a prokaryote initiator tRNA that functions in its nonformulated form, *Nature*, 252, 106, 1974.
268. **Baumstark, B. R., Spremulli, L. L., Raj Bhandary, U. L. and Brown, G. M.,** Initiation of protein synthesis without formylation in a mutant of *Escherichia coli* that grows in the abbsence of tetrahydrofolate, *J. Bacteriol.*, 129, 457, 1977.
269. **Marcu, K. B. and Dudock, B. S.,** Effect of ribothymidine in specific eukaryotic tRNAs on their efficiency in *in vitro* protein synthesis, *Nature*, 261, 159, 1976.
270. **Kersten, H., Albani, M., Männlein, E., Praisler, R., Wurmbach, P., and Nierhaus, K. H.,** On the role of ribosylthymine in prokaryotic tRNA function, *Eur. J. Biochem.*, 114, 451, 1981.
271. **Davenloo, P., Sprinzl, M., Watanabe, K., Albani, M., and Kersten, H.,** Role of ribothymidine in the termal stability of transfer RNA as monitored by proton magnetic resonance, *Nucleic Acid Res.*, 6, 1571, 1979.

# EPILOGUE

It is evident from all the chapters presented here that a great deal is already known about RNA processing even though it is a relatively young field of research. However, it is also obvious that much more is still missing. In general it is clear that much of the phenomenology is well defined, but very little of it is clearly understood. This is particularly true in eukaryotic cells. Moreover, about one of the more enigmatic phenomena, the prevalence of intervening sequences in so many of the genes of higher eukaryotic cells, it is not even clear whether good speculations about its raison d'être are available.

In prokaryotic cells it is clear that by and large the field is more advanced than in eukaryotic cells, and one has quite a bit of information concerning RNA processing where the substrate may be naked RNA. The enzymes and some of the enzymology are known. However, even in this field none of the enzymes has been analyzed in sufficient detail, and beyond the general statement that RNA processing enzymes recognize secondary and tertiary structure, there is not much more that can be said. Also, there is almost no knowledge regarding RNA processing where the substrate is obligatorily a ribonucleoprotein particle. While it is obvious that much has to be and can be learned about the mechanics of RNA processing, the biological significance of RNA processing is not very obvious in almost any of the systems studied.

While RNA processing could be clearly treated as a part and parcel of general cell biology, like protein synthesis or a membrane, it came as a surprise, at least to me, that it has a role and perhaps a significant one in viral and cellular development, and differentiation.

Thus it is easy to predict that the area of RNA processing is pregnant with problems the solution of which should be of great challenge and interest. It is also obvious that there is room for many more scientists to find challenge and satisfaction in this burgeoning field of research.

D.A.

# INDEX

## A



## B

## C

## D

## E

# I

## J



## K



## L



## M

## N

# O



# P

## Q



## R

## S

# T

Milton Keynes UK
Ingram Content Group UK Ltd.
UKHW051931141024
449569UK00027B/1447